Undergraduate Texts in Mathematics

Undergraduate Texts in Mathematics

Apostol: Introduction to Analytic Number Theory.

Armstrong: Basic Topology.

Bak/Newman: Complex Analysis.

Banchoff/Wermer: Linear Algebra Through Geometry.

Childs: A Concrete Introduction to Higher Algebra.

Chung: Elementary Probability Theory with Stochastic Processes.

Croom: Basic Concepts of Algebraic Topology.

Curtis: Linear Algebra: An Introductory Approach.

Dixmier: General Topology.

Driver: Why Math?

Ebbinghaus/Flum/Thomas Mathematical Logic.

Fischer: Intermediate Real Analysis.

Fleming: Functions of Several Variables. Second edition.

Foulds: Optimization Techniques: An Introduction.

Foulds: Combination Optimization for Undergraduates.

Franklin: Methods of Mathematical Economics.

Halmos: Finite-Dimensional Vector Spaces. Second edition.

Halmos: Naive Set Theory.

Iooss/Joseph: Elementary Stability and Bifurcation Theory.

Jänich: Topology.

Kemeny/Snell: Finite Markov Chains.

Klambauer: Aspects of Calculus.

Lang: Undergraduate Analysis.

Lang: A First Course in Calculus. Fifth Edition.

Lang: Calculus of One Variable. Fifth Edition.

Lang: Introduction to Linear Algebra. Second Edition.

Lax/Burstein/Lax: Calculus with Applications and Computing, Volume 1. Corrected Second Printing.

LeCuyer: College Mathematics with APL.

Lidl/Pilz: Applied Abstract Algebra.

Macki/Strauss: Introduction to Optimal Control Theory.

Malitz: Introduction to Mathematical Logic.

Marsden/Weinstein: Calculus I, II, III. Second edition.

continued after Index

Gabriel Klambauer

Aspects of Calculus

With 82 Illustrations

Springer-Verlag
New York Berlin Heidelberg Tokyo

Gabriel Klambauer
Department of Mathematics
University of Ottawa
Ottawa K1N 9B4 Ontario
Canada

AMS Subject Classification: 26-01, 26-A06

Library of Congress Cataloging in Publication Data
Klambauer, Gabriel.
  Aspects of calculus.
  (Undergraduate texts in mathematics)
  Bibliography: p.
  Includes index.
  1. Calculus.   I. Title.   II. Series.
QA303.K654   1986   515   85-30283

Typeset by Asco Trade Typesetting Ltd., Hong Kong.

9   8   7   6   5   4   3   2   1

ISBN-13:978-1-4613-9563-8   e-ISBN-13:978-1-4613-9561-4
DOI: 10.1007/978-1-4613-9561-4

To Agnes Catherine

Preface

This book is intended for students familiar with a beginner's version of differential and integral calculus stressing only manipulation of formulas and who are now looking for a closer study of basic concepts combined with a more creative use of information. The work is primarily aimed at students in mathematics, engineering, and science who find themselves in transition from elementary calculus to rigorous courses in analysis. In addition, this book may also be of interest to those preparing to teach a course in calculus.

Instead of exposing the reader to an excess of premature abstractions that so easily can degenerate into pedantry, I felt it more useful to stress instructive and stimulating examples. The book contains numerous worked out examples and many of the exercises are provided with helpful hints or a solution in outline. For further exercises the interested reader may want to consult a problem book by the author entitled *Problems and Propositions in Analysis* (New York: Marcel Dekker, 1979). For the history of calculus I recommend the book by C. B. Boyer, *The Concepts of the Calculus* (New York: Dover, 1949).

This book is made up of seven chapters and the Contents gives detailed information concerning the topics covered. The book begins with a study of the logarithmic and exponential functions. The treatment of these functions is geometric rather than arithmetic in nature and quickly leads to the evaluation of certain limits that are of crucial importance; the approach, which depends on a specific relation between hyperbolic segment and logarithmic function, goes back to A. A. de Sarasa (1618–1667). In the Bibliography at the end of the book the reader will find suitable references for further study.

I thank Professor P. R. Halmos, Indiana University, for the kind interest he has shown in my work, my son Peter who prepared the illustrations for this book, and my friends Dr. E. L. Cohen, University of Ottawa, and Dr.

G. K. R. Rao, University of Kenya, for valuable help and steadfast encouragement. I also wish to express my gratitude to the Board of Governors of the University of Ottawa for the benefit of a sabbatical leave during a portion of the writing of this book, and to the staff of Springer-Verlag for their fine cooperation.

                                                                    Gabriel Klambauer

Contents

CHAPTER 1

The Logarithmic and Exponential Functions

1. An Area Problem

The curve $y = 1/x$, for $x > 0$, is of special interest to us; it is located above the x-axis in the first quadrant of the x, y plane and it is seen to be symmetric with respect to the line $y = x$ (because the equation $xy = 1$ remains unchanged when x and y are interchanged). See Figure 1.1 for a display of the curve under consideration.

Definition. If $0 < a < b$, let $A_{a,b}$ denote the area of the region bounded above by the curve $y = 1/x$, bounded below by the x-axis, bounded to the left by the line $x = a$, and bounded to the right by the line $x = b$; if $0 < b < a$, let $A_{a,b} = -A_{b,a}$.

It is easily seen that $A_{a,a} = 0$, $A_{a,b} = -A_{b,a}$, and

$$A_{a,c} = A_{a,b} + A_{b,c} \tag{1.1}$$

for any points a, b, and c on the positive part of the x-axis [see Figure 1.2 and Figure 1.3 in connection with equation (1.1)]. Perhaps less obvious is the relation

$$A_{1,r} = A_{1/r,1} \tag{1.2}$$

for any $r > 0$. To verify equation (1.2) we proceed as follows. Suppose $r > 1$. (The case $r = 1$ needs no proof; the case $0 < r < 1$ reduces to the case under consideration when we interchange the roles of r and $1/r$.) In Figure 1.4 the region indicated by vertical cross-hatching has the same area as the region indicated by horizontal cross-hatching because the rectangle R_2 (with vertical cross-hatching) and the rectangle R_1 (with horizontal cross-hatching)

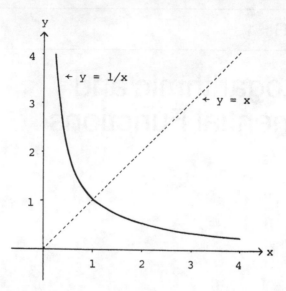

Figure 1.1

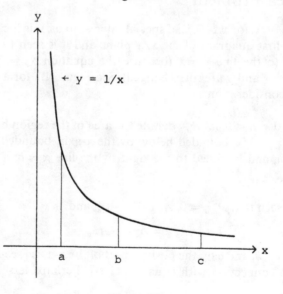

Figure 1.2

have the same area, namely, $1 - 1/r$. We shall see shortly that equation (1.2) is merely a special case of a more general relationship.

Let $0 < a < b$ and consider the *closed interval* $[a, b]$, that is, the set of all points x such that $a \leq x \leq b$. Then $y = 1/x$ with $a \leq x \leq b$ will assume its largest value for $x = a$ and its smallest value for $x = b$. On the interval $[a, b]$ we construct two rectangles, one with altitude $1/a$ and the other with altitude $1/b$. The larger rectangle has area $(b - a)/a$ and the smaller rectangle has area

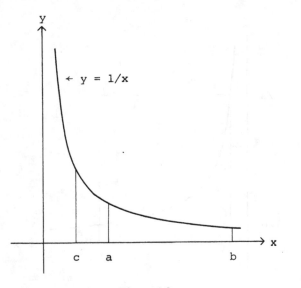

Figure 1.3

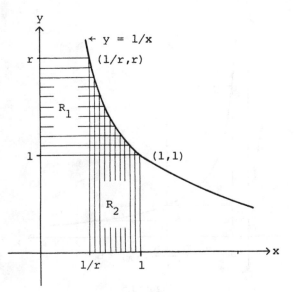

Figure 1.4

$(b - a)/b$ (see Figure 1.5). It is apparent that

$$\frac{b - a}{b} < A_{a,b} < \frac{b - a}{a}. \tag{1.3}$$

If $a = b$, each term in the foregoing inequality vanishes. Inequality (1.3) gives an estimate for $A_{a,b}$ which we shall find very useful later on.

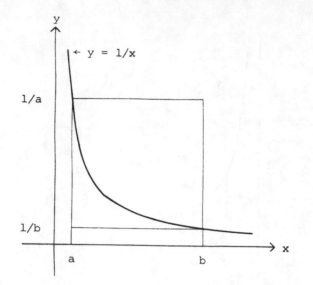

Figure 1.5

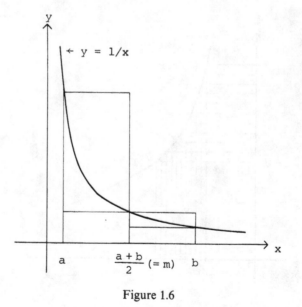

Figure 1.6

To refine the estimate for $A_{a,b}$ provided by inequality (1.3) we bisect the interval $[a, b]$ into two subintervals $[a, m]$ and $[m, b]$, where $m = (a + b)/2$, the midpoint of the interval $[a, b]$, and on each of these two subintervals we construct the corresponding larger and smaller rectangles to estimate $A_{a,m}$ and $A_{m,b}$ (see Figure 1.6). We obtain

$$\frac{m - a}{m} < A_{a,m} < \frac{m - a}{a} \quad \text{and} \quad \frac{b - m}{b} < A_{m,b} < \frac{b - m}{m}$$

or

$$\frac{m-a}{m}+\frac{b-m}{b} < A_{a,m}+A_{m,b}=A_{a,b} < \frac{m-a}{a}+\frac{b-m}{m}. \qquad (1.4)$$

Let

$$L_1 = \frac{m-a}{m}+\frac{b-m}{b} \quad \text{and} \quad U_1 = \frac{m-a}{a}+\frac{b-m}{m}.$$

We call L_1 a *lower approximating sum for* $A_{a,b}$ and U_1 an *upper approximating sum for* $A_{a,b}$; L_1 is the sum of the areas of the two inscribed rectangles and U_1 is the sum of the areas of the two circumscribed rectangles by which we seek to approximate $A_{a,b}$ and L_1 represents an underestimate and U_1 an overestimate of $A_{a,b}$. It can be seen that

$$\frac{b-a}{b} < L_1 \quad \text{and} \quad \frac{b-a}{a} > U_1; \qquad (1.5)$$

indeed, $(b-a)/b = (b-m)/b + (m-a)/b < (b-m)/b + (m-a)/m$ and $(b-a)/a = (b-m)/a + (m-a)/a > (b-m)/a + (m-a)/m$. Combining the inequalities (1.4) and (1.5) we obtain

$$\frac{b-a}{b} < L_1 < A_{a,b} < U_1 < \frac{b-a}{a}. \qquad (1.6)$$

Looking back, we have used the process of bisection of the interval $[a,b]$ to obtain from the estimate in inequality (1.3) the refinement expressed in inequality (1.6). This process can of course be continued by bisecting the subintervals $[a,m]$ and $[m,b]$ (see Figure 1.7) and obtaining a lower approxi-

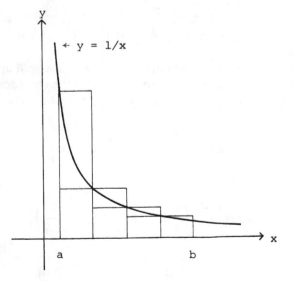

Figure 1.7

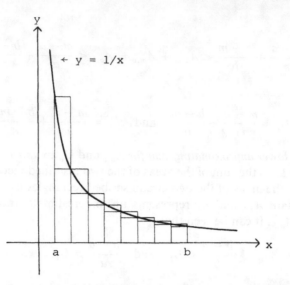

Figure 1.8

mating sum L_2 for $A_{a,b}$ and an upper approximating sum U_2 for $A_{a,b}$; L_2 simply stands for the sum of the areas of the four inscribed rectangles and U_2 stands for the sum of the areas of the four circumscribed rectangles by which we try to approximate $A_{a,b}$. In Figure 1.8 we display the next step of this process of continued bisection which will result in a lower approximating sum L_3 for $A_{a,b}$ and an upper approximating sum U_3 for $A_{a,b}$; here L_3 stands for the sum of the areas of the eight inscribed rectangles and U_3 denotes the sum of the areas of the eight circumscribed rectangles by which we seek to approximate $A_{a,b}$. It is clear that

$$\frac{b-a}{b} < L_1 < L_2 < L_3 < \cdots < A_{a,b} < \cdots < U_3 < U_2 < U_1 < \frac{b-a}{a}. \qquad (1.7)$$

At the nth step of bisection of the interval $[a,b]$ we have split up this interval into 2^n subintervals of equal length $(b-a)/2^n$; let the enumeration of these subintervals be $[a, t_1], [t_1, t_2], [t_2, t_3], \ldots, [t_{k-1}, t_k]$, where $k = 2^n$ and $t_k = b$. Then

$$L_n = \frac{b-a}{k}\left(\frac{1}{t_1} + \frac{1}{t_2} + \frac{1}{t_3} + \cdots + \frac{1}{b}\right)$$

and

$$U_n = \frac{b-a}{k}\left(\frac{1}{a} + \frac{1}{t_1} + \frac{1}{t_2} + \cdots + \frac{1}{t_{k-1}}\right)$$

and we have

$$U_n - L_n = \frac{b-a}{k}\left(\frac{1}{a} - \frac{1}{b}\right) = \frac{(b-a)^2}{ab} \cdot \frac{1}{2^n}. \qquad (1.8)$$

But a and b are fixed and n can be made as large as we please; therefore the difference $U_n - L_n$ tends to zero as n becomes arbitrarily large.

For $n = 1, 2, 3, \ldots$ let I_n denote the closed interval $[L_n, U_n]$. It is clear from inequality (1.7) that I_{n+1} is contained in I_n for $n = 1, 2, 3, \ldots$ and thus we are dealing with a sequence of nested closed intervals, where I_2 is a subset of I_1, I_3 is a subset of I_2, and so forth. Moreover, as n becomes arbitrarily large, the length of the interval I_n becomes as small as we please in view of equality (1.8). By an argument that we omit, it can be proved that there is one and only one point t that is common to all intervals I_n for $n = 1, 2, 3, \ldots$; this unique point t on the number line is precisely our quantity $A_{a,b}$. We now state the result that we have invoked.

Nested Interval Principle. For $n = 1, 2, 3, \ldots$ let J_n be closed intervals such that

(i) J_{n+1} is contained in J_n for each n,
(ii) the length of the intervals J_n tends to zero as n becomes arbitrarily large.

Then there is one and only one point that is common to all intervals J_n.

COMMENTS. It is easy to see that there can not be two (or more) points in common to all intervals J_n for this would violate condition (ii). Indeed, if two points, say, t and s with $t \neq s$, were common to all J_n then the distance between t and s (which could not be zero) would act as a barrier and no interval J_n could possibly have length less than this distance. It is also necessary that the intervals J_n be closed. Indeed, suppose J_n would be the interval $0 < x \leq 1/n$ for $n = 1, 2, 3, \ldots$. Clearly, the intervals J_n for $n = 1, 2, 3, \ldots$ would be nested and the length of J_n, being $1/n$, would tend to zero as n becomes arbitrarily large. However, there is no point t that is common to all J_n in this case. To see this note that the point 0 does not qualify; in fact, 0 is not a member of any J_n. Also, any t larger than 0 does not qualify because, by letting n grow, $1/n$ can be made less than any such fixed t. Finally, note that condition (ii) can also be expressed in the following way: For any $r > 0$ there is an interval J_n so that the length of J_n is less than r.

Proposition 1.1. Let $0 < a < b$ and $s > 0$; then

$$A_{sa,sb} = A_{a,b}. \tag{1.9}$$

PROOF. Let $k = 2^n$, where n is a positive integer, and let

$$a = t_0 < t_1 < t_2 < \cdots < t_{k-1} < t_k = b$$

be equally spaced points splitting the interval $[a, b]$ into k subintervals $[a, t_1], [t_1, t_2], \ldots, [t_{k-1}, b]$. On these subintervals we construct inscribed and circumscribed rectangles as indicated in Figure 1.9. The sum of the areas of the inscribed rectangles is

$$\frac{b-a}{k}\left(\frac{1}{t_1} + \frac{1}{t_2} + \cdots + \frac{1}{b}\right)$$

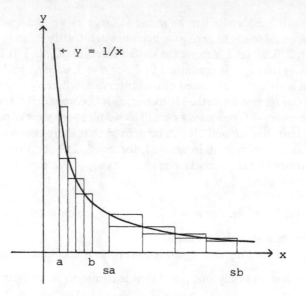

Figure 1.9

and the sum of the areas of the circumscribed rectangles is

$$\frac{b-a}{k}\left(\frac{1}{a}+\frac{1}{t_1}+\cdots+\frac{1}{t_{k-1}}\right);$$

we therefore have

$$\frac{b-a}{k}\left(\frac{1}{t_1}+\frac{1}{t_2}+\cdots+\frac{1}{b}\right) < A_{a,b} < \frac{b-a}{k}\left(\frac{1}{a}+\frac{1}{t_1}+\cdots+\frac{1}{t_{k-1}}\right). \quad (1.10)$$

Now let

$$sa = v_0 < v_1 < v_2 < \cdots < v_{k-1} < v_k = sb$$

be equally spaced points splitting the interval $[sa, sb]$ into k subintervals $[sa, v_1]$, $[v_1, v_2]$, ..., $[v_{k-1}, sb]$. On these subintervals we construct inscribed and circumscribed rectangles and obtain, in analogy to the case just considered,

$$\frac{sb-sa}{k}\left(\frac{1}{v_1}+\frac{1}{v_2}+\cdots+\frac{1}{sb}\right) < A_{sa,sb} < \frac{sb-sa}{k}\left(\frac{1}{sa}+\frac{1}{v_1}+\cdots+\frac{1}{v_{k-1}}\right).$$

But $v_j = st_j$ for $j = 1, 2, \ldots, k-1$ and so

$$\frac{b-a}{k}\left(\frac{1}{t_1}+\frac{1}{t_2}+\cdots+\frac{1}{b}\right) < A_{sa,sb} < \frac{b-a}{k}\left(\frac{1}{a}+\frac{1}{t_1}+\cdots+\frac{1}{t_{k-1}}\right). \quad (1.11)$$

Comparison of (1.10) and (1.11) shows that the *same* sequence of nested closed intervals produces both $A_{a,b}$ and $A_{sa,sb}$ and thus (1.9) is true. This completes the proof. $\square$

REMARK. Putting $a = 1$, $b = r$, and $s = 1/r$ in (1.9) yields (1.2).

Proposition 1.2. *Let $v > 0$ and $w > 0$; then*

$$A_{1,vw} = A_{1,v} + A_{1,w} \tag{1.12}$$

and

$$A_{1,v/w} = A_{1,v} - A_{1,w}. \tag{1.13}$$

PROOF. By (1.1) and (1.9) we have

$$A_{1,vw} = A_{1,v} + A_{v,vw} = A_{1,v} + A_{1,w}$$

and

$$A_{1,v/w} = A_{1,v} + A_{v,v/w} = A_{1,v} + A_{w,1} = A_{1,v} - A_{1,w}.$$

But this is what we wanted to show.   □

Proposition 1.3. *Let $v > 0$ and $r = n/m$, where n is an integer and m is a positive integer; then*

$$A_{1,v^r} = rA_{1,v}. \tag{1.14}$$

PROOF. It is easy to see that

$$A_{1,v^k} = kA_{1,v}$$

for any nonnegative integer k. Since $A_{1,1/v} = -A_{1,v}$ by (1.2), we see that

$$A_{1,v^{-k}} = -kA_{1,v}$$

for any nonnegative integer k. If m is a positive integer, we have

$$A_{1,v} = A_{1,(v^{1/m})^m} = mA_{1,v^{1/m}} \quad \text{or} \quad A_{1,v^{1/m}} = \frac{1}{m}A_{1,v}.$$

Thus, if n is any integer and m is any positive integer and $n/m = r$, then

$$A_{1,v^{n/m}} = A_{1,(v^{1/m})^n} = nA_{1,v^{1/m}} = \frac{n}{m}A_{1,v} \quad \text{or} \quad A_{1,v^r} = rA_{1,v}.$$

This finishes the proof.   □

2. The Natural Logarithm

Given that a and b are positive numbers, we recall having defined $A_{a,b}$ to denote the area of the region under the curve $y = 1/x$, above the x-axis and between the lines $x = a$ and $x = b$ provided that $a < b$; if $a > b$ we let $A_{a,b} = -A_{b,a}$. We are already familiar with a number of properties of $A_{a,b}$.

Definition. For $x > 0$, the function $L(x) = \ln x$ with $\ln x$ given by

$$\ln x = A_{1,x}$$

is called the (*natural*) logarithm.

It is clear that $\ln x$ is negative for $0 < x < 1$, $\ln 1 = 0$, and $\ln x$ is positive for $x > 1$. It is evident that $\ln x$ is an increasing function because $x_1 < x_2$ obviously implies that $\ln x_1 < \ln x_2$. From (1.12) and (1.13) we know that

$$\ln(xy) = \ln x + \ln y \quad \text{and} \quad \ln\left(\frac{x}{y}\right) = \ln x - \ln y$$

for any positive x and y. From (1.14) it follows that

$$\ln x^r = r \ln x$$

for any rational number r, that is, any number r of the form n/m, where n and m are integers and m is not zero. This formula also holds for irrational numbers r, that is, numbers that are not rational; here the main difficulty lies in defining x^r when r is irrational. We shall overcome this obstacle in the next section.

The curve $y = \ln x$ lies completely to the right of the y-axis and below the line $y = x$; the domain of definition of $\ln x$ is the positive part of the x-axis and $\ln x \leq x - 1$ by (1.3). Since

$$\ln x - \ln a = A_{1,x} - A_{1,a} = A_{a,x},$$

it can be seen that $\ln x$ tends to $\ln a$ as x tends to a because $A_{a,x}$ tends to zero as x tends to a. In the next chapter we shall see that this property of $y = \ln x$ means the continuity of the logarithmic function for $0 < x$. It also means that $y = \ln x$ can "skip" no values and that its range or set of values is an interval. To show that this interval covers the entire number line, we need only to show that this interval is unbounded above and unbounded below. We can do this by letting M be an arbitrary positive number and showing that $\ln x$ has values greater than M and values smaller than $-M$. Indeed, since $\ln 2$ is positive [we have $\frac{1}{2} < \ln 2 < 1$ by (1.3)], we know that some positive multiple of $\ln 2$ has to be larger than M; namely, we know that there is a positive integer n such that $n(\ln 2) > M$. [Here we are using the *Archimedean property*: If $a > 0$ and $b > 0$, then for some positive integer n we have $na > b$.] Multiplying this inequality by -1 yields $-n(\ln 2) < -M$. Since $n(\ln 2) = \ln(2^n)$ and $-n(\ln 2) = \ln(2^{-n})$, we have $\ln(2^n) > M$ and $\ln(2^{-n}) < -M$, verifying the unboundedness of $y = \ln x$.

From our discussion we see that $y = \ln x$ has domain of definition $0 < x < \infty$ and range $-\infty < y < \infty$. Moreover, $y = \ln x$ takes on every value between $-\infty$ and ∞ and it does so only once because it is an increasing function. In particular, there is one and only one number e such that $\ln e = 1$; this number e is called the *base of the natural logarithm*. We now consider the question of calculating the number e.

For $x > 1$ we have

$$\frac{1}{x} < \frac{\ln x}{x - 1} < 1$$

by (1.3). For $x = 1 + 1/n$, where $n = 1, 2, 3, \ldots$, we have

$$\frac{n}{n + 1} < \frac{\ln(1 + 1/n)}{1/n} < 1 \quad \text{or} \quad \frac{n}{n + 1} < \ln\left(1 + \frac{1}{n}\right)^n < 1.$$

As n becomes arbitrarily large, $n/(n + 1)$ tends to 1 and so $\ln(1 + 1/n)^n$ tends to 1 as well. This suggests that the calculation of the number e will depend on a closer examination of the sequence $(1 + 1/n)^n$, where $n = 1$, $2, 3, \ldots$.

Lemma. *Let* $0 \leq a < b$; *then*

$$\frac{b^{n+1} - a^{n+1}}{b - a} < (n + 1)b^n \qquad (1.15)$$

and

$$\frac{b^{n+1} - a^{n+1}}{b - a} > (n + 1)a^n \qquad (1.16)$$

for $n = 1, 2, 3, \ldots$.

PROOF. Consider the identity

$$b^{n+1} - a^{n+1} = (b^n + ab^{n-1} + a^2 b^{n-2} + \cdots + a^{n-1}b + a^n)(b - a).$$

Then

$$\frac{b^{n+1} - a^{n+1}}{b - a} = b^n + ab^{n-1} + a^2 n^{n-2} + \cdots + a^{n-1}b + a^n$$
$$< b^n + bb^{n-1} + b^2 b^{n-2} + \cdots + b^{n-1}b + b^n = (n + 1)b^n$$

and

$$\frac{b^{n+1} - a^{n+1}}{b - a} = b^n + ab^{n-1} + a^2 n^{n-2} + \cdots + a^{n-1} + a^n$$
$$> a^n + aa^{n-1} + a^2 a^{n-2} + \cdots + a^{n-1}a + a^n = (n + 1)a^n.$$

This completes the proof. □

Proposition 1.4. *For* $n = 1, 2, 3, \ldots$ *let*

$$a_n = \left(1 + \frac{1}{n}\right)^n \quad \text{and} \quad b_n = \left(1 + \frac{1}{n}\right)^{n+1}.$$

Then $a_n < a_{n+1}$ *and* $b_n > b_{n+1}$.

PROOF. We can rewrite the inequality (1.15) as

$$b^n[b - (n + 1)(b - a)] < a^{n+1}.$$

Putting $a = 1 + 1/(n + 1)$ and $b = 1 + 1/n$, the term in brackets reduces to 1 and we have

$$\left(1 + \frac{1}{n}\right)^n < \left(1 + \frac{1}{n+1}\right)^{n+1};$$

this shows that $a_n < a_{n+1}$. Next we put $a = 1$ and $b = 1 + 1/(2n)$. This time the term in brackets reduces to $\frac{1}{2}$, and we have

$$\left(1 + \frac{1}{2n}\right)^n < 2 \quad \text{or} \quad \left(1 + \frac{1}{2n}\right)^{2n} < 4.$$

Since $a_n < a_{n+1}$,

$$\left(1 + \frac{1}{n}\right)^n < \left(1 + \frac{1}{2n}\right)^{2n} < 4$$

for any positive integer n. Noting that $a_1 = 2$, we see that

$$2 \leq a_n \leq 4$$

for $n = 1, 2, 3, \ldots$.

We can rewrite (1.16) as

$$b^{n+1} > a^{n+1} + (n + 1)a^n(b - a).$$

Putting $a = 1 + 1/(n + 1)$ and $b = 1 + 1/n$, we get

$$\left(1 + \frac{1}{n}\right)^{n+1} > \left(1 + \frac{1}{n+1}\right)^{n+1} + \frac{1}{n}\left(1 + \frac{1}{n+1}\right)^n$$

or

$$\left(1 + \frac{1}{n}\right)^{n+1} > \left(1 + \frac{1}{n+1}\right)^n\left(1 + \frac{1}{n+1} + \frac{1}{n}\right). \qquad (1.17)$$

But

$$\left(1 + \frac{1}{n}\right)^n\left(1 + \frac{1}{n+1} + \frac{1}{n}\right) > \left(1 + \frac{1}{n+1}\right)^{n+2} \qquad (1.18)$$

because

$$1 + \frac{1}{n+1} + \frac{1}{n} > \left(1 + \frac{1}{n+1}\right)^2$$

or, equivalently,

$$\frac{1}{n} - \frac{1}{n+1} = \frac{1}{n(n+1)} > \frac{1}{(n+1)^2}.$$

Hence, by (1.17) and (1.18), $b_n > b_{n+1}$ and the proof is finished.   □

Proposition 1.5. *For* $n = 1, 2, 3, \ldots$ *let* a_n *and* b_n *be as in Proposition 1.4. Then the closed intervals* $[a_1, b_1]$, $[a_2, b_2]$, $[a_3, b_3]$, $\ldots$ *form a nested sequence of intervals and* $b_n - a_n$ *tends to zero as n becomes arbitrarily large. By the Nested Interval Principle there is one and only one point common to all these intervals; this point is e, the base of the natural logarithm. The (increasing) sequence* $a_1, a_2, a_3, \ldots$ *and the (decreasing) sequence* $b_1, b_2, b_3, \ldots$ *tends to e as n becomes arbitrarily large.*

PROOF. To see that $b_n - a_n$, the length of the interval $[a_n, b_n]$, tends to zero as n becomes arbitrarily large, we note that

$$b_n - a_n = \frac{1}{n} a_n. \tag{1.19}$$

But a_n was seen to satisfy $2 \le a_n \le 4$ for all positive integers n and $1/n$ tends to zero as n becomes arbitrarily large. Knowing that $a_n < a_{n+1}$, $b_n > b_{n+1}$, and $b_n - a_n = (1/n)a_n > 0$ shows that the interval $[a_{n+1}, b_{n+1}]$ is contained in the interval $[a_n, b_n]$. Indeed, all conditions of the Nested Interval Principle are fulfilled and the conclusion of the proposition follows. $\qquad\square$

COMMENTS. Since $b_5 < 3$, it is clear that $2 \le a_n < 3$ for all positive integers n; here we have used the fact that for a fixed m the point b_m has to be to the right of the point a_n for any positive integer n. Rewriting (1.19) gives

$$\left(1 + \frac{1}{n}\right)^{n+1} - \left(1 + \frac{1}{n}\right)^{n} = \frac{1}{n}\left(1 + \frac{1}{n}\right)^{n}; \tag{1.20}$$

we also know that $e < 3$ and

$$\left(1 + \frac{1}{n}\right)^{n} < e < \left(1 + \frac{1}{n}\right)^{n+1}.$$

Hence, (1.20) yields, for any positive integer n,

$$e - \left(1 + \frac{1}{n}\right)^{n} < \frac{3}{n}. \tag{1.21}$$

Inequality (1.21) provides us with an error estimate in the calculation of e by using the approximating sequence $(1 + 1/n)^n$ for $n = 1, 2, 3, \ldots$; for example, if $n = 30,000$, then e can exceed $(1 + 1/n)^n$ only by an amount less than 0.0001. The number e is an irrational number, $e = 2.718281\ldots$. In the study of infinite series we shall encounter a more convenient way of calculating e; it is also in the study of infinite series where we shall come across effective methods of calculating the logarithm of a positive number. At this stage we simply use suitable pocket calculators or other aids such as tables for numerical work in connection with logarithms. For a sketch of the graph of the logarithmic function see Figure 1.10.

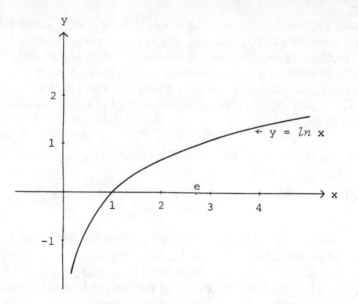

Figure 1.10

Proposition 1.6. *For any positive integers p and q, $p < q$, we have*

$$\ln\frac{q+1}{p} < \frac{1}{p} + \frac{1}{p+1} + \cdots + \frac{1}{q} < \ln\frac{q}{p-1}. \tag{1.22}$$

PROOF. Let $a = n$ and $b = n + 1$, for $n = 1, 2, 3, \ldots$, in (1.3); then

$$\frac{1}{n+1} < \ln\frac{n+1}{n} < \frac{1}{n}. \tag{1.23}$$

For $n = p, p + 1, \ldots, q$ the inequality (1.23) gives

$$\frac{1}{p+1} < \ln\frac{p+1}{p} < \frac{1}{p}, \qquad \frac{1}{p+2} < \ln\frac{p+2}{p+1} < \frac{1}{p+1}, \ldots,$$

$$\frac{1}{q+1} < \ln\frac{q+1}{q} < \frac{1}{q}.$$

Addition of the right halves of these $q - p$ inequalities gives the first half of (1.22) and the second half is obtained in like manner. □

APPLICATION. Let

$$H_n = \frac{1}{n+1} + \frac{1}{n+2} + \cdots + \frac{1}{2n}$$

for $n = 1, 2, 3, \ldots$ By (1.22) we obtain

$$\ln \frac{2n+1}{n+1} < H_n < \ln \frac{2n}{n} = \ln 2.$$

But we know that $\ln x$ tends to $\ln a$ as x tends to a and so it is clear that H_n tends to $\ln 2$ as n becomes arbitrarily large.

Let

$$T_n = 1 - \frac{1}{2} + \frac{1}{3} - \frac{1}{4} + \cdots - \frac{1}{2n}$$

for $n = 1, 2, 3, \ldots$. Then

$$T_n = 1 + \frac{1}{2} + \frac{1}{3} + \frac{1}{4} + \cdots + \frac{1}{2n} - 2\left(\frac{1}{2} + \frac{1}{4} + \cdots + \frac{1}{2n}\right)$$

$$= 1 + \frac{1}{2} + \frac{1}{3} + \frac{1}{4} + \cdots + \frac{1}{2n} - \left(1 + \frac{1}{2} + \cdots + \frac{1}{n}\right)$$

$$= H_n$$

and so T_n is seen to tend to $\ln 2$ as n becomes arbitrarily large.

Proposition 1.7. *For $n = 2, 3, 4, \ldots$ let*

$$x_n = 1 + \frac{1}{2} + \cdots + \frac{1}{n-1} - \ln n \tag{1.24}$$

and

$$y_n = 1 + \frac{1}{2} + \cdots + \frac{1}{n-1} + \frac{1}{n} - \ln n. \tag{1.25}$$

Then $x_{n+1} > x_n$, $y_{n+1} < y_n$, and $y_n - x_n = 1/n$. Hence, by the Nested Interval Principle, there is one and only one point C common to the closed intervals $[x_2, y_2]$, $[x_3, y_3]$, $[x_4, y_4]$, $\ldots$; this number C is called Euler's constant and it is known that $C = 0.5772156649\ldots$.

PROOF. By (1.23)

$$x_{n+1} - x_n = \frac{1}{n} - \ln(n+1) + \ln n = \frac{1}{n} - \ln \frac{n+1}{n} > 0$$

and

$$y_{n+1} - y_n = \frac{1}{n+1} - \ln(n+1) + \ln n = \frac{1}{n+1} - \ln \frac{n+1}{n} < 0.$$

It is evident that $y_n - x_n = 1/n$. We therefore see that the Nested Interval Principle is applicable, producing a unique number C. □

COMMENTS. To visualize Euler's constant, consider the following geometric situation. Given the closed interval $[1, n]$ and the curve $y = 1/x$ over this

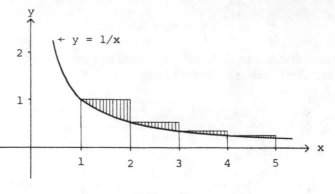

Figure 1.11

interval, the area of the region bounded above by the curve $y = 1/x$, bounded on the left by the line $x = 1$, bounded on the right by the line $x = n$, and bounded below by the x-axis is $\ln x$. On the other hand, dividing the interval $[1, n]$ into the $n - 1$ subinterval $[1, 2], [2, 3], \ldots, [n - 1, n]$ and erecting over $[1, 2]$ a rectangle of altitude 1, over $[2, 3]$ a rectangle of altitude $\frac{1}{2}$, ..., over $[n - 1, n]$ a rectangle of altitude $1/(n - 1)$, we can easily see that the region with vertical cross-hatching displayed in Figure 1.11 has area x_n [see (1.24)]. If we think of the $n - 1$ subregions with vertical cross-hatching as shifted to the left into the square bounded by the x-axis, the y-axis, and the lines $x = 1$ and $y = 1$ and, moreover, assume that n is becoming arbitrarily large, we can readily appreciate that C will be somewhat large than $\frac{1}{2}$ but definitely less than 1. A famous open question of mathematics asks: Is Euler's constant C a rational number (i.e., a ratio of two integers) or not?

APPLICATION. For $n = 1, 2, 3, \ldots$ let

$$S_n = 1 + \frac{1}{2} + \cdots + \frac{1}{n}. \tag{1.26}$$

Using a pocket calculator, we find

$$S_{10} = 2.9289683 \qquad S_{70} = 4.8328368$$

$$S_{20} = 3.5977397 \qquad S_{80} = 4.9654793$$

$$S_{30} = 3.9949871 \qquad S_{90} = 5.0825706$$

$$S_{40} = 4.278543 \qquad S_{100} = 5.1873775$$

$$S_{50} = 4.4992053 \qquad S_{110} = 5.2822346$$

$$S_{60} = 4.6798704 \qquad S_{120} = 5.3688683.$$

Let $z_n = (x_n + y_n)/2$, the midpoint of the interval $[x_n, y_n]$; x_n and y_n are given by (1.24) and (1.25). We have

$$z_{10} = 0.5763832 \qquad z_{70} = 0.5771987$$

$$z_{20} = 0.5770074 \qquad z_{80} = 0.5772027$$

$$z_{30} = 0.5771231 \qquad z_{90} = 0.5772054$$

$$z_{40} = 0.5771635 \qquad z_{100} = 0.5772073$$

$$z_{50} = 0.5771828 \qquad z_{110} = 0.5772088$$

$$z_{60} = 0.5771925 \qquad z_{120} = 0.5772099.$$

It is noteworthy how closely z_{120} approximates $C = 0.5772156649\ldots$.

Again, by pocket calculator we obtain $\ln 10 = 2.3025851$, $x_{10} = 0.5263832$, and $y_{10} = 0.6263832$. By Proposition 1.7, $S_n - \ln n = y_n$ has to be between x_{10} and y_{10} for $n > 10$; hence

$$23.55 < S_n < 23.66 \quad \text{for } n = 10^{10}.$$

Using the fact that $x_{120} = 0.5730432$ and $y_{120} = 0.5813765$, we get

$$23.598 < S_n < 23.608 \quad \text{for } n = 10^{10}.$$

In both of these estimates we have used the relation $S_n = y_n + \ln n$ together with the fact that $x_m < y_n < y_m$ for $n > m$. Both x_n and y_n tend to C as n becomes arbitrarily large. Evidently, S_n becomes arbitrarily large as n becomes arbitrarily large. However, S_n becomes arbitrarily large very slowly.

3. The Exponential Function

Rational powers of e have an established meaning: By $e^{n/m}$ we mean the mth root of e raised to the nth power. Moreover, we have seen that

$$\ln e^{n/m} = \frac{n}{m}. \tag{1.27}$$

Definition. If t is an irrational number, then by e^t we mean the unique number which has logarithm t:

$$\ln e^t = t. \tag{1.28}$$

Definition. The function

$$E(x) = e^x \quad \text{for all real numbers } x$$

is called the *exponential function*.

COMMENTS. Using (1.27) and (1.28) and writing

$$L(x) = \ln x \quad \text{and} \quad E(x) = e^x,$$

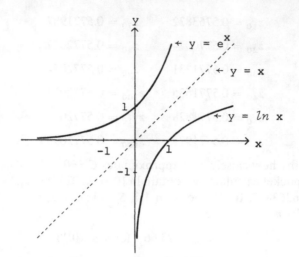

Figure 1.12

we obtain

$$L(E(x)) = x \quad \text{for all real numbers } x. \tag{1.29}$$

This means that *the exponential function is the inverse function of the logarithmic function.* Another way of expressing the relation that the exponential and the logarithmic functions are inverses of each other is

$$E(L(x)) = x \quad \text{for all positive real numbers.} \tag{1.30}$$

In more conventional notation (1.29) and (1.30) read:

$$\ln e^x = x \quad \text{for all real numbers } x \tag{1.31}$$

and

$$e^{\ln x} = x \quad \text{for all real } x > 0. \tag{1.32}$$

The graph of the exponential function appears in Figure 1.12; it can be obtained from the graph of the logarithmic function by reflection in the line $y = x$. It is clear that $x = e^y$ if and only if $y = \ln x$. Since the graph of the logarithmic function remains to the right of the y-axis, the graph of the exponential function remains above the x-axis; namely,

$$e^x > 0 \quad \text{for all real numbers } x. \tag{1.33}$$

Since the graph of the logarithmic function crosses the x-axis at $x = 1$, the graph of the exponential function crosses the y-axis at $y = 1$. Recalling that the logarithmic function is an increasing and continuous function and noting that the exponential function is the inverse of the logarithmic function, it will follow by general theory that the exponential function is an increasing and continuous function.

[Since it is easy to give a direct proof of the continuity of the exponential function, we shall do so now. Let $x = e^y$ and $x + h = e^{y+v}$. Then

$$v = A_{x, x+h}$$

and by (1.3) it follows that $|v| > h/(x + h)$ if $h > 0$ and $|v| > |h|/x$ if $h < 0$. This shows that if v is small in absolute value, then h must also be small in absolute value. Note that $|v| = v$ if $v \geq 0$ and $|v| = -v$ if $v \leq 0$.]

We will now make use of the logarithmic and the exponential functions to define the expression a^b for $a > 0$ and b denoting any real number; the elementary notion of exponent applies only when b is a rational number. We have already looked at the case $a = e$. If $a > 0$ and n/m is rational, we get

$$a^{n/m} = e^{(n/m)(\ln a)}. \tag{1.34}$$

To see that (1.34) is true, we need only to take logarithms on both sides of the equation. Next we note that the right-hand side of (1.34) is of the form

$$e^{b(\ln a)}$$

and has meaning for any real number b, be it rational or irrational. We are now ready to define the expression a^b for $a > 0$ and b denoting an irrational number.

Definition. If b is an irrational number and a is positive, we define the number a^b by setting

$$a^b = e^{b(\ln a)}. \tag{1.35}$$

Lemma. *Let $n = 2, 3, \ldots$; then $\sqrt[n]{n}$ tends to 1 as n becomes arbitrarily large.*

PROOF. Let $\sqrt[n]{n} = 1 + v_n$ for $n = 2, 3, \ldots$. Then $v_n > 0$ and

$$n = (1 + v_n)^n = 1 + nv_n + \frac{n(n-1)}{2} v_n^2 + \cdots + v_n^n.$$

Since all terms in the last sum are positive, we have

$$n > 1 + \frac{n(n-1)}{2} v_n^2 \quad \text{or} \quad 1 > \frac{n}{2} v_n^2.$$

Thus,

$$0 < v_n < \sqrt{\frac{2}{n}} \quad \text{for } n = 2, 3, \ldots$$

It is clear that $\sqrt{2/n} = \sqrt{2}/\sqrt{n}$ tends to 0 as n becomes arbitrarily large. Since v_n is wedged between 0 and $\sqrt{2/n}$ for $n = 2, 3, \ldots$, it can be seen that v_n tends to 0 as n becomes arbitrarily large and so $\sqrt[n]{n}$ must tend to 1 as n becomes arbitrarily large.   □

Proposition 1.8. *The function* ln *x tends to infinity with x, but more slowly than any positive power of* x. *In other words,* ln *x becomes arbitrarily large with x but* $(\ln x)/x^s$, *for any* $s > 0$, *tends to* 0 *as x becomes arbitrarily large.*

PROOF. Since ln t tends to ln a as t tends to a (as noted in Section 2), we get that ln $\sqrt[n]{n} = (1/n)\ln n$ tends to 0 as n becomes arbitrarily large by the Lemma. However, n takes the integer values $2, 3, \ldots$ in the Lemma. We must verify that $(\ln t)/t$ tends to 0 as t becomes arbitrarily large and t denoting any real number. We do this now. Let $[t]$ stand for the integer part of t. Then

$$0 < \frac{\ln t}{t} < \frac{\ln([t] + 1)}{[t]} = \frac{[t] + 1}{[t]} \cdot \frac{\ln([t] + 1)}{[t] + 1} \quad \text{for } t > 1$$

shows that $(\ln t)/t$ tends to 0 as t becomes arbitrarily large (note that $t - 1 < [t] \le t$). Letting $t = x^s$ for any $s > 0$, we see that

$$\frac{\ln x^s}{x^s} = s \cdot \frac{\ln x}{x^s}$$

tends to 0 as x becomes arbitrarily large. But s is positive and fixed.   □

Proposition 1.9. *The function* e^y *tends to infinity with y more rapidly than any power of* y, *or* y^t/e^y, *for any* $t > 0$, *tends to* 0 *when y becomes arbitrarily large for all values of t however great.*

PROOF. In Proposition 1.8 we saw that, for any positive value of s, $x^{-s}(\ln x)$ tends to 0 when x becomes arbitrarily large. Putting $t = 1/s$, we see that $x^{-1}(\ln x)^t$ tends to 0 as x becomes arbitrarily large for any $t > 0$. The desired result then follows on putting $x = e^y$.   □

DISCUSSION. In Section 2 we observed that the logarithmic function $L(x) = \ln x$ satisfies the functional equation

$$L(x_1 x_2) = L(x_1) + L(x_2) \quad \text{for any } x_1 > 0 \text{ and } x_2 > 0.$$

We now note that the exponential function $E(y) = e^y$ satisfies the functional equation

$$E(y_1 + y_2) = E(y_1)E(y_2) \quad \text{for any real numbers } y_1 \text{ and } y_2.$$

Indeed, let $y_1 = \ln x_1$ and $y_2 = \ln x_2$. Then $x_1 = e^{y_1}$, $x_2 = e^{y_2}$, and

$$y_1 + y_2 = \ln x_1 + \ln x_2 = \ln(x_1 x_2)$$

or

$$e^{y_1 + y_2} = e^{\ln(x_1 x_2)} = x_1 x_2 = e^{y_1} e^{y_2}.$$

More generally, for $a > 0$, we have

$$a^{y_1+y_2} = a^{y_1}a^{y_2}$$

because

$$a^{y_1+y_2} = e^{(y_1+y_2)(\ln a)} = e^{y_1(\ln a)}e^{y_2(\ln a)} = a^{y_1}a^{y_2}.$$

We also remark that

$$(a^{y_1})^{y_2} = (e^{y_1(\ln a)})^{y_2} = e^{y_1 y_2(\ln a)} = a^{y_1 y_2}.$$

If $a > 1$ then $a^x = e^{x(\ln a)} = e^{cx}$, where $c = \ln a$ is positive. The graph of a^x is in this case similar to that of e^x and a^x becomes infinite with x more rapidly than any power of x.

If $0 < a < 1$ then the graph of a^x is obtained from the graph of b^x, where $b = 1/a$ (and so $b > 1$), by reflecting the graph of $y = b^x$ with respect to the y-axis.

It is easy to express logarithms to a base other than e in terms of natural logarithms. If for a positive number a, with $a \neq 1$, the equation $x = a^y$ is satisfied, we write

$$y = \log_a x$$

and say that y is the logarithm of x with respect to the base a. Now $a^y = e^{y(\ln a)}$ so that $x = e^{y(\ln a)}$ or $y(\ln a) = \ln x$. It follows that

$$\log_a x = \frac{\ln x}{\ln a}.$$

Since logarithms to any base a, where $a > 0$ and $a \neq 1$, are proportional to natural logarithms, they satisfy the usual identities,

$$\log_a x + \log_a y = \log_a (xy),$$

$$\log_a \left(\frac{x}{y}\right) = \log_a x - \log_a y,$$

$$\log_a (x^c) = c(\log_a x),$$

for any positive real numbers x and y and any real number c.

Since $\log_A B = (\ln B)/(\ln A)$ for any positive real number B and any positive real number A with $A \neq 1$, we get at once that

$$\log_a x = (\log_a x)(\log_a b),$$

$$\log_b a = \frac{1}{\log_a b},$$

$$\log_a x = \frac{\log_b x}{\log_b a}$$

for any positive real numbers x, a, and b with $a \neq 1$ and $b \neq 1$. It is clear that $\log_a 1 = 0$ and $\log_a a = 1$. Moreover, we can write $\log_e x$ in place of $\ln x$.

Examples

1. Simplify $1/(\log_2 5) + 1/(\log_3 5) - (\log_5 3)/(\log_{30} 3)$.

SOLUTION. Since $1/(\log_2 5) = \log_5 2$ and $1/(\log_3 5) = \log_5 3$, we get that $1/(\log_2 5) + 1/(\log_3 5) = \log_5 2 + \log_5 3 = \log_5 6$. On the other hand, $(\log_5 3)/(\log_{30} 3) = (\log_5 3)(\log_3 30) = \log_5 30$. But $\log_5 6 - \log_5 30 = \log_5 (1/5) = -1$ and so the given expression simplifies to -1.

2. Let $\log_4 \log_3 \log_2 x = 0$. Find x.

SOLUTION. We have $\log_4 (\log_3 \log_2 x) = 0 = \log_4 1$ and so $\log_3 \log_2 x = 1$. But $\log_3 (\log_2 x) = 1 = \log_3 3$ implies that $\log_2 x = 3$ and hence $x = 8$.

3. Find x if $\log_{16} x + \log_4 x + \log_2 x = 7$.

SOLUTION. Since $\log_4 x = (\log_2 x)/(\log_2 4) = (\log_2 x)/2$ and $\log_{16} x = (\log_2 x)/(\log_2 16) = (\log_2 x)/4$, we see that $(1 + \frac{1}{2} + \frac{1}{4})(\log_2 x) = 7$ or $\log_2 x = 4$. We therefore have $x = 16$.

4. Express $\log_5 12$ in terms of $a = \log_{10} 2$ and $b = \log_{10} 3$.

SOLUTION. Let $\log_5 12 = x$; then $5^x = 12$ or $10^x(2^{-x}) = 12$. Taking logarithms to the base 10 gives $x(\log_{10} 10) - x(\log_{10} 2) = \log_{10} 12$. But $\log_{10} 12 = \log_{10} 3 + 2(\log_{10} 2) = b + 2a$ and so $x - xa = b + 2a$ or $x = (b + 2a)/(1 - a)$.

5. Consider a given geometric and arithmetic progression with positive terms:

$$G, G_1, G_2, \ldots, G_n, \ldots \quad \text{and} \quad A, A_1, A_2, \ldots, A_n, \ldots$$

The ratio of the geometric progression and the common difference of the arithmetic progression are positive. Show that there exists a system of logarithms for which

$$\log_c G_n - \log_c G = A_n - A$$

for any n and find the base c of this system.

SOLUTION. Let $G_n = Gq^n$ and $A_n = A + nd$. Then $\log_c G_n - \log_c G = n(\log_c q)$ and $A_n - A = nd$. Hence, $n(\log_c q) = nd$, that is, $\log_c q = d$. We therefore obtain $c^d = q$ or $c = q^{1/d}$.

Lemma. Let $b > 1$ and $n = 2, 3, \ldots$. Then $\sqrt[n]{b}$ tends to 1 as n becomes arbitrarily large.

PROOF. Since $b > 1$ we have $\sqrt[n]{b} > 1$ for $n = 2, 3, \ldots$. We set $\sqrt[n]{b} = 1 + h_n$. Then, for $n = 2, 3, \ldots$, we have $h_n > 0$ and

$$b = (1 + h_n)^n = 1 + nh_n + \frac{n(n-1)}{2} h_n^2 + \cdots + h_n^n.$$

Since all terms in the last sum are positive, we have

$$0 < h_n < \frac{b-1}{n}.$$

As n becomes arbitrarily large, h_n tends to 0 and $\sqrt[n]{b}$ tends to 1. $\qquad \square$

REMARK. An alternate way of proving the Lemma depends on observing that $\ln(\sqrt[n]{b}) = (1/n)(\ln b)$ tends to 0 as n becomes arbitrarily large.

Proposition 1.10. *Let $b > 1$. Then $n(\sqrt[n]{b} - 1)$, for $n = 2, 3, \ldots$, tends to $\ln b$ as n becomes arbitrarily large.*

PROOF. Let n be a positive integer larger than 1 and put $q = \sqrt[n]{b}$. Then

$$1 < q < q^2 < q^3 < \cdots < q^{n-1} < q^n = b.$$

Let $x_0 = 1$, $x_1 = q$, $x_2 = q^2$, $x_3 = q^3$, $\ldots$, $x_{n-1} = q^{n-1}$, $x_n = q^n = b$ and suppose that the interval $[1, b]$ is split up into the n subintervals

$$[x_0, x_1], \quad [x_1, x_2], \quad [x_2, x_3], \quad \ldots, \quad [x_{n-1}, x_n].$$

On the interval $[x_0, x_1]$ we construct the rectangle with altitude $1/x_0$, on $[x_1, x_2]$ the rectangle with altitude $1/x_1$, on $[x_2, x_3]$ the rectangle with altitude $1/x_2$, $\ldots$, and on $[x_{n-1}, x_n]$ the rectangle with altitude $1/x_{n-1}$. The sum of the areas of these n rectangles is

$$(q - 1) + \frac{1}{q}(q^2 - q) + \frac{1}{q^2}(q^3 - q^2) + \cdots + \frac{1}{q^{n-1}}(q^n - q^{n-1}) = n(q - 1).$$

The foregoing sum is an approximating sum for $A_{1,b} = \ln b$. We know that q tends to 1 as n becomes arbitrarily large; this means that we can make the length of the largest of the subintervals $[x_0, x_1]$, $[x_1, x_2]$, $[x_2, x_3]$, $\ldots$, $[x_{n-1}, x_n]$ as small as we please by taking n sufficiently large and our claim therefore follows. $\qquad \square$

COMMENTS. Let $a > 1$ and $b > 1$. Then the relation

$$n(\sqrt[n]{ab} - 1) = n(\sqrt[n]{a} - 1)\sqrt[n]{b} + n(\sqrt[n]{b} - 1)$$

naturally lends itself to deduce the functional equation for the logarithm

$$\ln(ab) = \ln a + \ln b.$$

Again let $b > 1$ and define $x_n = (b)^{1/2^n}$ for $n = 1, 2, 3, \ldots$. Then $x_{n+1}^2 = x_n$ and $x_{n+1} < x_n$. Moreover, x_n tends to 1 as n becomes arbitrarily large. We define

$$b_n = 2^n(x_n - 1) \quad \text{and} \quad a_n = 2^n\left(1 - \frac{1}{x_n}\right).$$

for $n = 1, 2, 3, \ldots$. We shall now verify that the sequence of closed intervals

$$[a_1, b_1], \quad [a_2, b_2], \quad [a_3, b_3], \ldots$$

satisfies the conditions of the Nested Interval Principle (See Section 1). To see that $b_n < b_{n-1}$ we note that $x_n > 1$ and so

$$(x_n + 1)(x_n - 1) = x_n^2 - 1 = x_{n-1} - 1$$

implies $2(x_n - 1) < x_{n-1} - 1$ and thus $2^n(x_n - 1) < 2^{n-1}(x_{n-1} - 1)$. To see that $a_n > a_{n-1}$ we note that $1/x_n < 1$ and so

$$\left(1 + \frac{1}{x_n}\right)\left(1 - \frac{1}{x_n}\right) = 1 - \frac{1}{x_n^2} = 1 - \frac{1}{x_{n-1}}$$

implies $2(1 - 1/x_n) > 1 - 1/x_{n-1}$ and thus $2^n(1 - 1/x_n) > 2^{n-1}(1 - 1/x_{n-1})$. To verify $a_n < b_n$ we observe that

$$x_n - 1 < x_n(x_n - 1)$$

because $x_n > 1$. Hence, $1 - 1/x_n < x_n - 1$ or $2^n(1 - 1/x_n) < 2^n(x_n - 1)$. Finally, to verify that $b_n - a_n$ tends to 0 as n becomes arbitrarily large we only need to realize that

$$x_n a_n = b_n$$

and that x_n tends to 1 as n becomes arbitrarily large. The unique point common to all intervals $[a_1, b_1], [a_2, b_2], [a_3, b_3], \ldots$ is $\ln b$.

Proposition 1.11. *Let $n = 1, 2, 3, \ldots$ and x be any real number. Then both*

$$\left(1 + \frac{x}{n}\right)^n \quad \text{and} \quad \left(1 - \frac{x}{n}\right)^{-n}$$

tend to e^x as n becomes arbitrarily large.

PROOF. By (1.3) we have that $(1/h)\ln(1 + xh) = (1/h)A_{1,1+xh}$ is between x and $x/(1 + xh)$. Thus, $(1/h)\ln(1 + xh)$ tends to x as h tends to 0. Putting $h = 1/k$, we see that $k\{\ln(1 + x/k)\}$ tends to x when k, taking integer values, tends to ∞ or $-\infty$. Since the exponential function is continuous,

$$\left(1 + \frac{x}{n}\right)^k = e^{k\{\ln(1 + x/k)\}}$$

tends to e^x as k tends to ∞ or $-\infty$.   □

REMARK. For $n = 2, 3, 4, \ldots$

$$\left(1 - \frac{1}{n}\right)^{-n} = \left(\frac{n-1}{n}\right)^{-n} = \left(\frac{n}{n-1}\right)^n = \left(1 + \frac{1}{n-1}\right)^n.$$

Proposition 1.12. *Let $x_1, x_2, x_3, \ldots$ be a sequence of numbers. If nx_n tends to K as n becomes arbitrarily large, then $(1 + x_n)^n$ tends to e^K.*

PROOF. If nx_n tends to K as n becomes arbitrarily large, we see that x_n tends to 0 and hence $\{\ln(1 + x_n)\}/x_n$ tends to 1 (see proof of Proposition 1.11). Writing $n\{\ln(1 + x_n)\}$ in the form

$$K\left(\frac{nx_n}{K}\right)\frac{\ln(1 + x_n)}{x_n}$$

we see that $n\{\ln(1 + x_n)\}$ tends to K as n becomes arbitrarily large.   □

APPLICATION. Let $a > 0$ and $b > 0$. We show that, for $n = 2, 3, 4, \ldots$,

$$\left(\frac{\sqrt[n]{a} + \sqrt[n]{b}}{2}\right)^n$$

tends to $\sqrt{ab}$ as n becomes arbitrarily large.

Indeed, let $x_n = (\sqrt[n]{a} + \sqrt[n]{b})/2 - 1$. Then

$$nx_n = \tfrac{1}{2}\{n(\sqrt[n]{a} - 1) + n(\sqrt[n]{b} - 1)\} \quad \text{and} \quad 1 + x_n = \frac{\sqrt[n]{a} + \sqrt[n]{b}}{2}.$$

By Proposition 1.10 we have that nx_n tends to $\tfrac{1}{2}(\ln a + \ln b) = \ln\sqrt{ab}$ as n becomes arbitrarily large. Hence, by Proposition 1.12, we have that $(1 + x_n)^n$ tends to

$$e^{\ln\sqrt{ab}} = \sqrt{ab}$$

and we have what we wanted to show.

Using the same method of proof we can show that for $a_1 > 0$, $a_2 > 0$, $\ldots$, $a_m > 0$ the sequence

$$\left(\frac{\sqrt[n]{a_1} + \sqrt[n]{a_2} + \cdots + \sqrt[n]{a_m}}{m}\right)^n \quad \text{for } n = 2, 3, 4, \ldots$$

tends to $\sqrt[m]{a_1 a_2 \ldots a_m}$ as n becomes arbitrarily large. Indeed, we let

$$x_n = \frac{\sqrt[n]{a_1} + \sqrt[n]{a_2} + \cdots + \sqrt[n]{a_m}}{m} - 1$$

and note that

$$nx_n = \frac{1}{m}\{n(\sqrt[n]{a_1} - 1) + n(\sqrt[n]{a_2} - 1) + \cdots + n(\sqrt[n]{a_m} - 1)\}.$$

We then observe that nx_n tends to $\ln(\sqrt[m]{a_1 a_2 \ldots a_m})$ and $(1 + x_n)^n$ tends to $\sqrt[m]{a_1 a_2 \ldots a_m}$ as n becomes arbitrarily large.

In Chapter 4 we shall establish some interesting properties of the function

$$f(x) = \left(\frac{a^x + b^x}{2}\right)^{1/x} \quad \text{for } x \neq 0,$$

$$= \sqrt{ab} \qquad\qquad \text{for } x = 0;$$

for details see Proposition 4.19.

4. The Hyperbolic Functions

Certain combinations of exponential functions (which are related to the hyperbola $x^2 - y^2 = 1$ in somewhat the same manner that trigonometric functions are related to the circle $x^2 + y^2 = 1$) appear so frequently in mathematics that they have been given special names. These functions are called the *hyperbolic functions* and their similarity to trigonometric functions is emphasized by calling them *hyperbolic sine, hyperbolic cosine, hyperbolic tangent*, and so on. They are defined as follows:

(i) $\sinh x = \dfrac{e^x - e^{-x}}{2}$.

(ii) $\cosh x = \dfrac{e^x + e^{-x}}{2}$.

(iii) $\tanh x = \dfrac{\sinh x}{\cosh x} = \dfrac{e^x - e^{-x}}{e^x + e^{-x}}$.

(iv) $\coth x = \dfrac{\cosh x}{\sinh x} = \dfrac{e^x + e^{-x}}{e^x - e^{-x}}$.

(v) $\operatorname{sech} x = \dfrac{1}{\cosh x} = \dfrac{2}{e^x + e^{-x}}$.

(vi) $\operatorname{csch} x = \dfrac{1}{\sinh x} = \dfrac{2}{e^x - e^{-x}}$.

The six hyperbolic functions satisfy identities that correspond to the usual trigonometric identities except for an occasional switch of plus and minus signs. For example, we have the following identities:

(1) $\cosh^2 x - \sinh^2 x = 1$.
(2) $1 - \tanh^2 x = \operatorname{sech}^2 x$.
(3) $\coth^2 x - 1 = \operatorname{csch}^2 x$.
(4) $\sinh(s \pm t) = (\sinh s)(\cosh t) \pm (\cosh s)(\sinh t)$.
(5) $\cosh(s \pm t) = (\cosh s)(\cosh t) \pm (\sinh s)(\sinh t)$.
(6) $\sinh 2x = 2(\sinh x)(\cosh x)$.
(7) $\cosh 2x = \cosh^2 x + \sinh^2 x = 2(\cosh^2 x) - 1 = 2(\sinh^2 x) + 1$.

The graphs of the six hyperbolic functions are shown in Figure 1.13.

Since the hyperbolic functions are defined in terms of exponential functions, it is not surprising to find that the inverse hyperbolic functions can be written in terms of logarithmic functions. The formulas for the inverse hyperbolic functions are:

(i) Inverse hyperbolic sine:

$$\sinh^{-1} x = \ln(x + \sqrt{x^2 + 1}) \quad \text{for any real number } x.$$

(ii) Inverse hyperbolic cosine:

$$\cosh^{-1} x = \ln(x + \sqrt{x^2 - 1}) \quad \text{for } x \geq 1.$$

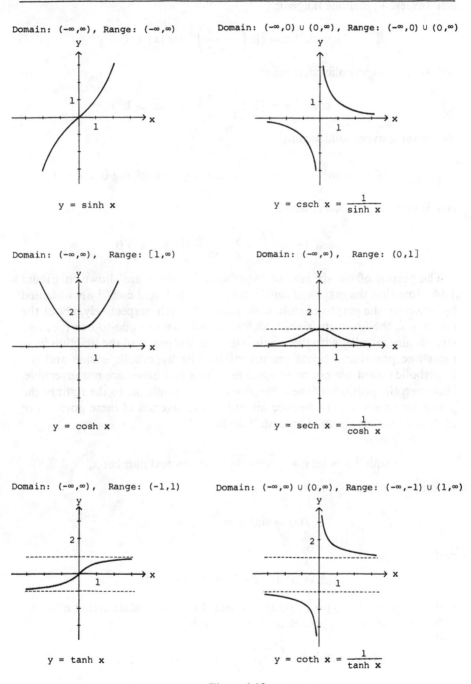

Domain: $(-\infty,\infty)$, Range: $(-\infty,\infty)$

$y = \sinh x$

Domain: $(-\infty,0) \cup (0,\infty)$, Range: $(-\infty,0) \cup (0,\infty)$

$y = \operatorname{csch} x = \dfrac{1}{\sinh x}$

Domain: $(-\infty,\infty)$, Range: $[1,\infty)$

$y = \cosh x$

Domain: $(-\infty,\infty)$, Range: $(0,1]$

$y = \operatorname{sech} x = \dfrac{1}{\cosh x}$

Domain: $(-\infty,\infty)$, Range: $(-1,1)$

$y = \tanh x$

Domain: $(-\infty,\infty) \cup (0,\infty)$, Range: $(-\infty,-1) \cup (1,\infty)$

$y = \coth x = \dfrac{1}{\tanh x}$

Figure 1.13

(iii) Inverse hyperbolic tangent:

$$\tanh^{-1} x = \tfrac{1}{2}\ln\left(\frac{1+x}{1-x}\right) \quad \text{for } |x| < 1.$$

(iv) Inverse hyperbolic cotangent:

$$\coth^{-1} x = \tfrac{1}{2}\ln\left(\frac{1+x}{1-x}\right) \quad \text{for } |x| > 1.$$

(v) Inverse hyperbolic secant:

$$\operatorname{sech}^{-1} x = \ln\left(\frac{1+\sqrt{1-x^2}}{x}\right) \quad \text{for any } x \text{ satisfying } 0 < x \le 1.$$

(vi) Inverse hyperbolic cosecant:

$$\operatorname{csch} x = \ln\left(\frac{1}{x} + \frac{\sqrt{1+x^2}}{|x|}\right) \quad \text{for } x \ne 0.$$

The graphs of the six inverse hyperbolic functions are shown in Figure 1.14. Note that the graphs of $\sinh^{-1}$, $\tanh^{-1}$, $\coth^{-1}$, and csch^{-1} are obtained by reflecting the graphs of sinh, tanh, coth, and csch, respectively, about the line $y = x$; the functions sinh, tanh, coth, and csch are one-to-one (i.e., any straight line parallel with the x-axis intersects the graph of the function in at most one point) and hence are invertible. The hyperbolic cosine and the hyperbolic secant are not one-to-one functions and hence are not invertible. However, the portions of these functions whose graphs lie to the right of the y-axis are one-to-one and hence invertible; the inverses of these portions of cosh and sech are denoted by $\cosh^{-1}$ and sech^{-1}.

To verify that

$$\sinh^{-1} x = \ln(x + \sqrt{x^2 + 1}) \quad \text{for any real number } x,$$

we only have to set

$$f(x) = \sinh x = \frac{e^x - e^{-x}}{2}$$

and

$$g(x) = \sinh^{-1} x = \ln(x + \sqrt{x^2 + 1})$$

and show that $f[g(x)] = g[f(x)] = x$. But this is a calculation that offers no difficulties. Another approach is to let $y = \sinh^{-1} x$ and so

$$x = \sinh y = \frac{e^y - e^{-y}}{2} \quad \text{or} \quad 2x = e^y - e^{-y}.$$

Multiplying both sides of the latter equation by e^y, we get

$$2xe^y = (e^y)^2 - 1 \quad \text{or} \quad (e^y)^2 - 2x(e^y) - 1 = 0.$$

Domain: $(-\infty,\infty)$,   Range: $(-\infty,\infty)$ Domain: $(-\infty,0) \cup (0,\infty)$, Range: $(-\infty,0) \cup (0,\infty)$

Domain: $[1,\infty)$, Range: $[0,\infty)$ Domain: $(0,1]$, Range: $[0,\infty)$

Domain: $(-1,1)$, Range: $(-\infty,\infty)$ Domain: $(-\infty,-1) \cup (1,\infty)$, Range: $(-\infty,0) \cup (0,\infty)$

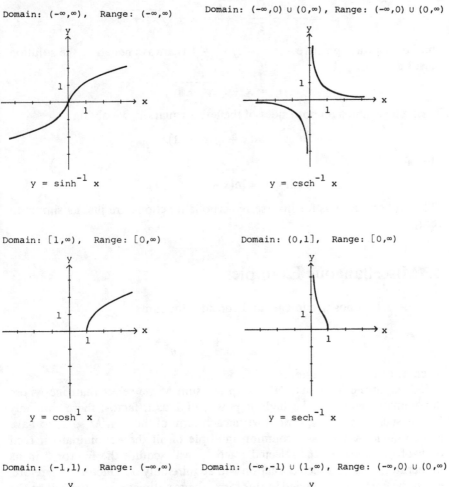
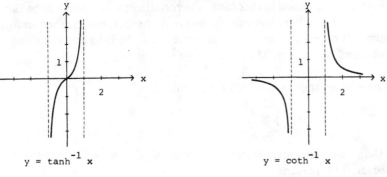

$y = \sinh^{-1} x$ $y = \operatorname{csch}^{-1} x$ $y = \cosh^{-1} x$ $y = \operatorname{sech}^{-1} x$ $y = \tanh^{-1} x$ $y = \coth^{-1} x$

Figure 1.14

Thus,

$$e^y = \frac{2x \pm \sqrt{4x^2 + 4}}{2} = x \pm \sqrt{x^2 + 1}.$$

Since e^y is always positive and $x - \sqrt{x^2 + 1}$ is always negative, the solution must be

$$e^y = x + \sqrt{x^2 + 1}.$$

Taking logarithms on both sides of the latter equation, we obtain

$$y = \ln(x + \sqrt{x^2 + 1}).$$

Therefore,

$$\sinh^{-1} x = \ln(x + \sqrt{x^2 + 1}).$$

The other formulas for inverse hyperbolic functions are just as simple to verify.

5. Miscellaneous Examples

1. Let n be a positive integer and consider the sum

$$M = \frac{1}{2} + \frac{1}{3} + \cdots + \frac{1}{n}.$$

Then M cannot be an integer.

Indeed, of the fractions making up the sum M we select that one whose denominator contains the highest power of 2 as a factor; there can only be one such term. Now, if we rewrite each term of the sum M so as to have as denominator the least common multiple of all the denominators, then each of them, except the selected fraction, will acquire the fractor 2 in its numerator, but the selected fraction will acquire only odd factors. Therefore, when the fractions are added in this form, the resulting numerator will be the sum of several even numbers and exactly one odd number, but the (common) denominator will be even. Hence, the numerator will be odd and the denominator even, and so the sum M cannot be an integer.

2. We know by Proposition 1.7 that the sum

$$1 + \frac{1}{2} + \frac{1}{3} + \cdots + \frac{1}{n}$$

is greater than any previously selected number N, if n is taken sufficiently large. However, if in the sum

$$1 + \frac{1}{2} + \frac{1}{3} + \cdots + \frac{1}{n}$$

we discard every term that contains the digit 9 in its denominator, then the sum of the remaining terms, for any n, will be less than 80.

Indeed, let n_k denote the number of undeleted fractions between $1/10^k$ and $1/10^{k+1}$. If the fraction $1/q$, lying between these two fractions, is one of the undeleted numbers, then of the numbers

$$\frac{1}{10q}, \quad \frac{1}{10q+1}, \quad \frac{1}{10q+2}, \quad \cdots, \quad \frac{1}{10q+8}, \quad \frac{1}{10q+9}$$

(all of which lie between $1/10^k$ and $1/10^{k+1}$), only the final fraction will be deleted when those containing a digit 9 in the denominator are crossed out. If $1/q$ is one of the deleted numbers, then all of the additional fractions

$$\frac{1}{10q}, \quad \frac{1}{10q+1}, \quad \cdots, \quad \frac{1}{10q+9}$$

will also be deleted. It follows that

$$n_k = 9n_{n-1}.$$

Of the fractions $1, \frac{1}{2}, \frac{1}{3}, \ldots, \frac{1}{8}, \frac{1}{9}$, only $\frac{1}{9}$ is deleted; hence $n_0 = 8$ and

$$n_1 = 8 \cdot 9 = 72,$$
$$n_2 = 8 \cdot 9^2,$$
$$\vdots$$
$$n_k = 8 \cdot 9^k.$$

Now consider, for $n < 10^{m+1}$, the sum

$$1 + \frac{1}{2} + \frac{1}{3} + \cdots + \frac{1}{n}.$$

We add up the sum

$$1 + \frac{1}{2} + \frac{1}{3} + \cdots + \frac{1}{10^{m+1} - 1},$$

after deleting all those fractions having a digit 9 in the denominator. Then we get

$$\left(1 + \frac{1}{2} + \frac{1}{3} + \cdots + \frac{1}{8}\right)$$

$$+ \left(\frac{1}{10} + \frac{1}{11} + \frac{1}{12} + \cdots + \frac{1}{18} + \frac{1}{20} + \cdots + \frac{1}{88}\right)$$

$$+ \left(\frac{1}{100} + \frac{1}{101} + \cdots + \frac{1}{888}\right) + \cdots + \left(\frac{1}{10^m} + \cdots + \frac{1}{88\ldots8}\right)$$

$$< 1 \cdot n_0 + \frac{1}{10} \cdot n_1 + \frac{1}{100} \cdot n_2 + \cdots + \frac{1}{10^{m-1}} \cdot n_{m-1} + \frac{1}{10^m} \cdot n_m$$

with $88\ldots8$ denoting the number made up of $m + 1$ digits 8. If we now replace each sum in parentheses by the product of the largest term contained therein and the number of terms in those parentheses, we obtain

$$1 \cdot n_0 + \frac{1}{10} \cdot n_1 + \frac{1}{100} \cdot n_2 + \cdots + \frac{1}{10^{m-1}} \cdot n_{m-1} + \frac{1}{10^m} \cdot n_m$$

$$= 8\left(1 + \frac{9}{10} + \frac{9^2}{10^2} + \cdots + \frac{9^{m-1}}{10^{m-1}} + \frac{9^m}{10^m}\right)$$

$$= 8 \cdot \frac{1 - (9/10)^{m+1}}{1 - 9/10} < 8 \cdot \frac{1}{1 - 9/10} = 8 \cdot 10 = 80.$$

This completes the proof of what was claimed at the start.

3. For any positive integer n the sum

$$1 + \frac{1}{2^2} + \frac{1}{3^2} + \cdots + \frac{1}{n^2}$$

is situated between the values

$$\left(1 - \frac{2}{n+1}\right)\left(1 - \frac{2}{2n+1}\right) \cdot \frac{\pi^2}{6} \quad \text{and} \quad \left(1 - \frac{1}{2n+1}\right)\left(1 + \frac{1}{2n+1}\right) \cdot \frac{\pi^2}{6}.$$

Letting $n \to \infty$, we readily see that

$$1 + \frac{1}{2^2} + \frac{1}{3^2} + \cdots = \frac{\pi^2}{6}.$$

Before commencing the proof, we recall two results of algebra: the first of these is *De Moivre's formula* and the second is *Vieta's formulas*.

De Moivre's Formula: Let n be a positive integer, α any real number and $i = \sqrt{-1}$; then

$$(\cos \alpha = i \sin \alpha)^n = \cos n\alpha + i \sin n\alpha.$$

Vieta's Formulas: Let the polynomial of degree n,

$$x^n + a_1 x^{n-1} + a_2 x^{n-2} + \cdots + a_{n-1} x + a_n = 0,$$

have the roots $x_1, x_2, \ldots, x_n$. Then for following relationship exists between the coefficients and the roots of the polynomial:

$$a_1 = -(x_1 + x_2 + \cdots + x_{n-1} + x_n),$$

$$a_2 = x_1 x_2 + x_1 x_3 + \cdots + x_{n-1} x_n,$$

$$a_3 = -(x_1 x_2 x_3 + \cdots + x_{n-2} x_{n-1} x_n),$$

$$\vdots$$

$$a_n = (-1)^n x_1 x_2 x_3 \ldots x_n.$$

Using De Moivre's formula and the binomial theorem, we have

$$\cos n\alpha + i \sin n\alpha = (\cos \alpha + i \sin \alpha)^n$$

$$= \left(\cos^n \alpha - \binom{n}{2} \cos^{n-2} \alpha \cdot \sin^2 \alpha + \binom{n}{4} \cos^{n-4} \alpha \cdot \sin^4 \alpha - \cdots \right)$$

$$+ i \left(\binom{n}{1} \cos^{n-1} \alpha \cdot \sin \alpha - \binom{n}{3} \cos^{n-3} \alpha \cdot \sin^3 \alpha + \cdots \right),$$

and so

$$\cos n\alpha = \cos^n \alpha - \binom{n}{2} \cos^{n-2} \alpha \cdot \sin^2 \alpha + \binom{n}{4} \cos^{n-4} \alpha \cdot \sin^4 \alpha - \cdots$$

and

$$\sin n\alpha = \binom{n}{1} \cos^{n-1} \alpha \cdot \sin \alpha - \binom{n}{3} \cos^{n-3} \alpha \cdot \sin^3 \alpha + \cdots .$$

Replacing n by $2n + 1$ in the foregoing formula for $\sin n\alpha$, we get

$$\sin(2n + 1)\alpha = (\sin^{2n+1} \alpha) \left(\binom{2n + 1}{1} \cot^{2n} \alpha - \binom{2n + 1}{3} \cot^{2n-2} \alpha + \cdots \right).$$

Thus, it follows that for

$$\alpha = \frac{\pi}{2n + 1}, \quad \frac{2\pi}{2n + 1}, \quad \cdots, \quad \frac{n\pi}{2n + 1}$$

the equation

$$\binom{2n + 1}{1} \cot^{2n} \alpha - \binom{2n + 1}{3} \cot^{2n-2} \alpha + \cdots = 0$$

holds. Therefore, the numbers

$$\cot^2 \frac{\pi}{2n + 1}, \quad \cot^2 \frac{2\pi}{2n + 1}, \quad \cdots, \quad \cot^2 \frac{n\pi}{2n + 1}$$

are roots of the polynomial

$$\binom{2n + 1}{1} x^n - \binom{2n + 1}{3} x^{n-1} + \cdots = 0$$

of degree n. But the sum of the roots of

$$x^n - \frac{\binom{2n + 1}{3}}{\binom{2n + 1}{1}} x^{n-1} + \cdots = 0$$

is equal to the negative of the coefficient of x^{n-1}; that is,

$$\cot^2 \frac{\pi}{2n + 1} + \cot^2 \frac{2\pi}{2n + 1} + \cdots + \cot^2 \frac{n\pi}{2n + 1} = \frac{n(2n - 1)}{3}. \qquad (1.36)$$

Since $\csc^2 \alpha = \cot^2 \alpha + 1$, equation (1.36) implies

$$\csc^2 \frac{\pi}{2n+1} + \csc^2 \frac{2\pi}{2n+1} + \cdots + \csc^2 \frac{n\pi}{2n+1} = \frac{2n(n+1)}{3}. \quad (1.37)$$

But $\sin \alpha < \alpha < \tan \alpha$ for $0 < \alpha < \pi/2$ [e.g., see (2.9) in Chapter 2] and so

$$\cot \alpha < \frac{1}{\alpha} < \csc \alpha \quad \text{for } 0 < \alpha < \frac{\pi}{2}.$$

It now follows from (1.36) and (1.37) that

$$\frac{n(2n-1)}{3} = \cot^2 \frac{\pi}{2n+1} + \cot^2 \frac{2\pi}{2n+1} + \cdots + \cot^2 \frac{n\pi}{2n+1}$$

$$< \left(\frac{2n+1}{\pi}\right)^2 + \left(\frac{2n+1}{2\pi}\right)^2 + \cdots + \left(\frac{2n+1}{n\pi}\right)^2 \quad (1.38)$$

$$< \csc^2 \frac{\pi}{2n+1} + \csc^2 \frac{2\pi}{2n+1} + \cdots + \csc^2 \frac{n\pi}{2n+1} = \frac{2n(n+1)}{3}.$$

By dividing all terms of (1.38) by $(2n+1)^2/\pi^2$, we obtain

$$\frac{2n}{2n+1} \cdot \frac{2n-1}{2n+1} \cdot \frac{\pi^2}{6} = \left(1 - \frac{1}{2n+1}\right)\left(1 - \frac{2}{2n+2}\right) \cdot \frac{\pi^2}{6}$$

$$< 1 + \frac{1}{2^2} + \frac{1}{3^2} + \cdots + \frac{1}{n^2}$$

$$< \frac{2n}{2n+1} \cdot \frac{2n+2}{2n+1} \cdot \frac{\pi^2}{6} = \left(1 - \frac{1}{2n+1}\right)\left(1 - \frac{1}{2n+1}\right) \cdot \frac{\pi^2}{6},$$

as was to be shown.

REMARK. See Proposition 7.19 (in Chapter 7) for another treatment of this sum.

4. Let j and n be positive integers and put

$$S_j = 1^j + 2^j + 3^j + \cdots + n^j.$$

Then

$$\binom{k+1}{1} S_1 + \binom{k+1}{2} S_2 + \cdots + \binom{k+1}{k} S_k = (n+1)^{k+1} - (n+1). \quad (1.39)$$

Indeed,

$$\sum_{p=1}^{n} (p+1)^{k+1} = \sum_{p=1}^{n} p^{k+1} + \binom{k+1}{1} \sum_{p=1}^{n} p^k + \binom{k+1}{2} \sum_{p=1}^{n} p^{k-1}$$

$$+ \cdots + \binom{k+1}{k} \sum_{p=1}^{n} p + n$$

by summing p from 1 to n in the identity

$$(p + 1)^{k+1} = p^{k+1} + \binom{k+1}{1}p^k + \binom{k+1}{2}p^{k-1} + \cdots + \binom{k+1}{k}p + 1.$$

However,

$$\sum_{p=1}^{n} (p + 1)^{k+1} - \sum_{p=1}^{n} p^{k+1} = (n + 1)^{k+1} - 1$$

and

$$\binom{k+1}{m} = \binom{k+1}{k+1-m} \quad \text{for } m = 1, 2, \ldots, k.$$

REMARK. The recursion formula (1.39) produces the following set of values:

$$S_1 = \frac{n(n+1)}{2}, \quad S_2 = \frac{n(n+1)(2n+1)}{6}, \quad S_3 = S_1^2, \ldots.$$

5. Let n and j be positive integers. Then

$$\frac{1}{j+1}n^{j+1} < 1^j + 2^j + 3^j + \cdots + n^j < \left(1 + \frac{1}{n}\right)^{j+1}\frac{1}{j+1}n^{j+1}. \quad (1.40)$$

Indeed, let $T = x^j + x^{j-1} + \cdots + x + 1$; if $x > 1$, then the first term is the largest, but if $0 < x < 1$, then the last term is the largest. Hence,

$$(j + 1)x^j > T > j + 1 \quad \text{if } x > 1$$

and

$$(j + 1)x^j < T < j + 1 \quad \text{if } 0 < x < 1.$$

If both sides of these inequalities are multiplied by $x - 1$, it is found that for $x \neq 1$

$$(j + 1)x^j(x - 1) > x^{j+1} - 1 > (j + 1)(x - 1)$$

[note that $T(x - 1) = x^{j+1} - 1$]. Assume now that $x = p/(p - 1)$; then we find

$$\frac{(j+1)p^j}{(p-1)^{j+1}} > \frac{p^{j+1} - (p-1)^{j+1}}{(p-1)^{j+1}} > \frac{(j+1)(p-1)^j}{(p-1)^{j+1}}.$$

Similarly, if we assume that $x = (p + 1)/p$, we obtain

$$\frac{(j+1)(p+1)^j}{p^{j+1}} > \frac{(p+1)^{j+1} - p^{j+1}}{p^{j+1}} > \frac{(j+1)p^j}{p^{j+1}}.$$

It follows that

$$(p + 1)^{j+1} - p^{j+1} > (j + 1)p^j > p^{j+1} - (p - 1)^{j+1},$$

or, letting p successively assume the values $1, 2, 3, \ldots, n$,

$$2^{j+1} - 1^{j+1} > (j + 1)1^j > 1^{j+1} - 0,$$

$$3^{j+1} - 2^{j+1} > (j + 1)2^j > 2^{j+1} - 1^{j+1},$$

$$4^{j+1} - 3^{j+1} > (j + 1)3^j > 3^{j+1} - 2^{j+1},$$

$$\vdots$$

$$(n + 1)^{j+1} - n^{j+1} > (j + 1)n^j > n^{j+1} - (n - 1)^{j+1}.$$

If these inequalities are added together we get

$$(n + 1)^{j+1} - 1 > (j + 1)(1^j + 2^j + 3^j + \cdots + n^j) > n^{j+1}. \qquad (1.41)$$

But (1.41) is easily seen to be equivalent to (1.40).

REMARK. A particular consequence of inequality (1.40) is the result

$$\lim_{n \to \infty} \frac{1^j + 2^j + 3^j + \cdots + n^j}{n^{j+1}} = \frac{1}{j + 1}. \qquad (1.42)$$

6. Show that the sum

$$S_j = 1^j + 2^j + 3^j + \cdots + n^j,$$

where n is an arbitrary positive integer and j is an *odd* positive integer, is divisible by S_1.

Indeed, $S_1 = n(n + 1)/2$ and we first note that for odd j, $a^j + b^j$ is divisible by $a + b$. Two cases are to be considered.

Case 1: Suppose n is even. Here the sum S_j is divisible by $n + 1$ because each of the sums

$$1^j + n^j, \quad 2^j + (n - 1)^j, \quad 3^j + (n - 2)^j, \quad \ldots, \quad \left(\frac{n}{2}\right)^j + \left(\frac{n}{2} + 1\right)^j$$

is divisible by

$$1 + n = 2 + (n - 1) = 3 + (n - 2) = \cdots = \frac{n}{2} + \left(\frac{n}{2} + 1\right).$$

The sum S_j is also divisible by $n/2$ because

$$1^j + (n - 1)^j, \quad 2^j + (n - 2)^j, \quad 3^j + (n - 3)^j,$$

$$\ldots, \quad \left(\frac{n}{2} - 1\right)^j + \left(\frac{n}{2} + 1\right)^j, \quad \left(\frac{n}{2}\right)^j, \quad n^j$$

are divisible by $n/2$.

Case 2: Suppose n is odd. Here the sum S_j is divisible by $(n + 1)/2$ because

$$1^j + n^j, \quad 2^j + (n - 1)^j, \quad 3^j + (n - 2)^j,$$

$$\ldots, \quad \left(\frac{n - 1}{2}\right)^j + \left(\frac{n + 3}{2}\right)^j, \quad \left(\frac{n + 1}{2}\right)^j$$

are all divisible by $(n + 1)/2$. Also, S_j is divisible by n because

$$1^j + (n-1)^j, \quad 2^j + (n-2)^j, \quad 3^j + (n-3)^j,$$

$$\cdots, \quad \left(\frac{n-1}{2}\right)^j + \left(\frac{n+1}{2}\right)^j, \quad n^j$$

are all divisible by n.

7. In the equality $N = N/2 + N/4 + N/8 + \cdots + N/2^n + \cdots$, where N is an arbitrary positive integer, every fraction may be replaced by the nearest whole number:

$$N = \left(\frac{N}{2}\right) + \left(\frac{N}{4}\right) + \left(\frac{N}{8}\right) + \cdots + \left(\frac{N}{2^n}\right) + \cdots.$$

Indeed, it is readily seen that $(a) = [a + \frac{1}{2}]$; here (a) denotes the nearest whole number to a and $[a + \frac{1}{2}]$ denotes the integer part of $a + \frac{1}{2}$, that is, the greatest integer less than or equal to $a + \frac{1}{2}$. Hence, we can put the equation which we wish to derive into the following form:

$$N = \left[\frac{N}{2} + \frac{1}{2}\right] + \left[\frac{N}{4} + \frac{1}{2}\right] + \left[\frac{N}{8} + \frac{1}{2}\right] + \cdots.$$

Now let

$$N = a_n \cdot 2^n + a_{n-1} \cdot 2^{n-1} + \cdots + a_1 \cdot 2 + a_0$$

($a_n, a_{n-1}, \ldots, a_1, a_0$ are either 0 or 1) be the expansion of N in powers of 2 as in the binary number system. We then have

$$\left[\frac{N}{2} + \frac{1}{2}\right] = \left[a_n \cdot 2^{n-1} + a_{n-1} \cdot 2^{n-2} + \cdots + a_1 + \frac{a_0 + 1}{2}\right]$$

$$= a_n \cdot 2^{n-1} + a_{n-1} \cdot 2^{n-2} + \cdots + a_1 + a_0,$$

$$\left[\frac{N}{4} + \frac{1}{2}\right] = \left[a_n \cdot 2^{n-2} + a_{n-1} \cdot 2^{n-3} + \cdots + \frac{a_1 + 1}{2} + \frac{a_0}{4}\right]$$

$$= a_n \cdot 2^{n-2} + a_{n-1} \cdot 2^{n-3} + \cdots + a_1,$$

$$\vdots$$

$$\left[\frac{N}{2^n} + \frac{1}{2}\right] = \left[a_n + \frac{a_{n-1} + 1}{2} + \frac{a_{n-2}}{4} + \cdots + \frac{a_0}{2^n}\right]$$

$$= a_n + a_{n-1},$$

$$\left[\frac{N}{2^{n+1}} + \frac{1}{2}\right] = \left[\frac{a_n + 1}{2} + \frac{a_{n-1}}{4} + \cdots + \frac{a_0}{2^{n+1}}\right]$$

$$= a_n,$$

and

$$\left[\frac{N}{2^{n+2}} + \frac{1}{2}\right] = \left[\frac{N}{2^{n+3}} + \frac{1}{2}\right] = \cdots = 0,$$

recalling that the a_i are either 0 or 1. Hence, we obtain

$$\left[\frac{N}{2}+\frac{1}{2}\right]+\left[\frac{N}{4}+\frac{1}{2}\right]+\cdots+\left[\frac{N}{2^{n+1}}+\frac{1}{2}\right]+\cdots$$

$$= a_n\{2^{n-1}+2^{n-2}+\cdots+1+1\}$$

$$+ a_{n-1}\{2^{n-2}+2^{n-3}+\cdots+1+1\}$$

$$+\cdots+a_1\{1+1\}+a_0$$

$$= a_n\cdot 2^n + a_{n-1}\cdot 2^{n-1}+\cdots+a_1\cdot 2 + a_0$$

$$= N$$

which is what we wished to verify.

8. The *Cauchy–Schwarz Inequality*: Let $a_1, a_2, \ldots, a_n$ and $b_1, b_2, \ldots, b_n$ be any real numbers. Then

$$(a_1b_1 + a_2b_2 + \cdots + a_nb_n)^2$$

$$\leq (a_1^2 + a_2^2 + \cdots + a_n^2)(b_1^2 + b_2^2 + \cdots + b_n^2) \tag{1.43}$$

with equality holding only if $a_1/b_1 = a_2/b_2 = \cdots = a_n/b_n$.

Indeed,

$$(xa_1 + b_1)^2 + (xa_2 + b_2)^2 + \cdots + (xa_n + b_n)^2 = Ax^2 + 2Bx + C, \tag{1.44}$$

where

$$A = a_1^2 + a_2^2 + \cdots + a_n^2, \qquad B = a_1b_1 + a_2b_2 + \cdots + a_nb_n,$$

$$C = b_1^2 + b_2^2 + \cdots + b_n^2.$$

By (1.44), $Ax^2 + 2Bx + C$ is the sum of squares and so $Ax^2 + 2Bx + C \geq 0$ for any real number x. Hence, putting $x = -B/A$, we get

$$A\frac{B^2}{A^2} - 2B\frac{B}{A} + C = \frac{AC - B^2}{A} \geq 0.$$

Since $A > 0$, we obtain $AC - B^2 \geq 0$ or $B^2 \leq AC$ and so (1.43) is seen to hold. The equality sign in (1.43) is possible only if

$$xa_1 + b_1 = xa_2 + b_2 = \cdots = xa_n + b_n = 0$$

or $b_1/a_1 = b_2/a_2 = \cdots = b_n/a_n \,(= -x)$.

9. Let $c_1, c_2, \ldots, c_n$ be positive real numbers. Then, for any real numbers $t_1, t_2, \ldots, t_n$ we have

$$(c_1t_1 + c_2t_2 + \cdots + c_nt_n)^2$$

$$\leq (c_1 + c_2 + \cdots + c_n)(c_1t_1^2 + c_2t_2^2 + \cdots + c_nt_n^2). \tag{1.45}$$

Indeed, putting $a_k = \sqrt{c_k}$ and $b_k = \sqrt{c_k}\cdot t_k$ for $k = 1, 2, \ldots, n$ in (1.43), we get the desired inequality (1.45).

REMARK. Putting $c_1 = \frac{1}{2}$, $c_2 = \frac{1}{3}$, and $c_3 = \frac{1}{6}$, inequality (1.45) yields

$$(\tfrac{1}{2}t_1 + \tfrac{1}{3}t_2 + \tfrac{1}{6}t_3)^2 \le \tfrac{1}{2}t_1^2 + \tfrac{1}{3}t_2^2 + \tfrac{1}{6}t_3^2$$

for any real numbers t_1, t_2, and t_3.

10. We have the identity

$$a + b(1 + a) + c(1 + a)(1 + b) + d(1 + a)(1 + b)(1 + c)$$
$$+ \cdots + q(1 + a)(1 + b) \cdots (1 + p) \qquad (1.46)$$
$$= (1 + a)(1 + b)(1 + c) \cdots (1 + q) - 1.$$

Indeed, adding 1 to the left side, we can write

$$[(1 + a) + b(1 + a)] + c(1 + a)(1 + b) + d(1 + a)(1 + b)(1 + c)$$
$$+ \cdots + q(1 + a)(1 + b) \cdots (1 + p)$$
$$= [(1 + a)(1 + b) + c(1 + a)(1 + b)] + d(1 + a)(1 + b)(1 + c)$$
$$+ \cdots + q(1 + a)(1 + b) \cdots (1 + p)$$
$$= (1 + a)(1 + b)(1 + c)(1 + d)$$
$$+ \cdots + q(1 + a)(1 + b) \cdots (1 + p)$$
$$= (1 + a)(1 + b)(1 + c)(1 + d) \cdots (1 + p)$$

REMARKS. If $a = b = c = \cdots = q$, then

$$a + a(1 + a) + a(1 + a)^2 + a(1 + a)^3 + \cdots + a(1 + a)^{n-1} = (1 + a)^n - 1,$$

where n is the number of integers $a, b, \ldots, q$; writing $1 + a = x$, we get

$$(x - 1)(1 + x + x^2 + \cdots + x^{n-1}) = x^n - 1$$

which is the formula for the sum of a geometric progression.

Letting $a = 1, b = 2, c = 3, \ldots, q = n$, we get

$$1 \cdot 1! + 2 \cdot 2! + 3 \cdot ! + \cdots + n \cdot n! = (n + 1)! - 1$$

and putting $a = (n + 1)/1$, $b = (n + 1)/2$, $c = (n + 1)/3$, $\ldots$, $q = (n + 1)/k$, we obtain

$$\binom{n + 1}{1} + \binom{n + 2}{2} + \binom{n + 3}{3} + \cdots + \binom{n + k}{k} = \binom{n + k + 1}{k} - 1.$$

11. Let $a > b > 0$ and n be a positive integer. Then

$$\frac{1 + a + a^2 + \cdots + a^{n-1} + a^n}{1 + a + a^2 + \cdots + a^{n-1}} > \frac{1 + b + b^2 + \cdots + b^{n-1} + b^n}{1 + b + b^2 + \cdots + b^{n-1}}. \qquad (1.47)$$

Indeed, since $a > b > 0$, we have

$$\frac{1 + a}{a^2} = \frac{1}{a^2} + \frac{1}{a} < \frac{1}{b^2} + \frac{1}{b} = \frac{1 + b}{b^2} \quad \text{or} \quad \frac{a^2}{1 + a} > \frac{b^2}{1 + b},$$

implying

$$\frac{1 + a + a^2}{1 + a} = 1 + \frac{a^2}{1 + a} > 1 + \frac{b^2}{1 + b} = \frac{1 + b + b^2}{1 + b}.$$

Similarly,

$$\frac{1 + a + a^2}{a^3} = \frac{1}{a^3} + \frac{1}{a^2} + \frac{1}{a} < \frac{1}{b^3} + \frac{1}{b^2} + \frac{1}{b} = \frac{1 + b + b^2}{b^3}$$

or

$$\frac{a^3}{1 + a + a^2} > \frac{b^3}{1 + b + b^2},$$

implying

$$\frac{1 + a + a^2 + a^3}{1 + a + a^2} = 1 + \frac{a^3}{1 + a + a^2} > 1 + \frac{b^3}{1 + b + b^2} = \frac{1 + b + b^2 + b^3}{1 + b + b^2}$$

and so forth.

12. We have $\cos(\sin x) > \sin(\cos x)$ for all real numbers x.
   Indeed, using the identity $\cos(A + B) = (\cos A)(\cos B) - (\sin A)(\sin B)$, we get

$$\cos\left(\frac{\pi}{2} + \cos x\right) = \left(\cos\frac{\pi}{2}\right)(\cos x) - \left(\sin\frac{\pi}{2}\right)\sin(\cos x)$$

$$= -\left(\sin\frac{\pi}{2}\right)\sin(\cos x) = -\sin(\cos x),$$

or,

$$\sin(\cos x) = -\cos\left(\frac{\pi}{2} + \cos x\right),$$

and so

$$\cos(\sin x) - \sin(\cos x) = \cos(\sin x) + \cos\left(\frac{\pi}{2} + \sin x\right).$$

But

$$\cos A + \cos B = 2\cos\frac{A + B}{2} \cdot \cos\frac{A - B}{2} \quad \text{and} \quad \cos A = \cos(-A),$$

and so

$$\cos(\sin x) - \sin(\cos x) = 2\cos\frac{\sin x + \pi/2 + \cos x}{2} \cdot \cos\frac{-\sin x + \pi/2 + \cos x}{2}.$$

Now,

$$|\cos x + \sin x| = (\cos^2 x + 2\{\cos x\}\{\sin x\} + \sin^2 x)^{1/2}$$
$$= \sqrt{1 + \sin 2x} \le \sqrt{2}$$

and

$$|\cos x - \sin x| = (\cos^2 x - 2\{\cos x\}\{\sin x\} + \sin^2 x)^{1/2}$$
$$= \sqrt{1 - \sin 2x} \le \sqrt{2}.$$

Since $\pi/2 = 1.57\ldots$ and $\sqrt{2} = 1.41\ldots$, we have

$$\frac{\pi}{2} > \frac{\pi/2 + \cos x + \sin x}{2} > 0 \quad \text{and} \quad \frac{\pi}{2} > \frac{\pi/2 + \cos x - \sin x}{2} > 0.$$

Therefore,

$$\cos \frac{\pi/2 + \cos x + \sin x}{2} > 0 \quad \text{and} \quad \cos \frac{\pi/2 + \cos x - \sin x}{2} > 0$$

and hence $\cos(\sin x) - \sin(\cos x) > 0$.

Exercises to Chapter 1

1.1. Show that $(1 - 1/n)^{-n}$, for $n = 2, 3, \ldots$, is a decreasing sequence tending to e.
   [Hint: If $n = k + 1$, then $(1 - 1/n)^{-n} = (1 + 1/k)^{k+1}$.]

1.2. Show that

$$\sinh x + \sinh 2x + \sinh 3x + \cdots + \sinh nx = (\sinh \tfrac{1}{2}nx)\frac{\sinh \tfrac{1}{2}(n + 1)x}{\sinh \tfrac{1}{2}x}$$

and

$$\tfrac{1}{2} + \cosh x + \cosh 2x + \cosh 3x + \cdots + \cosh nx = \frac{\sinh(n + \tfrac{1}{2})x}{2 \sinh \tfrac{1}{2}x}.$$

[Hint: We have

$$2(\sinh \tfrac{1}{2}nx)\frac{\sinh \tfrac{1}{2}(n + 1)x}{2 \sinh \tfrac{1}{2}x} = (e^{nx/2} - e^{-nx/2})\frac{e^{(n+1)x/2} - e^{-(n+1)x/2}}{e^{x/2} - e^{-x/2}}$$

$$= \frac{e^{(n+1)x} - e^x + e^{-nx} - 1}{e^x - 1}$$

$$= \frac{e^x(e^{nx} - 1)}{e^x - 1} - \frac{e^{-x}(e^{-nx} - 1)}{e^{-x} - 1}$$

$$= \sum_{r=1}^{n} e^{rx} - \sum_{r=1}^{n} e^{-rx} = 2\left(\sum_{r=1}^{n} \sinh rx\right)$$

and so forth.]

1.3. If $n = a + b + c + \cdots + z$, where $a, b, c, \ldots, z$ are positive integers, then

$$\frac{n!}{a!\,b!\,c! \cdots z!} \le \frac{n^n}{a^a b^b c^c \cdots z^z}.$$

[Hint: We have $(a + b + c + \cdots + z)^n = n^n$. Each term of the multinomial expansion of the left-hand side is positive and less than the sum of all terms. Hence, the particular term

$$\frac{n!}{a!\,b!\,c!\cdots z!}\, a^a b^b c^c \cdots z^z \leq n^n.]$$

1.4. Let $a_k > 0$ and $b_k > 0$, for $k = 1, 2, \ldots, n$. Verify that

$$(a_1^2 + b_1^2)^{1/2} + (a_2^2 + b_2^2)^{1/2} + \cdots + (a_n^2 + b_n^2)^{1/2}$$

$$\geq [(a_1 + a_2 + \cdots + a_n)^2 + (b_1 + b_2 + \cdots + b_n)^2]^{1/2}.$$

(1.48)

[Hint: Let P_0 be the point with coordinates $(0,0)$, P_1 with coordinates (a_1, b_1), P_2 with coordinates $(a_1 + a_2, b_1 + b_2)$, ..., and P_n with coordinates $(a_1 + a_2 + \cdots + a_n, b_1 + b_2 + \cdots + b_n)$. Then

$(a_1^2 + b_1^2)^{1/2}$ is the distance between P_0 and P_1,

$(a_2^2 + b_2^2)^{1/2}$ is the distance between P_1 and P_2,

$$\vdots$$

$(a_n^2 + b_n^2)^{1/2}$ is the distance between P_{n-1} and P_n,

while $[(a_1 + a_2 + \cdots + a_n)^2 + (b_1 + b_2 + \cdots + b_n)^2]^{1/2}$ is the distance between P_0 and P_n. But the shortest path between P_0 and P_n is the line segment connecting P_0 and P_n.]

1.5. (i) Is inequality (1.48) valid if a_k and b_k are not positive? (ii) Under what conditions will we have equality in inequality (1.48)?

1.6. Let $A = (1/2)(3/4)(5/6)\cdots(9999/10,000)$. Verify that $A < 1/100$.
    [Hint: Put $B = (2/3)(4/5)(6/7)\cdots(10,000/10,001)$. Then $A < B$ and so $A^2 < AB = 1/10,001$.]

1.7. Show that

$$\frac{1}{2}\cdot\frac{3}{4}\cdot\frac{5}{6}\cdots\frac{2n-1}{2n} < \frac{1}{\sqrt{2n+1}}.$$

[Hint: It is clear that

$$1\cdot 3 < 2^2, \quad 3\cdot 5 < 4^2, \quad \ldots, \quad (2n-1)(2n+1) < (2n)^2.$$

Thus,

$$1^2\cdot 3^2\cdot 5^2\cdots(2n-1)^2(2n+1) < 2^2\cdot 4^2\cdot 6^2\cdots(2n)^2.]$$

1.8. Show that

$$\frac{1\cdot 3\cdot 5\cdots(2n-1)}{2\cdot 4\cdot 6\cdots 2n} > \frac{\sqrt{n+1}}{2n+1}.$$

[Hint: If $a \neq b$, then $a + b > 2\sqrt{ab}$ and so $k + (k-1) > 2\sqrt{k(k-1)}$; hence, $(2k-1)/2(k-1) > \sqrt{k/(k-1)}$. For $k = 2, 3, \ldots, n+1$ we get

$$\left(\frac{3}{2}\right)1 > \sqrt{\frac{2}{1}}, \quad \left(\frac{5}{2}\right)2 > \sqrt{\frac{3}{2}}, \quad \left(\frac{7}{2}\right)3 > \sqrt{\frac{4}{3}}, \quad \ldots$$

$$\frac{2n-1}{2(n-1)} > \sqrt{\frac{n}{n-1}}, \quad \frac{2n+1}{2n} > \sqrt{\frac{n+1}{n}}.$$

Multiplying these n inequalities together we get

$$\frac{1\cdot3\cdot5\cdots(2n+1)}{2\cdot4\cdot6\cdots2n} > \sqrt{n+1}$$

and so

$$\frac{1\cdot3\cdot5\cdots(2n-1)}{2\cdot4\cdot6\cdots2n} > \frac{\sqrt{n+1}}{2n+1}.]$$

REMARK. Since $\sqrt{(n+1)}/(2n+1) > \sqrt{(n+1)}/(2n+2) > 1/2\sqrt{n+1}$ and by the result in Exercise 1.7, we may state that

$$\frac{1}{\sqrt{2n+1}} > \frac{1\cdot3\cdot5\cdots(2n-1)}{2\cdot4\cdot6\cdots2n} > \frac{1}{2\sqrt{n+1}}.$$

1.9. Real numbers $a_1, a_2, \ldots, a_n$, not all zero, are given, and $x_1, x_2, \ldots, x_n$ are real variables satisfying the equation $a_1 x_1 + a_2 x_2 + \cdots + a_n x_n = 1$. Show that the least value of $x_1^2 + x_2^2 + \cdots + x_n^2$ is

$$(a_1^2 + a_2^2 + \cdots + a_n^2)^{-1}.$$

[Hint: By the Cauchy–Schwarz Inequality (1.43),

$$(x_1^2 + x_2^2 + \cdots + x_n^2)(a_1^2 + a_2^2 + \cdots + a_n^2) \geq (a_1 x_1 + a_2 x_2 + \cdots + a_n x_n)^2.]$$

1.10. If $0 < a < x < b$, show that

$$\frac{1}{x} + \frac{1}{a+b-x} < \frac{1}{a} + \frac{1}{b}.$$

[Hint: We have $a - x < 0$ and $b - x > 0$ and so $(a-x)(b-x) < 0$. Hence, $ab - (a+b)x + x^2 < 0$, that is, $ab < x(a+b-x)$. It follows that

$$\frac{1}{ab} > \frac{1}{x(a+b-x)} \quad \text{or} \quad \frac{a+b}{ab} > \frac{a+b}{x(a+b-x)}.]$$

1.11. Let $n = 2, 3, \ldots$. Show that

$$\left(1 - \frac{1}{2n}\right)^{-1} = 1 + \frac{1}{2n} + \frac{A_n}{n^2}, \quad \text{where } A_n < \tfrac{1}{2}.$$

[Hint: Since

$$(1-x)^{-1} = 1 + x + \frac{x^2}{1-x},$$

we have

$$\left(1 - \frac{1}{2n}\right)^{-1} = 1 + \frac{1}{2n} + \frac{1}{4n^2 - 2n}.$$

But $4n^2 - 2n > 4n^2 - 2n^2 = 2n^2$ for $n > 1$. Hence, $A_n < \tfrac{1}{2}$.]

1.12. Let the function f be defined for all real x and y, and satisfy the relation $f(x+y) = f(x)f(y)$. Show that if f is not identically zero, then f is positive for all x and $f(0) = 1$. Show also that if f is not identically unity, then there is no M such that $f(x) < M$ for all x.

[Hint: Let $y = 0$. Then $f(x) = f(x)f(0)$ for all x. Hence, $f(0) = 1$. Put $x = y = \frac{1}{2}a$. Then $f(a) = [f(\frac{1}{2}a)]^2 \geq 0$. If $f(a) = 0$, then $f(x) = 0$ for all x (put $x = a$, $y = x - a$). Suppose $f(x) < M$ for all x, then $f(-x) = [f(x)]^{-1} > M$, a contradiction unless $f(x) = 1$ for all x.]

1.13. If $\log_a n = x$ and $\log_c n = y$, where $n \neq 1$, show that

$$\frac{x - y}{x + y} = \frac{\log_b c - \log_b a}{\log_b c + \log_b a}.$$

[Hint: We have

$$\frac{\log_a n - \log_c n}{\log_a n + \log_c n} = \frac{(\log_b n)(\log_a b) - (\log_b n)(\log_c b)}{(\log_b n)(\log_a b) + (\log_b n)(\log_c b)} = \frac{\log_a b - \log_c b}{\log_a b + \log_c b}$$

$$= \frac{1/(\log_b a) - 1/(\log_b c)}{1/(\log_b a) + 1/(\log_b c)} = \frac{\log_b c - \log_b a}{\log_b c + \log_b a}.]$$

1.14. If $x = \log_a(bc)$, $y = \log_b(ca)$ and $z = \log_c(ab)$, show that $x + y + z = xyz - 2$.
[Hint: We have

$$(\log_a b + \log_a c)(\log_b c + \log_b a)$$

$$= (\log_a b)(\log_b c) + (\log_a b)(\log_b a) + (\log_a c)(\log_b c) + (\log_a c)(\log_b a)$$

$$= (\log_a c) + 1 + (\log_a c)(\log_b c) + \log_b c$$

and

$$\{(\log_a c) + 1 + (\log_a c)(\log_b c) + (\log_b c)\}(\log_c a + \log_c b)$$

$$= (\log_c a)(\log_a c) + \log_c a + (\log_c a)(\log_a c)(\log_b c) + (\log_c a)(\log_b c)$$

$$\quad + (\log_c b)(\log_a c) + \log_c b + (\log_c b)(\log_a c)(\log_b c) + (\log_c b)(\log_b c)$$

$$= 1 + \log_c a + \log_b c + \log_b a + \log_a b + \log_c b + \log_a c + 1$$

$$= \log_a(bc) + \log_b(ca) + \log_c(ab) + 2.]$$

1.15. If c_r is the coefficient of x^r in the expansion of $(1 + x)^n$, where n is a positive integer, and $f(r) = c_0 c_r + c_1 c_{r+1} + \cdots + c_{n-r} c_n$, show that

(i) $f(r) = \dfrac{(2n)!}{(n + r)!(n - r)!}$,

(ii) $c_0 f(0) + c_1 f(1) + \cdots + c_n f(n) = \dfrac{(3n)!}{(n)!(2n)!}$.

[Hint: (i) Since $c_r = c_{n-r}$,

$$(c_0 + c_1 x + \cdots + c_r x^r + \cdots + c_n x^n)(c_n + c_{n-1} x + \cdots + c_{n-r} x^r + \cdots + c_0 x^n)$$

$$= (1 + x)^n (1 + x)^n = (1 + x)^{2n}.$$

The coefficient of x^{n+r} on the left-hand side is $f(r)$ while the coefficient of x^{n+r} in $(1 + x)^{2n}$ is $(2n)!/(n + r)!(n - r)!$ and (i) is established.
    (ii) It has been show in (i) that

$$(1 + x)^{2n} = \text{terms up to } x^{n-1} + f(0)x^n + f(1)x^{n+1} + \cdots + f(n)x^{2n}.$$

Multiplying by $(1 + x)^n = c_n + c_{n-1}x + \cdots + c_0 x^n$, it is clear that

$$c_0 f(0) + c_1 f(1) + \cdots + c_n f(n)$$

is the coefficient of x^{2n} in the expansion of $(1 + x)^{3n}$. This being

$$\frac{(3n)!}{(2n)!\,(n)!},$$

the validity of (ii) follows.]

1.16. Show that, if a and b are positive integers and $b > a$,

$$\lim_{n \to \infty} \left(\frac{1}{an + 1} + \frac{1}{an + 2} + \frac{1}{an + 3} + \cdots + \frac{1}{bn} \right) = \ln \frac{b}{a}.$$

[Hint: We have

$$\lim_{n \to \infty} \left(1 + \frac{1}{2} + \frac{1}{3} + \cdots + \frac{1}{bn} - \ln(bn + 1) \right) = C$$

and

$$\lim_{n \to \infty} \left(1 + \frac{1}{2} + \frac{1}{3} + \cdots + \frac{1}{an} - \ln(an + 1) \right) = C,$$

where C is Euler's constant. Subtracting these equations, the required result appears.]

1.17. If $0 < x < 1$ and $0 < y < 1$, show that $0 < x + y - xy < 1$.
  [Hint: Since $0 < (x - 1)(y - 1) < 1$, we have $0 < xy - x - y + 1 < 1$ and so $-1 < xy - x - y < 0$ or $1 > -xy + x + y > 0$.]

1.18. Show that the range of the function $y = (x^2 + x + 1)/(x + 1)$ does not contain the open interval $(-3, 1)$.
  [Hint: We have $x^2 + (1 - y)x + 1 - y = 0$. For x to be real

$$(1 - y)^2 \geq 4(1 - y) \quad \text{or} \quad (y - 1)(y + 3) \geq 0.$$

If y lies between -3 and 1, $y + 3 > 0$, and $y - 1 < 0$ giving $(y - 1)(y + 3) < 0$ and the above inequality $(1 - y)^2 \geq 4(1 - y)$ is not satisfied. Hence, there is no real value between -3 and 1.]

1.19. Let $E(x) = \lim_{n \to \infty} (1 + x/n)^n$. Show that $E(x) \cdot E(y) = E(x + y)$.
  [We have, using Proposition 1.12,

$$E(x) \cdot E(y) = \lim_{n \to \infty} \left(1 + \frac{x}{n} \right)^n \cdot \lim_{n \to \infty} \left(1 + \frac{y}{n} \right)^n = \lim_{n \to \infty} \left(\left(1 + \frac{x}{n} \right)\left(1 + \frac{y}{n} \right) \right)^n$$

$$= \lim_{n \to \infty} \left(1 + \frac{x + y}{n} + \frac{xy}{n^2} \right)^n = E(x + y).]$$

1.20. Estimate the magnitude of the sum

$$1 + \frac{1}{\sqrt{2}} + \frac{1}{\sqrt{3}} + \frac{1}{\sqrt{4}} + \cdots + \frac{1}{\sqrt{1{,}000{,}000}}.$$

[Hint: Let k be a positive number. Then

$$\sqrt{k+1} - \sqrt{k} = \frac{1}{\sqrt{k+1} + \sqrt{k}}$$

and so

$$2\sqrt{k+1} - 2\sqrt{k} < \frac{1}{\sqrt{k}} < 2\sqrt{k} - 2\sqrt{k-1}.$$

Hence,

$$2\sqrt{3} - 2\sqrt{2} < \frac{1}{\sqrt{2}} < 2\sqrt{2} - 2,$$

$$2\sqrt{4} - 2\sqrt{3} < \frac{1}{\sqrt{3}} < 2\sqrt{3} - 2\sqrt{2},$$

$$2\sqrt{5} - 2\sqrt{4} < \frac{1}{\sqrt{4}} < 2\sqrt{4} - 2\sqrt{3},$$

$$\vdots$$

$$2\sqrt{n+1} - 2\sqrt{n} < \frac{1}{\sqrt{n}} < 2\sqrt{n} - 2\sqrt{n-1}$$

and we get upon addition

$$2\sqrt{n+1} - 2\sqrt{2} < \frac{1}{\sqrt{2}} + \frac{1}{\sqrt{3}} + \frac{1}{\sqrt{4}} + \cdots + \frac{1}{\sqrt{n}} < 2\sqrt{n} - 2.$$

Since $2\sqrt{2} < 3$, and $\sqrt{n+1} > \sqrt{n}$, it follows that

$$2\sqrt{n} - 2 < 1 + \frac{1}{\sqrt{2}} + \frac{1}{\sqrt{3}} + \frac{1}{\sqrt{4}} + \cdots + \frac{1}{\sqrt{n}} < 2\sqrt{n} - 1.$$

But $n = 1,000,000$ and so

$$1998 < 1 + \frac{1}{\sqrt{2}} + \frac{1}{\sqrt{3}} + \frac{1}{\sqrt{4}} + \cdots + \frac{1}{\sqrt{1,000,000}} < 1999.]$$

1.21. Show the inequality

$$2\sqrt{n+1} - 2\sqrt{m} < \frac{1}{\sqrt{m}} + \frac{1}{\sqrt{m+1}} + \cdots + \frac{1}{\sqrt{n}} < 2\sqrt{n} - 2\sqrt{m-1}$$

and verify that

$$1800 < \frac{1}{\sqrt{10,000}} + \frac{1}{\sqrt{10,001}} + \cdots + \frac{1}{\sqrt{1,000,000}} < 1800.02.$$

[Hint: See the hint to Exercise 1.20.]

1.22. Let $a > 1$ and $b > 1$. Show that

$$\log_a b + \log_b a \geq 2.$$

[Hint: If $m > 0$ then $m + 1/m \geq 2$ with equality if and only if $m = 1$. Indeed, $(m - 1)^2 = m^2 - 2m + 1 \geq 0$ and division by m gives the desired result. But $\log_b a = 1/(\log_a b)$.]

1.23. Show that $1/\sqrt{n} < \sqrt{n + 1} - \sqrt{n - 1}$ for $n \geq 1$.

[Hint: We have

$$(\sqrt{n + 1} + \sqrt{n - 1})^2 = 2n + 2(n^2 - 1)^{1/2} < 2n + 2(n^2)^{1/2} = 4n$$

and so $\sqrt{n + 1} + \sqrt{n - 1} < 2\sqrt{n}$. Thus,

$$\frac{1}{2\sqrt{n}} < \frac{1}{\sqrt{n + 1} + \sqrt{n - 1}} = \frac{\sqrt{n + 1} - \sqrt{n - 1}}{2}$$

and the desired result follows.]

1.24. Let a and b denote numbers larger than zero. The *arithmetic*, *geometric*, and *harmonic mean of a and b* are, respectively,

$$A = \frac{a + b}{2}, \quad G = \sqrt{ab}, \quad \text{and} \quad H = \frac{2ab}{a + b}.$$

[Observe that $a - A = A - b$, $a/G = G/b$, $1/a - 1/H = 1/H - 1/b$, and $AH = G^2$.] If $0 < a < b$, verify that

(i) $A < G < H$,
(ii) $A - G > G - H$,
(iii) $A - G < (b - a)^2/8a$,
(iv) $A - H < (b - a)^2/4a$.

[Hints: We have

$$A - G = \tfrac{1}{2}(\sqrt{a} - \sqrt{b})^2 > 0 \quad \text{and} \quad H - G = -\frac{\sqrt{ab}}{a + b}(\sqrt{a} - \sqrt{b})^2 < 0,$$

proving (i). Next we note that

$$A - 2G + H = A - G - (G - H) = \frac{(\sqrt{a} - \sqrt{b})^4}{2(a + b)} > 0,$$

proving (ii). From

$$A - G = \frac{(a - b)^2}{2(\sqrt{a} + \sqrt{b})^2}$$

and the assumption that $a < b$, we obtain (iii). Finally, from

$$A - H = \frac{(a - b)^2}{2(a + b)}$$

and the assumption $a < b$ we get (iv).]

1.25. Show that if $x \geq 0$, then

$$\frac{1}{n} \leq \frac{1 + 2x + \cdots + nx^{n-1}}{1 + 2^2x + \cdots + n^2x^{n-1}} \leq 1 \quad \text{and} \quad \frac{1}{n} \leq \frac{1 + 2x + \cdots + nx^{n-1}}{n + (n - 1)x + \cdots + x^{n-1}} \leq n.$$

[Hint: Let $a_1/b_1, a_2/b_2, \ldots, a_n/b_n$ be n fractions with positive denominators. Then the fraction

$$\frac{a_1 + a_2 + \cdots + a_n}{b_1 + b_2 + \cdots + b_n}$$

is contained between the largest and the smallest of these fractions (see Worked Example 1 following Proposition 7.15 in Chapter 7). The inequalities in question now follow immediately.]

1.26. Let a, b, c, and d be positive numbers. Show that

$$\frac{(a^2 + a + 1)(b^2 + b + 1)(c^2 + c + 1)(d^2 + d + 1)}{abcd} \geq 81.$$

[Hint: If $a > 0$, then $a + (1/a) \geq 2$.]

CHAPTER 2

Limits and Continuity

1. Limits

Let x be any real number. By the *absolute value of* x, in notation $|x|$, we mean x if $x \geq 0$ and $-x$ if $x \leq 0$. If we picture x as a point on the number line, then $|x|$ can be viewed as the distance between the points 0 and x. It is obvious that $|-x| = |x|$.

Proposition 2.1. *Let a and b be any real numbers. Then*

$$|a + b| \leq |a| + |b|. \tag{2.1}$$

Proof. If we add the (trivial) inequalities

$$-|a| \leq a \leq |a| \quad \text{and} \quad -|b| \leq b \leq |b|,$$

we get

$$-(|a| + |b|) \leq a + b \leq |a| + |b|;$$

but the inequality $|A| \leq B$ is clearly equivalent to the inequality $-B \leq A \leq B$ for any two real numbers A and B. $\qquad\square$

Comments. The inequality in Proposition 2.1 is sometimes called the *triangle inequality*. We can use it to derive some other useful inequalities. For example, for any real numbers a and b

$$|a - b| = |a + (-b)| \leq |a| + |-b| = |a| + |b|. \tag{2.2}$$

Moreover, since $a = a + b - b$,

$$|a| \leq |a + b| + |-b| = |a + b| + |b|$$

or

$$|a| - |b| \le |a + b|.$$

By interchanging the role of a and b in the last inequality we obtain

$$|b| - |a| \le |a + b|.$$

Thus, we see that

$$||a| - |b|| \le |a + b|. \tag{2.3}$$

On the other hand, since $a = a - b + b$, we get

$$|a| \le |a - b| + |b|$$

or

$$|a| - |b| \le |a - b|.$$

By interchanging the role of a and b in the last inequality we get

$$|b| - |a| \le |b - a| = |a - b|.$$

Therefore,

$$||a| - |b|| \le |a - b|. \tag{2.4}$$

Combining inequalities (2.1) to (2.4) we get

$$||a| - |b|| \le |a \pm b| \le |a| + |b| \tag{2.5}$$

for any real numbers a and b.

Another simple consequence of (2.1) is that

$$|a + b + c| \le |a| + |b| + |c| \quad \text{for any real numbers } a, b, \text{ and } c.$$

Indeed, $|a + b + c| \le |a| + |b + c| \le |a| + |b| + |c|$. Instead of only taking three summands, we could of course have taken any finite number of summands.

Proposition 2.2. *Let a and b be any real numbers and let $\max\{a, b\}$ and $\min\{a, b\}$ denote the larger and the smaller of the two numbers a and b, respectively. Then*

$$\frac{a + b + |a - b|}{2} = \max\{a, b\}$$

and

$$\frac{a + b - |a - b|}{2} = \min\{a, b\}.$$

PROOF. Let x be any real number. Then

$$\frac{x + |x|}{2} = x \qquad \text{if } x \ge 0,$$
$$= 0 \qquad \text{if } x \le 0$$

and

$$\frac{x - |x|}{2} = 0 \qquad \text{if } x \geq 0,$$
$$= x \qquad \text{if } x \leq 0.$$

Replacing x by $a - b$ and then adding a to both sides gives what we have set
out to show.   □

DISCUSSION. Geometrically, $|x - a|$ means the distance between the two points
a and x. Let δ (delta) be larger than zero; then $|x - a| < \delta$ means that
$a - \delta < x < a + \delta$ and $|x| < \delta$ means that $-\delta < x < \delta$ for in the latter case
$a = 0$. Let $a < b$; then the inequality $a < x < b$ can be expressed in the form
$|x - A| < B$, where $A = (a + b)/2$ and $B = |a - b|/2$ (here A is the midpoint
between the points a and b and B is half the distance between the points a
and b).

If we know the curve $y = f(x)$, then the curve $y = |f(x)|$ is easy to picture.
To obtain the curve $y = |f(x)|$ from the curve $y = f(x)$, we leave unchanged
that portion of the curve which is above the x-axis, but reflect through the
x-axis the portion which is below the x-axis. For example, $y = x - 2$ is a
straight line with slope 1 intersecting the x-axis at $x = 2$; the curve $y = |x - 2|$
consists of two branches: for $x \geq 2$ we have $y = x - 2$, but for $x \leq 2$ we have
$y = -(x - 2)$. Figure 2.1 shows the curve $y = |x - 2|$.

The set of points (x, y) satisfying the equation $|x| + |y| = 1$ is the closed
curve that we get by connecting consecutively the points $(1, 0)$, $(0, 1)$, $(-1, 0)$,
$(0, -1)$, and $(1, 0)$ by line segments; the shape of the figure is a diamond. The
set of points (x, y) satisfying the equation $x - |x| = y - |y|$ consists of all
points making up the first quadrant, that is, all (x, y) satisfying $x \geq 0$ and
$y \geq 0$, and the points of the line $y = x$ in the third quadrant.

To find the smallest value of

$$f(x) = |x + 2| + |x - 1| + |x - 3|$$

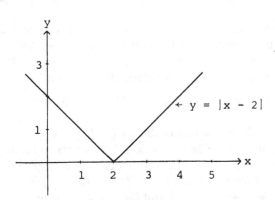

Figure 2.1

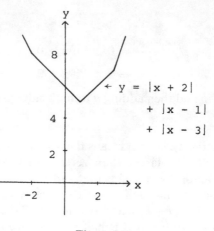

Figure 2.2

we proceed as follows. If x satisfies $-2 \leq x \leq 3$, then $f(x) = 5 + |x - 1|$ (because $|x + 2| + |x - 3| = 5$ when x is between -2 and 3); but $|x - 1| \geq 0$ with equality precisely when $x = 1$. Hence, $f(x)$ is smallest at $x = 1$ when x ranges over the closed interval $[-2, 3]$. For $x \geq 3$ we have $f(x) = 3x - 2$ (because $|x + 2| = x + 2$, $|x - 1| = x - 1$, and $|x - 3| = x - 3$ for $x \geq 3$) and for $x \leq -2$ we have $f(x) = -3x + 2$ (because $|x + 2| = -(x + 2)$, $|x - 1| = -(x - 1)$, and $|x - 3| = -(x - 3)$ for $x \leq -2$). Therefore, $f(x) \geq f(3) = 7$ for $x \geq 3$ and $f(x) \geq f(-2) = 8$ for $x \leq -2$. Hence, $f(x)$ is smallest at $x = 1$ for any real number x; $f(1) = 5$. Among the points $x = -2$, $x = 1$, and $x = 3$ the point $x = 1$ is the *median point* in the sense that $x = -2$ is to its left and $x = 3$ is to its right. The smallest value of

$$f(x) = |x + 2| + |x - 1| + |x - 3|$$

is assumed at $x = 1$, the median point of the set of three points $x = -2$, $x = 1$, and $x = 3$. Figure 2.2 shows the graph of $y = |x + 2| + |x - 1| + |x - 3|$.

Given the points $x = -2$, $x = 0$, $x = 2$, $x = 3$, and $x = 10$ the median point is $x = 2$ in the sense that the two points $x = -2$ and $x = 0$ are to its left and the two points $x = 3$ and $x = 10$ are to its right. By an argument similar to the one above we can conclude that the smallest value of

$$g(x) = |x + 2| + |x| + |x - 2| + |x - 3| + |x - 10|$$

is assumed at $x = 2$, the median of the set $\{-2, 0, 2, 3, 10\}$.

To solve the inequality $|2x - 7| < x + 1$ amounts to finding the set of all x for which the graph of $y = |2x - 7|$ is below the graph of $y = x + 1$. But $|2x - 7| = 2x - 7$ for $x \geq \frac{7}{2}$ and $|2x - 7| = -(2x - 7)$ for $x \leq \frac{7}{2}$. For $x \geq \frac{7}{2}$ we have $2x - 7 = x + 1$ or $x = 8$ and for $x \leq \frac{7}{2}$ we have $-(2x - 7) = x + 1$ or $x = 2$. The set of x satisfying $|2x - 7| < x + 1$ is the set of x with the

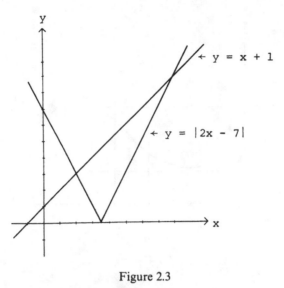

Figure 2.3

property $2 < x < 8$. Figure 2.3 shows the graphs of $y = |2x - 7|$ and $y = x + 1$.

Definition. Let a and b be two points on the number line and $a < b$; then the set of all points x satisfying $a < x < b$ is called an *open interval* and is denoted by (a, b). For any positive ε (epsilon), the interval $(a - \varepsilon, a + \varepsilon)$ is called a *neighborhood of a* (or, more precisely, an *ε-neighborhood of a*). We recall that the set of all points x satisfying $a \le x \le b$ is called a *closed interval* and is denoted by $[a, b]$.

Definition. Let a function f be defined on some neighborhood of a point a, except possibly at the point a itself, and the set of values or range of f be a set of real numbers. We say that *the limit of $f(x)$ as x tends to a is L*, or that *$f(x)$ tends to L as x tends to a*, and write

$$\lim_{x \to a} f(x) = L, \quad \text{or} \quad f(x) \to L \text{ as } x \to a,$$

if, given any $\varepsilon > 0$, there exists some $\delta > 0$, such that $|f(x) - L| < \varepsilon$ for any x satisfying $0 < |x - a| < \delta$.

DISCUSSION. A more compact phrasing of the defining condition of a functional limit is: if for any $\varepsilon > 0$ and some $\delta > 0$ we have $|f(x) - L| < \varepsilon$ for any x satisfying $0 < |x - a| < \delta$. To say that $f(x)$ *does not* tend to L as x tends to a means: if for some $\varepsilon > 0$ and any $\delta > 0$ we have $|f(x) - L| \ge \varepsilon$ for some x satisfying $0 < |x - a| < \delta$.

To visualize the defining condition of a functional limit, let us consider the

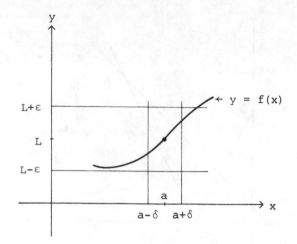

Figure 2.4

graph of the function $y = f(x)$. We take any $\varepsilon > 0$ and consider the lines $y = L - \varepsilon$ and $y = L + \varepsilon$. Then there has to be an open interval $(a - \delta, a + \delta)$ on the x-axis so that for all points x of this interval, with the exception of its midpoint $x = a$, we have that $f(x)$ is between $L - \varepsilon$ and $L + \varepsilon$. See Figure 2.4.

To make the defining condition of a functional limit even more accessible, picture the function f to be a gun which shoots from the point t on the x-axis and the bullet hits at the point $(t, f(t))$ in the x, y-plane. The strip between the lines $y = L - \varepsilon$ and $y = L + \varepsilon$ is the target that we are trying to hit. For any choice of $\varepsilon > 0$ we must find a δ-neighborhood of a so that the bullet will hit the target. This δ-neighborhood of a need not be the largest possible for a given ε and the point $x = a$ is completely excluded from consideration.

In the definition of functional limit we speak of "the" limit instead of "a" limit; the reason for this is the following Uniqueness Theorem.

Proposition 2.3. *If* $\lim_{x \to a} f(x) = L$ *and* $\lim_{x \to a} f(x) = M$, *then* $L = M$.

PROOF. We shall show that $L = M$ by proving that the assumption $L \neq M$ leads to the absurd conclusion $|L - M| < |L - M|$.

Let us assume that $L \neq M$. It follows that $\frac{1}{2}|L - M| > 0$. Since $f(x) \to L$ as $x \to a$, we know that there exists some $\delta_1 > 0$ such that

$$\text{if } 0 < |x - a| < \delta_1, \quad \text{then } |f(x) - L| < \tfrac{1}{2}|L - M|.$$

Since $f(x) \to M$ as $x \to a$, we know that there exists some $\delta_2 > 0$ such that

$$\text{if } 0 < |x - a| < \delta_2, \quad \text{then } |f(x) - M| < \tfrac{1}{2}|L - M|.$$

Let $\delta = \min\{\delta_1, \delta_2\}$. For t satisfying $0 < |t - a| < \delta$ we find that

$$|f(t) - L| < \tfrac{1}{2}|L - M| \quad \text{and} \quad |f(t) - M| < \tfrac{1}{2}|L - M|.$$

Using (2.1), it follows that

$$|L - M| = |[L - f(t)] + [f(t) - M]| \le |L - f(t)| + |f(t) - M|$$
$$< \tfrac{1}{2}|L - M| + \tfrac{1}{2}|L - M| = |L - M|.$$

Indeed, we have arrived at the absurdity $|L - M| < |L - M|$.   □

Definition. The sequence $x_1, x_2, x_3, \ldots$ of real numbers is said to *converge* if there is a real number x with the following property: For each $\varepsilon > 0$, there is a positive integer n_0, dependent on ε, such that $n > n_0$ implies $|x_n - x| < \varepsilon$. The number x is then called the *limit of the sequence* $x_1, x_2, x_3, \ldots$ and we say that the sequence *converges to* x, or x_n *tends to* x *as* n *becomes arbitrarily large*, and we write

$$\lim_{n \to \infty} x_n = x, \quad \text{or} \quad x_n \to x \text{ as } n \to \infty.$$

REMARKS. If a sequence has a limit, then this limit is unique; we may therefore speak of "the" limit of a sequence. Indeed, suppose the contrary: Let $x_n \to x$ and $x_n \to x'$ as $n \to \infty$ with $x \neq x'$. Let $\varepsilon = \tfrac{1}{2}|x - x'|$. Since $x_n \to x$ as $n \to \infty$, there exists a positive integer n_1 such that $|x_n - x| < \varepsilon$ for $n > n_1$. Similarly, since $x_n \to x'$ as $n \to \infty$, there exists a positive integer n_2 such that $|x_n - x'| < \varepsilon$ for $n > n_2$. Let $n_0 = \max\{n_1, n_2\}$. Then both $|x_n - x|$ and $|x_n - x'|$ are less than ε for $n > n_0$. Thus

$$|x - x'| = |(x_n - x') - (x_n - x)| \le |x_n - x'| + |x_n - x| < 2\varepsilon = |x - x'|$$

which shows that $|x - x'| < |x - x'|$. This contradiction establishes that $x = x'$.

Proposition 2.4. *Let a function f be defined on a neighborhood J of a point a, except possibly at the point a itself. Then*

$$\lim_{x \to a} f(x) = L$$

if and only if

$$\lim_{n \to \infty} f(x_n) = L$$

for every sequence $x_1, x_2, x_3, \ldots$ of points in J such that $x_n \neq a$ for any positive integer n and $x_n \to a$ as $n \to \infty$.

PROOF. Suppose $\lim_{x \to a} f(x) = L$. We choose a sequence $x_1, x_2, x_3, \ldots$ in J such that $x_n \neq a$ for $n = 1, 2, 3, \ldots$ and $x_n \to a$ as $n \to \infty$. Let $\varepsilon > 0$ be given. Then there is a $\delta > 0$ such that $|f(x) - L| < \varepsilon$ whenever x belongs to J and $0 < |x - a| < \delta$. Also, there exists an n_0 such that $n > n_0$ implies $0 < |x_n - a| < \delta$. Thus, for all $n > n_0$, we have $|f(x_n) - L| < \varepsilon$ and we see that $\lim_{n \to \infty} f(x_n) = L$ holds.

Conversely, suppose $\lim_{x \to a} f(x) = L$ fails. Then there exists some $\varepsilon > 0$ such that for any $\delta > 0$ there is a point x in the neighborhood J (depending on δ), for which $|f(x) - L| \geq \varepsilon$ but $0 < |x - a| < \delta$. Taking $\delta_n = 1/n$ for $n = 1, 2, 3, \ldots$, we thus find a sequence $x_1, x_2, x_3, \ldots$ in J satisfying $x_n \neq a$ for $n = 1, 2, 3, \ldots, x_n \to a$ as $n \to \infty$, for which $\lim_{n \to \infty} f(x_n) = L$ is false. This completes the proof.   □

REMARK. The following are equivalent statements:

(i) $\lim_{x \to a} f(x) = L$.   (ii) $\lim_{x \to a} [f(x) - L] = 0$.

(iii) $\lim_{x \to a} |f(x) - L| = 0$.   (iv) $\lim_{h \to 0} f(a + h) = L$.

Examples

1. Show that $\lim_{x \to 0} (x/x) = 1$.

SOLUTION. The function $f(x) = x/x = 1$ for $x \neq 0$, but is undefined for $x = 0$. Let $\varepsilon > 0$. Here we must find $\delta > 0$ such that

$$\text{if } 0 < |x| < \delta, \quad \text{then } |1 - 1| < \varepsilon.$$

Since $|1 - 1| = 0$, we always have $|1 - 1| < \varepsilon$ no matter how δ is chosen; in short, any positive number will do for δ.

2. Show that $\lim_{x \to 1} (3x - 8) = -5$.

SOLUTION. Let $\varepsilon > 0$. Here we must find $\delta > 0$ such that

$$\text{if } 0 < |x - 1| < \delta, \quad \text{then } |(3x - 8) - (-5)| < \varepsilon.$$

But $|(3x - 8) - (-5)| = 3|x - 1|$. Therefore, the condition $|(3x - 8) - (-5)| < \varepsilon$ is equivalent to $3|x - 1| < \varepsilon$, that is, $|x - 1| < \varepsilon/3$. Hence, we must determine a positive δ such that

$$\text{if } 0 < |x - 1| < \delta, \quad \text{then } |x - 1| < \varepsilon/3.$$

Obviously, $\delta = \varepsilon/3$ works. So does any *smaller* positive value of δ.

3. Show that $\lim_{x \to 3} x^2 = 9$.

SOLUTION. Let $\varepsilon > 0$. We must find $\delta > 0$ such that

$$\text{if } 0 < |x - 3| < \delta, \quad \text{then } |x^2 - 9| < \varepsilon.$$

But $|x^2 - 9| = |x - 3||x + 3|$. Let us first require that $\delta \leq 1$. Then $|x - 3| < \delta$ gives that $2 < x < 4$ and hence $5 < x + 3 < 7$. Now, $|x^2 - 9| = |x - 3| \cdot |x + 3|$ will be less than ε if simultaneously $|x - 3| < \varepsilon/7$ and $|x + 3| < 7$; to achieve this we only have to take δ to be the smaller of the two numbers 1 and $\varepsilon/7$, that is, $\delta = \min\{1, \varepsilon/7\}$. The graph of $\delta = \min\{1, \varepsilon/7\}$ is shown in Figure 2.5.

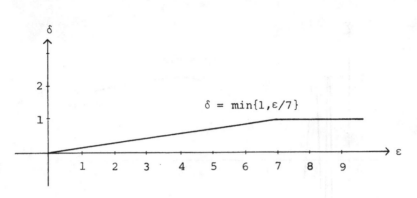

Figure 2.5

4. Show that $\lim_{x \to 2} [(x^2 - x + 18)/(3x - 1)] = 4$.

SOLUTION. We have

$$\left| \frac{x^2 - x + 18}{3x - 1} - 4 \right| = \left| \frac{x^2 - 13x + 22}{3x - 1} \right| = |x - 2| \cdot \left| \frac{x - 11}{3x - 1} \right|.$$

We now require that $\delta \le 1$. Then $|x - 2| < \delta$ gives $1 < x < 3$ and hence also $-10 < x - 11 < -8$ and $2 < 3x - 1 < 8$, so that $|x - 11| < 10$ and $|3x - 1| < 2$. Thus,

$$\left| \frac{x - 11}{3x - 1} \right| < \frac{10}{2} = 5 \quad \text{whenever } |x - 2| < 1.$$

But

$$|x - 2| \cdot \left| \frac{x - 11}{3x - 1} \right|$$

will be less than ε if simultaneously

$$|x - 2| < \frac{\varepsilon}{5} \quad \text{and} \quad \left| \frac{x - 11}{3x - 1} \right| < 5;$$

we only have to take $\delta = \min\{1, \varepsilon/5\}$.

5. Show that $\lim_{x \to 0} \sin(1/x)$ does not exist, that is, there is no real number L such that $\sin(1/x) \to L$ as $x \to 0$.

SOLUTION. We have $\sin(2n - 1/2)\pi = -1$ and $\sin(2n + 1/2)\pi = 1$ for $n = 1$, $2, \ldots$. Let

$$x_n = \frac{1}{(2n - 1/2)\pi} \quad \text{and} \quad t_n = \frac{1}{(2n + 1/2)\pi} \quad \text{for } n = 1, 2, \ldots$$

Then $x_n \to 0$ and $t_n \to 0$ as $n \to \infty$; however, $\sin(1/x_n) \to -1$ and $\sin(1/t_n) \to 1$ as $n \to \infty$ which is in violation of Proposition 2.3. The curve $y = \sin(1/x)$ is shown in Figure 2.6.

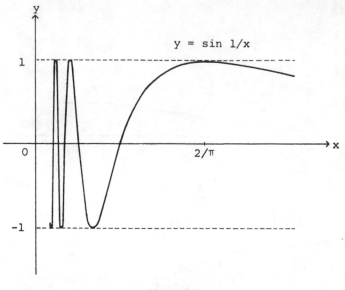

Figure 2.6

6. Show that $\lim_{x \to 0} x[\sin(1/x)] = 0$.

SOLUTION. Since $|\sin t| \le 1$ for any real number t, we have

$$0 \le \left| x \sin\frac{1}{x} \right| \le |x|.$$

Hence, given any $\varepsilon > 0$, we can take $\delta = \varepsilon$ to obtain the desired result. The curve $y = x\sin(1/x)$ is shown in Figure 2.7.

7. Let $f(x) = x$ if x is rational and $f(x) = -x$ if x is irrational. For what values of a does $\lim_{x \to a} f(x)$ exist?

SOLUTION. It is easily seen that the limit in question exists for $a = 0$. The limit does not exist if $a \ne 0$; note that any (nonempty) open interval on the number line contains both rational and irrational points.

Proposition 2.5. *If* $\lim_{x \to a} f(x) = L$ *and* $\lim_{x \to a} g(x) = M$, *then*

(i) $\lim_{x \to a} [f(x) + g(x)] = L + M$,

(ii) $\lim_{x \to a} [cf(x)] = cL$ *for each real number* c,

(iii) $\lim_{x \to a} [f(x)g(x)] = LM$.

PROOF. Let $\varepsilon > 0$. To establish (i) we must verify that there exists $\delta > 0$ such that

$$\text{if } 0 < |x - a| < \delta, \quad \text{then } |[f(x) + g(x)] - [L + M]| < \varepsilon.$$

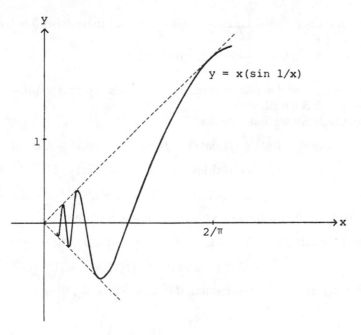

Figure 2.7

Since $f(x) \to L$ and $g(x) \to M$ as $x \to a$, we know that there exist positive numbers δ_1 and δ_2 such that

$$\text{if } 0 < |x - c| < \delta_1, \quad \text{then } |f(x) - L| < \frac{\varepsilon}{2}$$

and

$$\text{if } 0 < |x - a| < \delta_2, \quad \text{then } |g(x) - M| < \frac{\varepsilon}{2}.$$

Putting $\delta = \min\{\delta_1, \delta_2\}$, we observe:

$$\text{if } 0 < |x - a| < \delta, \quad \text{then } |f(x) - L| < \frac{\varepsilon}{2} \quad \text{and} \quad |g(x) - M| < \frac{\varepsilon}{2}.$$

Hence,

$$\text{if } 0 < |x - a| < \delta, \quad \text{then } |[f(x) + g(x)] - [L + M]| < \varepsilon$$

because

$$|[f(x) + g(x)] - [L - M]| = |[f(x) - L] - [g(x) - M]|$$
$$\leq |f(x) - L| + |g(x) - M|$$

by (2.1) and (i) is proved.

To establish (ii) we consider two cases: $c \neq 0$ and $c = 0$. If $c \neq 0$, then

$\varepsilon/|c| > 0$ and since $f(x) \to L$ as $x \to a$, we know that there exists $\delta > 0$ so that,

$$\text{if } 0 < |x - a| < \delta, \quad \text{then } |f(x) - L| < \frac{\varepsilon}{|c|}.$$

But $|f(x) - L| < \varepsilon/|c|$ implies that $|c||f(x) - L| < \varepsilon$ and thus $|cf(x) - cL| < \varepsilon$. The case $c = 0$ is trivial.

To establish (iii) we first note that

$$|f(x)g(x) - LM| \le |f(x)g(x) - f(x)M| + |f(x)M - LM|$$

$$= |f(x)||g(x) - M| + |M||f(x) - L|$$

$$\le |f(x)||g(x) - M| + (1 + |M|)|f(x) - L|.$$

Let $\varepsilon > 0$. Since $f(x) \to L$ and $g(x) \to M$ as $x \to a$, we know

(1) that there exists $\delta_1 > 0$ such that, if $0 < |x - a| < \delta_1$, then

$$|f(x) - L| < 1 \quad \text{and thus} \quad |f(x)| < 1 + |L|;$$

(2) that there exists $\delta_2 > 0$ such that, if $0 < |x - a| < \delta_2$, then

$$|g(x) - M| < \frac{\varepsilon}{2}\left(\frac{1}{1 + |L|}\right);$$

(3) that there exists $\delta_3 > 0$ such that, if $0 < |x - a| < \delta_3$, then

$$|f(x) - L| < \frac{\varepsilon}{2}\left(\frac{1}{1 + |M|}\right).$$

Now we set $\delta = \min\{\delta_1, \delta_2, \delta_3\}$ and note that if $0 < |x - a| < \delta$, then

$$|f(x)g(x) - LM| < (1 + |L|)\frac{\varepsilon}{2}\left(\frac{1}{1 + |L|}\right) + (1 + |M|)\frac{\varepsilon}{2}\left(\frac{1}{1 + |M|}\right) = \varepsilon.$$

This completes the proof.   □

Proposition 2.6. *If* $\lim_{x \to a} f(x) = L$ *and* $\lim_{x \to a} g(x) = M$ *with* $M \ne 0$, *then*

$$\lim_{x \to a} \frac{f(x)}{g(x)} = \frac{L}{M}.$$

PROOF. We show first that $\lim_{x \to a} g(x) = M$ with $M \ne 0$ implies

$$\lim_{x \to a} \frac{1}{g(x)} = \frac{1}{M}.$$

Indeed, for $g(x) \ne 0$,

$$\left|\frac{1}{g(x)} - \frac{1}{M}\right| = \frac{|g(x) - M|}{|M||g(x)|}.$$

Pick $\delta_1 > 0$ such that

$$\text{if } 0 < |x - a| < \delta_1, \quad \text{then } |g(x) - M| < \frac{|M|}{2}.$$

For such x we have

$$|g(x)| > \frac{|M|}{2}, \qquad \frac{1}{|g(x)|} < \frac{2}{|M|}$$

and therefore

$$\left| \frac{1}{g(x)} - \frac{1}{M} \right| \leq \frac{2}{|M|^2} |g(x) - M|.$$

Now let $\varepsilon > 0$ and pick $\delta_2 > 0$ such that

$$\text{if } 0 < |x - a| < \delta_2, \quad \text{then } |g(x) - M| < \frac{|M|^2}{2} \varepsilon.$$

Putting $\delta = \min\{\delta_1, \delta_2\}$, we get that

$$\text{if } 0 < |x - a| < \delta, \quad \text{then } \left| \frac{1}{g(x)} - \frac{1}{M} \right| < \varepsilon.$$

This shows that $1/g(x) \to 1/M$ as $x \to a$.

To prove the proposition we only note that

$$\frac{f(x)}{g(x)} = f(x) \frac{1}{g(x)}.$$

With $f(x) \to L$ and $1/g(x) \to 1/M$ as $x \to a$, part (iii) of Proposition 2.5 gives

$$\lim_{x \to a} \frac{f(x)}{g(x)} = L \frac{1}{M} = \frac{L}{M}. \qquad \square$$

REMARKS. The results in Proposition 2.5 and Proposition 2.6 are readily extended to any *finite* number of functions. Thus, it is easy to see that every polynomial

$$P(x) = c_n x^n + c_{n-1} x^{n-1} + \cdots + c_1 x + c_0$$

satisfies

$$\lim_{x \to a} P(x) = P(a). \tag{2.6}$$

Also, if P and Q are polynomials and $Q(a) \neq 0$, then

$$\lim_{x \to a} \frac{P(x)}{Q(x)} = \frac{P(a)}{Q(a)}. \tag{2.7}$$

If $\lim_{x \to a} f(x) = L$ with $L \neq 0$ and $\lim_{x \to a} g(x) = 0$, then $\lim_{x \to a} [f(x)/g(x)]$ does not exist. Indeed, suppose on the contrary that there exists a real number T such that

$$\lim_{x \to a} \frac{f(x)}{g(x)} = T.$$

Then

$$L = \lim_{x \to a} f(x) = \lim_{x \to a} \left(g(x) \cdot \frac{f(x)}{g(x)} \right) = \lim_{x \to a} g(x) \cdot \lim_{x \to a} \frac{f(x)}{g(x)} = 0 \cdot T = 0,$$

which contradicts the assumption that $L \neq 0$.

Proposition 2.7. *Suppose that there is a number $q > 0$ such that*

$$h(x) \leq f(x) \leq g(x)$$

for all x satisfying $0 < |x - a| < q$. If

$$\lim_{x \to a} h(x) = L \quad \text{and} \quad \lim_{x \to a} g(x) = L,$$

then

$$\lim_{x \to a} f(x) = L.$$

PROOF. Let $\varepsilon > 0$. Let $q > 0$ be such that

$$\text{if } 0 < |x - a| < q, \quad \text{then } h(x) \leq f(x) \leq g(x).$$

Pick $\delta_1 > 0$ such that

$$\text{if } 0 < |x - a| < \delta_1, \quad \text{then } L - \varepsilon < h(x) < L + \varepsilon$$

and pick $\delta_2 > 0$ such that

$$\text{if } 0 < |x - a| < \delta_2, \quad \text{then } L - \varepsilon < g(x) < L + \varepsilon.$$

Let $\delta = \min\{q, \delta_1, \delta_2\}$. For x satisfying $0 < |x - a| < \delta$, we have

$$L - \varepsilon < h(x) \leq f(x) \leq g(x) < L + \varepsilon$$

and so $|f(x) - L| < \varepsilon$.   □

APPLICATION. We show that

$$\lim_{x \to 0} \frac{\sin x}{x} = 1. \tag{2.8}$$

First we verify the inequality

$$\sin x < x < \tan x \quad \text{for } 0 < x < \frac{\pi}{2}. \tag{2.9}$$

Consider in a circle of radius R an acute angle $\angle AOB$, the chord AB and the tangent AC to the circle at the point A (see Figure 2.8). Let x be the radian measure of the angle $\angle AOB$; then the length of the circular arc AB equals

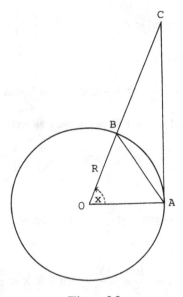

Figure 2.8

Rx. Evidently, the area of the triangle AOB (which equals $\frac{1}{2}R^2 \sin x$) is less than the area of the circular sector AOB which equals $\frac{1}{2}R^2 x$); moreover, the area of the circular sector AOB is less than the area of the triangle AOC (which equals $\frac{1}{2}R^2 \tan x$). Thus, upon division by $\frac{1}{2}R^2$, we get (2.9).

Since $\sin x$ is positive for $0 < x < \pi/2$, we can divide by $\sin x$ in (2.9) and obtain

$$1 < \frac{\sin x}{x} < \cos x \quad \text{or} \quad 0 < 1 - \frac{\sin x}{x} < 1 - \cos x \quad \text{for } 0 < x < \frac{\pi}{2}.$$

But $1 - \cos x = 2(\sin^2 x/2) < 2(\sin x/2)$ and so, by (2.9), $2(\sin x/2) < x$. Thus,

$$0 < 1 - \frac{\sin x}{x} < x \quad \text{for } 0 < x < \frac{\pi}{2}$$

or

$$0 < \left| \frac{\sin x}{x} - 1 \right| < |x| \quad \text{for } 0 < x < \frac{\pi}{2}.$$

Letting $f(x) = (\sin x)/x$ if $x \neq 0$, we see that $f(-x) = f(x)$; on the other hand, $|-x| = |x|$. Thus,

$$0 < \left| \frac{\sin x}{x} - 1 \right| < |x| \quad \text{for } 0 < |x| < \frac{\pi}{2}. \tag{2.10}$$

Applying Proposition 2.7 we get at once the desired result (2.8). In fact, for any $\varepsilon > 0$ and $\delta = \min\{\varepsilon, \pi/2\}$ we have:

$$\text{if } 0 < |x| < \delta, \quad \text{then} \quad \left| \frac{\sin x}{x} - 1 \right| < \varepsilon.$$

Returning to the estimate

$$0 < 1 - \frac{\sin x}{x} < 1 - \cos x \quad \text{for } 0 < x < \frac{\pi}{2},$$

we could have proceeded in the following way as well: Since

$$1 - \cos x = 2 \sin^2 \frac{x}{2} \quad \text{and} \quad \sin \frac{x}{2} < \frac{x}{2} \text{ by (2.9)},$$

we obtain

$$1 - \cos x < \frac{x^2}{2} \quad \text{for } 0 < x < \frac{\pi}{2} \tag{2.11}$$

and so

$$0 < 1 - \frac{\sin x}{x} < \frac{x^2}{2} \quad \text{for } 0 < x < \frac{\pi}{2}.$$

But replacing x by $-x$ does not alter the foregoing inequality and so

$$0 < \left| \frac{\sin x}{x} - 1 \right| < \frac{x^2}{2} \quad \text{for } 0 < |x| < \frac{\pi}{2}. \tag{2.12}$$

Applying Proposition 2.7 we get at once the desired result (2.8) from (2.12). Indeed, for any $\varepsilon > 0$ and $\delta = \min\{\sqrt{2\varepsilon}, \pi/2\}$ we have

$$\text{if } 0 < |x| < \delta, \quad \text{then} \quad \left| \frac{\sin x}{x} - 1 \right| < \varepsilon.$$

It is easy to prove that

$$\lim_{x \to 0} \frac{1 - \cos x}{x} = 0. \tag{2.13}$$

It is clear that the inequality (2.11) remains true if we replace x by $-x$. Thus,

$$1 - \cos x < \frac{x^2}{2} \quad \text{for } 0 < |x| < \frac{\pi}{2}$$

or

$$0 < \left| \frac{1 - \cos x}{x} \right| < \frac{|x|}{2} \quad \text{for } 0 < |x| < \frac{\pi}{2}$$

and (2.13) follows by applying Proposition 2.7.

Definition. Let f be a function defined at least on an open interval of the form (d, a); then the *left-side limit of $f(x)$ is L as x tends to a*, written as

$$\lim_{x \uparrow a} f(x) = L \quad \text{or} \quad f(a-) = L,$$

if, given any $\varepsilon > 0$, there exists some $\delta > 0$, such that $|f(x) - L| < \varepsilon$ for any x satisfying $a - \delta < x < a$.

Definition. Let f be a function defined at least on an open interval of the form (a, d); then the *right-side limit of $f(x)$ is L as x tends to a*, written as

$$\lim_{x \downarrow a} f(x) = L \quad \text{or} \quad f(a+) = L,$$

if, given any $\varepsilon > 0$, there exists some $\delta > 0$, such that $|f(x) - L| < \varepsilon$ for any x satisfying $a < x < a + \delta$.

REMARKS. Since $\sqrt{x}$ is not defined on both sides of 0, we can not consider the (two-sided) limit of $\sqrt{x}$ as x tends to 0. However, we have that

$$\lim_{x \downarrow 0} \sqrt{x} = 0$$

because, for any $\varepsilon > 0$ and $\delta = \varepsilon^2$, we have:

$$\text{if } 0 < x < \delta, \quad \text{then } |\sqrt{x} - 0| = \sqrt{x} < \varepsilon.$$

One-sided limits give us a simple method of deciding whether or not a (two-sided) limit exists:

$$\lim_{x \to a} f(x) = L \quad \text{if and only if} \quad \lim_{x \uparrow a} f(x) = L \quad \text{and} \quad \lim_{x \downarrow a} f(x) = L.$$

One-sided limits are of particular interest in connection with increasing and decreasing functions, as we shall see in the next section.

Definition. Let f be a function defined for all real numbers $x > d$, where d is some real number. We say that the *limit of $f(x)$ is L as x tends to infinity (or as x becomes arbitrarily large)*, and we write

$$\lim_{x \to \infty} f(x) = L,$$

if, given any $\varepsilon > 0$, there exists some $K > 0$, such that $|f(x) - L| < \varepsilon$ for any x satisfying $x > K$.

REMARK. There is a simple relation between limits at ∞ and one-sided limits at 0; setting $x = 1/t$, it is easily seen that

$$\lim_{x \to \infty} f(x) = L \quad \text{if and only if} \quad \lim_{t \downarrow 0} f\left(\frac{1}{t}\right) = L.$$

Definition. Let f be a function defined for all real numbers $x < d$, where d is some real number. We say that the *limit of $f(x)$ is L as x tends to negative infinity (or minus infinity)*, written as

$$\lim_{x \to -\infty} f(x) = L,$$

if, given any $\varepsilon > 0$, there exists some $K < 0$, such that $|f(x) - L| < \varepsilon$ for any x satisfying $x < K$.

REMARK. Propositions 2.3 to 2.7, suitably modified, are also valid for one-sided limits and limits at $\pm \infty$.

Examples

1. Let n be a positive integer. Show that

$$\lim_{x \to 1} \frac{x^n - 1}{x - 1} = n.$$

SOLUTION. For $x \neq 1$ we have

$$x^n - 1 = (x - 1)(x^{n-1} + x^{n-2} + \cdots + x + 1)$$

and so

$$\lim_{x \to 1} \frac{x^n - 1}{x - 1} = \lim_{x \to 1} (x^{n-1} + x^{n-2} + \cdots + x + 1) = n.$$

2. Let m and n be positive integers. Show that

$$\lim_{x \to 1} \frac{x^m - 1}{x^n - 1} = \frac{m}{n}.$$

SOLUTION. For $x \neq 1$ we have

$$\frac{x^m - 1}{x^n - 1} = \frac{(x - 1)(x^{m-1} + x^{m-2} + \cdots + x + 1)}{(x - 1)(x^{n-1} + x^{n-2} + \cdots + x + 1)}$$

and so

$$\lim_{x \to 1} \frac{x^m - 1}{x^n - 1} = \frac{\lim_{x \to 1}(x^{m-1} + x^{m-2} + \cdots + x + 1)}{\lim_{x \to 1}(x^{n-1} + x^{n-2} + \cdots + x + 1)} = \frac{m}{n}.$$

3. Let n and k be positive integers and $n > k$. Show that

$$\lim_{x \to 1} \frac{(x^n - 1)(x^{n-1} - 1) \cdots (x^{n-k+1} - 1)}{(x - 1)(x^2 - 1) \cdots (x^k - 1)} = \frac{n(n - 1) \cdots (n - k + 1)}{1 \cdot 2 \cdots k}.$$

SOLUTION. We have

$$\lim_{x \to 1} \frac{(x^n - 1)(x^{n-1} - 1) \cdots (x^{n-k+1} - 1)}{(x - 1)(x^2 - 1) \cdots (x^k - 1)}$$

$$= \lim_{x \to 1} \frac{x^n - 1}{x - 1} \lim_{x \to 1} \frac{x^{n-1} - 1}{x^2 - 1} \cdots \lim_{x \to 1} \frac{x^{n-k+1} - 1}{x^k - 1}$$

$$= \frac{n}{1} \cdot \frac{n - 1}{2} \cdots \frac{n - k + 1}{k}.$$

4. Let m and n be positive integers. Show that

$$\lim_{x \to 1}\left(\frac{m}{1 - x^m} - \frac{n}{1 - x^n}\right) = \frac{m - n}{2}.$$

SOLUTION. We have

$$\frac{m}{1 - x^m} - \frac{n}{1 - x^n} = \frac{n(x^m - 1) - m(x^n - 1)}{(x^m - 1)(x^n - 1)}$$

$$= \frac{n(x^{m-1} + x^{m-2} + \cdots + x + 1) - m(x^{n-1} + x^{n-2} + \cdots + x + 1)}{(x - 1)(x^{m-1} + x^{m-2} + \cdots + x + 1)(x^{n-1} + x^{n-2} + \cdots + x + 1)}.$$

But

$$\frac{n(x^{m-1} + x^{m-2} + \cdots + x + 1)}{x - 1}$$

$$= n\left(\frac{x^{m-1} - 1}{x - 1} + \frac{x^{m-2} - 1}{x - 1} + \cdots + \frac{x - 1}{x - 1} + \frac{1 - 1}{x - 1}\right) + \frac{nm}{x - 1}$$

and

$$\frac{m(x^{n-1} + x^{n-2} + \cdots + x + 1)}{x - 1}$$

$$= m\left(\frac{x^{n-1} - 1}{x - 1} + \frac{x^{n-2} - 1}{x - 1} + \cdots + \frac{x - 1}{x - 1} + \frac{1 - 1}{x - 1}\right) + \frac{nm}{x - 1}.$$

Thus,

$$\lim_{x \to 1}\frac{n(x^{m-1} + x^{m-2} + \cdots + x + 1) - m(x^{n-1} + x^{n-2} + \cdots + x + 1)}{x - 1}$$

$$= n\{(m - 1) + (m - 2) + \cdots + 1\} - m\{(n - 1) + (n - 2) + \cdots + 1\}$$

$$= n\frac{m(m - 1)}{2} - m\frac{n(n - 1)}{2}$$

$$= \frac{m^2 n - mn^2}{2}.$$

On the other hand,

$$\lim_{x \to 1}(x^{m-1} + x^{m-2} + \cdots + x + 1)(x^{n-1} + x^{n-2} + \cdots + x + 1) = mn.$$

But

$$\frac{m^2 n - mn^2}{2mn} = \frac{m - n}{2}$$

and so

$$\lim_{x \to 1}\left(\frac{m}{1 - x^m} - \frac{n}{1 - x^n}\right) = \frac{m - n}{2}.$$

5. Show that

$$\lim_{x \to \infty} \{\sqrt{(x + a)(x + b)} - x\} = \frac{a + b}{2}.$$

SOLUTION. We commence with an inequality: If A and B are any positive real numbers, then

$$\frac{2}{1/A + 1/B} \le \sqrt{AB} \le \frac{A + B}{2}. \tag{2.14}$$

Indeed, $0 \le (\sqrt{A} - \sqrt{B})^2 = A + B - 2\sqrt{AB}$ and so $\sqrt{AB} \le (A + B)/2$. This shows the validity of the second part of (2.14). Replacing in $\sqrt{AB} \le (A + B)/2$ the numbers A and B by $1/A$ and $1/B$, respectively, we get the first part of (2.14). Inequality (2.14) expresses the familiar fact that the harmonic mean of two positive numbers A and B is less than or equal to the geometric mean of A and B which in turn is less than or equal to the arithmetic mean of A and B; there is equality in (2.14) if and only if $A = B$.

Now, putting $A = x + a$ and $B = x + b$ in (2.14), we obtain

$$\frac{(a + b)x + 2ab}{2x + a + b} \le \sqrt{(x + a)(x + b)} - x \le \frac{a + b}{2}$$

whenever x is such that $x + a > 0$ and $x + b > 0$. Letting $x \to \infty$ in (2.14) yields the desired result.

6. If a regular polygon of n sides is inscribed in a circle of radius r, its area is

$$A(n) = \tfrac{1}{2}nr^2 \sin\frac{2\pi}{n}$$

and its circumference is

$$C(n) = 2nr \sin\frac{\pi}{n}.$$

Show that $\lim_{n \to \infty} A(n) = \pi r^2$ and $\lim_{n \to \infty} C(n) = 2\pi r$.

SOLUTION. We have $(\sin x)/x \to 1$ as $x \to 0$; but

$$A(n) = \pi r^2 \frac{\sin(2\pi/n)}{2\pi/n} \quad \text{and} \quad C(n) = 2\pi r \frac{\sin(\pi/n)}{\pi/n}.$$

2. Continuity

Definition. Let a function f be defined on some neighborhood of a point a (including the point a itself) and the set of values or range of f be a set of real numbers. We say that f is *continuous at* a if, given any $\varepsilon > 0$, there exists some $\delta > 0$, such that $|f(x) - f(a)| < \varepsilon$ for any x satisfying $|x - a| < \delta$.

REMARK. It follows immediately from the definition that f is continuous at a if and only if $\lim_{x \to a} f(x) = f(a)$.

Definition. If a function f is continuous at every point of an open interval (a, b), we say that *f is continuous on the interval (a, b)*. If a function f is defined on a closed interval $[a, b]$, we say that *f is continuous on $[a, b]$* if f is continuous on (a, b) and if, in addition,

$$\lim_{x \downarrow a} f(x) = f(a) \quad \text{and} \quad \lim_{x \uparrow b} f(x) = f(b).$$

Proposition 2.8. *Let f and g be functions defined on the same neighborhood of a point a and both functions be continuous at the point a. Then each of the following functions is continuous at a:*

(i) $(f + g)(x) = f(x) + g(x)$,
(ii) $(fg)(x) = f(x)g(x)$,
(iii) $\left(\dfrac{f}{g}\right)(x) = \dfrac{f(x)}{g(x)}$ $[if\ g(a) \neq 0]$,
(iv) $|f|(x) = |f(x)|$.

PROOF. Properties (i), (ii), and (iii) follow immediately from Propositions 2.5 and 2.6; property (iv) is a consequence of the inequality

$$\big||f(x)| - |f(a)|\big| \leq |f(x) - f(a)|$$

[see inequality (2.4)]. $\qquad\qquad\qquad\qquad\qquad\qquad\qquad\qquad\qquad\qquad\quad$ □

REMARK. If f and g are as in Proposition 2.8, then the functions

$$(f \vee g)(x) = \max\{f(x), g(x)\} \quad \text{and} \quad (f \wedge g)(x) = \min\{f(x), g(x)\}$$

are continuous at a. Indeed, using Propositions 2.2 and 2.8, the claim follows at once.

Proposition 2.9. *Let a function g (of the variable y) be defined on an interval Y, a function f (of the variable x) be defined on an interval X, and assume that the set of values $f(x)$, as x ranges over X, are contained in the interval Y. If f is continuous at a point x_0 of X and g is continuous at the corresponding point $y_0 = f(x_0)$ of Y, then the composite function*

$$h(x) = g[f(x)] \quad \text{for all } x \text{ in } X$$

is continuous at x_0.

PROOF. Let $\varepsilon > 0$ be given. Since g is continuous at $y = y_0$, there exists some $\sigma > 0$ such that

$$\text{if } |y - y_0| < \sigma, \quad \text{then } |g(y) - g(y_0)| < \varepsilon.$$

On the other hand, since f is continuous at $x = x_0$, for this σ (sigma) there exists some $\delta > 0$ such that

$$\text{if } |x - x_0| < \delta, \quad \text{then } |f(x) - f(x_0)| = |f(x) - y_0| < \sigma.$$

It therefore follows that

$$\text{if } |x - x_0| < \delta, \quad \text{then } |g[f(x)] - g(y_0)| = |g[f(x)] - g[f(x_0)]| < \varepsilon.$$

This shows the continuity of h at $x = x_0$. $\square$

REMARK. We can view the power function $h(x) = x^b$ with $x > 0$ and b a fixed real number as a composition of the logarithmic and the exponential functions; by (1.35) in Chapter 1 we have

$$x^b = e^{b(\ln x)}.$$

The continuity of the power function $h(x) = x^b$ with $x > 0$ and b fixed will follow by use of Proposition 2.9 as soon as we have established the continuity of the logarithmic and the exponential functions.

The function $h(x) = \cos x$ can be seen to be continuous by Proposition 2.9 once we know that $g(x) = \sin x$ and $f(x) = c - x$, where c is a constant, are continuous because $\cos x = \sin(\frac{1}{2}\pi - x)$.

Proposition 2.10. *Let f be a continuous function at a point a and suppose that $f(a)$ is positive. Then we can determine a positive number σ such that f is positive throughout the interval $(a - \sigma, a + \sigma)$.*

PROOF. Let $\varepsilon = \frac{1}{2}f(a)$ in the defining condition for continuity and denote the corresponding δ by σ. Then

$$|f(x) - f(a)| < \tfrac{1}{2}f(a)$$

for any x satisfying $|x - a| < \sigma$. But $|f(x) - f(a)| < \frac{1}{2}f(a)$ means

$$-\tfrac{1}{2}f(a) < f(x) - f(a) < \tfrac{1}{2}f(a) \quad \text{or} \quad \tfrac{1}{2}f(a) < f(x) < \tfrac{3}{2}f(a).$$

Thus $f(x) > \frac{1}{2}f(a) > 0$ for any x in the interval $(a - \sigma, a + \sigma)$. $\square$

REMARK. There is plainly a corresponding proposition referring to negative values of a continuous function f.

Proposition 2.11. *If a function f is continuous on a closed interval $[a, b]$ of finite length, it is bounded on that interval, that is, there are numbers m and M such that*

$$m \leq f(x) \leq M \quad \text{for all } x \text{ in } [a, b].$$

PROOF. Let f be continuous on $[a, b]$. We observe first that if x_0 is any point of $[a, b]$, there is *some* subinterval containing x_0 on which f is bounded. For, if we take $\varepsilon = 1$ in the defining condition for continuity and denote the corresponding δ by δ_1, we have

$$f(x_0) - 1 < f(x) < f(x_0) + 1$$

provided that x is a point of $[a, b]$ such that $x_0 - \delta_1 < x < x_0 + \delta_1$. Note that this subinterval extends to both sides of x_0 if $a < x_0 < b$, and one side of x_0 if $x_0 = a$ or $x_0 = b$. A second observation of importance is this: if f is bounded on two subintervals of $[a, b]$ whose union is $[a, b]$, then f is also bounded on $[a, b]$.

We now turn to the proof of the proposition at hand and suppose that our claim was false. In other words, suppose f was not bounded on $[a, b]$. Bisecting $[a, b]$ into $[a, d]$ and $[d, b]$, where $d = (a + b)/2$, the function f would not be bounded on at least one of these two subintervals. Let us call $[a_1, b_1]$ the subinterval on which f is not bounded. Next we bisect $[a_1, b_1]$, obtaining a subinterval $[a_2, b_2]$ on which f is not bounded and so forth. By repetition of this bisection process we generate a sequence of closed nested intervals on each of which f is not bounded; the length of the nth interval $[a_n, b_n]$ is $(b - a)/2^n$. By the Nested Interval Principle (see Section 1 of Chapter 1) there is one and only one point c common to all these intervals $[a_n, b_n]$. Clearly, the point c is in the interval $[a, b]$. Now, as shown at the beginning of our proof, f is bounded on *some* interval J containing the point c. Since c is in $[a_n, b_n]$ and since the length of $[a_n, b_n]$ tends to zero as n becomes arbitrarily large, it is clear that J must contain $[a_n, b_n]$ when n is sufficiently large. But this is a contradiction for f is *not* bounded on $[a_n, b_n]$ and it *is* bounded on J. Because of this contradiction, our initial claim that the proposition is false must be rejected and the proof is complete.   □

Proposition 2.12. *Let f be a continuous function on a closed interval $[a, b]$ of finite length and suppose that $f(a) \neq f(b)$. Then for every real number A between $f(a)$ and $f(b)$ there exists a point t such that $f(t) = A$, that is, f assumes all intermediate values between $f(a)$ and $f(b)$.*

PROOF. Suppose that $f(a) < f(b)$. Starting with the points $x_0 = a$ and $y_0 = b$, we construct by successive bisections a sequence of nested closed intervals $[x_n, y_n]$ for $n = 1, 2, 3, \ldots$ such that $f(x_n) \le A \le f(y_n)$ for $n = 1, 2, 3, \ldots$. To this end we only have to set

$$x_{n+1} = d_n \quad \text{and} \quad y_{n+1} = y_n \quad \text{in case} \quad f(d_n) \le A$$

and

$$x_{n+1} = x_n \quad \text{and} \quad y_{n+1} = d_n \quad \text{in case} \quad f(d_n) > A,$$

where $d_n = (x_n + y_n)/2$. Let t be the point determined by this sequence of nested closed intervals (see the Nested Interval Principle), that is, let t be the point common to all these nested closed intervals. Then $x_n \to t$ and $y_n \to t$ as $n \to \infty$. Since $f(x_n) \le A \le f(y_n)$ and f is continuous at t, we have that $f(t) = A$.

The case $f(b) < f(a)$ reduces to the case already discussed when we consider the function $-f$. The proof is finished.   □

APPLICATION. We wish to show that a polynomial function P of odd degree
has at least one real root.

Let $P(x) = c_0 + c_1 x + \cdots + c_n x^n$, where $c_n \neq 0$ and n is odd, that is, an
integer not divisible by 2. We may suppose that $c_n = 1$; otherwise we would
work with $(1/c_n)P$. It is clear that P is continuous everywhere because

$$\lim_{x \to a} P(x) = P(a)$$

for any real number a, as already noted in the remarks following Proposition 2.6. To apply Proposition 2.12 we need only to show that $P(x) > 0$ for
some x and $P(x) < 0$ for some other x. However, this is seen to be true
because

$$\lim_{x \to \infty} P(x) = +\infty \quad \text{and} \quad \lim_{x \to -\infty} P(x) = -\infty,$$

remembering that $c_n = 1$. We can avoid these limit notions by the following
argument. Observe that

$$P(x) = x^n D(x),$$

where

$$D(x) = 1 + \frac{c_0 + c_1 x + \cdots + c_{n-1} x^{n-1}}{x^n}.$$

Let $d = 1 + |c_0| + |c_1| + \cdots + |c_{n-1}|$. If $|x| > d$, then

$$|x| > 1 \quad \text{and} \quad |x| > |c_0| + |c_1| + \cdots + |c_{n-1}|$$

and so

$$|c_0 + c_1 x + \cdots + c_{n-1} x^{n-1}| \leq (|c_0| + |c_1| + \cdots + |c_{n-1}|)|x|^{n-1} < |x|^n$$

so that $D(x) > 0$ for $|x| > d$. Now, if $x > d$, then $x^n > 0$ and so $P(x) > 0$. And
if $x < -d$, then $x^n < 0$ (because n is odd) and so $P(x) < 0$.

Definition. Let S be a nonempty set of real numbers. We call M an *upper
bound of S* if $s \leq M$ for all s in S and we call m a *lower bound of S* if $s \geq m$ for
all s in S. If S has an upper bound, we say that S is *bounded above*; if S has a
lower bound, we say that S is *bounded below*. If S is both bounded above and
bounded below, then S is said to be *bounded*.

Definition. Let S be a nonempty set of real numbers which is bounded above.
Suppose that a real number M^* has the following two properties:

(i) M^* is an upper bound of S.
(ii) If $K < M^*$, then K is not an upper bound of S.

Then M^* is called the *least upper bound of S* [that there is at most one such
M^* is clear from (ii)], or the *supremum of S*; in notation, $M^* = \text{lub} S$ or
$M^* = \sup S$.

If S is a nonempty set of real numbers which is not bounded above, we put $\sup S = +\infty$.

REMARK. Let M^* be an upper bound of a nonempty set S of real numbers. Then M^* is the supremum of S if and only if, for any $\varepsilon > 0$, there is an s in S such that $s > M^* - \varepsilon$.

Indeed, let M^* be the supremum of S. Suppose that $\varepsilon > 0$ is given. Then $M^* - \varepsilon$ can not be the supremum of S, hence there is an s in S such that we have $s > M^* - \varepsilon$. On the other hand, suppose M^* is not the supremum of S. Then there is some $M_0^* < M^*$ such that M_0^* is the supremum of S. We have to find an $\varepsilon > 0$ such that $s \le M^* - \varepsilon$ for all s in S. But we only need to put $\varepsilon = M^* - M_0^*$.

Definition. Let S be a nonempty set of real numbers which is bounded below. Suppose that m_* is a real number satisfying the following two conditions:

 (i) m_* is a lower bound of S.
(ii) If $L > m_*$, then L is not a lower bound of S.

Then m_* is called the *greatest lower bound of S* [that there is at most one such m_* is clear from (ii)], or the *infimum of S*; in notation, $m_* = \text{glb}\, S$ or $m_* = \inf S$.

If S is a nonempty set of real numbers which is not bounded below, we put $\inf S = -\infty$.

Axiom of Completeness. Every nonempty set of real numbers that has a lower bound has a greatest lower bound. Also, every nonempty set of real numbers that has an upper bound has a least upper bound.

DISCUSSION. The Axiom of Completeness and the Nested Interval Principle can be derived from each other.

We assume the Axiom of Completeness and we show: If $J_1, J_2, \ldots, J_n, \ldots$ is a sequence of closed intervals, if $J_n \supset J_{n+1}$ for $n = 1, 2, 3, \ldots$, and if the length of J_n is less than any preassigned positive number for all large n, then there is one and only one point common to all intervals J_n.

Indeed, let $J_n = [a_n, b_n]$ and so the "nesting" property

$$a_n \le a_{n+1} < b_{n+1} \le b_n$$

holds for $n = 1, 2, 3, \ldots$. This shows that the set A of points that occur as left endpoints of intervals J_n is a bounded set and that the same is true for the set B of points that occur as right endpoints of the intervals J_n. Let

$$a = \sup A \quad \text{and} \quad b = \inf B.$$

It is clear that $a \le b_n$ and $b \ge a_n$ for $n = 1, 2, 3, \ldots$ and so $a \le b$ with both a and b belonging to J_n for $n = 1, 2, 3, \ldots$. But

$$b - a < b_n - a_n \quad \text{for } n = 1, 2, 3, \ldots$$

and $b_n - a_n \to 0$ as $n \to \infty$, implying that $a = b$. This is, however, what we wanted to show.

Now, we assume the Nested Interval Principle and we show: If S is a nonempty set of real numbers that is bounded above, then S has a least upper bound.

Indeed, let M be an upper bound of S and s belong to S; consider the closed interval $[s, M] = J_1$. If J_1 has only the single point s in common with S, then we are finished and s is the least upper bound of S. If J_1 has more than one point, namely, the left endpoint s, in common with the set S, then we bisect the interval J_1 and denote by J_2 the right-hand or the left-hand half of J_1 depending on whether or not there are points of S that belong to the right-hand half of J_1. According to the same rule we select one half of J_2 as J_3 and so forth. The intervals $J_1, J_2, J_3, \ldots$ have the property that to the right of each such interval there is no point of S, but that in each such interval there is at least one point of S. The point c common to all $J_1, J_2, J_3, \ldots$ satisfies the property that for any $\varepsilon > 0$ there is an s in S such that $s > c - \varepsilon$ and so $c = \sup S$.

Proposition 2.13. *Let f be a continuous function on a closed interval $[a, b]$ of finite length. By Proposition 2.11 the set S of all $f(x)$ as x ranges over the interval $[a, b]$ is a bounded set; let $m = \inf S$ and $M = \sup S$. Then f assumes the values M and m at least once each in the interval $[a, b]$.*

PROOF. Let M be the supremum of the set S of all $f(x)$ with x ranging over the interval $[a, b]$; M is called the *upper bound of f* on $[a, b]$. If $[a, b]$ is bisected, then it is possible to find a half $[a_1, b_1]$ on which the upper bound of f is also M. Proceeding by the method of repeated bisection, we construct a sequence of closed nested intervals

$$[a, b], \quad [a_1, b_1], \quad [a_2, b_2], \ldots$$

on each of which the upper bound of f is M. These intervals have one and only one point c in common and the sequences

$$a, a_1, a_2, \ldots \quad \text{and} \quad b, b_1, b_2, \ldots$$

tend to c. But $f(c) = M$. Suppose not, that is, suppose $f(c) = L \neq M$. Then $|L - M| = 2\varepsilon$ for some $\varepsilon > 0$. By the continuity of f at c, there is an open interval $(c - \delta, c + \delta)$ on which $f(x)$ can not differ from L by more than ε and hence M can not be the upper bound of f on $(c - \delta, c + \delta)$. But $(c - \delta, c + \delta)$ must contain $[a_n, b_n]$ for n larger than some n_0 because the length

$$b_n - a_n = \frac{b - a}{2^n}$$

tends to zero as n becomes arbitrarily large. Let n be such that $[a_n, b_n]$ is contained in $(c - \delta, c + \delta)$; then

$$M = \sup\{f(x): a_n \le x \le b_n\} \le \sup\{f(x): c - \delta < x < c + \delta\}.$$

But $[a,b]$ contains $(c - \delta, c + \delta)$ and so

$$\sup\{f(x): c - \delta < x < c + \delta\} \le \sup\{f(x): a \le x \le b\} = M.$$

We have a contradiction. [Trivial modifications are needed if $c = a$ or $c = b$.]

In a similar manner we can show that there is a point d in $[a,b]$ such that $f(d) = m$. $\qquad\square$

ALTERNATIVE PROOF OF PROPOSITION 2.13. Given any positive number σ, we can find a point x in $[a,b]$ for which

$$M - f(x) < \sigma \quad \text{or} \quad \frac{1}{M - f(x)} > \frac{1}{\sigma}.$$

Hence, $1/[M - f(x)]$ is not bounded, and therefore, by Proposition 2.11, not continuous. But $M - f(x)$ is a continuous function and so $1/[M - f(x)]$ is continuous at any point at which its denominator does not vanish. There must therefore be a point at which the denominator vanishes, and at which $f(x) = M$. Similarly, it may be shown that there is a point at which $f(x) = m$. $\qquad\square$

Definition. Let f be a bounded function on an interval $[a,b]$ and J be a subinterval of $[a,b]$. By the *oscillation of* f *on* J, denoted by $\omega(f,J)$, we mean

$$\omega(f,J) = \sup\{f(x): x \in J\} - \inf\{f(x): x \in J\},$$

that is, the difference between the upper bound and the lower bound of f on J.

Proposition 2.14. *Let* f *be a continuous function on a closed interval* $[a,b]$ *of finite length. Given* $\varepsilon > 0$, *there exists a partition of* $[a,b]$ *into a finite number of subintervals of equal length such that the oscillation of* f *on each of these subintervals does not exceed* ε.

PROOF. Suppose not, that is, suppose there was a continuous function f on $[a,b]$ and an $\varepsilon > 0$ for which no partition of $[a,b]$ of the desired type existed. For convenience, let us agree to say that the *interval* $[a,b]$ *has property* P_ε if there exists no partition of $[a,b]$ into a finite number of subintervals of equal length such that the oscillation of f on each of these subintervals is less than or equal to ε. In short, we suppose that $[a,b]$ has property P_ε. Bisecting $[a,b]$, we obtain a subinterval $[a_1, b_1]$ with property P_ε. We then bisect $[a_1, b_1]$ and obtain a subinterval $[a_2, b_2]$ having property P_ε and so forth. Proceeding in this manner we get a sequence of nested closed intervals

$$[a_1, b_1], \quad [a_2, b_2], \quad \ldots, \quad [a_n, b_n], \quad \ldots$$

with each interval of the sequence having property P_ε. In particular, this implies that $\omega(f, [a_n, b_n]) > \varepsilon$ for $n = 1, 2, 3, \ldots$. But the length of $[a_n, b_n]$

equals $(b - a)/2^n$ which tends to zero as n becomes arbitrarily large. By the Nested Interval Principle there is one and only one point c in common to all these intervals $[a_n, b_n]$ for $n = 1, 2, 3, \ldots$. But c is in $[a, b]$ and so f is continuous at c. Hence, we can find a positive δ such that $|f(x) - f(c)| < \varepsilon/2$ for any x in $[a, b]$ which satisfies $|x - c| < \delta$. If x_1 and x_2 are two such x, then

$$f(x_1) - f(x_2) = [f(x_1) - f(c)] + [f(c) - f(x_2)]$$

and so

$$|f(x_1) - f(x_2)| \leq |f(x_1) - f(c)| + |f(c) - f(x_2)| < \frac{\varepsilon}{2} + \frac{\varepsilon}{2} = \varepsilon$$

and so the oscillation on the interval $I = [a, b] \cap (c - \delta, c + \delta)$ is less than ε.

For sufficiently large n the interval $[a_n, b_n]$ is contained in the interval I and so the oscillation of f on such $[a_n, b_n]$ will be smaller than or equal to the oscillation of f on I. We have reached a contradiction because by our assumption and construction the oscillation of f on $[a_n, b_n]$ for $n = 1, 2, 3, \ldots$ is larger than ε. $\qquad\qquad\square$

Proposition 2.15. *Let f and $[a, b]$ be as in Proposition 2.14. Given $\varepsilon > 0$, there exists a positive δ such that on any subinterval J of $[a, b]$ having length less than δ the oscillation of f on J is smaller than ε.*

PROOF. By Proposition 2.14 we can pick m so large that the oscillation of f on each of the intervals

$$[a, a + \delta], \quad [a + \delta, a + 2\delta], \quad \ldots, \quad [a + (m - 1)\delta, b], \qquad (2.15)$$

each of which has length $\delta = (b - a)/m$, is less than $\varepsilon/2$. We consider any subinterval J of $[a, b]$ having length less than δ and we let x_1 and x_2 denote points of J at which f assumes largest and smallest values on J, respectively, and thus $f(x_1) - f(x_2)$ will denote the oscillation of f on J. These points x_1 and x_2 either belong to the same interval in (2.15) and then $f(x_1) - f(x_2)$ will be smaller than $\varepsilon/2$ or they belong to two abutting intervals

$$[a + (k - 1)\delta, a + k\delta] \quad \text{and} \quad [a + k\delta, a + (k + 1)\delta]$$

and in this case

$$f(x_1) - f(x_2) = [f(x_1) - f(a + k\delta)] + [f(a + k\delta) - f(x_2)] < \frac{\varepsilon}{2} + \frac{\varepsilon}{2} = \varepsilon.$$

On any subinterval J of length less than δ the oscillation of f is thus seen to be smaller than the prescribed ε. $\qquad\qquad\square$

Definition. A function f is said to be *uniformly continuous on* $[a, b]$ if, given any $\varepsilon > 0$, there exists some $\delta > 0$, dependent on ε only, such that for any points x_1 and x_2 in $[a, b]$ the inequality

$$|x_1 - x_2| < \delta \quad \text{implies} \quad |f(x_1) - f(x_2)| < \varepsilon.$$

REMARKS. Uniform continuity is readily expressed in terms of oscillation. The definition implies that $\omega(f, J) < \varepsilon$ if the length of the interval J is less than δ regardless where J is placed in the interval $[a, b]$. Or, we can say that $\omega(f, J) < \varepsilon$ holds uniformly with respect to all intervals J, provided the length of J is less than δ. Geometrically, the situation can be described as follows. A small rectangle of height 2ε and width 2δ is centered at a point on the curve. We have uniform continuity, if we can choose the width of the rectangle, whenever its height is given, so that while the rectangle slides along the curve $y = f(x)$ remaining parallel to itself, the curve projects from the vertical sides but does not touch the top or the bottom of the rectangle.

Proposition 2.16. *Let f be a continuous function on a closed interval $[a, b]$ of finite length. Then f is uniformly continuous on $[a, b]$.*

PROOF. The claim follows at once from Proposition 2.15. $\square$

Examples. The function $g(x) = 1/x$ is not uniformly continuous on the open interval $(0, 1)$. Indeed, suppose that g was uniformly continuous on $(0, 1)$. Then for any $\varepsilon > 0$ we should be able to find some δ, say between 0 and 1, such that $|g(x_1) - g(x_2)| < \varepsilon$ whenever $|x_1 - x_2| < \delta$ for any x_1 and x_2 in the interval. Now, let $x_1 = \delta$ and $x_2 = \delta/(1 + \varepsilon)$. Then $|x_1 - x_2| = [\varepsilon/(1 + \varepsilon)]\delta < \delta$. But $|1/x_1 - 1/x_2| = \varepsilon/\delta > \varepsilon$ (since $0 < \delta < 1$). Thus we have a contradiction and g is not uniformly continuous on $(0, 1)$.

The function $h(x) = \sin(1/x)$ is not uniformly continuous on the open interval $(0, 1/\pi)$. Figure 2.6 shows the graph of the function h. While the function h is bounded by 1 and -1, the oscillation $\omega(h, J) = 2$ on any interval J of the form $(0, a)$ no matter how small the positive number a is. This fact prevents uniform continuity of h on $(0, 1/\pi)$.

The function $w(x) = x^2$ is not uniformly continuous on $(0, \infty)$. Small changes in x can produce arbitrarily large changes in x^2 if only x is large enough. Indeed, suppose there was a $\delta > 0$ such that

$$|w(x) - w(a)| < 2 \quad \text{whenever} \quad |x - a| < \delta \quad \text{for all } a > 0.$$

We would than have for $x = a + \delta/2$

$$|w(x) - w(a)| = |x - a||x + a| = \frac{\delta}{2}\left|2a + \frac{\delta}{2}\right| < 2,$$

implying that $a\delta < 2$ for all $a > 0$ which is clearly false.

Let A be a nonempty bounded set of real numbers and define

$$v(x) = \inf\{|x - a|: a \in A\} \quad \text{for all real numbers } x.$$

Then v is uniformly continuous on $(-\infty, \infty)$. Indeed, for each $a \in A$ we have

$$v(x) \leq |x - a| \leq |x - y| + |y - a|$$

for any real number y. Taking the infimum over all $a \in A$, we obtain

$$v(x) \le |x - y| + v(y)$$

or $v(x) - v(y) \le |x - y|$. Exchanging the roles of x and y, we also obtain $v(y) - v(x) \le |x - y|$. This shows that

$$|v(x) - v(y)| \le |x - y|$$

and so v is seen to be uniformly continuous on $(-\infty, \infty)$.

3. Monotonic Functions

Definition. Let f be a real-valued function on an interval J and x and y be points of J. Then f is said to be *nondecreasing on J* if

$$x < y \quad \text{implies} \quad f(x) \le f(y);$$

f is said to be *strictly increasing on J* if

$$x < y \quad \text{implies} \quad f(x) < f(y).$$

Similarly, f is said to be *nonincreasing on J* if

$$x < y \quad \text{implies} \quad f(x) \ge f(y)$$

and f is said to be *strictly decreasing on J* if

$$x < y \quad \text{implies} \quad f(x) > f(x).$$

The class of *monotonic functions* consists of both the nondecreasing and the nonincreasing functions; the class of *strictly monotonic functions* consists of the strictly increasing and the strictly decreasing functions.

Definition. Let f be a real-valued function on an open interval (a, b). If f is not continuous at a point x of (a, b), then f is said to be *discontinuous at x*. If f is discontinuous at a point x of (a, b) and if the one-sided limits $f(x+)$ and $f(x-)$ exists, then f is said to have a *discontinuity of the first kind*, or a *simple discontinuity*, at x. Otherwise, the discontinuity is said to be of the *second kind*.

ILLUSTRATION. Consider the following three functions:

$$f(x) = 1 \qquad \text{for } x \ge 0,$$
$$= -1 \qquad \text{for } x < 0;$$
$$g(x) = 1 \qquad \text{for } x \ne 0,$$
$$= 0 \qquad \text{for } x = 0;$$
$$h(x) = \sin\frac{1}{x} \quad \text{for } x \ne 0,$$
$$= 0 \qquad \text{for } x = 0.$$

It is clear that $f(0+) = 1$ and $f(0-) = -1$; thus f has a discontinuity of the first kind at $x = 0$. For the function g we have $g(0+) = g(0-) = 1$, but $g(0) = 0$; thus g has a discontinuity of the first kind at $x = 0$. For the function h, neither $h(0+)$ nor $h(0-)$ exists [see Figure 2.6 and note that $h(-x) = -h(x)$]; thus h has a discontinuity of the second kind at $x = 0$.

REMARK. There are two ways in which a function f can have a simple discontinuity at x; either $f(x+) \neq f(x-)$, or $f(x+) = f(x-) \neq f(x)$. The following proposition shows that a monotonic function f has no discontinuities of the second kind since the one-sided limits $f(x+)$ and $f(x-)$ exist.

Proposition 2.17. *Let f be a nondecreasing function on an open interval (a,b). Then, for any x satisfying $a < x < b$,*

$$A = f(x-) \leq f(x) \leq f(x+) = B, \tag{2.16}$$

where $A = \sup\{f(t): a < t < x\}$ and $B = \inf\{f(t): x < t < b\}$. Moreover, if $a < x < y < b$, then

$$f(x+) \leq f(y-).$$

PROOF. By assumption, the set of numbers $f(t)$, where $a < t < x$, is bounded above by the number $f(x)$, and thus has a supremum which we have denoted by A. Evidently, $A \leq f(x)$. We verify that $A = f(x-)$.

Let $\varepsilon > 0$ be given. It follows from the definition of A as a supremum that there exists $\delta > 0$ such that $a < x - \delta < x$ and

$$A - \varepsilon < f(x - \delta) < A.$$

Since f is nondecreasing, we have $f(x - \delta) \leq f(t) \leq A$ for $x - \delta < t < x$. Thus, for $x - \delta < t < x$, we have $|f(t) - A| < \varepsilon$ and so $f(x-) = A$.

The second half of (2.16) is proved in the same way.

Next, if $a < x < y < b$, we get from (2.16) that

$$f(x+) = \inf\{f(t): x < t < b\} = \inf\{f(t): x < t < y\}.$$

The last equality in the foregoing is obtained by applying (2.16) to (a, y) in place of (a, b). Similarly,

$$f(y-) = \sup\{f(t): a < t < y\} = \sup\{f(t): x < t < y\}.$$

Thus, $f(x+) \leq f(y-)$ is seen to hold for $a < x < y < b$. $\quad\square$

Definition. Let X and Y be two nonempty sets; then the set of all ordered pairs (x, y) where $x \in X$, $y \in Y$, is called the *Cartesian product of X and Y* and is denoted by $X \times Y$. Here $(x_1, y_1) = (x_2, y_2)$ if and only if $x_1 = x_2$, $y_1 = y_2$. We refer to x as the *first coordinate of the pair* (x, y) and y as the *second coordinate*.

A *function from X to Y* is a nonempty subset of pairs (x, y) in $X \times Y$ such that no two distinct pairs have the same first coordinate. The sets

$$D[f] = \{x: x \in X, (x, y) \in f\},$$

$$R[f] = \{y: y \in Y, (x, y) \in f\},$$

are called the *domain of f* and the *range of f*, respectively.

We say that f is a *one-to-one* function if $(x_1, y_1) \in f$, $(x_2, y_2) \in f$, and $x_1 \neq x_2$ implies $y_1 \neq y_2$ for any x_1 and x_2 in $D[f]$.

Definition. Let f be a one-to-one function with domain $D[f]$ and range $R[f]$. Then f is said to be *invertible* and the *inverse function of f*, denoted by f^{-1}, is defined by the set of ordered pairs

$$\{(y, x): y \in R[f], x \in D[f]\};$$

we have

$$f^{-1}[f(x)] = x \quad \text{for all } x \in D[f],$$

$$f[f^{-1}(y)] = y \quad \text{for all } y \in R[f].$$

REMARKS. We are concerned with functions mapping a set of real numbers into a set of real numbers. The meaning of the terms just defined is simple enough to grasp. A function is defined as a set, namely, as the set of points making up the graph of the function. The function is one-to-one if any straight line parallel with the x-axis intersects the graph of the function in at most one point. Finally, if f is one-to-one, that is, if f is invertible, then the graph of f^{-1} is obtained from the graph of f by simply reflecting the graph of f about the line $y = x$. A strictly monotonic function is one-to-one, but a one-to-one function need not be monotonic; for example,

$$f(x) = \tfrac{1}{2}x \qquad \text{for } 0 \leq x \leq 2,$$

$$= -\tfrac{1}{2}x + 3 \quad \text{for } 2 < x < 4,$$

$$= \tfrac{1}{2}x \qquad \text{for } 4 \leq x \leq 6$$

maps the closed interval $[0, 6]$ onto the closed interval $[0, 3]$ in a one-to-one manner, but the function f is not monotonic on $[0, 6]$. Note also that f takes on all intermediate values between 0 and 3, but f is not continuous everywhere on $[0, 6]$ (having discontinuities at $x = 2$ and at $x = 4$). See Figure 2.9 for the graph of $y = f(x)$.

Proposition 2.18. *Let f be a continuous function on a closed interval $I = [a, b]$ of finite length. Then f is invertible if and only if f is strictly monotonic. Trivially, f^{-1} is increasing or decreasing depending on whether f is increasing or decreasing. Moreover, the inverse function f^{-1} is continuous on its interval of definition I^*.*

PROOF. It is clear that strict monotonicity is a sufficient condition for the invertibility of f. We only need to show that in the case of continuity it is also

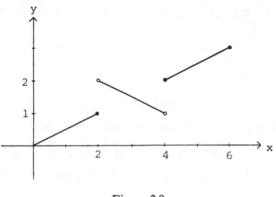

Figure 2.9

a necessary condition. For f to be invertible we need to have $f(x_1) \neq f(x_2)$ for any two distinct points x_1 and x_2 in I. Since f is assumed to be continuous, we have for any x between x_1 and x_2 that

$$\{f(x) - f(x_1)\}\{f(x) - f(x_2)\} < 0. \tag{2.17}$$

To see this, we suppose that x_1 and x_2 are so indexed that $f(x_1)$ is less than $f(x_2)$. Suppose now that $f(x) > f(x_2)$. By Proposition 2.12, the value $f(x_2)$ is attained between x_1 and x; however, this contradicts the invertibility of f since x_2 is not situated between x_1 and x. In the same way the assumption that $f(x)$ is less than $f(x_1)$ leads to a contradiction. Thus, $f(x_1) < f(x) < f(x_2)$ and we get (2.17).

Suppose now that f is not monotonic on I. Then there are two pairs of points y_1, y_2 and y_3, y_4 in I such that $y_1 < y_2$ and $y_3 < y_4$ satisfying

$$f(y_1) < f(y_2), \qquad f(y_3) > f(y_4). \tag{2.18}$$

Without loss of generality we may assume that none of the points y_1, y_2, y_3, and y_4 is an endpoint of the interval I. (Observe: if f is continuous at t and $f(t) > c$ [resp. $f(t) < c$] for a given real number c, then there is a neighborhood U of t such that $f(x) > c$ [resp. $f(x) < c$] for all x belonging to U and x being in I, the domain of definition of f. This is an easy consequence of Proposition 2.10 applied to the function $g(x) = f(x) - c$.) Then there are points v and w in I such that for $k = 1, 2, 3, 4$,

$$v < y_k < w.$$

By (2.17) we have

$$\{f(y_1) - f(v)\}\{f(y_1) - f(y_2)\} < 0$$

and

$$\{f(y_2) - f(y_1)\}\{f(y_2) - f(w)\} < 0;$$

adding the foregoing two inequalities we obtain

$$\{f(y_1) - f(y_2)\}^2 - \{f(v) - f(w)\}\{f(y_1) - f(y_2)\} < 0.$$

Thus,

$$\{f(v) - f(w)\}\{f(y_1) - f(y_2)\} > 0.$$

In the same way we obtain

$$\{f(v) - f(w)\}\{f(y_3) - f(y_4)\} > 0.$$

By multiplication of the last two inequalities, we see that

$$\{f(y_1) - f(y_2)\}\{f(y_3) - f(y_4)\} > 0.$$

But this violates the inequality (2.18). We have therefore shown that f is strictly monotonic.

We show that if f is strictly increasing, then f^{-1} is strictly increasing; the case of strictly decreasing is completely similar. Suppose that

$$y_1 < y_2 \quad \text{and} \quad x_1 = f^{-1}(y_1), \qquad x_2 = f^{-1}(y_2).$$

Then

$$y_1 = f(x_1) \quad \text{and} \quad y_2 = f(x_2).$$

If we had $x_1 > x_2$, then it would follow that $y_1 > y_2$ (because f is strictly increasing) and this violates our assumption. If we had $x_1 = x_2$, then it would follow that $y_1 = y_2$ which is also against our assumption. Hence, only $x_1 < x_2$ is possible and so f^{-1} is seen to be strictly increasing.

To complete the proof, we must show that f^{-1} is continuous on I^* which is as the range of f a closed interval of finite length by Propositions 2.12 and 2.13.

Let r be in the range of f and write $r = f(s)$ with $s \in I$. We want to prove that given any $\varepsilon > 0$, there exists some $\delta > 0$ such that, for any y in the range of f satisfying $|y - x| < \delta$, we have $|f^{-1}(y) - f^{-1}(r)| < \varepsilon$.

To this end, let $\varepsilon > 0$. By what we have proved already, f is strictly monotonic. We may without loss of generality suppose that f is increasing. Since $s - \varepsilon < s < s + \varepsilon$, we have

$$f(s - \varepsilon) < f(s) < f(s + \varepsilon).$$

Let δ be the smaller of the two numbers $f(s) - f(s - \varepsilon)$ and $f(s + \varepsilon) - f(s)$. Then

$$f(s - \varepsilon) \leq f(s) - \delta < f(s) + \delta \leq f(s + \varepsilon).$$

Hence, for any y satisfying $f(s) - \delta < y < f(s) + \delta$, we have

$$f(s - \varepsilon) < y < f(s + \varepsilon).$$

Since f is increasing, so is f^{-1}. Hence,

$$f^{-1}\{f(s - \varepsilon)\} < f^{-1}(y) < f^{-1}\{f(s + \varepsilon)\},$$

that is,

$$s - \varepsilon < f^{-1}(y) < s + \varepsilon.$$

But $s = f^{-1}(r)$ and so

$$|f^{-1}(r) - f^{-1}(y)| < \varepsilon.$$

We therefore have shown that, for any y satisfying $r - \delta < y < r + \delta$,

$$|f^{-1}(r) - f^{-1}(y)| < \varepsilon.$$

Hence, f^{-1} is continuous at r.   □

4. Miscellaneous Examples

1. The *logarithmic functions* $f(x) = \ln x$ and $g(x) = \log_a x$, where $a > 0$ and $a \neq 1$, are continuous on $(0, \infty)$.
   Indeed, we have by (1.3)

$$\frac{x - c}{x} < \ln x - \ln c < \frac{x - c}{c} \quad \text{for } x > c$$

and

$$\frac{c - x}{c} \ln c - \ln x < \frac{c - x}{x} \quad \text{for } x < c.$$

In either case, $\ln x \to \ln c$ as $x \to c$ for any point c in $(0, \infty)$. To verify the continuity of $g(x) = \log_a x$ we only need to recall that

$$\log_a x = (\log_a e)(\ln x).$$

2. The *exponential functions* $v(x) = e^x$ and $w(x) = a^x$, where $a > 0$ and $a \neq 1$, are continuous on $(-\infty, \infty)$.
   Indeed, the claim follows at once by use of Proposition 2.18. The function $v(x) = e^x$ is the inverse function of the continuous strictly increasing function $f(x) = \ln x$; the fact that $f(x) = \ln x$ is strictly increasing can be seen from the inequality

$$\frac{x - c}{x} < \ln x - \ln c \quad \text{for } x > c > 0,$$

already used in Example 1. In Section 3 of Chapter 1 there is a direct proof of the continuity of the exponential function $v(x) = e^x$.
   To verify the continuity of the exponential function with the base a, namely, $w(x) = a^x$, we may use Proposition 2.18 and treat $w(x) = a^x$ as the inverse function of the strictly monotonic and continuous function $g(x) = \log_a x$. Another approach consists in showing that $a^x \to a^c$ as $x \to c$. Since

$$a^x - a^c = a^c(a^{x-c} - 1),$$

we need to verify that $a^{x-c} \to 1$ as $x - c \to 0$ or, equivalently, $a^t \to 1$ as $t \to 0$. We shall do this now. It will be enough to consider the case $a > 1$ only.

Let $a > 1$. By the Lemma preceding Proposition 1.10, $a^{1/n} \to 1$ as $n \to \infty$. We wish to show the more general result $a^t \to 1$ as $t \to 0$. First we note that

$$\lim_{n \to \infty} a^{-1/n} = \lim_{n \to \infty} \frac{1}{a^{1/n}} = 1.$$

Thus, corresponding to any $\varepsilon > 0$, there exists some positive integer n_0 such that

$$1 - \varepsilon < a^{-1/n_0} < a^{1/n_0} < 1 + \varepsilon \quad \text{(for } a > 1)$$

holds. Now, if

$$|t| < \frac{1}{n_0} \quad \text{or} \quad -\frac{1}{n_0} < t < \frac{1}{n_0},$$

then

$$a^{-1/n_0} < a^t < a^{1/n_0};$$

from this it follows that

$$1 - \varepsilon < a^t < 1 + \varepsilon \quad \text{or} \quad |a^t - 1| < \varepsilon$$

and we have what we wanted to show.

3. The *power function* $h(x) = x^b$, where $x > 0$ and b denotes a fixed real number, is continuous on $(0, \infty)$.

Indeed, see the Remark following Proposition 2.9.

4. The *polynomial function* $P(x) = c_n x^n + c_{n-1} x^{n-1} + \cdots + c_1 x + c_0$, where $c_n, c_{n-1}, \ldots, c_1, c_0$ are fixed real numbers, is continuous on $(-\infty, \infty)$. The *rational function*

$$R(x) = \frac{P(x)}{Q(x)},$$

where P and Q are polynomial functions, is continuous on $(-\infty, \infty)$ with the exception of those points a for which $Q(a) = 0$.

Indeed, see the Remarks following Proposition 2.6.

5. The *trigonometric function* $s(x) = \sin x$ is continuous on $(-\infty, \infty)$. Indeed, by (2.9)

$$\sin x < x \quad \text{for } 0 < x < \frac{\pi}{2}.$$

From this it easily follows that

$$|\sin x| \le |x| \quad \text{for any real number } x$$

[for $|x| \geq \pi/2 > 1$ it follows from $|\sin x| \leq 1$; moreover, $s(-x) = -s(x)$]. But

$$\sin x - \sin c = 2 \sin \frac{x - c}{2} \cos \frac{x + c}{2}$$

and so

$$|\sin x - \sin c| = 2 \left| \sin \frac{x - c}{2} \right| \left| \cos \frac{x + c}{2} \right| \leq 2 \left| \sin \frac{x - c}{2} \right| \leq 2 \frac{|x - c|}{2},$$

that is,

$$|\sin x - \sin c| \leq |x - c| \quad \text{for any real numbers } x \text{ and } c. \qquad (2.19)$$

Thus, $\sin x \to \sin c$ as $x \to c$. [In fact, (2.19) shows uniform continuity on $(-\infty, \infty)$.]

From the Remark following Proposition 2.9 we can see that the cosine function is also continuous on $(-\infty, \infty)$. However, the functions

$$\tan x = \frac{\sin x}{\cos x}, \qquad \sec x = \frac{1}{\cos x}, \qquad \cot x = \frac{\cos x}{\sin x}, \qquad \csc x = \frac{1}{\sin x}$$

have certain points of discontinuity; $\tan x$ and $\sec x$ are continuous for any x such that $\cos x \neq 0$ and $\cot x$ and $\csc x$ are continuous for any x such that $\sin x \neq 0$. The points of discontinuity of $\tan x$ and $\sec x$ are of the form $(2k + 1)\pi/2$ and the points of discontinuity of $\cot x$ and $\csc x$ are of the form $k\pi$, where k denotes any integer.

6. Let f be a continuous function on the closed interval $[0, 2]$ and $f(0) = f(2)$. Then there are points x_1 and x_2 in $[0, 2]$ such that $|x_1 - x_2| = 1$ and $f(x_1) = f(x_2)$.

Indeed, let $g(x) = f(x + 1) - f(x)$ on $[0, 1]$. Then g is continuous on $[0, 1]$ and $g(0) = -g(1)$. If $g(0) = 0$, then $f(1) = f(0)$ and we are finished. If $g(0) \neq 0$, then $g(0)$ and $g(1)$ have opposite signs; by Proposition 2.12 there is a point t in $[0, 1]$ such that $g(t) = f(t + 1) - f(t) = 0$, that is, $f(t + 1) = f(t)$.

7. Let g be a continuous function mapping the closed interval $[0, 1]$ into itself. Then there is a point t in $[0, 1]$ such that $g(t) = t$.

Indeed, let $h(x) = g(x) - x$ on $[0, 1]$. Then h is continuous on $[0, 1]$. Since $h(0) = g(0) - 0 = g(0) \geq 0$ and $h(1) = g(1) - 1 \leq 1 - 1 = 0$, Proposition 2.12 shows that $h(t) = 0$ for some t in $[0, 1]$ and so we have $g(t) = t$.

8. The equation $x2^x = 1$ is satisfied for some x in $(0, 1)$.

Indeed, let $h(x) = x2^x$ on $[0, 1]$. Then $h(0) = 0$ and $h(1) = 2$. By Proposition 2.12, there is some point t between 0 and 1 such that $h(t) = 1$.

9. Let f be a continuous function on an interval J of finite length and x_1, x_2, and x_3 be points of J and p_1, p_2, and p_3 be positive real numbers. Then

there is a point t in J such that

$$f(t) = \frac{p_1 f(x_1) + p_2 f(x_2) + p_3 f(x_3)}{p_1 + p_2 + p_3}.$$

Indeed, let $a = \min\{x_1, x_2, x_3\}$ and $b = \max\{x_1, x_2, x_3\}$ and set

$$v = \min\{f(x_1), f(x_2), f(x_3)\} \quad \text{and} \quad V = \max\{f(x_1), f(x_2), f(x_3)\}.$$

Then f is continuous on the closed interval $[a, b]$ of finite length and f assumes all values between v and V. But

$$v(p_1 + p_2 + p_3) \le p_1 f(x_1) + p_2 f(x_2) + p_3 f(x_3) \le V(p_1 + p_2 + p_3)$$

or

$$v \le \frac{p_1 f(x_1) + p_2 f(x_2) + p_3 f(x_3)}{p_1 + p_2 + p_3} \le V.$$

[Clearly, the foregoing result can be extended from the case of three points to the case of any finite number of points.]

10. A rational number x can be written in the form $x = p/q$, where $q > 0$, and p and q are integers without common divisor. When $x = 0$, we take $q = 1$. Consider the function f on the closed interval $[0, 1]$ defined by

$$f(x) = 0 \quad \text{for } x \text{ irrational,}$$

$$= \frac{1}{q} \quad \text{for } x = \frac{p}{q}.$$

Then f is continuous at every irrational point of $(0, 1)$ and discontinuous at every rational point of $(0, 1)$.

Indeed, let t be any point of $(0, 1)$. Given $\varepsilon > 0$, there is only a finite number of positive integers q that are not larger than $1/\varepsilon$; this means that in $(0, 1)$ there are only finitely many rational points p/q for which $f(p/q) = 1/q \ge \varepsilon$. Thus, one may construct around the point t a neighborhood $(t - \delta, t + \delta)$ with $\delta > 0$ such that in this neighborhood there is no point x for which $f(x) \ge \varepsilon$ (except possibly the point t itself). Thus, if $0 < |x - t| < \delta$, then for both rational and irrational x we have $|f(x)| < \varepsilon$. Thus, the one-sided limits for every point t in $(0, 1)$ satisfy

$$f(t+) = f(t-) = 0.$$

If t is irrational, then $f(t) = 0$, that is, f is continuous at t; if t is rational, then $f(t) \ne 0$, that is, f is discontinuous at t.

11. Let g be defined on $(-\infty, \infty)$ as follows:

$$g(x) = 0 \quad \text{for } x \text{ irrational,}$$

$$g(x) = x \quad \text{for } x \text{ rational.}$$

Then g is continuous at $x = 0$ and discontinuous everywhere else.

Indeed, $\lim_{x \to 0} g(x) = 0$; but $\lim_{x \to a} g(x)$ does not exist for $a \neq 0$ because any (nonempty) open interval on the number line contains both rational and irrational points.

12. **A strictly increasing (or strictly decreasing) function on an interval that assumes all intermediate values is continuous.**

Indeed, let g be a strictly increasing function on an interval J such that the set of $g(x)$ as x runs through J is an interval I. Consider a point t in J and suppose that t is not an endpoint of J; only minor changes in the proof are needed if t is an endpoint of J. Then $g(t)$ is not an endpoint of I and so there is some $\sigma > 0$ such that the open interval $(g(t) - \sigma, g(t) + \sigma)$ is contained in I. Take any $\varepsilon > 0$ with $\varepsilon < \sigma$. Then there exist points x_1 and x_2 in J such that

$$g(x_1) = g(t) - \varepsilon \quad \text{and} \quad g(x_2) = g(t) + \varepsilon.$$

It is clear that $x_1 < t < x_2$. Moreover, if $x_1 < x < x_2$, then

$$g(x_1) < g(x) < g(x_2) \quad \text{or} \quad g(t) - \varepsilon < g(x) < g(t) + \varepsilon$$

and hence $|g(x) - g(t)| < \varepsilon$. Now, if we set $\delta = \min\{x_2 - t, t - x_1\}$, then

$$|x - t| < \delta \quad \text{implies} \quad x_1 < x < x_2$$

and hence $|g(x) - g(t)| < \varepsilon$.

If g is strictly decreasing, then $-g$ is strictly increasing; thus the case when g is strictly decreasing does not require a separate proof.

13. Let a and b denote given real numbers. Then

$$\lim_{x \to \infty} \{\sqrt{(x + a)(x + b)} - x\} = \frac{a + b}{2}.$$

Indeed, we have

$$\sqrt{(x + a)(x + b)} - x = \frac{(x + a)(x + b) - x^2}{\sqrt{(x + a)(x + b)} + x} = \frac{(a + b)x + ab}{\sqrt{(x + a)(x + b)} + x}$$

$$= \frac{a + b + ab/x}{\sqrt{(1 + a/x)(1 + b/x)} + 1}.$$

Using the continuity of the power function (see Example 3), we get

$$\sqrt{\left(1 + \frac{a}{x}\right)\left(1 + \frac{b}{x}\right)} \to 1 \quad \text{as } x \to \infty$$

and the desired result follows.

14. Let $a_1, a_2, \ldots, a_k$ denote given real numbers and put

$$S(x) = \sqrt[k]{(x + a_1)(x + a_2) \cdots (x + a_k)} - x.$$

Then $\lim_{x \to \infty} S(x) = (a_1 + a_2 + \cdots + a_k)/k$.

Indeed, putting

$$y = \sqrt[k]{(x + a_1)(x + a_2) \cdots (x + a_k)} \quad \text{and} \quad z = x$$

in the identity

$$y - z = \frac{y^k - z^k}{y^{k-1} + y^{k-2}z + \cdots + z^{k-1}},$$

we obtain

$$S(x) = \frac{(x + a_1) \cdots (x + a_k) - x^k}{(\sqrt[k]{\cdots})^{k-1} + x(\sqrt[k]{\cdots})^{k-2} + \cdots + x^{k-1}}$$

$$= \frac{(a_1 + \cdots + a_k) + (a_1 a_2 + \cdots + a_{k-1} a_k)/x + \cdots + (a_1 a_2 \cdots a_k)/x^{k-1}}{\{\sqrt[k]{(1 + a_1/x) \cdots (1 + a_k/x)}\}^{k-1} + \cdots + 1}.$$

Since $\sqrt[k]{(1 + a_1/x) \cdots (1 + a_k/x)} \to 1$ as $x \to \infty$, we have

$$\lim_{x \to \infty} S(x) = \frac{a_1 + a_2 + \cdots + a_k}{k}.$$

15. Let m be a positive integer; then

$$\lim_{x \to 1} \frac{\sqrt[m]{x} - 1}{x - 1} = \frac{1}{m}.$$

Indeed, let $x = t^m$. Then $t = x^{1/m}$ is continuous for $x > 0$ (see Example 3) and $t \to 1$ as $x \to 1$. Thus,

$$\lim_{x \to 1} \frac{\sqrt[m]{x} - 1}{x - 1} = \lim_{t \to 1} \frac{t - 1}{t^m - 1} = \lim_{t \to 1} \frac{1}{t^{m-1} + t^{m-2} + \cdots + t + 1} = \frac{1}{m}.$$

16. Let m and n be positive integers; then

$$\lim_{x \to 1} \frac{\sqrt[m]{x} - 1}{\sqrt[n]{x} - 1} = \frac{n}{m}.$$

Indeed, we have (using Example 15)

$$\lim_{x \to 1} \frac{\sqrt[m]{x} - 1}{\sqrt[n]{x} - 1} = \lim_{x \to 1} \frac{\sqrt[m]{x} - 1}{x - 1} \cdot \lim_{x \to 1} \frac{x - 1}{\sqrt[n]{x} - 1} = \frac{1}{m} \cdot \frac{n}{1} = \frac{n}{m}.$$

17. Let $p, q, r,$ and s be positive integers; then

$$\lim_{x \to 1} \frac{x^{p/q} - 1}{x^{r/s} - 1} = \frac{ps}{qr}.$$

Indeed, let $x = t^q$. Then $t = x^{1/q}$ is continuous for $x > 0$ (see Example 3) and $t \to 1$ as $x \to 1$. Thus, by Example 16,

$$\lim_{x \to 1} \frac{x^{p/q} - 1}{x - 1} = \lim_{t \to 1} \frac{t^p - 1}{t^q - 1} = \frac{p}{q}.$$

On the other hand,

$$\lim_{x \to 1} \frac{x - 1}{x^{r/s} - 1} = \frac{s}{r}.$$

Thus, the desired result follows.

18. Let a and b be given real numbers; then

$$\lim_{x \to 0} \frac{\sin ax}{\sin bx} = \frac{a}{b}.$$

Indeed, we have

$$\frac{\sin ax}{\sin bx} = \frac{\sin ax}{ax} \cdot \frac{bx}{\sin bx} \cdot \frac{a}{b}.$$

But $(\sin t)/t \to 1$ as $t \to 0$ by (2.8).

19. We have

$$\lim_{x \to \infty} \{\ln(x + 1) - \ln x\} = 0.$$

Indeed, since

$$\ln(x + 1) - \ln x = \ln \frac{x + 1}{x} \quad \text{and} \quad \ln t \to 0 \text{ as } t \to 1,$$

the desired result follows.

20. We have

$$\lim_{x \to 0} \frac{\ln(1 + x)}{x} = 1. \tag{2.20}$$

Indeed, by (1.3)

$$\frac{1}{1 + x} < \frac{\ln(1 + x)}{x} < 1 \quad \text{for } x > 0$$

and

$$1 < \frac{\ln(1 + x)}{x} < \frac{1}{1 + x} \quad \text{for } -1 < x < 0;$$

the claim follows immediately.

21. Let $a > 0$, $a \neq 1$; then

$$\lim_{x \to 0} \frac{\log_a(1 + x)}{x} = \log_a e. \tag{2.21}$$

Indeed, (2.21) follows from (2.20) by noting that

$$\log_a(1 + x) = (\log_a e)\ln(1 + x).$$

22. Let $a > 0$, $a \neq 1$; then

$$\lim_{x \to 0} \frac{a^x - 1}{x} = \ln a. \tag{2.22}$$

Indeed, let us set $a^x - 1 = y$. Then $y \to 0$ as $x \to 0$ because the exponential function is continuous. Moreover, $x = \log_a(1 + y)$ and so, by (2.21),

$$\lim_{x \to 0} \frac{a^x - 1}{x} = \lim_{y \to 0} \frac{y}{\log_a(1 + x)} = \frac{1}{\log_a e} = \ln a.$$

Note that (2.22) implies $n(\sqrt[n]{a} - 1) \to \ln a$ as $n \to \infty$ when we let $x = 1/n$ in (2.22); we have already come across this result in Proposition 1.10.

23. Let c be a given real number; then

$$\lim_{x \to 0} \frac{(1 + x)^c - 1}{x} = c. \tag{2.23}$$

Indeed, let $(1 + x)^c - 1 = y$. Then $y \to 0$ as $x \to 0$ because the power function is continuous (see Example 3). But $(1 + x)^c = 1 + y$ gives $c[\ln(1 + x)] = \ln(1 + y)$ and so

$$\frac{(1 + x)^c - 1}{x} = \frac{y}{x} = \frac{y}{\ln(1 + y)} \cdot c \cdot \frac{\ln(1 + x)}{x}.$$

By (2.20)

$$\frac{y}{\ln(1 + y)} \to 1 \quad \text{for } y \to 0 \quad \text{and} \quad \frac{\ln(1 + x)}{x} \to 1 \quad \text{for } x \to 0;$$

the desired result (2.23) follows.

24. Let

$$f(x) = \lim_{n \to \infty} \frac{x^{2n} - 1}{x^{2n} + 1} \quad \text{and} \quad g(x) = \lim_{n \to \infty} \frac{x^{2n+1} + x^2}{x^{2n} + 1}.$$

Then

$$f(x) = 1 \quad \text{for } |x| > 1,$$
$$= -1 \quad \text{for } |x| < 1,$$
$$= 0 \quad \text{for } x = \pm 1.$$

Indeed, for $|x| > 1$ we have $x^{2n} \to \infty$ as $n \to \infty$. Thus, for $|x| > 1$,

$$\frac{x^{2n} - 1}{x^{2n} + 1} = \frac{1 - 1/x^{2n}}{1 + 1/x^{2n}} \to 1 \quad \text{as } n \to \infty.$$

For $|x| < 1$ we have $x^{2n} \to 0$ as $n \to \infty$ and so, for $|x| < 1$,

$$\frac{x^{2n} - 1}{x^{2n} + 1} \to -1 \quad \text{as } n \to \infty.$$

For $x = \pm 1$ we have $x^{2n} = 1$ and so $f(\pm 1) = 0$.

In an entirely similar way we obtain $g(x) = x$ for $|x| > 1$, $g(x) = x^2$ for $|x| < 1$, $g(1) = 1$, and $g(-1) = 0$. Thus, g is discontinuous at $x = -1$ only.

25. We have

$$\lim_{x \to 0} \frac{1 - \cos x}{x^2} = \frac{1}{2}. \tag{2.24}$$

Indeed, putting $x = 2t$, we get

$$\frac{1 - \cos x}{x^2} = \frac{1 - \cos 2t}{4t^2} = \frac{\sin^2 t}{2t^2} = \frac{1}{2} \frac{\sin t}{t} \frac{\sin t}{t}$$

and the desired result easily follows from (2.8).

26. We have

$$\lim_{x \to 0} \frac{\cos mx - \cos nx}{x^2} = \frac{n^2 - m^2}{2}.$$

Indeed,

$$\frac{1 - \cos nx}{x^2} - \frac{1 - \cos mx}{x^2} = n^2 \frac{1 - \cos nx}{(nx)^2} - m^2 \frac{1 - \cos mx}{(mx)^2}$$

and we can use (2.24) to deduce the result in question.

27. We have

$$\lim_{x \to 0} \frac{\tan x - \sin x}{x^3} = \frac{1}{2}.$$

Indeed, we have

$$\frac{\sin x - (\sin x)(\cos x)}{x^3 \cos x} = \frac{\sin x}{x} \frac{1 - \cos x}{x^2} \frac{1}{\cos x}$$

and we can use (2.24) to get the desired result.

28. We have

$$\lim_{x \to 0} \frac{\sqrt{1 - \cos x}}{x} = \frac{1}{\sqrt{2}}.$$

Indeed,

$$\frac{\sqrt{1-\cos x}}{x} = \frac{\sqrt{1-\cos x}}{x}\frac{\sqrt{1+\cos x}}{\sqrt{1+\cos x}} = \frac{\sqrt{1-\cos^2 x}}{x\sqrt{1+\cos x}} = \frac{\sin x}{x\sqrt{1+\cos x}}$$

and we use (2.8) to get what we are after.

29. We have

$$\lim_{x\to 0}\frac{x-\sin 2x}{x+\sin 3x} = -\frac{1}{4}.$$

Indeed, since

$$\frac{x-\sin 2x}{x+\sin 3x} = \frac{2(x-\sin 2x)/2x}{3(x+\sin 3x)/3x} = \frac{2\frac{1}{2}-(\sin 2x)/2x}{3\frac{1}{3}+(\sin 3x)/3x}$$

we see that

$$\lim_{x\to 0}\frac{x-\sin 2x}{x+\sin 3x} = \frac{2\frac{1}{2}-1}{3\frac{1}{3}+1} = -\frac{1}{4}.$$

30. We have

$$\lim_{x\to 3}\frac{\sqrt{x^2-2x+6}-\sqrt{x^2+2x-6}}{x^2-4x+3} = -\frac{1}{3}.$$

Indeed,

$$\frac{\sqrt{x^2-2x+6}-\sqrt{x^2+2x-6}}{x^2-4x+3}$$

$$= \frac{-4(x-3)}{(x^2-4x+3)(\sqrt{x^2-2x+6}+\sqrt{x^2+2x-6})}$$

$$= \frac{-4}{(x-1)(\sqrt{x^2-2x+6}+\sqrt{x^2+2x-6})}$$

and so the limit in question equals

$$\frac{-4}{2(\sqrt{9}+\sqrt{9})} \quad \text{or} \quad -\frac{1}{3}.$$

31. We have

$$\lim_{n\to\infty}\frac{2^{n+1}+3^{n+1}}{2^n+3^n} = 3.$$

Indeed,

$$\frac{2^{n+1}+3^{n+1}}{2^n+3^n} = \frac{2(\frac{2}{3})^n+3}{(\frac{2}{3})^n+1}$$

and the desired result follows.

32. Let $f(-2) = -1, f(-1) = 0, f(1) = 1$, and

$$f(x) = \frac{x^2 + x - 2}{x^3 + 2x^2 - x - 2} \quad \text{for } x \neq -2, -1, 1.$$

Then f is discontinuous for $x = \pm 1$, but continuous for $x = -2$.
Indeed, for $x \neq -2, -1, 1$ we have

$$f(x) = \frac{(x+2)(x-1)}{(x+2)(x-1)(x+1)} = \frac{1}{x+1}.$$

But

$$\lim_{x \downarrow -1} \frac{1}{x+1} = \infty \quad \text{and} \quad \lim_{x \uparrow -1} \frac{1}{x+1} = -\infty.$$

In addition,

$$\lim_{x \to 1} \frac{1}{x+1} = \frac{1}{2}.$$

On the other hand,

$$\lim_{x \to -2} \frac{1}{x+1} = -1.$$

33. We have

$$\lim_{x \to \infty} x^{3/2}(\sqrt{x+1} + \sqrt{x-1} - 2\sqrt{x}) = -\tfrac{1}{4}.$$

Indeed, since

$$\sqrt{x+1} + \sqrt{x-1} - 2\sqrt{x} = (\sqrt{x+1} - \sqrt{x}) - (\sqrt{x} - \sqrt{x-1})$$

$$= \frac{1}{\sqrt{x+1} + \sqrt{x}} - \frac{1}{\sqrt{x} + \sqrt{x-1}}$$

$$= \frac{\sqrt{x-1} - \sqrt{x+1}}{(\sqrt{x+1} + \sqrt{x})(\sqrt{x} + \sqrt{x-1})},$$

we have

$$x^{3/2}(\sqrt{x+1} + \sqrt{x-1} - 2\sqrt{x})$$

$$= \frac{-2(\sqrt{x})^3}{(\sqrt{x+1} + \sqrt{x})(\sqrt{x} + \sqrt{x-1})(\sqrt{x-1} + \sqrt{x+1})}$$

$$= \frac{-2}{(\sqrt{1+1/x} + 1)(1 + \sqrt{1-1/x})(\sqrt{1-1/x} + \sqrt{1+1/x})} \to -\frac{1}{4}$$

as $x \to \infty$.

34. Let A and B be two bounded sets of real numbers and let C be defined by

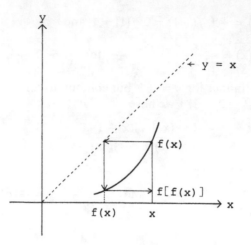

Figure 2.10

$$C = \{x: x = a + b \text{ where } a \in A, b \in B\}.$$

Then C is bounded above and $\sup C = \sup A + \sup B$.

Indeed, for any $x \in C$, $x = a + b$ where $a \in A$ and $b \in B$. Now $a \le \sup A$ and $b \le \sup B$. Thus, $x \le \sup A + \sup B$. This holds for all $x \in C$; thus $\sup A + \sup B$ is an upper bound for C, C is bounded above, and $\sup C \le \sup A + \sup B$.

By the Remark following the definition of supremum, for any $\varepsilon > 0$ there is an $a \in A$ such that $a > \sup A - \frac{1}{2}\varepsilon$ and an element $b \in B$ such that $b > \sup B - \frac{1}{2}\varepsilon$. But $a + b \in C$. Hence, $\sup C \ge a + b > \sup A + \sup B - \varepsilon$. This is true for all $\varepsilon > 0$ and thus $\sup C \ge \sup A + \sup B$. Thus, finally, $\sup C = \sup A + \sup B$.

35. Let A and B be two bounded sets of real numbers. Then

$$\sup(A \cup B) = \max\{\sup A, \sup B\}.$$

Indeed, $A \cup B \supset A$. Thus, $\sup(A \cup B)$ is an upper bound of A. Hence, $\sup(A \cup B) \ge \sup A$ and similarly $\sup(A \cup B) \ge \sup B$. If $a \in A \cup B$, then $a \in A$ or $a \in B$. Thus, $a \le \sup A$ or $a \le \sup B$ and in either case $a \le \max\{\sup A, \sup B\}$. Hence, we have $\sup(A \cup B) = \max\{\sup A, \sup B\}$.

36. Knowing the graph of the function $y = f(x)$, Figure 2.10 illustrates how to plot the graph of the function $y = f[f(x)]$.

EXERCISES TO CHAPTER 2

2.1. The functions f and g are defined as follows:

$$f(x) = -\frac{1}{2(1 + x)} \quad \text{if } -1 < x \le -\tfrac{1}{2},$$

$$= 2x \quad \text{if } -\tfrac{1}{2} \le x \le \tfrac{1}{2},$$

$$= \frac{1}{2(1 - x)} \quad \text{if } \tfrac{1}{2} \le x < 1,$$

$$= 0 \quad \text{if } x \ge 1 \text{ or } x \le -1,$$

and

$$g(x) = \frac{2x - 1}{2x} \quad \text{if } x \ge 1,$$

$$= \frac{x}{2} \quad \text{if } -1 \le x \le 1,$$

$$= -\frac{2x + 1}{2x} \quad \text{if } x \le -1.$$

Show that $f[g(x)] = x$ for all values of x but $g[f(x)]$ is different from x outside the interval $(-1, 1)$.

[Hint: If $\tfrac{1}{2} \le x < 1$, then $g[f(x)] = x$; if $-\tfrac{1}{2} \le x \le \tfrac{1}{2}$, $g[f(x)] = x$; if $x \ge 1$, $g[f(x)] = g(0) = 0 \ne x$; if $-1 < x \le -\tfrac{1}{2}$, $g[f(x)] = x$; and if $x \le -1$, $g[f(x)] = 0 \ne x$. Note that g is increasing but its values are confined to the interval $(-1, 1)$, and so $g[f(x)] = x$ only for x in $(-1, 1)$.

On the other hand, if $x \ge 1$, then

$$f[g(x)] = f\left(\frac{2x - 1}{2x}\right) = \frac{1}{2}\left[1 - \left(1 - \frac{1}{2x}\right)\right] = x;$$

if $-1 \le x \le 1$,

$$f[g(x)] = f(\tfrac{1}{2}x) = 2(\tfrac{1}{2}x) = x;$$

if $x \le -1$,

$$f[g(x)] = f\left(-\frac{2x + 1}{2x}\right) = -\frac{1}{2}\left(1 - \frac{2x + 1}{2x}\right) = x.$$

Hence, $f[g(x)] = x$ for all x.]

2.2. The function f is continuous in (a, b) and $f(x)$ is not zero for any x in (a, b). Show that f is of constant sign in (a, b).

[Hint: For any positive t, and any A, B in (a, b), $[tf(A) + f(B)]/(t + 1)$ lies between $f(A)$ and $f(B)$ and therefore, since f is continuous, is a value of f for an x between A and B and so different from zero. Suppose that $f(A) > 0$; take $t = |f(B)|/f(A)$, then $f(B) \ne -|f(B)|$ and so $f(B) > 0$. In the same way it follows that if $f(A) < 0$, then $f(B) < 0$.]

2.3. Let

$$d(x, y) = \frac{|x - y|}{1 + |x - y|}.$$

Show that $d(a, b) \le d(a, c) + d(b, c)$ for any real numbers a, b, and c.

[Hint: Since $a - b = (a - c) + (c - b)$, we have

$$|a - b| \leq |a - c| + |c - b|.$$

Evidently $|c - b| = |b - c|$. But the function

$$f(t) = \frac{t}{1 + t} = 1 - \frac{1}{1 + t}$$

is increasing for $t \geq 0$ and so

$$\frac{|a - b|}{1 + |a - b|} \leq \frac{|a - c| + |b - c|}{1 + |a - c| + |b - c|}.$$

However, it is clear that

$$\frac{|a - c| + |b - c|}{1 + |a - c| + |b - c|} \leq \frac{|a - c|}{1 + |a - c|} + \frac{|b - c|}{1 + |b - c|}.]$$

2.4. A function f is said to be *concave up on* (A, B) if

$$f[tx + (1 - t)y] \leq tf(x) + (1 - t)fy$$

whenever $A < x < B, A < y < B, 0 < t < 1$. Geometrically, this means that if P, Q, and R are any three points on the graph of f with Q between P and R, then Q is on or below chord PR.

Show that if f is concave up on (A, B) and $[a, b]$ is any closed subinterval of (A, B), then f is continuous on $[a, b]$.

[Hint: We show that the set of all values $f(x)$ as x ranges over the interval $[a, b]$ is bounded above and below and then establish that there exists a constant K so that for any two points x, y of $[a, b]$ we have

$$|f(x) - f(y)| \leq K|x - y|.$$

We observe that $M = \max\{f(a), f(b)\}$ is an upper bound for f on $[a, b]$, since for any point $z = ta + (1 - t)b$ in $[a, b]$

$$f(z) \leq tf(a) + (1 - t)f(b) \leq tM + (1 - t)M = M.$$

But f is also bounded from below because, writing an arbitrary point in the form $(a + b)/2 + s$, we have

$$f\left(\frac{a + b}{2}\right) \leq \frac{1}{2}f\left(\frac{a + b}{2} + s\right) + \frac{1}{2}f\left(\frac{a + b}{2} - s\right)$$

or

$$f\left(\frac{a + b}{2} - s\right) \leq 2f\left(\frac{a + b}{2}\right) + f\left(\frac{a + b}{2} - s\right).$$

Using M as an upper bound,

$$-f\left(\frac{a + b}{2} + s\right) \geq -M,$$

so

$$f\left(\frac{a + b}{2} + s\right) \geq 2f\left(\frac{a + b}{2}\right) - M = m.$$

This shows that M and m are upper and lower bounds of f on $[a, b]$, respectively.

We next pick $h > 0$ so that $a - h$ and $b + h$ belong to (A, B), and let m and M be the lower and upper bounds for f on $[a - h, b + h]$. If x and y are distinct points of $[a, b]$, set

$$z = y + \frac{h}{|y - x|}(y - x), \qquad t = \frac{|y - x|}{h + |y - x|}.$$

Then z belongs to the closed interval $[a - h, b + h]$, $y = tz + (1 - t)x$, and we have

$$f(y) \le tf(z) + (1 - t)f(x) = t[f(z) - f(x)] + f(x),$$

$$f(y) - f(x) \le t(M - m) < \frac{|y - x|}{h}(M - m) = K|y - x|,$$

where $K = (M - m)/h$. Since this is true for any x and y in $[a, b]$, we conclude that $|f(y) - f(x)| \le K|y - x|$ as desired.]

2.5. Show that f is concave up on (A, B) if and only if for all sets of distinct x_1, x_2, x_3 belonging to (A, B),

$$\frac{(x_3 - x_2)f(x_1) + (x_1 - x_3)f(x_2) + (x_2 - x_1)f(x_3)}{(x_1 - x_2)(x_2 - x_3)(x_3 - x_1)}$$

is nonnegative.

[Hint: Let $x_1 < x_2 < x_3$ and consider the determinant

$$\begin{vmatrix} 1 & x_1 & f(x_1) \\ 1 & x_2 & f(x_2) \\ 1 & x_3 & f(x_3) \end{vmatrix}$$

Interpret this determinant geometrically.]

2.6. Let f_1, f_2, and f_3 be continuous functions on $[a, b]$ and let $f(x)$ denote that one of the three values $f_1(x)$, $f_2(x)$, and $f_3(x)$ that lies between the other two. Show that f is continuous on $[a, b]$.

[Hint: Note that

$$f(x) = f_1(x) + f_2(x) + f_3(x) - \max\{f_1(x), f_2(x), f_3(x)\}$$

$$- \min\{f_1(x), f_2(x), f_3(x)\}.]$$

2.7. Any continuous function f mapping the set of real numbers into itself such that $f(x + y) = f(x) + f(y)$ for any real numbers x and y is of the form $f(x) = cx$, where c is a constant. Verify this claim.

[Hint: For any positive integer n, we have, by induction on n, $f(nx) = nf(x)$; replacing x by x/n, we see that $f(x/n) = (1/n)f(x)$. On the other hand, $f(0 + x) = f(0) + f(x)$, hence $f(0) = 0$; and $f[x + (-x)] = f(x) + f(-x) = f(0) = 0$, hence $f(-x) = -f(x)$. Thus, for any pair of integers p, q such that $q > 0$, we have $f(px/q) = (p/q)f(x)$; in other words, $f(rx) = rf(x)$ for any rational number r. If we let $x = 1$ and denote $f(1)$ by c, we obtain $f(r) = cr$.

Let ρ be any irrational number. We choose for ρ an approximating sequence r_1, r_2, r_3, ... of rational numbers, for example, the sequence of finite decimal fractions approximating ρ. Then, for $n = 1, 2, 3, \ldots$, $f(r_n) = cr_n$. Since f is assumed continuous on the entire number line, we obtain, on passing to the limit as $n \to \infty$,

$$f(\rho) = \lim_{n \to \infty} f(r_n) = \lim_{n \to \infty} cr_n = c\rho.]$$

2.8. Show that if g is any continuous function mapping the set of real numbers into itself such that $g[(x + y)/2] = [g(x) + g(y)]/2$ for any real numbers x and y, then g is of the form $g(x) = cx + a$, where c and a are constants.

[Hint: Indeed, for $y = 0$, we obtain $g(x/2) = [g(x) + g(0)]/2 = [g(x) + a]/2$, where $a = g(0)$; thus,

$$\frac{g(x) + g(y)}{2} = g\left(\frac{x + y}{2}\right) = \frac{g(x + y) + a}{2},$$

that is, $g(x + y) = g(x) + g(y) - a$, and, with $f(x) = g(x) - a$,

$$f(x + y) = f(x) + f(y).$$

But g is continuous; hence f is continuous and so, by Exercise 2.7, f is of the form $f(x) = cx$. This shows that g is of the form $g(x) = cx + a$, where c and a are constants.]

2.9. Let $[a, b]$ be a closed interval of finite length and f be a continuous function on $[a, b]$. If for each x in $[a, b]$ there exists a y in $[a, b]$ such that

$$|f(y)| \leq \tfrac{1}{2}|f(x)|,$$

show that there is a point t in $[a, b]$ for which $f(t) = 0$.

[Hint: The function f^2 is continuous on $[a, b]$ and hence achieves a minimum value c at some point t of $[a, b]$. But there exists an s in $[a, b]$ such that $|f(s)| \leq \tfrac{1}{2}|f(t)| = \tfrac{1}{2}\sqrt{c}$. Thus, f^2 takes the value $\tfrac{1}{4}c$ on $[a, b]$ and hence $c \leq \tfrac{1}{4}c$. Since $c \geq 0$, it follows that $c = 0$.]

2.10. Let the numbers of the open interval $(0, 1)$ be expressed as finite or infinite decimals $x = 0.a_1 a_2 a_3 \ldots a_n \ldots$, and let $f(x) = 0.0a_1 0a_2 0a_3 \ldots$. Is f discontinuous for every value of x represented by a finite decimal?

2.11. Show that a continuous function on a closed interval J of finite length which takes on no value more than twice must take on some value exactly once.

[Hint: Let M be the maximum value which the function f takes on J. If $f(x) = M$ nowhere else on J, we are done; if not, let m be the minimum which f takes on between the two points where the value of f is M. Then evidently between these two points f takes on all the values $m < a < M$ at least twice, and thus, by assumption, exactly twice. But the value m is taken only once, since for it to be taken twice, either within or without the interval between the two maxima, f would have to take on some of the values $m < a < M$ still more times.]

2.12. A function f is said to be *periodic*, with period a, if $f(x) = f(x + a)$ for all values of x for which $f(x)$ is defined. Show that no periodic function can be a rational function, that is, a quotient of two polynomials, unless it is a constant.

[Hint: Let f be a periodic function with period a and suppose that f is a rational function so that $f(x) = P(x)/Q(x)$, where P and Q are polynomials. If $f(0) = c$, then $P(x)/Q(x) = c$ when $x = 0, a, 2a, 3a, \ldots$. Thus, whatever the degree n of the equation $P(x) - cQ(x) = 0$, it is satisfied by more than n values of x and this is only possible if $P(x) - cQ(x) = 0$ for all values of x, that is, if $f(x) = c$ (constant).]

2.13. Show that if $P(x) = x^m + a_1 x^{m-1} + a_2 x^{m-2} + \cdots + a_m$, where m is a positive integer and $a_1, a_2, \ldots, a_m$ are real numbers, then there exists a number X such that, when $x > X$,

$$\tfrac{1}{2} x^m < P(x) < \tfrac{3}{2} x^m.$$

[Hint: For $x > 1$,

$$|a_1 x^{m-1} + a_2 x^{m-2} + \cdots + a_m| \leq |a_1 x^{m-1}| + |a_2 x^{m-2}| + \cdots + |a_m| \leq A x^{m-1}$$

where $A = |a_1| + |a_2| + \cdots + |a_m|$. Thus,

$$|a_1 x^{m-1} + a_2 x^{m-2} + \cdots + a_m| \leq \tfrac{1}{2} x^m \quad \text{if } x > X = \max\{1, 2A\}.$$

That is,

$$-\tfrac{1}{2} x^m \leq a_1 x^{m-1} + a_2 x^{m-2} + \cdots + a_m \leq \tfrac{1}{2} x^m,$$

and the claim follows on addition of x^m.]

2.14. Let

$$f(x) = \lim_{n \to \infty} \frac{\ln(x + 2) - x^{2n}(\sin x)}{1 + x^{2n}} \quad \text{for } 0 \leq x \leq \tfrac{1}{2}\pi.$$

Describe the graph of f in the interval $[0, \tfrac{1}{2}\pi]$ and note that f does not vanish anywhere in this interval, although $f(0)$ and $f(\tfrac{1}{2}\pi)$ are of opposite sign.

[Hint: Note that

$$\begin{aligned}
f(x) &= \ln(x + 2) & \text{for } 0 \leq x < 1, \\
&= \tfrac{1}{2}(\ln 3 - \sin 1) & \text{for } x = 1, \\
&= -\sin x & \text{for } 1 < x \leq \tfrac{1}{2}\pi.]
\end{aligned}$$

2.15. Show that a function f which, in a given interval $[a, b]$, possesses either of the properties

(i) it attains its largest and smallest values for any closed subinterval $[a', b']$ of $[a, b]$ at least once in the subinterval;

(ii) it attains at least once in any subinterval $[a', b']$ of $[a, b]$ every value between $f(a')$ and $f(b')$;

does not necessarily possess the other.

Show further that a function which possesses both these properties in $[a, b]$ is not necessarily continuous on $[a, b]$.

[Hint: On the interval $[a, b]$ with $a = -1$ and $b = 1$, consider the functions

$$\begin{aligned}
f(x) &= 0 & x \text{ rational,} \\
&= 1 & x \text{ irrational;} \\
g(x) &= (1 - x^2)\sin(x^{-2}) & \text{for } x \neq 0, \\
&= 0 & \text{for } x = 0; \\
h(x) &= \sin(x^{-1}) & \text{for } x \neq 0, \\
&= 0 & \text{for } x = 0.
\end{aligned}$$

Then f is an example of a function showing that (i) does not imply (ii), g is an example of a function showing that (ii) does not imply (i), and, lastly, h is an example of a function that shows that (i) and (ii) in combination are not enough to secure continuity throughout $[a, b]$. Note especially that while 1 is the upper bound of $g(x)$ on $[-1, 1]$, 1 is never attained and that the graph of g is contained between the parabolic arcs $y = \pm(1 - x^2)$ for $-1 \leq x \leq 1$.]

2.16. Let $p > 1$, $q > 1$, and $y_1, y_2, \ldots, y_n$ be functions defined by the relations

$$y_1 = \{x(x)^{1/q}\}^{1/p}, \quad y_2 = \{x(xy_1)^{1/q}\}^{1/p}, \quad \ldots, \quad y_n = \{x(xy_{n-1})^{1/q}\}^{1/p}.$$

Find $y = \lim_{n \to \infty} y_n$.

[Hint: Let $a_1, a_2, \ldots, a_n$ be the exponents of x in $y_1, y_2, \ldots, y_n$, respectively. Then

$$a_1 = \frac{1}{p}\left(1 + \frac{1}{q}\right), \qquad a_2 = a_1\left(1 + \frac{1}{pq}\right),$$

$$a_3 = a_1 + \frac{a_2}{pq} = a_1\left(1 + \frac{1}{pq} + \frac{1}{p^2 q^2}\right),$$

$$\vdots$$

$$a_n = a_1\left(1 + \frac{1}{pq} + \frac{1}{p^2 q^2} + \cdots + \frac{1}{p^{n-1} q^{n-1}}\right).$$

Hence,

$$a_n \to \frac{q+1}{pq-1} \quad \text{as } n \to \infty$$

and so

$$y = x^{(q+1)/(pq-1)}.]$$

2.17. If f is defined on $(0, \infty)$, has inverse f^{-1}, and satisfies the relation

$$f(x) + f(y) = f(xy) \quad \text{for all } x > 0 \text{ and } y > 0,$$

find a corresponding relation satisfied by f^{-1}.

[Hint: We must have $f^{-1}\{f(x) + f(y)\} = f^{-1}\{f(xy)\} = xy$. Put $f(x) = u$ and $f(y) = v$. Then $x = f^{-1}(u)$, $y = f^{-1}(v)$, and $f^{-1}(u + v) = f^{-1}(u)f^{-1}(v)$. The last equation is the relaltion we desire.]

2.18. Let $f(x) = (ax + b)/(cx + d)$, where we assume that $ad - bc \neq 0$. Show that $f^{-1}(x) = (b - dx)/(cx - a)$. In particular, $f = f^{-1}$ when $a = -d$.

2.19. Let $f(x) = 1 - |x|$ if $|x| \leq 1$ and $f(x) = 0$ if $|x| \geq 1$. Sketch the graph of $y = f(x)f(a - x)$ if (i) $a = 0$, (ii) $a = 1$, and (iii) $a = 2$.

2.20. Find

$$\lim_{n \to \infty}\left(1 + \frac{1}{2}\cos x + \frac{1}{4}\cos 2x + \cdots + \frac{1}{2^n}\cos nx\right).$$

[Hint: Let

$$S = 1 + a(\cos x) + a^2(\cos 2x) + \cdots + a^k(\cos kx)$$

and

$$T = a(\sin x) + a^2(\sin 2x) + \cdots + a^k(\sin kx).$$

Then

$$S + iT = 1 + (\cos x + i\sin x) + a^2(\cos 2x + i\sin 2x) + \cdots$$
$$+ a^k(\cos kx + i\sin kx).$$

Putting $A = \cos x + i\sin x$, we obtain

$$S + iT = 1 + aA + a^2A^2 + \cdots + a^kA^k$$

$$= \frac{a^{k+1}A^{k+1} - 1}{aA - 1} = \frac{a^{k+1}A^{k+1} - 1}{aA - 1} \cdot \frac{aA^{-1} - 1}{aA^{-1} - 1}$$

$$= \frac{a^{k+2}A^k - a^{k+1}A^{k+1} - aA^{-1} + 1}{a^2 - a(A + A^{-1}) + 1}.$$

Hence,

$$S = \frac{a^{k+2}(\cos kx) - a^{k+1}[\cos(k + 1)x] - a(\cos x) + 1}{a^2 - 2a(\cos x) + 1}.$$

Finally, letting $k = n$ and $a = \frac{1}{2}$, we see that

$$\lim_{n\to\infty}\left(1 + \frac{1}{2}\cos x + \frac{1}{4}\cos 2x + \cdots + \frac{1}{2^n}\cos nx\right) = \frac{2(2 - \cos x)}{5 - 4\cos x}.]$$

CHAPTER 3

Differentiation

1. Basic Rules of Differentiation

Definition. Let f be a real-valued function defined on an open interval (a, b) containing the point s. We form the quotient

$$\frac{f(t) - f(s)}{t - s}$$

with $a < t < b$ and $t \neq s$. If the limit

$$\lim_{t \to s} \frac{f(t) - f(s)}{t - s}$$

exists (as a finite real number), its value is denoted by $f'(s)$; in this case $f'(s)$ is called the *derivative of f at s* and f is said to be *differentiable at s*. If f is differentiable at every point of a set S, then f is said to be *differentiable on S*.

REMARK. Consider the graph of the function f and let $(s, f(s))$ and $(t, f(t))$ be two points on the graph of f. The straight line containing the points $(s, f(s))$ and $(t, f(t))$ is a *secant line* to the graph of f and has the slope

$$\frac{f(t) - f(s)}{t - s}.$$

At $t \to s$, the slope $[f(t) - f(s)]/(t - s)$ of the secant line approaches the slope m of the *tangent line* to the graph of f at the point $(s, f(s))$ and $m = f'(s)$. See Figure 3.1.

Definition. Let f be a real-valued function defined on a closed interval $[a, b]$. Then f is said to be *differentiable at the endpoints a and b* if

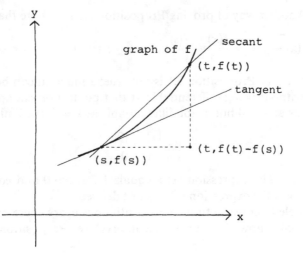

Figure 3.1

$$f'(a) = \lim_{t\downarrow a}\frac{f(t) - f(a)}{t - a} \quad \text{and} \quad f'(b) = \lim_{t\uparrow b}\frac{f(t) - f(b)}{t - b}$$

exist (as finite real numbers); the notation $t \downarrow a$ signifies that t approaches a from above, that is, $t > a$, and $t \uparrow b$ signifies that t approaches b from below, that is, $t < b$.

Proposition 3.1. *Let f be defined on an interval J. If f is differentiable at some point s of J, then f is continuous at s.*

PROOF. For all points x of J, $x \neq s$, sufficiently close to s the expression

$$\left|\frac{f(x) - f(s)}{x - s}\right|$$

has bound M, that is,

$$|f(x) - f(s)| \le M|x - s|. \tag{3.1}$$

Indeed, for all points x of J, $x \neq s$, let

$$h(x) = \frac{f(x) - f(s)}{x - s}.$$

But $\lim_{x\to s} h(x) = L$, where L is a finite real number. Let $\varepsilon = 1$ in the defining condition of limit. Then there is a subinterval I of J, with s not in I, such that

$$L - 1 < h(x) < L + 1$$

for any x in I. Now, take $M = \max\{|L - 1|, |L + 1|\}$.

As $x \to s$, the expression on the right-hand side of inequality (3.1) tends to zero; hence $f(x) \to f(s)$ as $x \to s$ and the continuity of f at s is established. $\square$

REMARKS. Another way of proving Proposition 3.1 is to note that

$$f(x) - f(s) = \frac{f(x) - f(s)}{x - s}(x - s) \to f'(s) \cdot 0 = 0 \quad \text{as } x \to s.$$

The converse of Proposition 3.1 is not true; a function can be continuous at a point without being differentiable at that point. For example, $f(x) = |x|$ is continuous at $x = 0$ but is not differentiable at $x = 0$; the limit

$$\lim_{x \to 0} \frac{|x| - |0|}{x - 0}$$

does not exist. The expression $|x|/x$ equals 1 for $x > 0$ and equals -1 for $x < 0$; for $x = 0$ the expression $|x|/x$ is not defined.

It is possible to construct functions continuous everywhere on an interval but differentiable nowhere on the same interval; see Proposition 7.65.

Proposition 3.2. *Let f and g be functions defined on $[a, b]$ and be differentiable at a point x of $[a, b]$ and let k be a fixed real number. Then kf, $f + g$, fg, and f/g [where $g(x) \neq 0$] are differentiable at x, and*

(i) $(kf)'(x) = kf'(x)$;
(ii) $(f + g)'(x) = f'(x) + g'(x)$;
(iii) $(fg)'(x) = f'(x)g(x) + f(x)g'(x)$;
(iv) $\left(\dfrac{f}{g}\right)'(x) = \dfrac{g(x)f'(x) - g'(x)f(x)}{g^2(x)}$.

PROOF. Parts (i) and (ii) are easy consequences of Proposition 2.5. To verify part (iii), we let $h = fg$ and note that

$$h(t) - h(x) = f(t)[g(t) - g(x)] + g(x)[f(t) - f(x)].$$

Dividing by $t - x$ and observing that f and g are continuous at x (by Proposition 3.1), we let $t \to x$ and see that part (iii) follows.

Finally, let $w = f/g$. Then

$$\frac{w(t) - w(x)}{t - x} = \frac{1}{g(t)g(x)}\left(g(x)\frac{f(t) - f(x)}{t - x} - f(x)\frac{g(t) - g(x)}{t - x}\right).$$

Letting $t \to x$, we obtain (iv). $\square$

Proposition 3.3 (Chain Rule). *Let f be a function defined on an interval J and f be differentiable at some point x of J; let g be defined on an interval I which contains the range of f, and let g be differentiable at the point $f(x)$. If, for t in J,*

$$h(t) = g[f(t)],$$

then h is differentiable at x and

$$h'(x) = g'[f(x)]f'(x). \tag{3.2}$$

PROOF. Let $y = f(x)$. By the definition of derivative, we have

$$f(t) - f(x) = (t - x)[f'(x) + u(t)],$$

$$g(s) - g(y) = (s - y)[g'(y) + v(s)],$$

where $t \in J$, $s \in I$, and $u(t) \to 0$ as $t \to x$, $v(s) \to 0$ as $s \to y$. We let $s = f(t)$. Then

$$h(t) - h(x) = g[f(t)] - g[f(x)] = [f(t) - f(x)][g'(y) + v(s)]$$

$$= (t - x)[f'(x) + u(t)][g'(y) + v(s)],$$

or, if $t \neq x$,

$$\frac{h(t) - h(x)}{t - x} = [g'(y) + v(s)][f'(x) + u(t)].$$

Letting $t \to x$, we see that $s \to y$, by the continuity of f at x (see Proposition 3.1); thus, as $t \to x$,

$$\frac{h(t) - h(x)}{t - x} \to g'(y)f'(x). \qquad \square$$

Proposition 3.4. *Let f be continuous and strictly monotonic on an interval I; moreover, let f be differentiable at some point x of I with $f'(x) \neq 0$. Then the inverse function f^{-1} is differentiable at $y = f(x)$ and*

$$(f^{-1})'(y) = \frac{1}{f'(x)}. \tag{3.3}$$

PROOF. Let I^* be the interval on which f^{-1} is defined (see Proposition 2.18). Let $w \in I^*$, $w \neq y$, and $w \to y$. Then [with $v = f^{-1}(w), v \in I, v \neq x, v \to x$]

$$\frac{f^{-1}(w) - f^{-1}(y)}{w - y} = \frac{v - x}{f(v) - f(x)} = \frac{1}{[f(v) - f(x)]/(v - x)} = \frac{1}{f'(x)}. \qquad \square$$

REMARKS. The relation (3.3) is easy to remember. Since $h(x) = x$ obviously yields $h'(x) = 1$, we get by Proposition 3.3 that $f^{-1}[f(x)] = x$ implies

$$(f^{-1})'(y)f'(x) = 1.$$

Figure 3.2 illustrates formula (3.3) geometrically. The graphs of f and f^{-1} are reflections of one another in the line $y = x$. The tangent lines l_1 and l_2 are also reflections of one another in the line $y = x$; we have

$$f'(x) = \text{slope of } l_1 = \frac{f(x) - b}{x - b} \quad \text{and} \quad (f^{-1})'(y) = \text{slope of } l_2 = \frac{x - b}{f(x) - b}$$

are reciprocals of one another.

Let $y = f(x)$. Then $f'(x)$ can also be written

$$\frac{dy}{dx}, \quad \frac{d}{dx}f(x), \quad D_x y, \quad D_x f(x), \quad \text{and } y'.$$

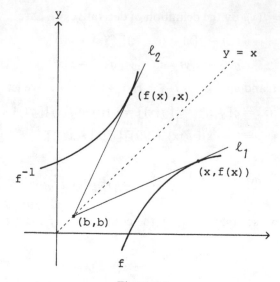

Figure 3.2

For $w = g(y)$ and $y = f(x)$ we can write (3.2) in the simple form

$$\frac{dw}{dx} = \frac{dw}{dy} \cdot \frac{dy}{dx} \quad \text{or} \quad D_x w = D_y w \cdot D_x y. \tag{3.4}$$

Similarly, for $y = f(x)$ and $x = f^{-1}(y)$ we can write (3.3) in the simple form

$$\frac{dx}{dy} = \frac{1}{dy/dx} \quad \text{or} \quad D_y x = \frac{1}{D_x y}. \tag{3.5}$$

Definition. A function f defined on an open interval (a, b) is said to be *twice differentiable at s* in (a, b) if f is differentiable on a neighborhood of s and if f' has a derivative at s. The derivative of f' is called the *second derivative of f* or the *second order derivative of f* and its value at s is denoted by $f''(s)$. If this situation prevails at all points s of (a, b), then f is said to be *twice differentiable on (a, b)* and we denote the second derivative of f by f''. Instead of saying that f is twice differentiable at s we can also say that f has a second derivative at s or that the second derivative of f exists at s. We say that the *third derivative of f exists at s* if f' is differentiable on a neighborhood of s and if f'' has a derivative at s; the third derivative at s is denoted by $f'''(s)$. Proceeding inductively, we say that the *nth derivative of f exists at s* if the derivative of order $n - 2$ of f is differentiable in a neighborhood of s and if the derivative of order $n - 1$ of f has a derivative at s. We denote by $f^{(n)}$ the nth order derivative of f.

Let $y = f(x)$, then $f^{(n)}(x)$ can also be written

$$\frac{d^n y}{dx^n}, \quad \frac{d^n}{dx^n} f(x), \quad D_x^n y, \quad D_x^n f(x), \quad \text{and} \quad y^{(n)}.$$

2. Derivatives of Basic Functions

1. Let $f(x) = c$, where c is a fixed real number. Then $f'(x) = 0$.
   Indeed,

$$f'(x) = \lim_{t \to x} \frac{f(t) - f(x)}{t - x} = \lim_{t \to x} \frac{c - c}{t - x} = \lim_{t \to x} 0 = 0.$$

2. Let $g(x) = x^n$, where n is a positive integer. Then $g'(x) = nx^{n-1}$.
   Indeed, for $t \neq x$ we have

$$\frac{t^n - x^n}{t - x} = t^{n-1} + t^{n-2}x + \cdots + tx^{n-2} + x^{n-1}.$$

Thus,

$$g'(x) = \lim_{t \to x} \frac{t^n - x^n}{t - x} = \lim_{t \to x} (t^{n-1} + t^{n-2}x + \cdots + tx^{n-2} + x^{n-1})$$

$$= nx^{n-1}.$$

3. For $x > 0$, let $f(x) = \ln x$. Then $f'(x) = 1/x$.
   Indeed, by (1.3)

$$\frac{1}{t} < \frac{\ln t - \ln x}{t - x} < \frac{1}{x} \quad \text{for } x > t$$

and

$$\frac{1}{x} < \frac{\ln x - \ln t}{x - t} < \frac{1}{t} \quad \text{for } x < t.$$

In either case

$$f'(x) = \lim_{t \to x} \frac{\ln t - \ln x}{t - x} = \frac{1}{x}.$$

REMARK. Another method for obtaining the derivative of $\ln x$ is to note that

$$\frac{\ln t - \ln x}{t - x} = \frac{1}{x} \frac{\ln[1 + (t - x)/x]}{(t - x)/x}$$

and then use (2.20).

4. For $x > 0$, let $g(x) = \log_a x$, where a is a fixed positive real number different from 1. Then $g'(x) = (\log_a e)(1/x)$.
   Indeed, $\log_a x = (\log_a e)(\ln x)$ and so

$$\frac{d}{dx}(\log_a x) = (\log_a e)\frac{d}{dx}(\ln x) = (\log_a e)\frac{1}{x}.$$

5. Let $h(x) = a^x$, where a is a fixed positive real number different from 1 and x is any real number. Then $h'(x) = (\ln a)a^x$.

Indeed, for $-\infty < x < \infty$ and $0 < y < \infty$

$$y = a^x \quad \text{if and only if} \quad x = \log_a y.$$

Thus, by (3.5),

$$D_x y = \frac{1}{D_y x} = (\log_e a)y = (\ln a)a^x$$

In particular,

$$\frac{d}{dx}(e^x) = e^x.$$

REMARK. Another method for obtaining the derivative of a^x is to note that

$$\frac{a^t - a^x}{t - x} = a^x \frac{a^{t-x} - 1}{t - x}$$

and then use (2.22).

6. For $x > 0$, let $w(x) = x^b$, where b is a fixed real number. Then $w'(x) = bx^{b-1}$.

Indeed, since

$$x^b = e^{b(\ln x)},$$

we get, letting $w = e^y$ and $y = b(\ln x)$,

$$\frac{dw}{dx} = \frac{dw}{dy} \cdot \frac{dy}{dx} = e^y \left(\frac{b}{x}\right) = x^b \left(\frac{b}{x}\right) = bx^{b-1}.$$

REMARK. Another way of getting the derivative of the power function x^b is to observe that

$$\frac{t^b - x^b}{t - x} = x^{b-1} \frac{[1 + (t - x)/x]^b - 1}{(t - x)/x}$$

and then use (2.23)

7. For $x > 0$, let $v(x) = x^x$. Then $v'(x) = x^x(1 + \ln x)$.

Indeed, $x^x = e^{x(\ln x)}$ and so, putting $v = e^y$ and $y = x(\ln x)$, we get

$$\frac{dv}{dx} = \frac{dv}{dy} \cdot \frac{dy}{dx} = e^y(1 + \ln x) = x^x(1 + \ln x).$$

8. Let $f(x) = \sin x$. Then $f'(x) = \cos x$.

Indeed, since

$$\sin t - \sin x = 2 \sin \frac{t - x}{2} \cos \frac{t + x}{2},$$

we have

$$f'(x) = \lim_{t \to x} \frac{\sin t - \sin x}{t - x} = \lim_{t \to x} \frac{2}{t - x} \sin \frac{t - x}{2} \cos \frac{t + x}{2}$$

$$= \lim_{t \to x} \frac{\sin[(t - x)/2]}{(t - x)/2} \cdot \lim_{t \to x} \cos \left[\frac{(t + x)}{2} \right] = \cos x$$

by (2.8).

9. Let $g(x) = \cos x$. Then $g'(x) = -\sin x$.
   Indeed, since $\cos x = \sin(\frac{1}{2}\pi - x)$, we have by Proposition 3.3

$$\frac{d}{dx}(\cos x) = \frac{d}{dx}[\sin(\frac{1}{2}\pi - x)] = -\cos(\frac{1}{2}\pi - x) = -\sin x.$$

REMARK. Since

$$\tan x = \frac{\sin x}{\cos x}, \quad \cot x = \frac{\cos x}{\sin x}, \quad \sec x = \frac{1}{\cos x}, \quad \text{and } \csc x = \frac{1}{\sin x}$$

it is easily seen by using part (iv) of Proposition 3.2 that

$$\frac{d}{dx}(\tan x) = \sec^2 x, \quad \frac{d}{dx}(\cot x) = -\csc^2 x,$$

$$\frac{d}{dx}(\sec x) = (\sec x)(\tan x), \quad \text{and} \quad \frac{d}{dx}(\csc x) = -(\csc x)(\cot x).$$

10. The *inverse sine function*, denoted by $\sin^{-1}$, or arc sin, is defined as follows:

$$y = \sin^{-1} x \quad \text{if and only if} \quad x = \sin y \text{ and } -\frac{\pi}{2} \leq y \leq \frac{\pi}{2}.$$

(The domain of $\sin^{-1}$ is the closed interval $[-1, 1]$, and the range is the closed interval $[-\pi/2, \pi/2]$.)
   For $-1 < x < 1$ we have

$$\frac{d}{dx}(\sin^{-1} x) = \frac{1}{\sqrt{1 - x^2}}.$$

Indeed, $D_y(\sin y) = \cos y > 0$ for any y in the open interval $(-\pi/2, \pi/2)$. In this case the derivative $D_x y$ exists and

$$D_x y = \frac{1}{D_y x} = \frac{1}{\cos y} = \frac{1}{\sqrt{1 - \sin^2 y}} = \frac{1}{\sqrt{1 - x^2}};$$

the root is positive because $\cos y > 0$.

REMARKS. The *inverse cosine function*, denoted by $\cos^{-1}$, or arc cos, is defined as follows:

$$y = \cos^{-1} x \quad \text{if and only if} \quad x = \cos y \text{ and } 0 \le y \le \pi.$$

(The domain of $\cos^{-1}$ is the closed interval $[-1, 1]$, and the range is the closed interval $[0, \pi]$. Moreover, $\cos^{-1} x = (\pi/2) - \sin^{-1} x$ for $-1 \le x \le 1$.)

It is not difficult to show that

$$\frac{d}{dx}(\cos^{-1} x) = -\frac{1}{\sqrt{1 - x^2}} \quad \text{for } -1 < x < 1.$$

11. The *inverse tangent function*, denoted by $\tan^{-1}$, or arc tan, is defined as follows:

$$y = \tan^{-1} x \quad \text{if and only if} \quad x = \tan y \text{ and } -\frac{\pi}{2} < y < \frac{\pi}{2}.$$

(The domain of $\tan^{-1}$ is the set of all real numbers, and the range is the open interval $(-\pi/2, \pi/2)$.)

For $-\infty < x < \infty$ we have

$$\frac{d}{dx}(\tan^{-1} x) = \frac{1}{1 + x^2}.$$

Indeed, for any real number x

$$D_x y = \frac{1}{D_y x} = \frac{1}{\sec^2 y} = \frac{1}{1 + \tan^2 y} = \frac{1}{1 + x^2}.$$

REMARKS. The *inverse cotangent function*, denoted by $\cot^{-1}$, or arc cot, is defined by

$$y = \cot^{-1} x \quad \text{if and only if} \quad x = \cot y \text{ and } 0 < y < \pi.$$

(The domain of $\cot^{-1}$ is the set of all real numbers and the range is the open interval $(0, \pi)$. Moreover, $\cot^{-1} x = (\pi/2) - \tan^{-1} x$ for all real numbers x.)

It is not difficult to see that

$$\frac{d}{dx}(\cot^{-1} x) = -\frac{1}{1 + x^2} \quad \text{for } -\infty < x < \infty.$$

12. The *inverse secant function*, denoted by $\sec^{-1}$, or arc sec, is defined as follows:

$$y = \sec^{-1} x \quad \text{if and only if} \quad x = \sec y, \, y \ne \pi/2, \text{ and } 0 \le y \le \pi.$$

(The domain of $\sec^{-1}$ is $(-\infty, -1]$ together with $[1, \infty)$ and its range is $[0, \pi/2)$ together with $(\pi/2, \pi]$. Moreover, $\sec^{-1} x = \cos^{-1}(1/x)$ for $|x| \ge 1$.)

For $|x| > 1$ we can see that

$$\frac{d}{dx}(\sec^{-1} x) = \frac{1}{|x|\sqrt{x^2 - 1}}.$$

Note: The *inverse cosecant function*, denoted by $\csc^{-1}$, or arc csc, is defined by

$$y = \csc^{-1} x \quad \text{if and only if} \quad x = \csc y, \ y \neq 0, \text{ and } -\frac{\pi}{2} \leq y \leq \frac{\pi}{2}.$$

(The domain of $\csc^{-1}$ is $(-\infty, -1]$ together with $[1, \infty)$ and its range is $[-\pi/2, 0)$ together with $(0, \pi/2]$. Moreover, $\csc^{-1} x = \sin^{-1}(1/x)$ for $|x| \geq 1$.)

For $|x| > 1$ we can see that

$$\frac{d}{dx}(\csc^{-1} x) = -\frac{1}{|x|\sqrt{x^2 - 1}}.$$

REMARK. The graphs of the six inverse trigonometric functions are obtained by reflecting each of the graphs in Figure 3.3 about the line $y = x$; Figure 3.4 shows the resulting graphs. Figure 3.5 shows the graphs of some additional functions that are of interest.

13. The hyperbolic and inverse hyperbolic functions were introduced in Section 4 of Chapter 1. It is easily seen that

$$\frac{d}{dx}(\sinh x) = \cosh x, \quad \frac{d}{dx}(\cosh x) = \sinh x, \quad \frac{d}{dx}(\tanh x) = \operatorname{sech}^2 x,$$

$$\frac{d}{dx}(\coth x) = -\operatorname{csch}^2 x, \quad \frac{d}{dx}(\operatorname{sech} x) = -(\operatorname{sech} x)(\tanh x),$$

$$\frac{d}{dx}(\operatorname{csch} x) = -(\operatorname{csch} x)(\coth x),$$

and

$$\frac{d}{dx}(\sinh^{-1} x) = \frac{1}{\sqrt{1 + x^2}},$$

$$\frac{d}{dx}(\cosh^{-1} x) = \frac{1}{\sqrt{x^2 - 1}} \quad \text{if } x > 1,$$

$$\frac{d}{dx}(\tanh^{-1} x) = \frac{1}{1 - x^2} \quad \text{if } |x| < 1,$$

$$\frac{d}{dx}(\coth^{-1} x) = \frac{1}{1 - x^2} \quad \text{if } |x| > 1,$$

$$\frac{d}{dx}(\operatorname{sech}^{-1} x) = \frac{-1}{x\sqrt{1 - x^2}} \quad \text{if } 0 < x < 1,$$

$$\frac{d}{dx}(\operatorname{csch}^{-1} x) = \frac{-1}{|x|\sqrt{1 + x^2}} \quad \text{if } x \neq 0.$$

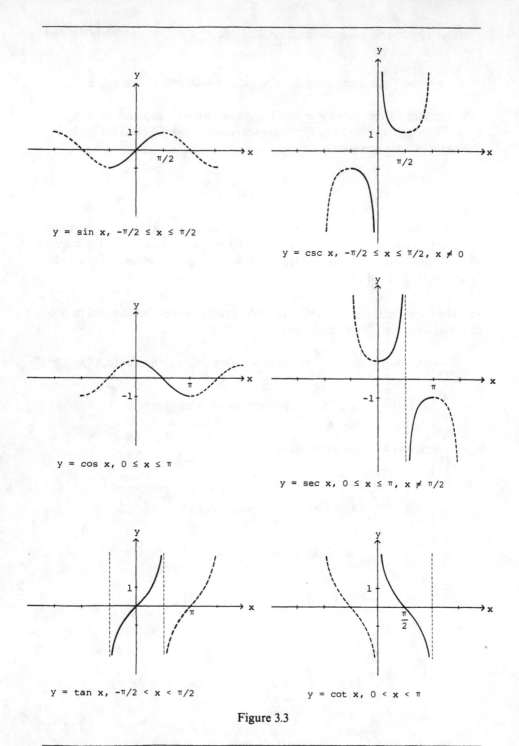

$y = \sin x,\ -\pi/2 \le x \le \pi/2$

$y = \csc x,\ -\pi/2 \le x \le \pi/2,\ x \ne 0$

$y = \cos x,\ 0 \le x \le \pi$

$y = \sec x,\ 0 \le x \le \pi,\ x \ne \pi/2$

$y = \tan x,\ -\pi/2 < x < \pi/2$

$y = \cot x,\ 0 < x < \pi$

Figure 3.3

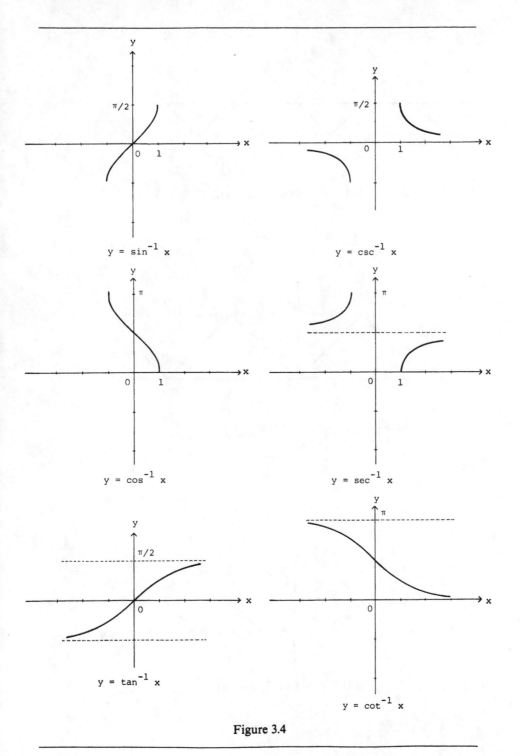

Figure 3.4

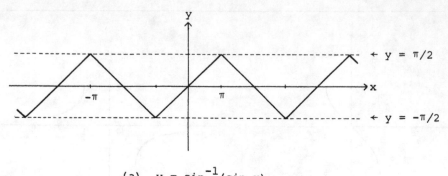

(a) $y = \sin^{-1}(\sin x)$

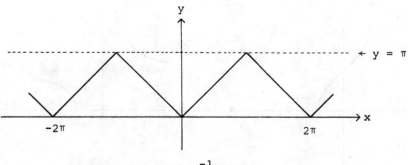

(b) $y = \cos^{-1}(\cos x)$

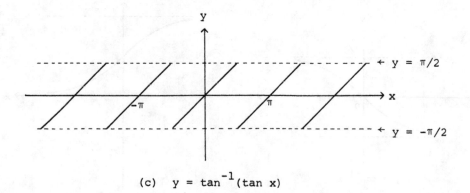

(c) $y = \tan^{-1}(\tan x)$

Figure 3.5

3. Mechanics of Differentiation

We tabulate the various differentiation formulas discussed so far in this chapter. It is assumed that c and n are fixed real numbers and that the functions u, v, w, and y are differentiable.

(1) $D_x(c) = 0$.

(2) $D_x(x) = 1$.

(3) $D_x(x^n) = nx^{n-1}$.

(4) $D_x(cu) = cD_x(u)$.

(5) $D_x(u + v) = D_x(u) + D_x(v)$.

(6) $D_x(uv) = uD_x(v) + vD_x(u)$.

(7) $D_x\left(\dfrac{u}{v}\right) = \dfrac{vD_x(u) - uD_x(v)}{v^2}$ $(v \neq 0)$.

(8) $D_x w = D_y w \cdot D_x y$ (*Chain Rule*).

(9) $D_y x = \dfrac{1}{D_x y}$ $(D_x y \neq 0)$.

(10) $D_x(\ln x) = \dfrac{1}{x}$ $(x > 0)$.

(11) $D_x(\log_a x) = (\log_a e)\dfrac{1}{x} = \dfrac{1}{(\ln a)x}$ $(x > 0)$.

(12) $D_x(e^x) = e^x$.

(13) $D_x(a^x) = (\ln a)a^x$.

(14) $D_x(x^x) = x^x(1 + \ln x)$ $(x > 0)$.

(15) $D_x(\sin x) = \cos x$.

(16) $D_x(\cos x) = -\sin x$.

(17) $D_x(\tan x) = \sec^2 x$.

(18) $D_x(\cot x) = -\csc^2 x$.

(19) $D_x(\sec x) = (\sec x)(\tan x)$.

(20) $D_x(\csc x) = -(\csc x)(\cot x)$.

(21) $D_x(\sin^{-1} x) = \dfrac{1}{\sqrt{1 - x^2}}$ $(|x| < 1)$.

(22) $D_x(\cos^{-1} x) = -\dfrac{1}{\sqrt{1 - x^2}}$ $(|x| < 1)$.

(23) $D_x(\tan^{-1} x) = \dfrac{1}{1 + x^2}$.

(24) $D_x(\cot^{-1} x) = -\dfrac{1}{1 + x^2}$.

(25) $D_x(\sec^{-1} x) = \dfrac{1}{|x|\sqrt{x^2 - 1}} \qquad (|x| > 1)$.

(26) $D_x(\csc^{-1} x) = -\dfrac{1}{|x|\sqrt{x^2 - 1}} \qquad (|x| > 1)$.

(27) $D_x(\sinh x) = \cosh x$.

(28) $D_x(\cosh x) = \sinh x$.

(29) $D_x(\tanh x) = \operatorname{sech}^2 x$.

(30) $D_x(\coth x) = -\operatorname{csch}^2 x$.

(31) $D_x(\operatorname{sech} x) = -(\operatorname{sech} x)(\tanh x)$.

(32) $D_x(\operatorname{csch} x) = -(\operatorname{csch} x)(\coth x)$.

(33) $D_x(\sinh^{-1} x) = \dfrac{1}{\sqrt{1 + x^2}}$.

(34) $D_x(\cosh^{-1} x) = \dfrac{1}{\sqrt{x^2 - 1}} \qquad (x > 1)$.

(35) $D_x(\tanh^{-1} x) = \dfrac{1}{1 - x^2} \qquad (|x| < 1)$.

(36) $D_x(\coth^{-1} x) = \dfrac{1}{1 - x^2} \qquad (|x| > 1)$.

(37) $D_x(\operatorname{sech}^{-1} x) = \dfrac{-1}{x\sqrt{1 - x^2}} \qquad (0 < x < 1)$.

(38) $D_x(\operatorname{csch}^{-1} x) = \dfrac{-1}{|x|\sqrt{1 + x^2}} \qquad (x \neq 0)$.

The logarithmic function is very useful in certain types of arithmetic calculations; it is equally useful in the computation of derivatives of some types of function. For example, suppose that

$$F(x) = f_1(x)f_2(x)\cdots f_n(x), \tag{3.6}$$

where each $f_k(x)$ with $k = 1, 2, \ldots, n$ is defined as a positive differentiable function in the same interval (a, b). Then $F(x)$ is seen to have the same properties; we note that

$$\ln F(x) = \ln f_1(x) + \ln f_2(x) + \cdots + \ln f_n(x)$$

and use of the chain rule of differentiation gives

$$\frac{F'(x)}{F(x)} = \frac{f_1'(x)}{f_1(x)} + \frac{f_2'(x)}{f_2(x)} + \cdots + \frac{f_n'(x)}{f_n(x)}. \tag{3.7}$$

This method of differentiation is referred to as *logarithmic differentiation*; the name derives from the fact that

$$\frac{d}{dx}(\ln F(x)) = \frac{F'(x)}{F(x)}$$

and so the logarithmic derivative is not some kind of new derivative but merely the derivative of the logarithm of a function. Logarithmic differentiation is often time-saving and brings out properties of the derivative which may otherwise be easily overlooked.

Logarithmic differentiation is particularly important when powers are involved. Suppose that $f(x)$ and $g(x)$ are differentiable functions in an interval (a, b) and that $f(x)$ is positive. We wish to consider

$$G(x) = [f(x)]^{g(x)}.$$

Taking logarithms we get

$$\ln G(x) = g(x)\{\ln f(x)\}$$

and differentiation gives

$$\frac{G'(x)}{G(x)} = g'(x)\{\ln f(x)\} + g(x)\frac{f'(x)}{f(x)}.$$

Thus,

$$G'(x) = [f(x)]^{g(x)}\left(g'(x)\{\ln f(x)\} + g(x)\frac{f'(x)}{f(x)}\right).$$

We now illustrate the foregoing remarks with some examples.

EXAMPLE 1. Find y' if

$$y = \sqrt[3]{x^2}\frac{1-x}{1+x^2}(\sin^4 x)(\cos^2 x) \quad \text{for } x < 1.$$

SOLUTION. Taking logarithms we have

$$\ln y = \tfrac{2}{3}(\ln x) + \ln(1-x) - \ln(1+x^2) + 4\{\ln(\sin x)\} + 2\{\ln(\cos x)\}$$

and differentiation yields

$$\frac{1}{y}y' = \frac{2}{3}\frac{1}{x} - \frac{1}{1-x} - \frac{2x}{1+x^2} + 4\cot x - 2\tan x;$$

hence,

$$y' = y\left(\frac{2}{3x} - \frac{1}{1-x} - \frac{2x}{1+x^2} + 4\cos x - 2\tan x\right).$$

EXAMPLE 2. Find y' if $y = (\tan x)^{\cos x}$, where $0 < x < \pi/2$.

SOLUTION. We have $\ln y = (\cos x)\{\ln(\tan x)\}$ and so

$$\frac{1}{y}y' = \csc x - (\sin x)\{\ln(\tan x)\}.$$

Therefore, $y' = (\tan x)^{\cos x}[\csc x - (\sin x)\{\ln(\tan x)\}]$.

EXAMPLE 3. Let $f(x)$ and $g(x)$ be positive on an interval (a, b). Moreover, let f and g be differentiable on (a, b). Using logarithmic differentiation, show that

$$\left(\frac{f}{g}\right)' = \frac{gf' - fg'}{g^2}.$$

SOLUTION. Let $h = f/g$ on (a, b). Then $\ln h = \ln f - \ln g$ and so

$$\frac{1}{h}h' = \frac{1}{f}f' - \frac{1}{g}g' \quad \text{or} \quad h' = h\left(\frac{1}{f}f' - \frac{1}{g}g'\right).$$

But $h = f/g$ and thus we get the desired formula.

COMMENTS. It is clear that the function $\ln(-x)$ is defined for $x < 0$. Using the chain rule of differentiation, we get

$$D_x\{\ln(-x)\} = \frac{1}{-x}(-1) = \frac{1}{x} \quad (x < 0).$$

Combining this with the relation

$$D_x(\ln x) = \frac{1}{x} \quad (x > 0),$$

we observe that

$$D_x(\ln|x|) = \frac{1}{x} \quad (x \neq 0). \tag{3.8}$$

We can therefore see that if $f(x)$ is differentiable on an interval and is not zero at any point of this interval, then on this interval

$$\frac{f'(x)}{f(x)} = D_x\{\ln|f(x)|\} \tag{3.9}$$

exists; we call it the *logarithmic derivative of* $f(x)$.

Letting $F(x)$ be as in (3.6), where each $f_k(x)$ with $k = 1, 2, \ldots, n$ is a differentiable function in the same interval (a, b) and is not zero at any point of this interval, then formula (3.7) is valid because

$$\frac{F'(x)}{F(x)} = D_x\{\ln|F(x)|\} = D_x\{\ln|f_1(x)\cdots f_n(x)|\}$$

and

$$D_x\{\ln|f_1(x)| + \cdots + \ln|f_n(x)|\} = \frac{f_1'(x)}{f_1(x)} + \cdots + \frac{f_n'(x)}{f_n(x)}.$$

We know that the formula

$$D_x(x^n) = nx^{n-1} \qquad (3.10)$$

holds if $x > 0$ and n is any fixed real number. Now, let $n > 0$ and define

$$f(x) = x^n \quad \text{for } x > 0,$$

$$= 0 \quad \text{for } x = 0.$$

(Note that x^n is not defined for $x = 0$ and $n \leq 0$.) Then, for $h > 0$, we have

$$\frac{f(0 + h) - f(0)}{h} = \frac{h^n}{h} = h^{n-1}.$$

For $n > 1$ and $h > 0$ we have

$$h^{n-1} \to 0 \quad \text{as } h \to 0;$$

for $0 < n < 1$ and $h > 0$ we have

$$h^{n-1} \to +\infty \quad \text{as } h \to 0.$$

If $n = 1$, then $f(x) = x$. However, we know that $D_x(x) = 1$.

We have also seen that formula (3.10) holds if x is any real number and n is a fixed positive integer. If $n = 0$, then $x^n = 1$ provided $x \neq 0$ and $D_x(1) = 0$. It is easy to see that formula (3.10) remains valid if n is a fixed negative integer and $x \neq 0$. For example,

$$y = x^{-2} = \frac{1}{x^2} \qquad (x \neq 0)$$

has the derivative

$$y' = \frac{-2x}{x^4} = -2x^{-3} \qquad (x \neq 0).$$

Suppose that $n > 0$ is a rational number, that is, $n = p/q$ with p and q being relatively prime positive integers. If q is an odd number, then $x^{p/q}$ is defined as the (uniquely determined) solution of the equation $y^q = x^p$. In this case the function $y = x^{p/q} = x^n$ is also defined for negative x and we have for negative x

$$x^{1/q} = -(-x)^{1/q} \quad \text{and} \quad x^n = (x^{1/q})^p = (-1)^p(-x)^n.$$

We now verify that formula (3.10) is valid in the case of the function $y = x^n = x^{p/q}$, where p and q are relatively prime positive integers, q is odd, and $x \neq 0$. Indeed,

$$|x^n| = |x|^n \quad \text{or} \quad \ln|x^n| = n(\ln|x|).$$

But (3.9) gives

$$D_x[\ln|x^n|] = \frac{D_x(x^n)}{x^n} \qquad (x \neq 0)$$

and (3.8) gives

$$D_x[n(\ln|x|)] = n\frac{1}{x} \qquad (x \neq 0);$$

therefore,

$$\frac{D_x(x^n)}{x^n} = n\frac{1}{x} \quad \text{or} \quad D_x(x^n) = nx^{n-1} \quad \text{for } x \neq 0.$$

To motivate the next proposition, we consider the following example. Let

$$y = \sqrt{1 - x^2} \qquad (-1 < x < 1).$$

Then

$$D_x y = -\frac{x}{\sqrt{1 - x^2}}.$$

If we put $x = \sin t$, where $-\pi/2 < t < \pi/2$, then $y = (1 - \sin^2 t)^{1/2} = \cos t$ and we get

$$D_t x = \cos t, \qquad D_t y = -\sin t;$$

thus,

$$D_x y = \frac{-\sin t}{\cos t} = \frac{D_t y}{D_t x}. \tag{3.11}$$

This suggests the possibility of obtaining $D_x y$ in terms of $D_t y$ and $D_t x$.

Proposition 3.5. *Let f and g be continuous functions on a closed interval $[a, b]$ and be differentiable on the open interval (a, b). Suppose, moreover, that $f'(t) \neq 0$ for any t satisfying $a < t < b$ and f is strictly monotonic on $[a, b]$. Then the parametric equations*

$$x = f(t) \quad \text{and} \quad y = g(t) \quad \text{for } a \leq t \leq b \tag{3.12}$$

define y as a differentiable function of x, and

$$D_x y = \frac{D_t y}{D_t x} \quad \text{for } a < t < b.$$

PROOF. By Proposition 2.18 we see that f has a continuous inverse f^{-1} such that $t = f^{-1}(x)$ for all x on the closed interval whose endpoints are $f(a)$ and $f(b)$. Thus,

$$y = g(t) = g\{f^{-1}(x)\} = F(x), \tag{3.13}$$

where $F = g(f^{-1})$ is a continuous function whose domain is the closed interval with endpoints $f(a)$ and $f(b)$. Hence, the parametric equations (3.12) define y as a continuous function of x, whose rule of correspondence is given by

(3.13). By substituting $x = f(t)$ into $g(t) = F(x)$ from (3.13), we obtain the identity in t:

$$g(t) = F\{f(t)\}. \tag{3.14}$$

If we differentiate both members of (3.14) with respect to t by the chain rule (see Proposition 3.3), we obtain

$$g'(t) = F'[f(t)]f'(t),$$

which in view of equations (3.12) can be written

$$D_t y = D_x F \cdot D_t x.$$

Therefore, since $D_t x = f'(t) \neq 0$,

$$D_x F = \frac{D_t y}{D_t x}. \qquad \qquad \square$$

REMARK. Note that if $f'(t) \neq 0$ on (a, b) and f' is continuous on $[a, b]$, then f will be strictly monotonic on $[a, b]$.

DISCUSSION. Consider the equation

$$x^2 + y^2 = R^2; \tag{3.15}$$

it represents a circle of radius R with center at the origin $(0, 0)$ in the x, y plane. We look at two parametrizations of this circle.

First Parametric Representation. Let θ denote the (polar) angle between the line segment connecting the origin $(0, 0)$ with the point $(R, 0)$ and the line segment connecting the origin $(0, 0)$ with the point (x, y) on the circle; the angle θ is measured in radians and in the counterclockwise direction (see Figure 3.6). Then

$$x = R\cos\theta, \quad y = R\sin\theta, \quad \text{and} \quad 0 \leq \theta < 2\pi. \tag{3.16}$$

As θ goes from 0 to 2π the point (x, y) traces out the circle, starting at the point $(R, 0)$ and moving in the counterclockwise direction.

Second Parametric Representation. We now use instead of the (polar) angle θ the magnitude t, where

$$t = \tan\frac{\theta}{2},$$

as parameter; note that t represents the slope of the line connecting the points $(-R, 0)$ and (x, y) on the circle $x^2 + y^2 = R^2$ (see Figure 3.7). From Figure 3.8 we can see that

$$\sin\frac{\theta}{2} = \frac{t}{\sqrt{1 + t^2}}, \qquad \cos\frac{\theta}{2} = \frac{1}{\sqrt{1 + t^2}}.$$

But

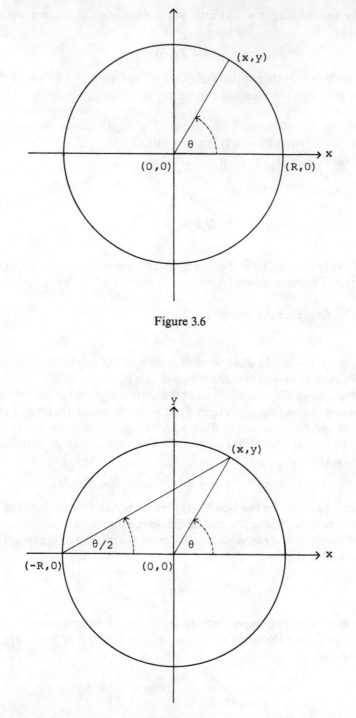

Figure 3.6

Figure 3.7

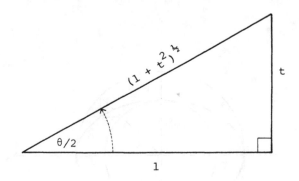

Figure 3.8

$$\sin \theta = 2 \sin \frac{\theta}{2} \cos \frac{\theta}{2} = \frac{2t}{1 + t^2},$$

$$\cos \theta = \cos^2 \frac{\theta}{2} - \sin^2 \frac{\theta}{2} = \frac{1 - t^2}{1 + t^2},$$

and so we obtain

$$x = R\frac{1 - t^2}{1 + t^2}, \quad y = R\frac{2t}{1 + t^2}, \quad \text{and} \quad -\infty < t < \infty \qquad (3.17)$$

as the parametric representation of $x^2 + y^2 = R^2$ in terms of the parameter t. As t goes from $-\infty$ to ∞ the angle $\theta/2 = \tan^{-1} t$ goes from $-\pi/2$ to $\pi/2$ and hence θ goes from $-\pi$ to π and the point (x, y) traces out the circle (3.15), starting from the point $(-R, 0)$ and moving in the counterclockwise direction. The point $(-R, 0)$ itself is obtained as a limit point for $t \to \pm\infty$.

From (3.16) we get (for $0 < \theta < \pi$ and $\pi < \theta < 2\pi$)

$$\frac{D_\theta y}{D_\theta x} = \frac{r \cos \theta}{-r \sin \theta} = -\cot \theta = -\frac{x}{y},$$

and from (3.17) we obtain (for $-\infty < t < 0$ and $0 < t < \infty$)

$$\frac{D_t y}{D_t x} = \frac{2r(1 - t^2)}{(1 + t^2)} \div \frac{-4rt}{(1 + t^2)^2} = -\frac{1 - t^2}{2t} = -\frac{x}{y}.$$

Hence, both parametrizations lead to the same answer, namely,

$$y' = -\frac{x}{y} \quad \text{for } y \neq 0. \qquad (3.18)$$

At the points $(\pm R, 0)$ the tangent line to the circle (3.15) is parallel with the y-axis and the derivative at these points has to be infinite.

We mention in passing two parametric representations for the ellipse

$$\frac{x^2}{a^2} + \frac{y^2}{b^2} = 1; \qquad (3.19)$$

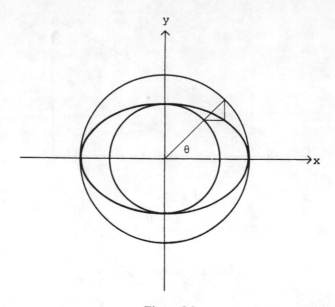

Figure 3.9

they are

$$x = a\cos\theta, \quad y = b\sin\theta, \quad \text{and} \quad 0 \le \theta < 2\pi \tag{3.20}$$

(see Figure 3.9) and

$$x = a\frac{1 - t^2}{1 + t^2}, \quad y = b\frac{2t}{1 + t^2}, \quad \text{and} \quad -\infty < t < \infty. \tag{3.21}$$

We note three parametric representations for the hyperbola

$$\frac{x^2}{a^2} - \frac{y^2}{b^2} = 1; \tag{3.22}$$

they are

$$x = a\cosh s, \quad y = b\sinh s, \quad \text{and} \quad -\infty < s < \infty; \tag{3.23}$$

$$x = a\sec v, \quad y = b\tan v, \quad \text{and} \quad 0 \le v < 2\pi, \quad v \ne \frac{\pi}{2}, \quad v \ne \frac{3\pi}{2}; \tag{3.24}$$

$$x = a\frac{1 + w^2}{1 - w^2}, \quad y = b\frac{2w}{1 - w^2}, \quad \text{and} \quad -\infty < w < \infty, w \ne \pm 1. \tag{3.25}$$

In the case of the ellipse (3.19) we obtain

$$y' = -\frac{b^2 x}{a^2 y} \quad \text{for } y \ne 0 \tag{3.26}$$

and in the case of the hyperbola (3.22) we get

$$y' = \frac{b^2 x}{a^2 y} \quad \text{for } y \ne 0. \tag{3.27}$$

In the case of the curves (3.15), (3.19), and (3.22) we used explicit parametric representations to calculate the derivative y'. However, the lack of such an explicit representation need not prevent us from finding y'. Let a curve C be given by an equation of the form

$$F(x, y) = 0, \tag{3.28}$$

which is not solved for y. *Assume* that in some interval or other there is at least one function f such that

$$F[x, f(x)] = 0 \tag{3.29}$$

holds for all values of x in the interval in question. *Suppose*, moreover, that F is a differentiable function of each of its two arguments x and y, and that f is a differentiable function of x. Then the derivative of f can be found by straightforward differentiation of (3.29). This does not exclude the possibility that there are points where f has no derivative or an infinite one (the latter meaning that there is a vertical tangent). The process by which $f'(x)$ is computed is called *implicit differentiation* and $f(x)$ is known as an *implicit function* since f is not given explicitly.

Before turning to examples of implicit differentiation, let us note clearly that our present purpose is not to examine any of the following questions: Under what conditions is a solution $y = f(x)$ of $F(x, y) = 0$ possible? What can be said about the differentiability of the function f in terms of what may be known about the function F? The answers to these and related questions belong to the study of multivariable calculus.

Examples

1. Consider the circle $x^2 + y^2 = R^2$ and compute y' by using implicit differentiation.

SOLUTION. We have

$$2x + 2y \cdot D_x y = 0 \quad \text{or} \quad y' = -\frac{x}{y}.$$

2. An *astroid* satisfies the equation

$$x^{2/3} + y^{2/3} = a^{2/3},$$

where a is a fixed positive number. See Figure 3.10 for the graph of the astroid. Using implicit differentiation, find y'.

SOLUTION. We have

$$\tfrac{2}{3}x^{-1/3} + \tfrac{2}{3}y^{-1/3} \cdot D_x y = 0 \quad \text{or} \quad y' = -\sqrt[3]{\frac{y}{x}}.$$

REMARKS. A simple calculation shows that $x^{2/3} + y^{2/3} = a^{2/3}$ implies

$$(x^2 + y^2 - a^2)^3 + 27a^2 x^2 y^2 = 0,$$

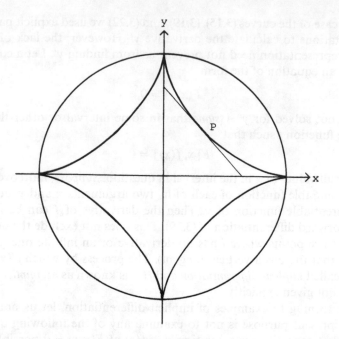

Figure 3.10

showing that the astroid is inside the circle $x^2 + y^2 = a^2$ and is symmetric with respect to both coordinate axes. It is therefore sufficient to study the curve only in the first quadrant of the x, y plane. It is easy to establish that the length of the line segment intercepted on the tangent line by the coordinate axes is independent of the point of contact P on the curve; the length equals a. This geometric property is useful in the construction of an astroid. Finally, an astroid has the parametric representation

$$x = a\cos^3 t, \quad y = a\sin^3 t, \quad \text{and} \quad 0 \le t < 2\pi.$$

3. The equation of the *folium of Descartes* is

$$x^3 + y^3 = 3axy,$$

where a is a fixed positive number. See Figure 3.11 for the graph of the folium of Descartes. Using implicit differentiation, find y'.

SOLUTION. We have

$$3x^2 + 3y^2 \cdot D_x y = 3a(x \cdot D_x y + y)$$

or

$$y' = \frac{ay - x^2}{y^2 - ax}.$$

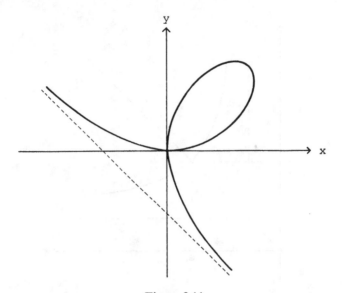

Figure 3.11

REMARKS. The straight line $y = tx$, where t is an arbitrary real number, intersects the folium of Descartes in three points, counting the origin $(0,0)$ as two; the third point has the coordinates

$$x = \frac{3at}{1+t^3}, \quad y = \frac{3at^2}{1+t^3}, \quad \text{and} \quad -\infty < t < \infty; t \neq -1.$$

This is a parametric representation of the folium of Descartes. As t goes from $-\infty$ to -1, the point (x, y) on the curve travels, starting at the origin $(0,0)$, along the right-side branch of the curve to infinity. As t goes from -1 to 0, the point on the curve goes from infinity to the origin $(0,0)$ along the left-side branch of the curve. As t goes from 0 to ∞, the point on the curve moves along the loop of the curve in the counterclockwise direction. The folium of Descartes is situated in the half-plane bounded by the line $x + y + a = 0$ containing the origin $(0,0)$. The line $x + y + a = 0$ is indicated by the ashed line in Figure 3.11. The highest point of the loop has coordinates $(2^{1/3}a, 2^{2/3}a)$. Three distinct points on the folium of Descartes given by the parametric values t_1, t_2, and t_3 are colinear if and only if the product $t_1 t_2 t_3$ equals -1.

4. Consider the curve $x^y = y^x$, where $x > 0$ and $y > 0$. It consists of two branches, namely, the line $y = x$ in the first quadrant of the x, y plane and the curve given by the parametric representation

$$x = \left(1 + \frac{1}{u}\right)^u \quad \text{and} \quad y = \left(1 + \frac{1}{u}\right)^{u+1}$$

obtained by first putting $y = tx$ into $x^y = y^x$ and then setting $t - 1 = 1/u$. See

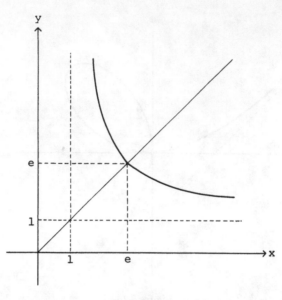

Figure 3.12

Figure 3.12 for the curve $x^y = y^x$, where $x > 0$ and $y > 0$. Calculate y' when $x \neq y$.

SOLUTION. For $x > 0$ and $y > 0$ the equation $x^y = y^x$ is equivalent to

$$y(\ln x) = x(\ln y).$$

Using implicit differentiation yields

$$y\frac{1}{x} + y'(\ln x) = \ln y + x\frac{1}{y}y' \quad \text{for} \quad y' = \left(\frac{x(\ln y) - y}{y(\ln x) - x}\right)\frac{y}{x}.$$

REMARKS. The representation $x = (1 + 1/u)^u$, $y = (1 + 1/u)^{u+1}$ permits us to find easily points on the curved part of the graph of $x^y = y^x$ for $x > 0$, $y > 0$. Note, for example, that

if $u = 1$, then $x = 2$ and $y = 4$;
if $u = 2$, then $x = \frac{9}{4}$ and $y = \frac{27}{8}$;
if $u = 3$, then $x = \frac{64}{27}$ and $y = \frac{256}{81}$.

The solution $2^4 = 4^2$ is the *only* solution of $n^m = m^n$ in positive integers n and m with $n \neq m$. Indeed, suppose $m < n$ and we write $m = n + r$, where r is a positive integer. Substituting into $m^n = n^m$, we find that $(n + r)^n = n^{n+r}$ or

$$\left(1 + \frac{r}{n}\right)^n = n^r < e^r,$$

by Proposition 1.11 in Chapter 1. Hence, $n = 1$ or $n = 2$. If $n = 1$, then $m = 1$ and this case is excluded because $n \neq m$. The case $n = 2$ yields $m = 4$.

5. Let x and y be connected by the equation

$$\ln \sqrt{x^2 + y^2} = \tan^{-1} \frac{y}{x}.$$

Using implicit differentiation, calculate y' and y''.

SOLUTION. Since $\frac{1}{2}\ln(x^2 + y^2) = \tan^{-1} y/x$ yields

$$\frac{1}{2} \frac{2x + 2yy'}{x^2 + y^2} = \frac{1}{1 + (y/x)^2} \frac{xy' - y}{x^2} \quad \text{or} \quad x + yy' = xy' - y,$$

we obtain

$$y' = \frac{x + y}{x - y}.$$

Implicit differentiation of $x + yy' = xy' - y$ gives $1 + (y')^2 + yy'' = xy''$ and so (substituting for y' the expression obtained)

$$y'' = \frac{1 + (y')^2}{x - y} = 2\frac{x^2 + y^2}{(x - y)^3}.$$

6. Let $x = y - \alpha \sin y$, where α is fixed and $0 < \alpha < 1$. Find y' and y''.

SOLUTION. We have $1 = y' - \alpha(\cos y)y'$. Thus,

$$y' = \frac{1}{1 - \alpha \cos y} \quad \text{and} \quad y'' = \frac{-\alpha(\sin y)y'}{(1 - \alpha \cos y)^2} = \frac{-\alpha \sin y}{(1 - \alpha \cos y)^3}.$$

REMARK. In $x = y - \alpha \sin y$, where $0 < \alpha < 1$, x is a strictly increasing function of y. Indeed, let $x_k = y_k - \alpha \sin y_k$ for $k = 1, 2$; then

$$x_2 - x_1 = (y_2 - y_1) - \alpha(\sin y_2 - \sin y_1).$$

But $|\sin y_2 - \sin y_1| < y_2 - y_1$ whenever $y_2 > y_1$ [see inequality (2.19) in Chapter 2]. It follows therefore that $y_2 > y_1$ implies $x_2 > x_1$. The inverse function of

$$x = y - \alpha \sin y \qquad (0 < \alpha < 1)$$

exists, but we can't express it in terms of elementary functions. Incidentally, $D_y x > 0$ for any real y and we could have used Proposition 3.4 to find y'.

7. Let x and y be connected by the equation

$$y^5 e^y - (2x^3 + 3)(\sin y) + x^2 y^2 - x \cos x = 0. \tag{3.30}$$

It is clear that the origin $(0, 0)$ is a point of the graph of (3.30). However, we would be hard up to verify that for an x close to $x = 0$ there corresponds a y close to $y = 0$ satisfying (3.30) because we can not solve equation (3.30). Yet straightforward implicit differentiation gives, at the point $(x, y) = (0, 0)$,

$$y' = -\frac{1}{3}, \qquad y'' = 0, \qquad y''' = \frac{26}{27}. \tag{3.31}$$

REMARK. While we have no way of sketching the graph of equation (3.30), the result expressed in (3.31) reveals that (3.30) has a point of inflection (a concept to be studied later on) at the point $(0, 0)$.

We complete this section with a few observations concerning higher order derivatives. Suppose that f and g are functions having derivatives of order n on the same interval. Then the product function fg satisfies

$$(fg)' = fg' + f'g,$$
$$(fg)'' = fg'' + 2f'g' + f''g,$$
$$(fg)''' = fg''' + 3f'g'' + 3f''g' + f'''g, \qquad (3.32)$$
$$\vdots$$
$$(fg)^{(n)} = fg^{(n)} + \binom{n}{1}f'g^{(n-1)} + \binom{n}{2}f''g^{(n-2)}$$
$$+ \cdots + \binom{n}{n-1}f^{(n-1)}g' + f^{(n)}g,$$

where

$$\binom{n}{k} = \frac{n!}{k!(n-k)!} \quad \text{and} \quad m! = 1 \cdot 2 \cdot 3 \cdots m.$$

The relation (3.32) is known as the *formula of Leibniz*. To verify (3.32) we note that it is fairly obvious that we must have a formula of the type

$$(fg)^{(n)} = fg^{(n)} + C_{n,1}f'g^{(n-1)} + C_{n,2}f''g^{(n-2)}$$
$$+ \cdots + C_{n,n-1}f^{(n-1)}g' + f^{(n)}g,$$

where the coefficients $C_{n,k}$ are positive integers independent of the choice of f and g. The values of the coefficients can be found by substituting suitable special functions. We take, for example,

$$f(x) = x^k, \qquad g(x) = x^{n-k},$$

so that the left-hand side becomes

$$(fg)^{(n)} = D_x^n(x^k x^{n-k}) = D_x^n(x^n) = n!.$$

On the right-hand side the derivatives of f occur in increasing orders while those of g are in decreasing orders. This gives

$$D_x^j(x^k) = k! \quad \text{if } j = k,$$
$$= 0 \quad \text{if } j > k,$$

and

$$D_x^{n-j}(x^{n-k}) = 0 \qquad \text{if } j = 0, 1, \ldots, k-1,$$
$$= (n-k)! \quad \text{if } j = k.$$

Hence, all terms on the right-hand side vanish with only one exception, so that we get the equation

$$n! = C_{n,k} k! (n-k)!$$

and $C_{n,k}$ is seen to have the value claimed in (3.32).

Next we consider composite functions for higher derivatives. We assume, for example, that f and g have derivatives of order 4 and we suppose that

$$F = f[g]$$

exists; then

$$F' = f'[g]g', \qquad F'' = f''[g](g')^2 + f'[g]g'',$$

$$F''' = f'''[g](g')^3 + 3f''[g]g'g'' + f'[g]g''',$$

$$F^{(4)} = f^{(4)}[g](g')^4 + 6f'''[g](g')^2 g'' + f''[g]\{3(g'')^2 + 4g'g'''\} + f'[g]g^{(4)}.$$

We now look at the inverse function and we suppose that f is strictly monotone, three times differentiable, and $f'(x) \neq 0$. Then f^{-1} satisfies

$$(f^{-1})'(y) = \frac{1}{f'(x)}, \qquad (f^{-1})''(y) = -\frac{f''(x)}{[f'(x)]^3},$$

$$(f^{-1})'''(y) = \frac{3[f''(x)]^3 - f'(x)f'''(x)}{[f'(x)]^5}.$$

Finally, we consider the parametric representation $x = f(t)$ and $y = g(t)$ with $a \leq t \leq b$; we assume that the functions f and g are two times differentiable on the open interval (a,b), that f and g together with their first order derivatives with respect to t are continuous on the closed interval $[a,b]$, that $f'(t) \neq 0$ for any t satisfying $a < t < b$, and that f is strictly monotone on the closed interval $[a,b]$. Then y is a twice differentiable function of x and

$$D_x y = \frac{D_t y}{D_t x}, \qquad D_x^2 y = \frac{D_t x \cdot D_t^2 y - D_t y \cdot D_t^2 x}{[D_t x]^3} \quad \text{for } a < t < b.$$

4. Asymptotes

Definition. A straight line is said to be an *asymptote* of an infinite branch of a curve, if, as the point P recedes to an infinite distance from the origin along the branch, the perpendicular distance of P from the straight line tends to zero.

REMARKS. The coordinate axes are asymptotes of $y = 1/x$. The x-axis is an asymptote of $y = e^x$. The straight line $y = x$ is clearly an asymptote of $y = x + 1/x$ because $1/x \to 0$ as $x \to \pm\infty$. The straight lines $y = \pm\pi/2$ are asymptotes of $y = \arctan x$.

Discussion. We proceed to the determination of asymptotes and first consider the case of *oblique asymptotes*, that is, asymptotes not parallel to the coordinate axes, having an equation of the form

$$y + Ax + B. \tag{3.33}$$

The abscissa, x, must tend to infinity as the point P with coordinates (x, y) recedes to infinity along the branch. We shall determine A and B so that the straight line (3.33) may be an asymptote of the given curve. The perpendicular distance of any point P with coordinates (x, y) on an infinite branch of a given curve from the line (3.33) is

$$d = \frac{|y - Ax - B|}{\sqrt{1 + A^2}}.$$

But $d \to 0$ as $x \to \infty$ and so $\lim_{x\to\infty}(y - Ax - B) = 0$, which means that

$$\lim_{x\to\infty} (y - Ax) = B.$$

Since $y/x - A = (y - Ax)(1/x)$,

$$\lim_{x\to\infty}\left(\frac{y}{x} - A\right) = \left(\lim_{x\to\infty}(y - Ax)\right)\left(\lim_{x\to\infty}\left(\frac{1}{x}\right)\right) = A \cdot 0 = 0$$

or

$$\lim_{x\to\infty}\left(\frac{y}{x}\right) = A.$$

Hence,

$$A = \lim_{x\to\infty}\frac{y}{x} \quad \text{and} \quad B = \lim_{x\to\infty}(y - Ax); \tag{3.34}$$

A is the slope and B is the y-intercept of the asymptote $y = Ax + B$.

Similar considerations have to be carried out for $x \to -\infty$.

Example 1. We are given the hyperbola

$$\frac{x^2}{a^2} - \frac{y^2}{b^2} = 1 \quad \text{or} \quad y = \pm\frac{b}{a}\sqrt{x^2 - a^2}.$$

For $x \to \infty$ we get by (3.34)

$$\frac{y}{x} = \pm\frac{b}{a}\frac{\sqrt{x^2 - a^2}}{x} = \pm\frac{b}{a}\sqrt{1 - a^2/x^2} \to \pm\frac{b}{a},$$

and then

$$y \mp \frac{b}{a}x = \pm\frac{b}{a}(\sqrt{x^2 - a^2} - x) = \mp\frac{ab}{x + \sqrt{x^2 - a^2}} \to 0;$$

thus,

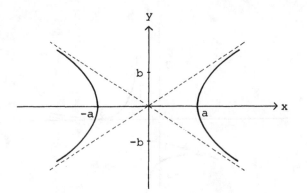

Figure 3.13

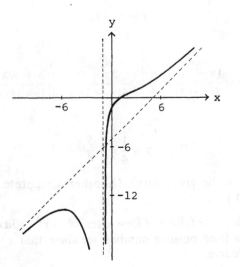

Figure 3.14

$$y = \pm\frac{b}{a}x$$

are the desired asymptotes. See Figure 3.13.

EXAMPLE 2. Let $y = (x - 1)^3/(x + 1)^2$ (see Figure 3.14). For $x \to \pm\infty$ we get

$$\frac{y}{x} \to 1, \qquad y - x = \frac{-5x^2 + 2x - 1}{(x + 1)^2} \to -5.$$

Therefore, $y = x - 5$ is an asymptote of the given curve. (Another asymptote is $x = -1$ because $y \to -\infty$ as $x \to -1$.)

EXAMPLE 3. Let $y = (x - 3)^2/4(x - 1)$ (see Figure 3.15). For $x \to \pm\infty$ we get

$$\frac{y}{x} = \frac{(x - 3)^2}{4(x - 1)x} = \frac{(1 - 3/x)^2}{4(1 - 1/x)} \to \frac{1}{4}$$

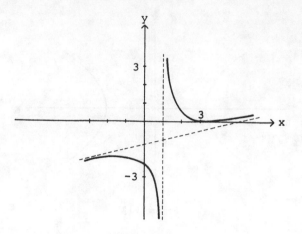

Figure 3.15

and

$$y - \frac{1}{4}x = \frac{(x-3)^2}{4(x-1)} - \frac{x}{4} = \frac{-5x+p}{4(x-1)} = \frac{-5+9/x}{4(1-1/x)} \to -\frac{5}{4}.$$

Thus,

$$y = \frac{1}{4}x - \frac{5}{4}$$

is an asymptote of the given curve. (Another asymptote is $x = 1$ because $y \to \pm\infty$ as $x \to 1$.)

EXAMPLE 4. Consider the folium of Descartes $x^3 + y^3 - 3axy = 0$ (see Figure 3.10), where a is a fixed positive number; we show that $x + y + a = 0$ is an asymptote of the curve.

Indeed, dividing the equation of the curve by x^3, we get

$$\left(\frac{y}{x}\right)^3 = 3a \cdot \frac{y}{x} \cdot \frac{1}{x} - 1.$$

Hence, for $|x| > 3a$, the quantity $|y/x|$ remains bounded; from this we get

$$\frac{y}{x} \to -1 \quad \text{as } x \to \pm\infty.$$

But, for $x \to \pm\infty$,

$$y + x = \frac{3axy}{x^2 - xy + y^2} = \frac{3a(y/x)}{1 - y/x + (y/x)^2} \to -a.$$

Definition. Let $P(x, y)$ be a polynomial in x and y with real coefficients, say

$$P(x, y) = \sum a_{jk} x^j y^k;$$

in other words, $P(x, y)$ is made up of a linear combination of terms of the form $x^j y^k$, where j and k are nonnegative integers, and the coefficients a_{jk} in the linear combination are real numbers. The degree of the polynomial $P(x, y)$ is

$$n = \max\{j + k\}.$$

We assume that $P(x, y)$ can not be factored into polynomial factors. The set of points (x, y) in the plane such that $P(x, y) = 0$ forms an *algebraic curve C of order n*.

REMARKS. In the foregoing Examples 1 and 3 we had algebraic curves of order 2; in Examples 2 and 4 we had algebraic curves of order 3. The astroid (see Figure 3.9) satisfies the equation $x^{2/3} + y^{2/3} = a^{2/3}$, where a is a fixed positive number; its equation can be written in the form

$$(x^2 + y^2 - a^2)^3 + 27a^2 x^2 y^2 = 0,$$

showing that the astroid is an algebraic curve of order 6.

DISCUSSION. Let the equation of an algebraic curve of order n be arranged in homogeneous sets of terms and expressed as

$$a_{n,0} x^n + a_{n-1,1} x^{n-1} y + a_{n-2,2} x^{n-2} y^2 + \cdots + a_{0,n} y^n$$
$$+ a_{n-1,0} x^{n-1} + a_{n-2,1} x^{n-2} y + \cdots + a_{0,n-1} y^{n-1}$$
$$+ a_{n-2,0} x^{n-2} + \cdots + a_{0,n-2} y^{n-2}$$
$$+ \cdots$$
$$+ a_{0,0} = 0$$

or

$$x^n H_n\left(\frac{y}{x}\right) + x^{n-1} H_{n-1}\left(\frac{y}{x}\right) + x^{n-2} H_{n-2}\left(\frac{y}{x}\right) + \cdots + H_0\left(\frac{y}{x}\right) = 0, \qquad (3.35)$$

where $H_k(y/x)$ is a polynomial of degree k in the unknown y/x. Dividing by x^n, we get

$$H_n\left(\frac{y}{x}\right) + \frac{1}{x} H_{n-1}\left(\frac{y}{x}\right) + \frac{1}{x^2} H_{n-2}\left(\frac{y}{x}\right) + \cdots + \frac{1}{x^n} H_0\left(\frac{y}{x}\right) = 0.$$

Letting $x \to \pm\infty$, we see that

$$H_n\left(\frac{y}{x}\right) = 0,$$

that is, y/x must tend to a number A and this A satisfies the equation

$$H_n(A) = 0. \qquad (3.36)$$

Hence, in the case of an algebraic curve we obtain the slope A of an oblique asymptote by finding the real roots of the polynomial (of highest degree) H_n. An algebraic curve of order n can therefore have at most n oblique asymptotes. Having found A by solving equation (3.36), we then substitute $y = Ax + B$ into the equation of the curve and in the resulting expression we let the coefficient of the highest power of x be zero; this procedure supplies for each value of A the corresponding value of B in most cases. We shall illustrate what goes on with the help of some examples and comments.

EXAMPLE 5. We wish to determine the asymptotes of the algebraic curve

$$P(x, y) = 9x^2 - 4y^2 - 5x + 2y + 1 = 0.$$

The algebraic curve under consideration has order 2. It is clear that $H_2(A) = 9 - 4A^2$. Thus, $H_2(A) = 0$ implies $A = \pm\frac{3}{2}$. Putting $y = Ax + B$ into the equation of the curve and noting that $9 - 4A^2 = 0$, we get

$$-8ABx - 4B^2 - 5x + 2Ax + 2B + 1 = 0.$$

Setting the coefficients of the highest power of x equal to zero gives

$$-8AB - 5 + 2A = 0 \quad \text{or} \quad B = \frac{2A - 5}{8A}.$$

Hence, if $A = \frac{3}{2}$, then $B = -\frac{1}{6}$ and if $A = -\frac{3}{2}$, then $B = \frac{2}{3}$. The two asymptotes are

$$y = \tfrac{3}{2}x - -\tfrac{1}{6} \quad \text{and} \quad y = -\tfrac{3}{2}x + \tfrac{2}{3}.$$

EXAMPLE 6. The asymptotes of the algebraic curve

$$2x^3 - x^2y - 2xy^2 + y^3 + 2x^2 + xy - y^2 + x + y + 1 = 0$$

of order 3 are

$$y = x + 1, \quad y = -x, \quad \text{and} \quad y = 2x.$$

Here $H_3(A) = A^3 - 2A^2 - A + 2 = (A - 1)(A + 1)(A - 2)$ and

$$B = \frac{A^2 - A - 2}{3A^2 - 4A - 1}.$$

EXAMPLE 7. We wish to find the asymptotes of the algebraic curve

$$(y - x)^2 x - 3y(y - x) + 2x = 0.$$

Here we have $xy^2 - 2x^2y + x^3 - 3y^2 + 3xy + 2x = 0$ and so $H_3(A) = (A - 1)^2$. $H_3(A) = 0$ produces the double root $A = 1$. We can of course write the equation of the curve in the form

$$(y - x)^2 - 3(y - x)\frac{y}{x} + 2 = 0$$

and make use of the fact that $y/x \to A = 1$ as $x \to \pm\infty$. Doing so we get

$$(y - x)^2 - 3(y - x) + 2 = 0$$

which shows that $y - x = 1$ and $y - x = 2$ are asymptotes; thus, B is either 1 or 2.

However, if we try to find B by substituting $y = Ax + B$ into the equation of the curve, we obtain (after noting that $A^2 - 2A + 1 = 0$)

$$(2AB - 2B - 3A^2 + 3A)x^2 + (B^2 - 6AB + 3B + 2)x - 3B^2 = 0.$$

The coefficients of the highest power of x is

$$2AB - 2B - 3A^2 + 3A;$$

putting $A = 1$ into this expression, we get $2B - 2B - 3 + 3$. Therefore, we can not determine B by setting the coefficient of x^2 equal to 0. But the coefficient of the next highest power of x, namely,

$$B^2 - 6AB + 3B + 2$$

will give us the expression

$$B^2 - 3B + 2 = (B - 1)(B - 2)$$

when $A = 1$ and so we see that B is either 1 or 2 when we set the coefficient of x equal to 0. Figure 3.16 illustrates the curve under discussion. (Incidentally, the curve also has a vertical asymptote, namely, $x = 3$.)

COMMENTS. We recall that in order to find the slope A of the asymptote $y = Ax + B$ we had to solve equation (3.36) for A. To find B (once A is known), we substituted $y = Ax + B$ into the equation of the curve and in the resulting expression we let the coefficient of the highest power of x be 0 (in Example 7 we let the coefficient of the second highest power of x be 0) and solved for B. If A is a simple real root of equation (3.36), we can obtain B by use of the formula

$$B \cdot H_n'(A) + H_{n-1}(A) = 0. \tag{3.37}$$

If $H_n'(A) = 0 = H_{n-1}(A)$ and $H''(A) \neq 0$, then the following formula replaces formula (3.37) for the determination of B:

$$\frac{B^2}{2} H_n''(A) + B \cdot H_{n-1}'(A) + H_{n-2}(A) = 0. \tag{3.38}$$

The reason for formulas (3.37) and (3.38) is, briefly, the following. Replacing y/x by $A + B/x$ in (3.35), we get

$$x^n[H_n(A)] + x^{n-1}[B \cdot H_n'(A) + H_{n-1}(A)]$$
$$+ x^{n-2}[\tfrac{1}{2}B^2 \cdot H_n''(A) + B \cdot H_{n-1}'(A) + H_{n-2}(A)] + \cdots = 0 \tag{3.39}$$

when we arrange terms according to descending powers of x; the details of calculation are somewhat tedious, but can be avoided if one makes use of

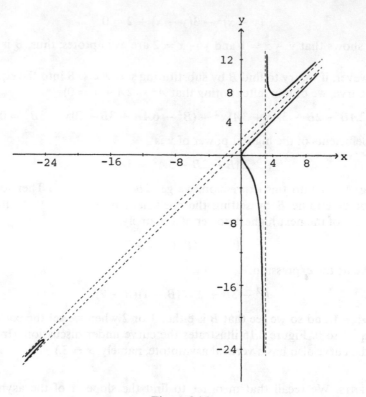

Figure 3.16

Taylor's Theorem (to be taken up in Section 2 of Chapter 4). We observe that upon putting $H_n(A) = 0$ in (3.39) and then dividing by x^{n-1} we obtain formula (3.37) when we let $x \to \pm\infty$; moreover, we can solve for B if $H_n'(A) \neq 0$. If $H_n'(A) = 0$ but $H_{n-1}(A) \neq 0$, then (3.40) does not determine any (finite) value of B and, thus, there is no asymptote corresponding to the slope A. However, if $H_n'(A) = 0 = H_{n-1}(A)$, (3.37) becomes an identity and we have to reexamine equation (3.39) which now becomes

$$x^{n-2}[\tfrac{1}{2}B^2 \cdot H''(A) + B \cdot H_{n-1}'(A) + H_{n-2}(A)] + \cdots = 0.$$

Division by x^{n-2} and then taking the limit as $x \to \pm\infty$ lead to formula (3.38) which determines two values of B provided that $H_n''(A) \neq 0$. The conditions

$$H_n(A) = 0 = H_n'(A), \quad H''(A) \neq 0, \quad \text{and } H_{n-1}(A) = 0$$

signify that A is a double root and so we have two parallel asymptotes. The exceptional case encountered in Example 7 was of this type because

$$H_3(A) = (A - 1)^2, \quad H_2(A) = -3A^2 + 3A, \quad \text{and } H_1(A) = 2;$$

applying formula (3.39) yields $B^2 - 3B + 2 = 0$ implying that B is either 1 or 2.

DISCUSSION. We now consider the determination of the asymptotes parallel to the coordinate axes. First we look at the case of asymptotes parallel to the y-axis, that is, vertical asymptotes.

Let $x = k$ be an asymptote of the curve and we have to determine k. Here y alone tends to infinity as a point P with coordinates (x, y) recedes to infinity along the curve. The distance of any point P with coordinates (x, y) on the curve from the line $x = k$ is equal to $|x - k|$. Hence, $x \to k$ when y tends to infinity.

Thus, to find the asymptotes parallel to the y-axis, we find the definite value or values k_1, k_2, and so on, to which x tends as y tends to infinity. Then $x = k_1$, $x = k_2$, and so on are the required asymptotes.

We will now obtain a simple rule for finding vertical asymptotes of an algebraic curve. We arrange the equation of the curve in descending powers of y, so that it takes the form

$$y^m Q(x) + y^{m-1} Q_1(x) + y^{m-2} Q_2(x) + \cdots = 0, \qquad (3.40)$$

where $Q(x), Q_1(x), Q_2(x), \ldots$ are polynomials in x. Dividing equation (3.40) by y^m, we get

$$Q(x) + \frac{1}{y} Q_1(x) + \frac{1}{y^2} Q_2(x) + \cdots = 0. \qquad (3.41)$$

Letting y tend to infinity and writing $x \to k$ as y tends to infinity, equation (3.41) gives $Q(k) = 0$ so that k is a root of the equation $Q(x) = 0$. Let k_1, k_2, and so on be the real roots of $Q(x) = 0$. Then the asymptotes parallel to the y-axis are $x = k_1$, $x = k_2$, and so on. Hence, we have the rule: The asymptotes parallel to the y-axis are obtained by setting equal to zero the real linear factors of the coefficient $Q(x)$ of the highest power of y in the equation of the curve.

In the same manner it can be shown that the horizontal asymptotes of an algebraic curve can be obtained by setting equal to zero the real linear factors of the coefficient of the highest power of x in the equation of the curve. In general, $y = r$ is an asymptote parallel to the x-axis if $y \to r$ when x tends to infinity in the equation of the curve.

EXAMPLE 8. The curve $x^2 y - 3x^2 - 5xy + 6y + 24 = 0$ has the asymptotes $x = 2$, $x = 3$, and $y = 3$.

5. Tangent to a Conic Section

From analytic geometry it is known that any conic section (e.g., circle, ellipse, hyperbola, parabola) can be represented by an equation of the form

$$Ax^2 + 2Bxy + Cy^2 + 2Dx + 2Ey + F = 0, \qquad (3.42)$$

where A, B, C, D, E, and F are constants. There is a very simple way of writing down the equation of the tangent to such a curve at a point (x_1, y_1) of the curve. Indeed, using implicit differentiation, we get from (3.42) that

$$y' = -\frac{Ax + By + D}{Bx + Cy + E}.$$

The equation of the tangent at (x_1, y_1) is therefore

$$y - y_1 = -\frac{Ax_1 + By_1 + D}{Bx_1 + Cy_1 + E}(x - x_1)$$

or

$$Ax_1x + B(x_1y + y_1x) + Cy_1y + D(x_1 + x) + E(y_1 + y) + F = 0. \quad (3.43)$$

If we compare this equation with equation (3.42) from which it was derived, we see that the equation of the tangent at a point (x_1, y_1) to any curve defined by an equation of the second degree in x and y can be obtained from the equation of the curve by replacing x^2 by x_1x, $2xy$ by $(x_1y + xy_1)$, y^2 by y_1y, $2x$ by $(x_1 + x)$, and $2y$ by $(y_1 + y)$. This method, of course, is applicable *only* in finding tangents to curves of second order.

Let $P_3 = (x_3, y_3)$ be a given point not on the conic section (3.42) such that two tangents can be drawn to the conic section (3.42) from P_3; let $P_1 = (x_1, y_1)$ and $P_2 = (x_2, y_2)$ be the points of contact of these two tangents (see Figure 3.17). We wish to determine the equation of the chord of contact joining P_1 and P_2.

By (3.43), the equations of the two tangents are

$$Ax_1x + B(x_1y + y_1x) + Cy_1y + D(x_1 + x) + E(y_1 + y) + F = 0 \quad (3.44)$$

and

$$Ax_2x + B(x_2y + y_2x) + Cy_2y + D(x_2 + x) + E(y_2 + y) + F = 0, \quad (3.45)$$

where x_1, y_1, x_2, y_2, of course, are not given at the outset. Since the two tangents are to pass through $P_3 = (x_3, y_3)$ equations (3.44) and (3.45) must be satisfied if we set $x = x_3$ and $y = y_3$. That is,

$$Ax_1x_3 + B(x_1y_3 + y_1x_3) + Cy_1y_3 + D(x_1 + x_3) + E(y_1 + y_3) + F = 0$$

and

$$Ax_2x_3 + B(x_2y_3 + y_2x_3) + Cy_2y_3 + D(x_2 + x_3) + E(y_2 + y_3) + F = 0.$$

These equations show that the coordinates of P_1 and the coordinates of P_2 satisfy the equation

$$Ax_3x + B(x_3y + y_3x) + Cy_3y + D(x_3 + x) + E(y_3 + y) + F = 0. \quad (3.46)$$

But (3.46) is linear in x and y, and so represents a straight line. Since (3.46) is satisfied by the coordinates of P_1 and by those of P_2, this line is the line through P_1 and P_2. Therefore, equation (3.46) is the equation of the chord of contact of the tangents drawn to the conic section (3.42) from the point

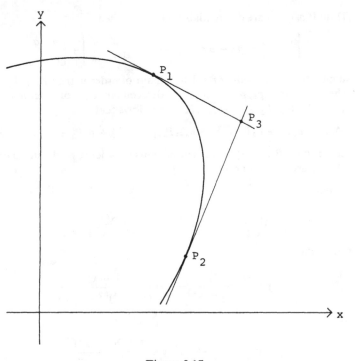

Figure 3.17

$P_3 = (x_3, y_3)$. It is easily remembered from the form of the equation of the tangent to the conic at a point of the conic.

EXAMPLE 9. The equation of the chord of contact of the tangents drawn to the ellipse

$$x^2 + 4y^2 - 18 = 0$$

from the external point $(2, 2)$ is, by (3.46),

$$2x + 4 \cdot 2y - 18 = 0 \quad \text{or} \quad x + 4y = 9.$$

From this we easily obtain the points of tangency

$$\left(\tfrac{3}{5}, \tfrac{21}{10}\right) \quad \text{and} \quad \left(3, \tfrac{3}{2}\right)$$

and the equations of the two tangents

$$x + 2y - 6 = 0 \quad \text{and} \quad x + 14y - 30 = 0.$$

EXERCISES TO CHAPTER 3

3.1. If the entries of a determinant are differentiable functions, show that the derivative of the determinant is the sum of all the determinants formed by differentiating one row, leaving the other rows unchanged.

[Hint: If p, q, r, s are differentiable functions, then

$$\begin{bmatrix} p & q \\ r & s \end{bmatrix}' = p's - q'r + ps' - qr' = \begin{bmatrix} p' & q' \\ r & s \end{bmatrix} + \begin{bmatrix} p & q \\ r' & s' \end{bmatrix}.$$

Assuming the proposition for a determinant of order n, then if $P_1, P_2, \ldots, P_{n+1}$ are the cofactors of $p_1, p_2, \ldots, p_{n+1}$ in a determinant Δ_{n+1} of order $n + 1$, so that $\Delta_{n+1} = p_1 P_1 + p_2 P_2 + \cdots + p_{n+1} P_{n+1}$, it follows that

$$\Delta'_{n+1} = p'_1 P_1 + p'_2 P_2 + \cdots + p'_{n+1} P_{n+1} + p_1 P'_1 + p_2 P'_2 + \cdots + p_{n+1} P'_{n+1};$$

hence, since $P_1, P_2, \ldots, P_{n+1}$ are determinants of order n, we deduce the proposition for determinants of order $n + 1$.]

3.2. Show that

$$\frac{d}{dx} \ln \left(\frac{(1 + x^4)^{1/2} + \sqrt{2}x}{1 - x^2} \right) = \frac{\sqrt{2}(1 + x^2)}{(1 - x^2)(1 + x^4)^{1/2}}.$$

[Hint: Since

$$y = \frac{(1 + x^4) + \sqrt{2}x}{1 - x^2} = \frac{1 - x^2}{(1 + x^4)^{1/2} - \sqrt{2}x},$$

we have

$$y - \frac{1}{y} = \frac{2x\sqrt{2}}{1 - x^2}, \quad y + \frac{1}{y} = \frac{2(1 + x^4)^{1/2}}{1 - x^2}$$

and therefore

$$\left(y + \frac{1}{y} \right) \frac{1}{y} y' = \left(1 + \frac{1}{y^2} \right) y' = 2\sqrt{2} \left(\frac{1 + x^2}{(1 - x^2)^2} \right),$$

implying that

$$\frac{d}{dx} (\ln y) = \frac{1}{y} \frac{dy}{dx} = 2\sqrt{2} \left(\frac{1 + x^2}{(1 - x^2)^2} \right) \frac{1 - x^2}{2(1 + x^4)} = \frac{\sqrt{2}(1 + x^2)}{(1 - x^2)(1 + x^4)^{1/2}}.]$$

3.3. If P is a polynomial and $P(a) = P'(a) = 0$, show that $P(x)$ has factor $(x - a)^2$. More generally, if P is a polynomial and $P(a) = P'(a) = \cdots = P^{(n)}(a) = 0$, show that $P(x)$ is divisible by $(x - a)^{n+1}$.

[Hint: Since $P(a) = 0$, $P(x)$ is divisible by $x - a$. Let $P(x) = (x - a)Q(x)$, where Q is a polynomial. Then $P'(x) = (x - a)Q'(x) + Q(x)$. But $P'(a) = 0$ and so $Q(a) = 0$; hence, $Q(x)$ is divisible by $x - a$. Now use induction.]

3.4. We say that a function f is *periodic with period a* if $f(x + a) = f(x)$ for all x. Show that if f is differentiable and periodic with period a, then f' is periodic with period a.

[Hint: By periodicity,

$$\frac{f(t + a) - f(x + a)}{(t + a) - (x + a)} = \frac{f(t) - f(x)}{t - x}.]$$

3.5. Consider the parabola $y = x^2$ and the straight line $x - y - 2 = 0$. It is required to connect these two curves by a line segment of least length.

[Hint: Evidently only the point $(\frac{1}{2}, \frac{1}{4})$ on the parabola $y = x^2$ has a tangent line parallel to the line $x - y - 2 = 0$. The perpendicular from the point $(\frac{1}{2}, \frac{1}{4})$ to the line $x - y - 2 = 0$ intersects the line at $(\frac{11}{8}, -\frac{5}{8})$. The line segment of least length connecting $y = x^2$ and $x - y - 2 = 0$ therefore connects the points $(\frac{1}{2}, \frac{1}{4})$ and $(\frac{11}{8}, -\frac{5}{8})$ and has length $7\sqrt{2}/8$.]

3.6. Let $f(x) = Ax^2 + Bx + C$ be a parabola and a and b denote given numbers. Then $c = (a + b)/2$ is the only number strictly between a and b such that

$$f(b) - f(a) = (b - a)f'(c).$$

Note that $c = (a + b)/2$ is the midpoint of the interval $[a, b]$.
[Hint: We have $f(b) - f(a) = A(b^2 - a^2) + B(b - a)$ and so

$$\frac{f(b) - f(a)}{b - a} = A(b + a) + B.$$

On the other hand, $f'(c) = 2Ac + B$ and so $c = (a + b)/2$.]

3.7. If f and g are differentiable three times and if $f'(x)g'(x) = 1$ and $h(x) = f(x)g(x)$, show that

$$\frac{h'''(x)}{h(x)} = \frac{f'''(x)}{f(x)} + \frac{g'''(x)}{g(x)}.$$

[Hint: Since $f'g' = 1$, $f'g'' + f''g' = 0$ and so $h''' = f'''g + 3f''g' + 3f'g'' + g'''f = f'''g + g'''f$.]

3.8. Let $f(x) = \{x(x + 1)(x + 2)(x + 3)(x + 4)\}^{-1}$. Find $f^{(n)}$.
[Hint: Putting

$$\frac{1}{x(x + 1)(x + 2)(x + 3)(x + 4)} = \frac{A}{x} + \frac{B}{x + 1} + \frac{C}{x + 2} + \frac{D}{x + 3} + \frac{E}{x + 4},$$

we obtain

$$1 = A(x + 1)(x + 2)(x + 3)(x + 4) + Bx(x + 2)(x + 3)(x + 4)$$
$$+ Cx(x + 1)(x + 3)(x + 4) + Dx(x + 1)(x + 2)(x + 4)$$
$$+ Ex(x + 1)(x + 2)(x + 3).$$

Letting $x = 0, -1, -2, -3$, and -4 in succession, we obtain $A = \frac{1}{24}$, $B = -\frac{1}{6}$, $C = \frac{1}{4}$, $D = -\frac{1}{6}$, and $E = \frac{1}{24}$. Thus,

$$f(x) = \frac{1}{24x} - \frac{1}{6(x + 1)} + \frac{1}{4(x + 2)} - \frac{1}{6(x + 3)} + \frac{1}{24(x + 4)}$$

and the required nth derivative is therefore

$$(-1)^n n! \left(\frac{1}{24x^{n+1}} - \frac{1}{6(x + 1)^{n+1}} + \frac{1}{4(x + 2)^{n+1}} - \frac{1}{6(x + 3)^{n+1}} + \frac{1}{24(x + 4)^{n+1}} \right).]$$

3.9. Show that the polynomial

$$P_n(x) = \frac{d^n}{dx^n}(x^2 - 1)^n$$

satisfies the equation

$$(1 - x^2)P'_n(n) - 2xP'_n(x) + n(n + 1)P_n(x) = 0.$$

[Hint: Let $g = (x^2 - 1)^n$ so that, by logarithmic differentiation,

$$\frac{g'}{g} = \frac{2nx}{x^2 - 1}$$

and so $(1 - x^2)g' + 2nxg = 0$. Differentiating again

$$(1 - x^2)g'' + 2(n - 1)xg' + 2ng = 0.$$

Using the formula of Leibniz (3.32) to differentiate n times, we obtain

$$(1 - x^2)g^{(n+2)} + ng^{(n+1)}(-2x) + \frac{n(n + 1)}{2}g^{(n)}(-2)$$

$$+ 2(n - 1)[xg^{(n+1)} + ng^{(n)}] + 2ng^{(n)} = 0,$$

giving

$$(1 - x^2)g^{(n+2)} - 2xg^{(n+1)} + n(n + 1)g^{(n)} = 0.$$

But $g^{(n)} = P_n$.]

3.10. If $y = u/x$, where u is a function of x, verify that

$$\frac{d^n y}{dx^n} = \frac{(-1)^n n!}{x^{n+1}}\left(u - \frac{x}{1!}\frac{du}{dx} + \frac{x^2}{2!}\frac{d^2 u}{dx^2} - \frac{x^3}{3!}\frac{d^3 u}{dx^3} + \cdots + (-1)^n\frac{x^n}{n!}\frac{d^n u}{dx^n}\right).$$

Conclude, by putting $u = x^m$, that

$$\binom{m}{0} - \binom{m}{1} + \binom{m}{2} - \cdots + (-1)^n\binom{m}{n} = (-1)^n\binom{m - 1}{n},$$

where m is any rational number.

3.11. If $x^4 - (a + b)x^3 + (a - b)x - 1 = 0$ has a root of multiplicity two, show that $a^{4/3} - b^{4/3} = 2^{2/3}$.

[Hint: By Exercise 3.3, the equations

$$f(x) = x^4 - (a + b)x^3 + (a - b)x - 1 = 0$$

and

$$f'(x) = 4x^3 - 3(a + b)x^2 + a - b = 0$$

will have a common root and the required result will be obtained by eliminating x between these two equations. Multiplying the second equation by x and subtracting the first gives $a + b = (3x^4 + 1)/2x^3$; multiplying the second equation by x and subtracting three times the first equation gives $a - b = (x^4 + 3)/2x$. Adding and subtracting the relations giving $a + b$ and $a - b$, we get

$$2a = \frac{1}{2}\left(x + \frac{1}{x}\right)^3, \qquad 2b = -\frac{1}{2}\left(x - \frac{1}{x}\right)^3.$$

Thus, $x + 1/x = 2^{2/3}a^{2/3}$ and $x - 1/x = -2^{2/3}b^{2/3}$; by squaring and subtracting, the required result follows.]

3.12. Show that the length of the portion of the tangent to the astroid (see Figure 3.10)

$$x^{2/3} + y^{2/3} = a^{2/3}$$

intercepted between the coordinate axes is constant.

[Hint: Differentiation gives

$$\frac{2}{3}x^{-1/3} + \frac{2}{3}y^{-1/3}\frac{dy}{dx} = 0, \quad \text{that is,} \quad \frac{dy}{dx} = -\left(\frac{y}{x}\right)^{1/3}.$$

Therefore, the equation of the tangent at any point (x_1, y_1) is

$$y - y_1 = -\left(\frac{y_1}{x_1}\right)^{1/3}(x - x_1).$$

The x-intercept of the tangent at (x_1, y_1) is $(x_1^{1/3}a^{2/3}, 0)$ and the y-intercept is $(0, y_1^{1/3}a^{2/3})$; the distance between these two points is a.]

3.13. Show that the sum of the intercepts on the coordinate axes of any tangent to $\sqrt{x} + \sqrt{y} = \sqrt{a}$ is constant.

3.14. Find the equations for the tangents to the ellipse $4x^2 + y^2 = 72$ that pass through the point $(4, 4)$.

[Answer: $2x + y - 12 = 0$ and $14x + y - 60 = 0$.]

3.15. Let $P(x)$ be a polynomial and $y = mx + c$ be tangent to $y = P(x)$ at $x = x_0$. Show that the polynomial $Q(x) = P(x) - mx - c$ is divisible by $(x - x_0)^2$.

[Hint: Since $y = mx + c$ is tangent to $y = P(x)$ at $x = x_0$ we have $P(x_0) = mx_0 + c$, $P'(x_0) = m$, and so $Q(x_0) = Q'(x_0) = 0$ and the result follows by Exercise 3.3.]

3.16. Find the equations of the asymptotes of $y^2(x - 1) - x^3 = 0$

[Answer: $x - 1 = 0$, $y = x + \frac{1}{2}$, $y = -x - \frac{1}{2}$.]

3.17. Find the equations of the asymptotes of $x^3 - 2x^2y - y^2 = 0$.

[Answer: $4x - 8y - 1 = 0$.]

3.18. If $y = 1/(a^2 + x^2)$, show that

$$y^{(n)} = \frac{(-1)^n n!(\sin^{n+1} t)[\sin(n + 1)t]}{a^{n+2}},$$

where $t = \cot^{-1}(x/a)$.

[Hint: Put $i = \sqrt{-1}$. We have

$$y = \frac{1}{(x + ia)(x - ia)} = \frac{1}{2ia}\left(\frac{1}{x - ia} - \frac{1}{x + ia}\right)$$

and so

$$y^{(n)} = \frac{(-1)^n n!}{2ia}\left(\frac{1}{(x - ia)^{n+1}} - \frac{1}{(x + ia)^{n+1}}\right).$$

Now let r and t be such that $x = r\cos t$ and $a = r\sin t$; then

$$y^{(n)} = \frac{(-1)^n n!}{2iar^{n+1}}[(\cos t - i\sin t)^{-n-1} - (\cos t + i\sin t)^{-n-1}].$$

Therefore, since

$$(\cos t - i\sin t)^{-n-1} = \cos(n+1)t + i\sin(n+1)t$$

and

$$(\cos t + i\sin t)^{-n-1} = \cos(n+1)t - i\sin(n+1)t,$$

we have

$$y^{(n)} = \frac{(-1)^n n!}{ar^{n+1}}\sin(n+1)t = \frac{(-1)^n n!}{a^{n+2}}(\sin^{n+1} t)[\sin\{n+1)t].]$$

3.19. If $y = x/(a^2 + x^2)$, show that

$$y^{(n)} = \frac{(-1)^n n!(\sin^{n+1} t)[\cos(n+1)t]}{a^{n+1}},$$

where $t = \cot^{-1}(x/a)$.
    [Hint: See Exercise 3.18.]

3.20. If $y = \sin x$, show that $y^{(n)} = \sin(x + n\pi/2)$.
    [Hint: We have $y' = \cos x = \sin(x + \pi/2)$,

$$y'' = \frac{d}{dx}\sin\left(x + \frac{\pi}{2}\right) = \frac{d}{d(x + \pi/2)}\sin\left(x + \frac{\pi}{2}\right) = \sin\left(x + \frac{2\pi}{2}\right),$$

and so forth.]

3.21. If $y = \cos x$, show that $y^{(n)} = \cos(x + n\pi/2)$.

3.22. Show that

$$\frac{d^n}{dx^n}e^{ax}\sin bx = r^n e^{ax}\sin(bx + nt)$$

and

$$\frac{d^n}{dx^n}e^{ax}\cos bx = r^n e^{ax}\cos(bx + nt),$$

where $r^2 = a^2 + b^2$ and $t = \tan^{-1} b/a$.
    [Hint: If $y = e^{ax}\sin bx$, then

$$y' = ae^{ax}\sin bx + be^{ax}\cos bx = e^{ax}(a\sin bx + b\cos bx).$$

Let $a = r\cos t$ and $b = r\sin t$, so that

$$r^2 = a^2 + b^2 \quad\text{and}\quad \tan t = \frac{b}{a};$$

then we get

$$y' = re^{ax}[(\cos t)(\sin bx) + (\sin t)(\cos bx)] = re^{ax}\sin(bx + t).$$

Again,

$$y'' = re^{ax}[a\sin(bx+t) + b\cos(bx+t)]$$
$$= r^2 e^{ax}\{(\cos t)[\sin(bx+t)] + (\sin t)[\cos(bx+t)]\}$$
$$= r^2 e^{ax}\sin(bx+2t)$$

and so forth.]

3.23. If $y = (\sin x)(\sin 2x)$, show that

$$y^{(n)} = \frac{1}{2}\left[\cos\left(x + \frac{n\pi}{2}\right) - 3^n\cos\left(3x + \frac{n\pi}{2}\right)\right].$$

[Hint: Note that $y = \frac{1}{2}(\cos x - \cos 3x)$.]

3.24. If $y = e^x(\sin^2 x)(\sin 2x)$, show that

$$y^{(n)} = \frac{1}{2}(5)^{n/2}e^x\sin(2x + n\tan^{-1}2) - \frac{1}{4}(17)^{n/2}e^x\sin(4x + n\tan^{-1}4).$$

[Hint: Note that

$$y = (e^x/2)(1 - \cos 2x)(\sin 2x) = \frac{1}{2}e^x(\sin 2x) - \frac{1}{4}e^x(\sin 4x).]$$

3.25. Show that the determinant

$$\Delta(x) = \begin{bmatrix} \cos(x+a) & \sin(x+a) & 1 \\ \cos(x+b) & \sin(x+b) & 1 \\ \cos(x+c) & \sin(x+c) & 1 \end{bmatrix}$$

satisfies $\Delta'(x) = 0$.

[Hint: Take the transpose of the determinant $\Delta(x)$ and differentiate (see Exercise 3.1).]

3.26. Simplify the expression $f(x) = \cos\{\tan^{-1}[\sin(\cot^{-1}x)]\}$.

[Hint: Let $\alpha = \cot^{-1}x$ and $\beta = \tan^{-1}y$. Then

$$\sin\alpha = \sin(\cot^{-1}x) = (1+x^2)^{-1/2}, \qquad \cos\beta = \cos\{\tan^{-1}y\} = (1+y^2)^{-1/2}.$$

But $y = \sin(\cot^{-1}x) = (1+x^2)^{-1/2}$ and $f(x) = \cos\{\tan^{-1}y\} = (1+y^2)^{-1/2}$ so that

$$f(x) = \frac{(x^2+1)^{1/2}}{(x^2+2)^{1/2}}.]$$

3.27. Let $y = (ax+b)/(cx+d)$. Show that

$$y^{(n)} = (-1)^n\frac{n!\,c^{n-1}}{(cx+d)^{n+1}}(bc - ad).$$

[Hint: Note that

$$\frac{ax+b}{cx+d} = \frac{a}{c} + \frac{bc-ad}{c(cx+d)}.]$$

3.28. If $y = 1/(x^4 - a^4)$, show that

$$y^{(n)} = \frac{(-1)^n n!}{4a^3}\left(\frac{1}{(x-a)^{n+1}} - \frac{1}{(x+a)^{n+1}} - \frac{2}{a^{n+1}}(\sin^{n+1}t)[\sin(n+1)t]\right),$$

where $t = \cot^{-1}(x/a)$.

[Hint: We have

$$\frac{1}{x^4 - a^4} = \frac{1}{4a^3}\left(\frac{1}{x-a} - \frac{1}{x+a} - 2a\frac{1}{x^2+a^2}\right).$$

For the differentiation of $1/(x^2 + a^2)$ see Exercise 3.18.]

3.29. If $y = \tanh^{-1}(x/a)$, show that

$$y^{(n)} = \frac{(n-1)!}{2}\left(\frac{(-1)^{n-1}}{(a+x)^n} + \frac{1}{(a-x)^n}\right).$$

[Hint: Recall that (see Section 4 of Chapter 1)

$$\tanh^{-1}u = \tfrac{1}{2}\ln\left(\frac{1+u}{1-u}\right)\quad\text{for }|u| < 1.]$$

3.30. If $y = 1/[(x^2 + a^2)(x^2 + b^2)]$, show that

$$y^{(n)} = \frac{(-1)^n n!}{a^2 - b^2}\left(\frac{(\sin^{n+1} t)[\sin(n+1)t]}{b^{n+2}} - \frac{(\sin^{n+1} s)[\sin(n+1)s]}{a^{n+2}}\right),$$

where $x = b\cot t = a\cot s$.
[Hint: We have

$$\frac{1}{(x^2 + a^2)(x^2 + b^2)} = \frac{1}{a^2 - b^2}\left(\frac{1}{x^2+b^2} - \frac{1}{x^2+a^2}\right).$$

For the differentiation of $1/(x^2 + c^2)$ see Exercise 3.18.]

3.31. If $y = 1/(x^2 - a^2)$, show that

$$y^{(n)} = \frac{(-1)^n n!}{2a}\left(\frac{1}{(x-a)^{n+1}} - \frac{1}{(x+a)^{n+1}}\right).$$

3.32. Prove the following result: If f is continuous on $[a, b]$ and if at every point t in $[a, b]$

$$\frac{f(t+h) - f(t-h)}{h} \to 0\quad\text{as }h \to 0,$$

then f is constant on $[a, b]$.

COMMENTS. It should first be observed that the convergence, as $h \to 0$, of the special incremental ratio

$$\frac{f(t+h) - f(t-h)}{h}$$

does not secure the existence of $f'(t)$. To see this, it is enough to consider the functions

$$|x|, \quad \sin^2\left(\frac{1}{x}\right)$$

at $x = 0$. Neither function is differentiable at $x = 0$, yet, the special incremental ratio for each of them vanishes identically and so certainly converges as $h \to 0$.

We can see that the result is not necessarily true for a discontinuous function by considering the function

$$f(x) = 1 \quad \text{for } x \neq 0,$$

$$= 0 \quad \text{for } x = 0.$$

Here, if $t \neq 0$,

$$f(t + h) - f(t - h) = 0, \quad \text{if } |h| < |t|,$$

while $f(h) - f(-h) = 0$ for every h. Thus, the limit of the special incremental ratio is everywhere zero, but the function is not constant in any interval which contains the origin.

[Hint: To prove the result we state the condition of convergence in the form

$$|f(x + h) - f(x - h)| \leq h\varepsilon, \quad \text{if } 0 < h < \text{some } \delta(\varepsilon, x), \quad (3.47)$$

where it is sufficient to consider h positive, since the expression on the left is not affected by changing h to $-h$. We verify that the numbers δ are unrestricted, except, of course, by the condition that $x \pm h$ lie in the stated interval $[a, b]$. Suppose, on the contrary, that for given x, ε the numbers δ have a largest number or an upper bound $\delta_0(\delta, x)$. Then

$$|f(x + h) - f(x - h)| \leq h\varepsilon, \quad \text{if } 0 < h < \delta_0(\varepsilon, x), \quad (3.48)$$

$$|f(x + h') - f(x - h')| > h'\varepsilon, \quad \text{for some } h' < \text{any } \delta_0 + \varepsilon', \quad (3.49)$$

where, of course, $h' \geq \delta_0$ by (3.48).

Now, by (3.47),

$$|f(x + \delta_0 + h) - f(x + \delta_0 - h)| \leq h\varepsilon, \quad \text{if } 0 < h < \delta(\varepsilon, x + \delta_0), \quad (3.50)$$

$$|f(x - \delta_0 + h) - f(x - \delta_0 - h)| \leq h\varepsilon, \quad \text{if } 0 < h < \delta(\varepsilon, x - \delta_0). \quad (3.51)$$

Choose as ε' the smallest of δ_0 and $\delta(\varepsilon, x \pm \delta_0)$. Then, by (3.49), we can find a positive $h_1 < \varepsilon' \leq \delta_0$ such that

$$|f(x + \delta_0 + h_1) - f(x - \delta_0 - h_1)| > (\delta_0 + h_1)\varepsilon. \quad (3.52)$$

But by (3.50) and (3.51)

$$|f(x + \delta_0 + h_1) - f(x + \delta_0 - h_1)| \leq h_1\varepsilon, \quad \text{since } 0 < h_1 < \delta(\varepsilon, x + \delta_0),$$

$$|f(x - \delta_0 + h_1) - f(x - \delta_0 - h_1)| \leq h_1\varepsilon, \quad \text{since } 0 < h_1 < \delta(\varepsilon, x - \delta_0),$$

and by (3.48)

$$|f(x_0 + \delta_0 - h_1) - f(x - \delta_0 + h_1)| \leq (\delta_0 - h_1)\varepsilon, \quad \text{since } 0 < \delta_0 - h_1 < \delta_0.$$

By addition of the last three inequalities we have

$$|f(x + \delta_0 + h_1) - f(x - \delta_0 - h_1)| \leq (\delta_0 + h_1)\varepsilon,$$

which contradicts (3.52).

The preceding argument is fallacious, however, if, in (3.49), h' is always δ_0, that is, if

$$|f(x + \delta_0) - f(x - \delta_0)| > \delta_0\varepsilon > 0,$$

but

$$|f(x + h) - f(x - h)| \le h\varepsilon$$

in every open neighborhood of $h = \delta_0$. We now require the condition that f be continuous, in virtue of which we can take the limit $h \to \delta_0$. This gives

$$|f(x + \delta_0) - f(x - \delta_0)| \le \delta_0\varepsilon,$$

and so rules out our assumption.

The original assumption, then, that there is a largest $\delta(\varepsilon, x)$ is also disproved and we therefore have that

$$|f(x + h) - f(x - h)| \le h\varepsilon,$$

provided only that $x \pm h$ lie in $[a, b]$. Thus, ε is independent of x, h, and we may take the limit $\varepsilon \to 0$, giving

$$f(x + h) = f(x - h) \quad \text{for every } h.$$

In particular, if we put $x = h + c$, where c is a constant in $[a, b]$, we get

$$f(x + 2c) = f(c),$$

that is, f is a constant throughout $[a, b]$.]

3.33. Determine

$$\lim_{x \to 0} \frac{e^x - e^{\sin x}}{x - \sin x}.$$

[Hint: Note that

$$\frac{e^x - e^{\sin x}}{x - \sin x} = e^{\sin x} \frac{e^{x - \sin x} - 1}{x - \sin x}$$

and that

$$\lim_{t \to 0} \frac{e^t - 1}{t} = \frac{d}{dt}(e^t)\Big|_{t=0} = 1.]$$

3.34. If $(1 - x^2)^{1/2} + (1 - y^2)^{1/2} = a(x - y)$, show that $y' = (1 - y^2)^{1/2}/(1 - x^2)^{1/2}$.

[Indeed, putting $x = \sin t$ and $y = \sin s$, we get

$$\cos t + \cos s = a(\sin t - \sin s).$$

It follows that $a = \cot \frac{1}{2}(t - s)$, implying that $t - s = 2(\cot^{-1} a)$ or

$$\sin^{-1} x - \sin^{-1} y = 2(\cot^{-1} a).$$

Differentiating with respect to x gives the desired expression for y'.]

CHAPTER 4

Applications of Differentiation

1. Mean Value Theorems

Definition. Let (a, b) be an open interval contained in the domain of definition of a real-valued function f and c be a point of (a, b). The number $f(c)$ is said to be a *relative maximum of f at c* if there is some $\delta > 0$ such that the open interval $(c - \delta, c + \delta)$ is contained in (a, b) and $f(x) \leq f(c)$ for any point x in $(c - \delta, c + \delta)$. Similarly, the number $f(c)$ is said to be a *relative minimum of f at c* if there is some $\delta > 0$ such that the open interval $(c - \delta, c + \delta)$ is contained in (a, b) and $f(x) \geq f(c)$ for any point x in $(c - \delta, c + \delta)$. By a *relative extremum* we mean a relative maximum or a relative minimum. By an *absolute maximum of a function on an interval* we simply mean the largest value of that function on the given interval; by an *absolute minimum of a function on an interval* we mean the smallest value of that function on the given interval. Absolute maxima and absolute minima are often only referred to as *maxima* and *minima*, respectively.

COMMENTS. By Proposition 2.13, every function continuous on a closed interval $[a, b]$ of finite length has an absolute maximum and an absolute minimum on that interval. It can happen that a continuous function has an absolute maximum without having a relative maximum; a case in point is the function $f(x) = x$ on $[0, 1]$. By the definition of the relative extremum it must, if it is to exist at all, occur at an inner point rather than endpoint of the interval.

Proposition 4.1. *If a continuous function f on a closed interval $[a, b]$ of finite length takes on some value twice, then it has a relative extremum.*

PROOF. If f is constant, that is, if $f(x_1) = f(x_2)$ for any x_1 and x_2 in $[a, b]$, the proposition holds trivially. If f is not constant, we can assume without loss of generality that $f(a) = f(b)$ and that for some x inside the interval we have $f(x) > f(a)$. The absolute maximum of the function is then assumed at an inner point of the interval, say, at x_0. The proposition is then proved, for such an absolute maximum must be a relative maximum. If $f(x)$ had been less than $f(a)$, the same argument would have given us a minimum.   □

Proposition 4.2 (Fermat's Theorem). *Let f be defined on $[a, b]$; if f has a relative extremum at an inner point x of $[a, b]$, and if the derivative of f at x exists, then $f'(x) = 0$.*

PROOF. Suppose that we have a relative maximum at x. Let δ be chosen in accordance with the definition of a relative maximum so that

$$a < x - \delta < x < x + \delta < b.$$

If $x - \delta < t < x$, then

$$\frac{f(t) - f(x)}{t - x} \geq 0.$$

Letting $t \to x$, we see that $f'(x) \geq 0$. If $x < t < x + \delta$, then

$$\frac{f(t) - f(x)}{t - x} \leq 0.$$

which shows that $f'(x) \leq 0$. Hence, $f'(x) = 0$. The same argument shows that the derivative also vanishes at relative minima.   □

REMARKS. Proposition 4.2 shows that the relative extrema of a differentiable function are to be found among the zeros of the derivative. The absolute maximum must be the largest of the values of the function at these points and at the boundaries of the interval; the absolute minimum must be the smallest of the values of the function at these points and at the boundaries of the interval. The converse of Proposition 4.2 is not true; the function $f(x) = x^3$ has derivative zero at $x = 0$, but has no relative extremum at $x = 0$.

The graph of the function $g(x) = |x|$ has a corner at $x = 0$; g is not differentiable at $x = 0$, but g has a relative minimum at $x = 0$. The graph of the function $h(x) = x^{2/3}$ has a cusp at $x = 0$; h is not differentiable at $x = 0$, but h has a relative minimum at $x = 0$. Figure 4.1 shows the graph of the function g and Figure 4.2 the graph of h.

Proposition 4.3 (Rolle's Theorem). *If f is a continuous real-valued function on a closed interval $[a, b]$ of finite length and is differentiable on the open interval (a, b), and if $f(a) = f(b)$, then there is a point x in the open interval (a, b) such that $f'(x) = 0$.*

PROOF. The proposition is a consequence of Propositions 4.1 and 4.2.   □

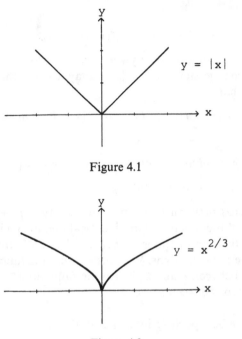

Figure 4.1

Figure 4.2

REMARKS. Proposition 4.3 shows that if $f'(x) \neq 0$ for each point x in (a, b), then $f(x_1) \neq f(x_2)$ for any two distinct points x_1 and x_2 in (a, b).

Proposition 4.3 includes the statement that between two real zeros of a function the derivative must have at least one real zero. If f is a polynomial with real coefficients such that $f(a) = f(b) = 0$ and $f(x) \neq 0$ for any x in the open interval (a, b), then f' has an odd number of real zeros in the open interval (a, b) and hence at least one real zero between a and b; the zeros of course are counted by their multiplicity. This result is a special case of Proposition 4.4, to be considered next.

Proposition 4.4 (Waring's Theorem). *Let f be a polynomial with real coefficients such that $f(a) = f(b) = 0$ and $f(x) \neq 0$ for any x in the open interval (a, b); moreover, let r be any fixed real number. Then $f' + rf$ has an odd number of zeros (hence at least one zero) in the open interval (a, b). It is understood here that the zeros are counted according to their multiplicity.*

PROOF. We recall from algebra the following: Let A and B be real numbers and P be a polynomial with real coefficients. If the numbers $P(A)$ and $P(B)$ have opposite signs, then there is an odd number of zeros of P between A and B (hence at least one zero). However, if the numbers $P(A)$ and $P(B)$ have the same sign, then there is either no zero or an even number of zeros of P between A and B. Here each zero is counted according to its multiplicity.

Turning to the proof of the proposition, we assume that $a < b$ and we suppose that a is a zero of f of multiplicity p and b a zero of f of multiplicity

q; thus,

$$f(x) = (x - a)^p(x - b)^q g(x),$$

where g does not become zero anywhere on the closed interval $[a, b]$ and hence does not change sign on $[a, b]$; we may assume that g is positive on $[a, b]$. It follows that

$$f'(x) + rf(x) = (x - a)^{p-1}(x - b)^{q-1}h(x),$$

where

$$h(x) = p(x - b)g(x) + q(x - a)g(x) + (x - a)(x - b)g'(x)$$
$$+ r(x - a)(x - b)g(x).$$

Therefore, the zeros between a and b of the function $f' + rf$ and the zeros between a and b of the function h are identical and also with regard to their multiplicity. But the sign of $h(a)$ is that of $a - b$ and of $h(b)$ that of $b - a$; so $h(a)$ and $h(b)$ have opposite signs. We thus see that the function h has an odd number of zeros between a and b; hence the function $f' + rf$ must have an odd number of zeros between a and b.   □

APPLICATION. Let n be a positive integer and $n! = 1 \cdot 2 \cdot 3 \ldots n$; the function

$$f_n(x) = 1 + \frac{x}{1!} + \frac{x^2}{2!} + \cdots + \frac{x^n}{n!}$$

has no real zero or one real zero according to whether n is even or odd.

Indeed, it is enough to show that f_n does not have two consecutive zeros; it is clear that f_n can not have positive zeros and, if n is odd, f_n has at least one zero (see the Application following Proposition 2.12). Now, suppose a and b were two such consecutive zeros [i.e., $f_n(a) = f_n(b) = 0$, $f_n(x) \neq 0$ for $a < x < b$], then we would have

$$f_n(a) = f_n'(a) + \frac{a^n}{n!} = 0, \qquad f_n(b) = f_n'(b) + \frac{b^n}{n!} = 0,$$

$$\operatorname{sgn} f_n'(a) = \operatorname{sgn} f_n'(b) \neq 0,$$

where sgn A denotes the sign of the number A. Since the sign of $f_n'(a)$ and the sign of $f_n'(b)$ are the same, f_n' would have an even number of zeros in the open interval (a, b); however, by Proposition 4.4 (with $r = 0$), f_n' would have an odd number of zeros in (a, b). Therefore, the assumption that f_n has two consecutive negative zeros leads to a contradiction. This proves the assertion.

Proposition 4.5 (Mean Value Theorem). *If f is continuous on a closed interval $[a, b]$ of finite length and is differentiable on the open interval (a, b), then there is a point x in the open interval (a, b) such that*

$$f'(x) = \frac{f(b) - f(a)}{b - a}.$$

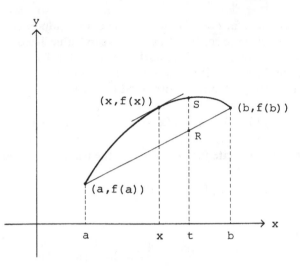

Figure 4.3

PROOF. Geometrically speaking, the Mean Value Theorem tells us that there exists at least one value x between a and b such that the tangent to the curve at the point $(x, f(x))$ is parallel to the chord joining $(a, f(a))$ with $(b, f(b))$; see Figure 4.3. The proof of the Mean Value Theorem is obtained by applying Proposition 4.3 to a properly chosen auxiliary function F. To this end, define

$$F(t) = f(t) - \left(f(a) + \frac{f(b) - f(a)}{b - a}(t - a) \right) \quad \text{for } a \le t \le b.$$

Note that $F(t)$ can be interpreted geometrically; it measures the separation between the point $S = (t, f(t))$ on the curve and the point R with abscissa t on the chord joining $(a, f(a))$ with $(b, f(b))$, as can be seen in Figure 4.3. Now F is clearly continuous on $[a, b]$ and differentiable on (a, b) and $F(a) = F(b) = 0$. So F satisfies the conditions of Proposition 4.3. There is accordingly a point x in the open interval (a, b) for which $F'(x) = 0$. But this implies

$$f'(x) - \frac{f(b) - f(a)}{b - a} = 0$$

and the proof is finished. $\qquad\qquad\qquad\qquad\qquad\qquad\qquad\qquad\qquad\qquad\qquad\square$

COMMENTS. To see that the assumptions in the Mean Value Theorem are necessary, consider the function

$$f(0) = 0 \quad \text{and} \quad f(t) = \frac{1}{t} \quad \text{for } t \ne 0.$$

If $t = 0$ belongs to the open interval (a, b), then the conclusion of the Mean Value Theorem would give the impossible relation $x^2 = ab$, where a and b have opposite signs; but the Mean Value Theorem is not applicable because

f is not differentiable for $t = 0$. If $t = 0$ is one of the endpoints of the interval $[a, b]$, say $a = 0$, then the conclusion of the Mean Value Theorem would give the impossible relation $x^2 = -b^2$; but the Mean Value Theorem is not applicable here either because f is not continuous for $t = 0$.

Another interesting situation is provided by the function

$$g(0) = 0 \quad \text{and} \quad g(x) = x^2 \sin\frac{1}{x} \quad \text{for } x \neq 0.$$

For $x \neq 0$ we can calculate the derivative in the usual fashion and obtain

$$g'(x) = 2x \sin\frac{1}{x} - \cos\frac{1}{x};$$

for $x = 0$ we get

$$g'(0) = \lim_{h \to 0} \frac{g(h) - g(0)}{h} = \lim_{h \to 0} h \sin\frac{1}{h} = 0$$

(because $|\sin 1/h| \leq 1$). Since $\lim_{x \to 0} \cos(1/x)$ does not exist, we see that $g'(x)$ does not tend to a limit as $x \to 0$ and so the function g' is not continuous at $x = 0$. This shows that a function can be differentiable everywhere but its derivative may fail to be continuous everywhere. Remarkable for us is the following circumstance: Let h be a positive real number and consider the closed interval $[0, h]$. The function g is continuous on $[0, h]$ and differentiable on the open interval $(0, h)$. By the Mean Value Theorem there is a point x in $(0, h)$ such that

$$g'(x) = \frac{g(h) - g(0)}{h}, \quad \text{that is,} \quad 2x \sin\frac{1}{x} - \cos\frac{1}{x} = h \sin\frac{1}{h}.$$

Now for each $h > 0$ there is an $x > 0$ by the Mean Value Theorem; moreover, since $0 < x < h$, it is clear that $x \to 0$ as $h \to 0$. We have

$$\lim_{h \to 0} g'(x) = 0 \quad \text{but} \quad \lim_{x \to 0} g'(x) \text{ does not exist}$$

and

$$\lim_{h \to 0} \cos\frac{1}{x} = 0 \quad \text{but} \quad \lim_{x \to 0} \cos\frac{1}{x} \text{ does not exist.}$$

This suggests that x must be a discontinuous function of h, being undefined at vastly more points than defined in any open interval of the form $(0, b)$.

A further example of this type is provided by the function

$$w(0) = 0 \quad \text{and} \quad w(x) = x \sin(\ln x) \quad \text{for } x > 0.$$

Here we have, by virtue of the Mean Value Theorem,

$$\sqrt{2} \sin\left(\frac{\pi}{4} + \ln x\right) = \sin(\ln h),$$

noting that

$$\sin(\ln x) + \cos(\ln x) = \sqrt{2} \sin\left(\frac{\pi}{4} + \ln x\right).$$

Since $\sin^2(\pi/4 + \ln x)$ can not be larger than $\frac{1}{2}$, there are infinitely many intervals in $(0, h)$, regardless of how small h is, that do not contain the point x. These intervals belong to the sequence

$$(q, q^2), \quad (q^3, q^4), \quad (q^5, q^6), \quad \dots, \quad \text{where } q = e^{-\pi/2}.$$

Proposition 4.6. (i) *Let f be a differentiable function on (a, b) such that $f'(x) = 0$ for all x in (a, b). Then f is a constant function on (a, b).*

(ii) *Let f and g be differentiable functions on (a, b) such that $f' = g'$ on (a, b). Then there is a constant c such that $f(x) = g(x) + c$ for all x in (a, b).*

PROOF. If f is not constant on the interval (a, b), then there are two points x_1 and x_2 such that $a < x_1 < x_2 < b$ and $f(x_1) \neq f(x_2)$. By the Mean Value Theorem, for some x satisfying $x_1 < x < x_2$ we have

$$f'(x) = \frac{f(x_2) - f(x_1)}{x_2 - x_1} \neq 0,$$

a contradiction. This establishes part (i) of the proposition.

To prove part (ii) of the proposition, we merely have to apply part (i) of the proposition to the function $f - g$. $\qquad\square$

Proposition 4.7. *Let f be a differentiable function on an interval (a, b). Then*

(i) *f is strictly increasing if $f'(x) > 0$ for all x in (a, b);*
(ii) *f is strictly decreasing if $f'(x) < 0$ for all x in (a, b);*
(iii) *f is nondecreasing if $f'(x) \geq 0$ for all x in (a, b);*
(iv) *f is nonincreasing if $f'(x) \leq 0$ for all x in (a, b).*

PROOF. To prove part (i), consider points x_1 and x_2, where $a < x_1 < x_2 < b$. By the Mean Value Theorem (see Proposition 4.5), for some x satisfying $x_1 < x < x_2$ we have

$$\frac{f(x_2) - f(x_1)}{x_2 - x_1} = f'(x) > 0.$$

Since $x_2 - x_1 > 0$, we get that $f(x_2) - f(x_1) > 0$ or $f(x_2) > f(x_1)$. The remaining cases are equally easy to establish. $\qquad\square$

COMMENTS. The function $f(x) = x^3$ is strictly increasing, but $f'(x) = 3x^2$ is zero for $x = 0$. This shows that f can be strictly increasing on an interval (a, b) while $f'(x) = 0$ for some isolated points x of the interval.

If f is a differentiable function on an interval (a, b) and if $f'(c) > 0$ for some point c of (a, b), then there is a subinterval (a_1, c) of (a, c) such that $f(x) < f(c)$

for any x in (a_1, c) and there exists a subinterval (c, b_1) of (c, b) such that $f(x) > f(c)$ for any x in (c, b_1). Indeed, from $f'(c) > 0$ it follows that for all x in (a, b) that are sufficiently close to c we have

$$\frac{f(x) - f(c)}{x - c} > 0,$$

that is, for $x < c$ {resp. $x > c$} we have $f(x) < f(c)$ {resp. $f(x) > f(c)$}. A similar statement holds if $f'(c) < 0$.

Proposition 4.8 (Darboux's Theorem). *Let f be a differentiable function on a closed interval $[a, b]$ of finite length and suppose that $f'(a) < y_0 < f'(b)$. Then there is a point x_0 in the open interval (a, b) such that $f'(x_0) = y_0$.*

PROOF. We first consider the special case where $f'(a) < 0$, $f'(b) > 0$ and show that there is an x in (a, b) such that $f'(x) = 0$.

We note that since f is differentiable it must be continuous (see Proposition 3.1). It accordingly attains its smallest value on $[a, b]$ (see Proposition 2.13). Since $f'(a) < 0$ there are points x_1 in (a, b) with $f(x_1) < f(a)$; similarly, since $f'(b) > 0$ there are points x_2 in (a, b) with $f(x_2) < f(b)$ (see Comments following Proposition 4.7). Thus, the least value of f in $[a, b]$ is attained at an x in (a, b). But then $f'(x) = 0$ (by Proposition 4.2).

Now suppose f is differentiable and only that $f'(a) < y_0 < f'(b)$. We show that there is an x_0 in (a, b) such that $f'(x_0) = y_0$.

Consider the auxiliary function $g(t) = f(t) - y_0 t$. Then

$$g'(a) = f'(a) - y_0 < 0 \quad \text{and} \quad g'(b) = f'(b) - y_0 > 0.$$

Since g satisfies the conditions of the special case already considered, there is an x_0 in (a, b) for which $g'(x_0) = 0$. But $f'(x_0) = g'(x_0) + y_0 = y_0$. $\square$

REMARKS. In place of the inequality $f'(a) < y_0 < f'(b)$ we could also have used the inequality $f'(a) > y_0 > f'(b)$ in Proposition 4.8; in the proof we would merely have to replace the function f by the function $-f$.

From the Comments following Proposition 4.5 we already know that the derivative of a differentiable function need not be a continuous function. Continuous functions have the property that they assume all intermediate values (see Proposition 2.12); by Proposition 4.8, derivatives of differentiable functions share this property. We now give an example of two functions f and g that have the intermediate value property but $f + g$ does not. Let

$$F(0) = 0 \quad \text{and} \quad F(t) = t^2 \sin\frac{1}{t} \quad \text{for } t \neq 0,$$

$$G(0) = 0 \quad \text{and} \quad G(t) = t^2 \cos\frac{1}{t} \quad \text{for } t \neq 0.$$

Now $F'(0) = G'(0) = 0$; for $t \neq 0$, $F'(t) = 2t \sin(1/t) - \cos(1/t)$ and $G'(t) =$

$2t\cos(1/t) + \sin(1/t)$. Putting $f(t) = \{F'(t)\}^2$ and $g(t) = \{G'(t)\}^2$, then f and g have the intermediate value property because F' and G' do, but

$$(f + g)(t) = 4t^2 + 1 \quad \text{for } t \neq 0$$

$$= 0 \qquad \text{for } t = 0.$$

Proposition 4.9 (Generalized Mean Value Theorem). (i) *Let f and g be continuous functions on a closed interval $[a,b]$ of finite length and both be differentiable on the open interval (a,b), then there is a point x in the open interval (a,b) such that*

$$[f(b) - f(a)]g'(x) = [g(b) - g(a)]f'(x). \tag{4.1}$$

(ii) *If, in addition, $g(a) \neq g(b)$, and $f'(t)$ and $g'(t)$ are not both zero for the same value of t in the open interval (a,b), then*

$$\frac{f(b) - f(a)}{g(b) - g(a)} = \frac{f'(x)}{g'(x)} \tag{4.2}$$

for some x in (a,b).

(iii) *If $g'(t)$ is not zero for any t in (a,b) the additional assumptions (ii) necessarily hold, so that equation (4.2) follows.*

PROOF. To prove part (i) we apply Proposition 4.3 to the auxiliary function F given by

$$F(t) = [f(b) - f(a)]g(t) - [g(b) - g(a)]f(t).$$

Then F is continuous on $[a,b]$ and differentiable on (a,b) and $F(a) = F(b) = f(b)g(a) - f(a)g(b)$. Thus, by Proposition 4.3, there is a point x in the open interval (a,b) for which $F'(x) = 0$. It therefore follows that

$$[f(b) - f(a)]g'(x) = [g(b) - g(a)]f'(x).$$

We next suppose that the additional assumptions of part (ii) hold. Then, if $f'(x) = 0$, it follows that $g'(x) \neq 0$. On the other hand, if $f'(x) \neq 0$, since $g(b) - g(a) \neq 0$, the right-hand member of equation (4.1) is not zero. Therefore, the left-hand member of equation (4.1) is not zero, and again $g'(x) \neq 0$. Thus, we may divide by $g'(x)$ and $g(b) - g(a)$ and so deduce equation (4.2).

Finally, we want to see that we may replace the additional assumptions of part (ii) by assuming that $g'(t) \neq 0$ for any t in (a,b), as was asserted in part (iii). Indeed, if $g'(t) \neq 0$ for any t in (a,b), then $f'(t)$ and $g'(t)$ can not both vanish for the same t in (a,b). Moreover, by the Mean Value Theorem,

$$g(b) - g(a) = (b - a)g'(s) \neq 0,$$

completing the proof.   □

COMMENTS. Taking $g(t) = t$, we see that Proposition 4.5 is a special case of Proposition 4.9. We can give Proposition 4.9 a geometric interpretation that is similar to the geometric interpretation of Proposition 4.5. We consider a curve given by the parametric representation

$$x = f(t), \qquad y = g(t) \quad \text{for } \alpha \le t \le \beta.$$

Then the formula

$$\frac{f(\beta) - f(\alpha)}{g(\beta) - g(\alpha)} = \frac{f'(\gamma)}{g'(\gamma)} \tag{4.3}$$

for some γ in the open interval (α, β) can be interpreted as follows: The left-hand side of (4.3) stands for the slope of the chord connecting the two endpoints of the curve and the right-hand side of (4.3) stands for the slope of the tangent line to the curve for $t = \gamma$.

Proposition 4.10 (L'Hôpital's Rules). *Let J be an open interval with an "endpoint" c, where c may be either finite or infinite. Assume that f and g are two functions satisfying the properties:*

  (i) *f and g are differentiable on J;*
 (ii) *$g(x) \ne 0$ and $g'(x) \ne 0$ for any x in J;*
(iii) *$\lim_{x \to c} f'(x)/g'(x) = L$, where L may be either finite or infinite.*

Moreover, assume that either

$$\text{(A)} \quad \lim_{x \to c} f(x) = 0 = \lim_{x \to c} g(x)$$

or

$$\text{(B)} \quad \lim_{x \to c} |g(x)| = \infty.$$

Then

$$\lim_{x \to c} \frac{f(x)}{g(x)} = L.$$

PROOF. Let $c < d$ and $J = (c, d)$. For any x in J define the functions

$$m(x) = \inf\left\{ \frac{f'(v)}{g'(v)} : c < v < x \right\} \quad \text{and} \quad M(x) = \sup\left\{ \frac{f'(v)}{g'(v)} : c < v < x \right\}.$$

Let s be any point between c and x. By Proposition 4.9 we have

$$\frac{f(x) - f(s)}{g(x) - g(s)} = \frac{f'(t)}{g'(t)},$$

where t is a number between s and x. Hence, we have

$$m(x) \le \frac{f(x) - f(s)}{g(x) - g(s)} \le M(x)$$

for *each* s between c and x. But clearly

$$m(x) \le \frac{f(x) - f(s)}{g(x) - g(s)} = \frac{f(x)/g(x) - f(s)/g(x)}{1 - g(s)/g(x)} \le M(x) \qquad (4.4)$$

and

$$m(x) \le \frac{f(x) - f(s)}{g(x) - g(s)} = \frac{f(s)/g(s) - f(x)/g(s)}{1 - g(x)/g(s)} \le M(x). \qquad (4.5)$$

Suppose now that condition (A) of the assumption is satisfied. Let x be fixed. Since both $f(s)$ and $g(s)$ tend to 0 as $s \to c$, it then follows from (4.4) that $m(x) \le f(x)/g(x) \le M(x)$. If case (B) holds and x is fixed, then letting $s \to c$, we see from (4.5) that

$$m(x) \le \lim_{s \to c} \frac{f(s)}{g(s)} \le M(x).$$

To complete the proof, it is enough to note that by (iii) both $m(x)$ and $M(x)$ tend toward L as $x \to c$ and that s is trapped between c and x.

If x does not tend to a finite real number c, we can use the foregoing to settle matters as well. Suppose, for example, that $x \to \infty$. Making the change of variable $y = 1/x$, we see that y tends to 0 from the right when $x \to \infty$. If we define

$$F(y) = f\left(\frac{1}{y}\right) = f(x) \quad \text{and} \quad G(y) = g\left(\frac{1}{y}\right) = g(x),$$

we have

$$F'(y) = -y^{-2} f'\left(\frac{1}{y}\right) = -x^2 f'(x) \quad \text{and} \quad G'(y) = -x^2 g'(x)$$

and so

$$\frac{F'(y)}{G'(y)} = \frac{f'(x)}{g'(x)}.$$

This shows that if $\lim_{x \to \infty} f'(x)/g'(x) = L$, then we have

$$\lim_{x \to \infty} \frac{f(x)}{g(x)} = \lim_{y \downarrow 0} \frac{F(y)}{F(y)} = \lim_{y \downarrow 0} \frac{F'(y)}{F'(y)} = \lim_{x \to \infty} \frac{f'(x)}{f'(x)} = L. \qquad \square$$

COMMENTS. If $f'(x)/g'(x)$ does not approach a limit as $x \to c$, we can not conclude that $f(x)/g(x)$ also has no limit as $x \to c$. Consider, for example,

$$f(x) = x - \sin x \quad \text{and} \quad g(x) = x + \sin x$$

Then

$$\lim_{x \to \infty} \frac{x - \sin x}{x + \sin x} = \lim_{x \to \infty} \frac{1 - (\sin x)/x}{1 + (\sin x)/x} = 1$$

but

$$\lim_{x \to \infty} \frac{1 - \cos x}{1 + \cos x} = \lim_{x \to \infty} \tan^2 \frac{x}{2}$$

does not exist. Another example of this type is the following: Let

$$f(0) = 0, \qquad f(x) = x^2 \sin \frac{1}{x} \quad \text{for } x \neq 0, \qquad \text{and } g(x) = \sin x.$$

Then, as $x \to 0$,

$$\frac{f(x)}{g(x)} = \frac{x}{\sin x} \left(x \sin \frac{1}{x} \right) \to 0,$$

but

$$\frac{f'(x)}{g'(x)} = \frac{2x \sin(1/x) - \cos(1/x)}{\cos x}$$

approaches no limit whatever as $x \to 0$.

The fact that $g'(x) \neq 0$ for x "near" c is important: Let

$$f(x) = x + \sin x \cos x \quad \text{and} \quad g(x) = (x + \sin x \cos x)e^{\sin x}.$$

Then, as $x \to \infty$, $f(x)/g(x)$ oscillates forever between e and $1/e$ and has no limit. Now

$$f'(x) = 2(\cos x)^2 \quad \text{and} \quad g'(x) = e^{\sin x}(\cos x)(x + 2\cos x + \sin x \cos x).$$

But, as $x \to \infty$,

$$\frac{f'(x)}{g'(x)} = \frac{2e^{-\sin x} \cos x}{x + (2 + \sin x)\cos x} \to 0$$

because the numerator remains bounded and the denominator becomes arbitrarily large. Note that both $f'(x) = 0$ and $g'(x) = 0$ whenever $\cos x = 0$.

From the Remarks to Proposition 4.3 we know that if g is a differentiable function on an open interval (c, d) and $g'(x) \neq 0$ for each point x in (c, d), then $g(x_1) \neq g(x_2)$ for any two distinct points x_1 and x_2 in (c, d). We add, if $g'(x) \neq 0$ for each point x in (c, d), then either $g'(x) > 0$ for each x in (c, d) or $g'(x) < 0$ for each x in (c, d) and hence the function g is seen to be strictly monotonic on (c, d) by Proposition 4.7. Indeed, if for two distinct points x_1 and x_2 we had $g'(x_1)g'(x_2) < 0$, then by the intermediate value property of differentiable functions (see Proposition 4.8) there would exist a point x_3 between x_1 and x_2 such that $g'(x_3) = 0$.

Discussion and Examples. We frequently come upon limits of the form

$$\lim_{x \to c} \frac{f(x)}{g(x)}, \tag{4.6}$$

where the limit can be one- or two-sided and where c is finite or infinite. The limit (4.6) exists and is simply

$$\frac{\lim_{x \to c} f(x)}{\lim_{x \to c} g(x)} \tag{4.7}$$

provided the limits $\lim_{x \to c} f(x)$ and $\lim_{x \to c} g(x)$ exist and are finite and provided $\lim_{x \to c} g(x) \neq 0$; see Proposition 2.6. When we calculate the derivative of a function h at a point x we consider the limit

$$\lim_{t \to x} \frac{h(t) - h(x)}{t - x},$$

provided that this limit exists. Applying the rule that "the limit of a quotient is the quotient of the limits" would only result in the indeterminate form of type $0/0$, which has no meaning, and would show that we forgot the important condition that the limit of the denominator must be different from zero for this rule to apply. If the expression (4.7) leads to an indeterminate form of the type $0/0$ or ∞/∞, then L'Hôpital's Rules (see Proposition 4.10) can frequently be used to evaluate the limit (4.6). In addition, other indeterminate forms, such as $0 \cdot \infty$, $\infty - \infty$, 0^0, ∞^0, or 1^∞, can usually be reformulated so as to take the form $0/0$ or ∞/∞.

Indeed, if the product fg of two functions presents itself in the form $0 \cdot \infty$ as $x \to c$, then we try to apply L'Hôpital's Rules to the quotient

$$\frac{f}{1/g}.$$

If the difference $f - g$ presents itself in the form $\infty - \infty$ as $x \to c$, then we consider the product

$$f(1 - g/f),$$

which will be of the type $0 \cdot \infty$ provided that g/f tends to 1 as $x \to c$. If

$$F(x) = [f(x)]^{g(x)}$$

produces an indeterminate of the type 0^0, ∞^0, or 1^∞ as $x \to c$, then

$$\ln F(x) = g(x)[\ln f(x)]$$

gives rise to a form of the type $0 \cdot \infty$. If $\ln\{F(x)\} \to K$ as $x \to c$, then $F(x) \to e^K$ as $x \to c$ by the continuity of the exponential function.

We now look at some examples.

1. We wish to find the limit

$$\lim_{x \to 0} \frac{a^x - b^x}{x},$$

where a and b are fixed positive real numbers.

Here $f(x) = a^x - b^x$ and $g(x) = x$; as $x \to 0$ we get an indeterminate of type $0/0$. Now

$$\lim_{x \to 0} \frac{f'(x)}{g'(x)} = \lim_{x \to 0} \{a^x(\ln a) - b^x(\ln b)\} = \ln \frac{a}{b}$$

and Proposition 4.10 is applicable. We obtain

$$\lim_{x \to 0} \frac{a^x - b^x}{x} = \ln \frac{a}{b}.$$

2. To determine the limit

$$\lim_{x \downarrow a} \frac{x^{1/2} - a^{1/2} + (x - a)^{1/2}}{(x^2 - a^2)^{1/2}},$$

where a is a fixed positive real number, we again use Proposition 4.10.
Here $f(x) = x^{1/2} - a^{1/2} + (x - a)^{1/2}$ and $g(x) = (x^2 - a^2)^{1/2}$ so that

$$\frac{f'(x)}{g'(x)} = \frac{(x^2 - a^2)^{1/2} + x^{1/2}(x + a)^{1/2}}{2xx^{1/2}} \to \frac{1}{(2a)^{1/2}} \quad \text{as } x \downarrow a.$$

Once again we had an indeterminate of type 0/0. By Proposition 4.10 we conclude that the limit in question equals $(2a)^{-1/2}$.

3. Consider the limit

$$\lim_{x \to 0} \frac{e^x + e^{-x} - 2}{1 - \cos x}.$$

With $f(x) = e^x + e^{-x} - 2$ and $g(x) = 1 - \cos x$, we see that

$$\frac{f'(x)}{g'(x)} = \frac{e^x - e^{-x}}{\sin x} \quad \text{and} \quad \frac{f''(x)}{g''(x)} = \frac{e^x + e^{-x}}{\cos x}.$$

Now $f(x)/g(x)$ and $f'(x)/g'(x)$ are indeterminates of type 0/0; however,

$$\lim_{x \to 0} \frac{f''(x)}{g''(x)} = \lim_{x \to 0} \frac{e^x - e^{-x}}{\cos x} = 2.$$

Thus, by Proposition 4.10,

$$\lim_{x \to 0} \frac{f'(x)}{g'(x)} = \lim_{x \to 0} \frac{e^x - e^{-x}}{\sin x} = 2$$

and, by Proposition 4.10 applied for a second time,

$$\lim_{x \to 0} \frac{f(x)}{g(x)} = \lim_{x \to 0} \frac{e^x + e^{-x} - 2}{1 - \cos x} = 2.$$

4. If $f'(x)$ exists on an interval containing c, then

$$\lim_{h \to 0} \frac{f(c + h) - f(c - h)}{2h} = \lim_{h \to 0} \frac{f'(c + h) + f'(c - h)}{2} = f'(c),$$

by Proposition 4.10. If $f''(x)$ exists on an interval containing c, then

$$\lim_{h \to 0} \frac{f(c + h) + f(c - h) - 2f(c)}{h^2} = \lim_{h \to 0} \frac{f'(c + h) - f'(c - h)}{2h} = f''(c).$$

5. Let n be a positive integer. We claim

$$\lim_{x \to \infty} (\sqrt[n]{x^n + a_1 x^{n-1} + a_2 x^{n-2} + \cdots + a_n} - x) = \frac{a_1}{n}.$$

Indeed, let us replace x by $1/y$ so that $y \downarrow 0$ corresponds to $x \to \infty$. Our problem now becomes the evaluation of the limit

$$\lim_{y \downarrow 0} \frac{\sqrt[n]{1 + a_1 y + a_2 y^2 + \cdots + a_n y^n} - 1}{y}.$$

Here both numerator and denominator tend to 0 as $y \downarrow 0$. Passing from the quotient $f(y)/g(y)$ to the quotient $f'(y)/g'(y)$, we must seek to evaluate the limit

$$\lim_{y \downarrow 0} \frac{a_1 + 2a_2 y + \cdots + na_n y^{n-1}}{n(1 + a_1 y + \cdots + a_n y^n)^{1 - 1/n}}.$$

However, this latter limit is easily seen to be a_1/n. Invoking Proposition 4.10 gives us what we had claimed.

6. From Example 5 we can immediately deduce that

$$\lim_{x \to \infty} (\sqrt[n]{x^n + a_1 x^{n-1} + \cdots + a_n} - \sqrt[n]{x^n + b_1 x^{n-1} + \cdots + b_n}) = \frac{a_1 - b_1}{n},$$

where n denotes a positive integer. In both Example 5 and the present example the indeterminate is of type $\infty - \infty$.

7. Let b denote a positive real number. We wish to show that

$$\lim_{x \to \infty} x(b^{1/x} - 1) = \ln b.$$

Indeed, letting $f(x) = b^{1/x} - 1$ and $g(x) = 1/x$, we see that

$$\frac{f'(x)}{g'(x)} = \frac{b^{1/x}(\ln b)(-1/x^2)}{-1/x^2} \to \ln b \quad \text{as } x \to \infty.$$

The rest follows by Proposition 4.10. (Look up Proposition 1.10 for comparison.)

8. The function

$$F(0) = 0, \qquad F(x) = e^{-1/x^2} \quad \text{for } x \neq 0$$

has the remarkable property that its derivatives of all orders are zero at

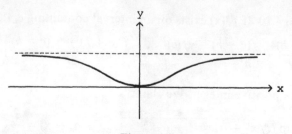

Figure 4.4

$x = 0$. In the graph of $y = F(x)$ (see Figure 4.4) this property accounts for the extreme flattening in the neighborhood of the origin. Note that the graph of $y = F(x)$ is symmetric about the y-axis because $F(-x) = F(x)$.

We now show that the derivatives of all orders of F vanish at $x = 0$. If we put $u = 1/x$ and use the notation $\exp(t)$ for e^t, we obtain

$$F'(0) = \lim_{x \to 0} \frac{F(x)}{x} = \lim_{u \to \pm\infty} \frac{u}{\exp(u^2)} = 0$$

by using Proposition 4.10. Since $F(x) = \exp(-u^2)$, $u' = -u^2$, we have by the chain rule for differentiation

$$F''(x) = \exp(-u^2)[-2uu'] = 2u^3 \exp(-u^2) \quad \text{for } x \neq 0.$$

Now

$$F''(0) = \lim_{x \to 0} \frac{F'(x)}{x} = \lim_{u \to \pm\infty} \frac{2u^4}{\exp(u^2)} = 0$$

by using Proposition 4.10 repeatedly and

$$F'''(x) = \exp(-u^2)[6u^2u' - 4u^4u'] = \exp(-u^2)[4u^6 - 6u^4] \quad \text{for } x \neq 0.$$

Again

$$F'''(0) = \lim_{x \to 0} \frac{F''(x)}{x} = \lim_{u \to \pm\infty} \frac{4u^7 - 6u^5}{\exp(u^2)} = 0$$

by using Proposition 4.10 several times. Continuing in this manner, for any positive integer n we find that

$$F^{(n)}(0) = \lim_{u \to \pm\infty} \frac{P_n(u)}{\exp(u^2)} = 0,$$

where P_n is a polynomial.

If $F^{(n)}(x) = \exp(-u^2) P_n(u)$, we obtain the recursion formula

$$P_{n+1}(u) = 2u^3 P_n(u) - u^2 P_n'(u).$$

Thus, with $P_0(u) = 1$, we have

$$P_1(u) = 2u^3, \quad P_2(u) = 4u^6 - 6u^4, \quad P_3(u) = 8u^9 - 36u^7 + 24u^5, \ldots$$

It is clear that the polynomial P_n has degree $3n$ and that $P_n(u)$ is made up of a linear combination of terms of the form u^m, where m is an integer satisfying $0 < m \leq 3n$. Thus,

$$\lim_{u \to \pm\infty} \frac{u^m}{\exp(u^2)} = 0 \quad \text{implies} \quad \lim_{u \to \pm\infty} \frac{P_n(u)}{\exp(u^2)} = 0.$$

We prove that

$$\lim_{u \to \pm\infty} \frac{u^m}{\exp(u^2)} = 0$$

by repeated use of Proposition 4.10:

$$\lim_{u \to \pm\infty} \frac{u^m}{\exp(u^2)} = \lim_{u \to \pm\infty} \frac{mu^{m-1}}{2u \exp(u^2)} = \frac{m}{2}\left(\lim_{u \to \pm\infty} \frac{u^{m-2}}{\exp(u^2)} \right);$$

after a finite number of steps the exponent in the numerator will no longer be positive, and then the limit is seen to be 0.

9. We show that

$$\lim_{x \to \infty} (1 + x)^{1/x} = 1.$$

Indeed, let $y = (1 + x)^{1/x}$; then $\ln y = (1/x)\ln(1 + x)$. But

$$\lim_{x \to \infty} \frac{\ln(1 + x)}{x} = \lim_{x \to \infty} \frac{1}{1 + x} = 0.$$

by Proposition 4.10. Since $\ln y \to 0$ as $x \to \infty$, we see that $y \to 1$ as $x \to \infty$.

10. We verify that

$$\lim_{x \downarrow 0} \frac{(1 + x)^{1/x} - e}{x} = -\frac{e}{2}.$$

Indeed, putting $y = (1 + x)^{1/x}$, hence $\ln y = (1/x)\ln(1 + x)$, we can see that

$$\lim_{x \downarrow 0} (1 + x)^{1/x} = e.$$

By Proposition 4.10,

$$\lim_{x \downarrow 0} \frac{(1 + x)^{1/x} - e}{x} = \lim_{x \downarrow 0} (1 + x)^{1/x} \frac{x - (1 + x)\ln(1 + x)}{x^2(1 + x)}$$

and two applications of Proposition 4.10 give us that

$$\lim_{x \to 0} \frac{x - (1 + x)\ln(1 + x)}{x^2(1 + x)} = -\frac{1}{2}.$$

11. We have

$$\lim_{x \to 0} \left(\frac{\tan x}{x} \right)^{1/x^2} = e^{1/3}.$$

Indeed, since

$$\frac{\tan x}{x} = \frac{\sin x}{x}\frac{1}{\cos x},$$

we see that the indeterminate form under investigation is of type 1^∞. We put

$$y = \left(\frac{\tan x}{x}\right)^{1/x^2}$$

and show that $\ln y \to \frac{1}{3}$ as $x \to \infty$ and hence that $y \to e^{1/3}$ as $x \to \infty$. Now

$$\ln y = \frac{\ln(\tan x) - \ln x}{x^2}$$

and applying Proposition 4.10 gives

$$\lim_{x\to 0} \ln y = \lim_{x\to 0} \frac{(\sec^2 x)/(\tan x) - 1/x}{2x} = \lim_{x\to 0} \frac{2x - \sin(2x)}{2x^2\sin(2x)}.$$

However, applying Proposition 4.10 three times in succession yields

$$\lim_{x\to 0} \frac{2x - \sin(2x)}{2x^2\sin(2x)} = \frac{1}{3}.$$

12. We show that

$$\lim_{x\to 0}\left(\frac{1}{x} - \cot x\right) = 0.$$

Indeed,

$$\frac{1}{x} - \cot x = \frac{\sin x - x\cos x}{x\sin x}$$

and, by Proposition 4.10,

$$\lim_{x\to 0} \frac{\sin x - x\cos x}{x\sin x} = \lim_{x\to 0} \frac{x\sin x}{\sin x + x\cos x}.$$

Another application of Proposition 4.10 gives

$$\lim_{x\to 0} \frac{x\sin x}{\sin x + x\cos x} = \lim_{x\to 0} \frac{\sin x + x\cos x}{2\cos x - x\sin x} = 0.$$

However, we could have avoided application of Proposition 4.10 in the last step by observing that $(\sin x)/x \to 1$ as $x \to 0$ [see (2.8)] and hence

$$\lim_{x\to 0} \frac{x\sin x}{\sin x + x\cos x} = \lim_{x\to 0} \frac{\sin x}{(\sin x)/x + \cos x} = 0.$$

13. We show that

$$\lim_{x\to 0} \frac{xe^{2x} + xe^x - 2e^{2x} + 2e^x}{(e^x - 1)^3} = \frac{1}{6}.$$

Indeed, we use Proposition 4.10 three times in succession. After the first application of Proposition 4.10 we cancel the common factor e^x in the numerator and denominator and after the second application of Proposition 4.10 we replace the factor e^x in the denominator by 1 (because $e^x \to 1$ as $x \to 0$). The actual calculation simplifies a lot by this procedure. The details are as follows:

$$\lim_{x \to 0} \frac{xe^{2x} + xe^x - 2e^{2x} + 2e^x}{(e^x - 1)^3} = \lim_{x \to 0} \frac{2xe^{2x} + e^{2x} + xe^x + e^x - 4e^{2x} + 2e^x}{3e^x(e^x - 1)^2}$$

$$= \lim_{x \to 0} \frac{2xe^x - 3e^x + 3 + x}{3(e^x - 1)^2}$$

$$= \frac{1}{3} \cdot \lim_{x \to 0} \frac{2xe^x + 2e^x - 3e^x + 1}{2e^x(e^x - 1)}$$

$$= \frac{1}{6} \cdot \lim_{x \to 0} \frac{-e^x + 2xe^x + 1}{e^x - 1}$$

$$= \frac{1}{6} \cdot \lim_{x \to 0} \frac{2xe^x + e^x}{e^x} = \frac{1}{6}.$$

14. Let

$$y = \left(\frac{\pi}{2} - \arctan x\right)^{1/(\ln x)}.$$

For $x \to \infty$ we obtain an indeterminate form of the type 0^0; we wish to show that $y \to 1/e$ as $x \to \infty$.

Indeed,

$$\ln y = \frac{\ln(\pi/2 - \arctan x)}{\ln x}$$

is an indeterminate form of the type ∞/∞; we apply Proposition 4.10 and obtain

$$\lim_{x \to \infty} \ln y = \lim_{x \to \infty} \frac{-\dfrac{1}{\pi/2 - \arctan x} \dfrac{1}{1 + x^2}}{1/x} = \lim_{x \to \infty} \frac{x/(1 + x^2)}{\arctan x - \pi/2}$$

$$= \lim_{x \to \infty} \frac{(1 - x^2)/(1 + x^2)^2}{1/(1 + x^2)} = \lim_{x \to \infty} \frac{1 - x^2}{1 + x^2} = -1.$$

15. We have

$$\lim_{x \to 0} \frac{x \cos x - \sin x}{x \sin^2 x} = -\frac{1}{3}.$$

From this follows in particular that

$$\lim_{x\to 0}\left(\cot^2 x - \frac{1}{x^2}\right) = -\frac{2}{3} \quad \text{and} \quad \lim_{x\to 0}\left(\frac{\sin x}{x}\right)^{1/(1-\cos x)} = e^{-1/3}.$$

Indeed, by Proposition 4.10,

$$\lim_{x\to 0}\frac{x\cos x - \sin x}{x\sin^2 x} = \lim_{x\to 0}\frac{-x\sin x}{\sin^2 x + 2x(\sin x)(\cos x)}$$

$$= \lim_{x\to 0}\frac{-1}{(\sin x)/x + 2\cos x} = -\frac{1}{3},$$

establishing the first assertion. Since

$$\cot^2 x - \frac{1}{x^2} = \frac{x^2\cos^2 x - \sin^2 x}{x^2\sin^2 x} = \frac{x\cos x + \sin x}{x}\frac{x\cos x - \sin x}{x\sin^2 x}$$

with

$$\frac{x\cos x + \sin x}{x} = \cos x + \frac{\sin x}{x} \to 2 \quad \text{as } x \to 0$$

the second assertion follows from the first. Finally, putting $y = f(x)$, where

$$y = \left(\frac{\sin x}{x}\right)^{1/(1-\cos x)},$$

we have

$$\ln y = \frac{\ln(\sin x) - \ln x}{1 - \cos x} \quad \text{for } x > 0.$$

The assumption $x > 0$ causes no restriction because $f(-x) = f(x)$. Using Proposition 4.10, we obtain

$$\lim_{x\to 0}\ln y = \lim_{x\to 0}\frac{(\cos x)/(\sin x) - 1/x}{\sin x} = \lim_{x\to 0}\frac{x\cos x - \sin x}{x\sin^2 x} = -\frac{1}{3},$$

establishing the third assertion in terms of the first.

16. Find a value of the fixed number c such that

$$\lim_{x\to\infty}\left(\frac{x+c}{x-c}\right)^x = 4.$$

We put

$$y = \left(\frac{x+c}{x-c}\right)^x \quad \text{or} \quad \ln y = \frac{\ln(x+c) - \ln(x-c)}{1/x}$$

and apply Proposition 4.10; we get

$$\lim_{x \to \infty} \ln y = \lim_{x \to \infty} \frac{\ln(x + c) - \ln(x - c)}{1/x} = \lim_{x \to \infty} \frac{1/(x + c) - 1/(x - c)}{-1/x^2}$$

$$= \lim_{x \to \infty} \frac{2cx^2}{x^2 - c^2} = 2c.$$

We therefore see that $2c = 2(\ln 2)$ or $c = \ln 2$.

17. Let a and b denote fixed numbers and suppose that $a > 0$. Then

$$\lim_{x \to \infty} a^x \cdot \sin \frac{b}{a^x} = b \quad \text{if } a > 1$$

$$= 0 \quad \text{if } 0 < a < 1$$

Indeed, let $a > 1$; then $a^x \to \infty$ as $x \to \infty$ and so

$$a^x \cdot \sin \frac{b}{a^x} = b \cdot \frac{\sin(b/a^x)}{b/a^x} \to b \quad \text{as } x \to \infty.$$

If $0 < a < 1$, then $a^x \to 0$ as $x \to \infty$; moreover, $|\sin t| \leq 1$ for any real number t.

18. We have

$$\lim_{x \to 0} \left(\frac{1}{\sin^2 x} - \frac{1}{x^2} \right) = \frac{1}{3}.$$

Indeed, we have

$$\frac{1}{\sin^2 x} - \frac{1}{x^2} = \frac{x^2 - \sin^2 x}{x^2 \sin^2 x} = \frac{x^2 - \sin^2 x}{x^4} \cdot \frac{x^2}{\sin^2 x}.$$

Applying Proposition 4.10 four times in succession yields

$$\lim_{x \to 0} \frac{x^2 - \sin^2 x}{x^4} = \frac{1}{3}.$$

On the other hand, $x^2/(\sin^2 x) \to 1$ as $x \to 0$.

2. Taylor's Theorem

Proposition 4.11 (Taylor's Theorem). *Suppose that f is a real-valued function on a closed interval $[a, b]$ of finite length and n denotes a nonnegative integer. Let f and its first n derivatives be continuous on $[a, b]$ and let the $(n + 1)$st derivative $f^{(n+1)}$ exist (i.e., $f^{(n)}$ be differentiable) on the open interval (a, b). Assume, moreover, that α and β are distinct points of $[a, b]$ and put*

$$P(t) = f(\alpha) + \frac{f'(\alpha)}{1!}(t - \alpha) + \cdots + \frac{f^{(n)}(\alpha)}{n!}(t - \alpha)^n. \tag{4.8}$$

Then there exists a point x between α and β such that

$$f(\beta) = P(\beta) + \frac{f^{(n+1)}(x)}{n!q}(\beta - \alpha)^q(\beta - x)^{n-q+1}, \qquad (4.9)$$

where q is a fixed number ≥ 1.

PROOF. Let K be a number defined by

$$f(\beta) = P(\beta) + K(\beta - \alpha)^q \qquad (4.10)$$

and put, for t in $[a, b]$,

$$g(t) = f(t) + \frac{f'(t)}{1!}(\beta - t) + \cdots + \frac{f^{(n)}(t)}{n!}(\beta - t)^n + K(\beta - t)^q,$$

where β is held fixed. It is clear that $g(\beta) = f(\beta)$; by (4.10) we have that $g(\alpha) = f(\beta)$. Thus, by Proposition 4.3, $g'(x) = 0$ for some x between α and β. But

$$g'(t) = f'(t) + \left(\frac{f''(t)}{1!}(\beta - t) - f'(t)\right)$$

$$+ \left(\frac{f'''(t)}{2!}(\beta - t)^2 - \frac{f''(t)}{1!}(\beta - t)\right)$$

$$+ \left(\frac{f^{(4)}(t)}{3!}(\beta - t)^3 - \frac{f'''(t)}{2!}(\beta - t)^2\right) + \cdots$$

$$+ \left(\frac{f^{(n+1)}(t)}{n!}(\beta - t)^n - \frac{f^{(n)}(t)}{(n-1)!}(\beta - t)^{n-1}\right) - qK(\beta - t)^{q-1}$$

and (4.9) follows.   □

REMARKS. The polynomial P in (4.8) is called the *Taylor polynomial of order n for f at α*. The expression

$$R_n = \frac{f^{(n+1)}(x)}{n!q}(\beta - \alpha)^q(\beta - x)^{n-q+1} \qquad (4.11)$$

that appears in (4.9) is called *Schlömilch's form of the remainder*. If we let $q = n + 1$ in (4.11), then we obtain

$$R_n = \frac{f^{(n+1)}(x)}{(n+1)!}(\beta - \alpha)^{n+1}$$

and we call it *Lagrange's form of the remainder* Putting $q = 1$ in (4.11) yields

$$R_n = \frac{f^{(n+1)}(x)}{n!}(\beta - \alpha)(\beta - x)^n$$

and we call it *Cauchy's form of the remainder*.

For $n = 0$ and using Lagrange's form of the remainder, Proposition 4.11 reduces to Proposition 4.5. In general, Proposition 4.11 shows that f may be approximated by a polynomial of degree n; and (4.9) allows us to estimate the error of the approximation, if we know bounds on $|f^{(n+1)}(x)|$.

In Example 8 following Proposition 4.10 we studied the function

$$F(0) = 0, \qquad F(t) = e^{-1/t^2} \quad \text{for } t \neq 0.$$

It is noteworthy to point out that the Taylor polynomials of all orders for F at 0 vanish because the derivatives of all orders of F are zero at $t = 0$. In the case of this function Proposition 4.11 does not yield an approximating polynomial (except for the trivial zero polynomial) and the function F is merely reproduced in the remainder. However, in the case of some other functions, Proposition 4.11 provides very useful information indeed. We consider some examples next.

Exponential Function. Let $f(t) = e^t$. Then $f^{(k)}(t) = e^t$ for $k = 1, 2, \dots$. Taking $\alpha = 0$ and choosing for R_n Lagrange's form of the remainder, we get

$$e^\beta = 1 + \frac{\beta}{1!} + \frac{\beta^2}{2!} + \cdots + \frac{\beta^n}{n!} + R_n(\beta),$$

where

$$R_n(\beta) = e^x \frac{\beta^{n+1}}{(n+1)!}$$

with x between 0 and β. We note that

$$|R_n(\beta)| < \frac{|\beta|^{n+1} e^{|\beta|}}{(n+1)!}$$

and, for fixed β,

$$\lim_{n \to \infty} R_n(\beta) = 0. \tag{4.12}$$

To verify (4.12), let m be a positive integer $> 2|\beta|$. Then for $n \geq m$ we have

$$\frac{|\beta|^{n+1} e^{|\beta|}}{(n+1)!} = e^{|\beta|} \cdot \frac{|\beta|^m}{m!} \cdot \left(\frac{|\beta|}{m+1}\right) \cdot \left(\frac{|\beta|}{m+2}\right) \cdots \left(\frac{|\beta|}{n+1}\right) < e^{|\beta|} \cdot \frac{|\beta|^m}{m!} \cdot \left(\frac{1}{2}\right)^{n-m+1}.$$

Thus, for $n \geq m$,

$$|R_n(\beta)| < M \cdot \left(\frac{1}{2}\right)^n,$$

where M is the fixed number

$$M = 2^{m-1} \cdot e^{|\beta|} \cdot \frac{|\beta|^m}{m!}.$$

But $(1/2)^n \to 0$ as $n \to \infty$ and so $M \cdot (1/2)^n \to 0$ as $n \to \infty$ with M a fixed number. Therefore, for fixed β we can calculate e^β by use of the relation

$$e^\beta = 1 + \frac{\beta}{1!} + \frac{\beta^2}{2!} + \cdots + \frac{\beta^n}{n!} + \frac{\beta^{n+1}}{(n+1)!} e^{\theta\beta}, \quad \text{where } 0 < \theta < 1 \quad (4.13)$$

with any degree of accuracy. In particular, for $\beta = 1$ we get from (4.13) that

$$e = 1 + \frac{1}{1!} + \frac{1}{2!} + \cdots + \frac{1}{n!} + R_n(1), \quad \text{where } 0 < R_n(1) < \frac{3}{(n+1)!}. \quad (4.14)$$

Taking $n = 13$ in (4.14) we see that $e = 2.718281828\ldots$, accurate to nine decimal places.

We observe that e is not a rational number, that is, e is not representable as the ratio of two integers. Indeed, from (4.13) we have

$$e = 1 + \frac{1}{1!} + \frac{1}{2!} + \cdots + \frac{1}{n!} + \frac{e^\theta}{(n+1)!}, \quad \text{where } 0 < \theta < 1$$

for any positive integer n; the number θ depends on n. Suppose now that $e = p/q$, where p and q denote positive integers. We only need to convince ourselves that

$$\frac{p}{q} = 1 + \frac{1}{1!} + \frac{1}{2!} + \cdots + \frac{1}{n!} + \frac{e^\theta}{(n+1)!} \quad (4.15)$$

can not be true if $n \geq q$ and ≥ 3 and $0 < \theta < 1$. Now (4.15) is equivalent to

$$n! \frac{p}{q} - n! \left(1 + \frac{1}{1!} + \frac{1}{2!} + \cdots + \frac{1}{n!} \right) = \frac{e^\theta}{n+1}. \quad (4.16)$$

But (4.16) can not be true because on the left-hand side of the equation we have an integer while on the right-hand side we have a quantity larger than 0 but less than $3/(n + 1)$ which in turn is less than 1 because $n \geq 3$ by assumption; however, there is no integer between 0 and 1.

Sine and Cosine Functions. Let $f(t) = \sin t$. If m is a positive integer, then

$$f^{(2m-1)}(t) = (-1)^{m-1}\cos t \quad \text{and} \quad f^{(2m)}(t) = (-1)^m \sin t.$$

Taking $\alpha = 0$ and choosing for R_n Lagrange's form of the remainder, we obtain upon setting $n = 2m$

$$\sin \beta = \beta - \frac{\beta^3}{3!} + \frac{\beta^5}{5!} - \frac{\beta^7}{7!} + \cdots + (-1)^{m-1} \frac{\beta^{2m-1}}{(2m-1)!} + R_n(\beta), \quad (4.17)$$

where

$$R_n(\beta) = (-1)^m \frac{\beta^{2m+1}}{(2m+1)!} \cos x$$

with x being some point between 0 and β. It is clear that

$$|R_n(\beta)| \leq \frac{|\beta|^{2m+1}}{(2m+1)!}$$

and so, just as in the case of the exponential function, $R_n(\beta) \to 0$ as $n \to \infty$ for any fixed β. Thus, $\sin \beta$ can be calculated by (4.17) with any degree of accuracy we please for any fixed β.

In a completely similar way we obtain

$$\cos \beta = 1 - \frac{\beta^2}{2!} + \frac{\beta^4}{4!} - \frac{\beta^6}{6!} + \cdots + (-1)^{m-1} \frac{\beta^{2m-2}}{(2m-2)!} + R_n(\beta), \quad (4.18)$$

where

$$R_n(\beta) = (-1)^m \frac{\beta^{2m}}{(2m)!} \cos v$$

with v being some point situated between 0 and β. Since

$$|R_n(\beta)| \leq \frac{|\beta|^{2m}}{(2m)!},$$

it follows that $R_n(\beta) \to 0$ as $n \to \infty$ for fixed β. Hence, (4.18) can be used to compute $\cos \beta$ with any degree of accuracy we please.

Logarithmic Function. Let $f(t) = \ln(1 + t)$. Then, for $k = 1, 2, \ldots$,

$$f^{(k)}(t) = (-1)^{k-1} \frac{(k-1)!}{(1+t)^k}.$$

We have $f(0) = 0$ and

$$\frac{1}{k!} f^{(k)}(0) = \frac{(-1)^{k-1}}{k} \quad \text{for } k = 1, 2, \ldots.$$

Taking $\alpha = 0$ and choosing for R_n Lagrange's form of the remainder, we get

$$\ln(1 + \beta) = \beta - \frac{\beta^2}{2} + \frac{\beta^3}{3} - \frac{\beta^4}{4} + \cdots + (-1)^{n-1} \frac{\beta^n}{n} + R_n(\beta), \quad (4.19)$$

where

$$R_n(\beta) = (-1)^n \frac{\beta^{n+1}}{n+1} \frac{1}{(1+x)^{n+1}}$$

with x being a point between 0 and β. We can write R_n also in the form

$$R_n(\beta) = (-1)^n \frac{\beta^{n+1}}{n+1} \frac{1}{(1+\theta\beta)^{n+1}},$$

where θ denotes some number between 0 and 1. For fixed β such that $0 \leq \beta \leq 1$ we have

$$|R_n(\beta)| \le \frac{1}{n+1} \to 0 \quad \text{as } n \to \infty$$

and so we see that $\ln(1 + \beta)$ can be computed with any prescribed degree of accuracy by use of (4.19) provided that β satisfies $0 \le \beta \le 1$.

Since for positive values of β the expression

$$\frac{1}{(1 + \theta\beta)^{n+1}}$$

as well as θ are between 0 and 1, we can write in place of (4.19)

$$\ln(1 + \beta) = \beta - \frac{\beta^2}{2} + \frac{\beta^3}{3} - \frac{\beta^4}{4} + \cdots + (-1)^{n-1}\frac{\beta^n}{n} + (-1)^n\theta_1\frac{\beta^{n+1}}{n+1}, \quad (4.20)$$

where $0 \le \beta \le 1$ and $0 < \theta_1 < 1$.

We now wish to investigate what happens when $-1 < \beta \le 0$. In place of Lagrange's form of the remainder, we now consider Cauchy's form of the remainder and get

$$\ln(1 + \beta) = \beta - \frac{\beta^2}{2} + \frac{\beta^3}{3} - \frac{\beta^4}{4} + \cdots + (-1)^{n-1}\frac{\beta^n}{n} + R_n(\beta), \quad (4.21)$$

where

$$R_n(\beta) = (-1)^n\frac{\beta^{n+1}}{1 + \omega\beta}\left(\frac{1 - \omega}{1 + \omega\beta}\right)^n,$$

where $0 < \omega < 1$ and $-1 < \beta \le 0$ (note that we have replaced x by $\omega\beta$ in the formula for Cauchy's form of the remainder). This form of the remainder makes it possible to see that $R_n(\beta)$ tends to 0 as $n \to \infty$ when β satisfies $-1 < \beta \le 0$. Indeed, if $-1 < \beta \le 0$, then (since $0 < \omega < 1$)

$$1 - \omega < 1 + \omega\beta$$

and so, for any $n \ge 0$,

$$0 < \left(\frac{1 - \omega}{1 + \omega\beta}\right)^n \le 1.$$

Moreover, if $-1 < \beta \le 0$ and $0 < \omega < 1$, then

$$0 < \frac{1}{1 + \omega\beta} \le \frac{1}{1 + \beta}.$$

Hence, we can write in place of (4.21)

$$\ln(1 + \beta) = \beta - \frac{\beta^2}{2} + \frac{\beta^3}{3} - \frac{\beta^4}{4} + \cdots + (-1)^{n-1}\frac{\beta^n}{n} + (-1)^n\theta_2\frac{\beta^{n+1}}{1 + \beta}, \quad (4.22)$$

where $-1 < \beta \le 0$ and $0 < \theta_2 < 1$, and it is also clear that

$$R_n(\beta) = (-1)^n\theta_2\frac{\beta^{n+1}}{1 + \beta}$$

tends to 0 as $n \to \infty$. Thus, (4.22) can be used to calculate $\ln(1 + \beta)$ when β satisfies $-1 < \beta \le 0$ with any prescribed degree of accuracy. Of course, we do not need such powerful methods for the trivial case $\beta = 0$. Since

$$\ln a = -\ln\frac{1}{a},$$

we can use (4.22) to compute $\ln a$ in the case $a > 0$; however, such a direct approach is often not advantageous.

Replacing β by $-\beta$ in (4.22) and multiplying both sides by -1, we obtain

$$\ln\frac{1}{1 - \beta} = \beta + \frac{\beta^2}{2} + \frac{\beta^3}{3} + \cdots + \frac{\beta^n}{n} + \theta_2\frac{\beta^{n+1}}{1 - \beta}, \qquad (4.23)$$

where $0 \le \beta < 1$ and $0 < \theta_2 < 1$.

Since

$$\frac{1 + \beta}{1 - \beta} = \ln(1 + \beta) + \ln\frac{1}{1 - \beta},$$

we can combine the results in (4.20) and (4.23); picking n to be the even number $2k$, we have

$$\frac{1 + \beta}{1 - \beta} = 2\left(\beta + \frac{\beta^3}{3} + \cdots + \frac{\beta^{2k-1}}{2k - 1}\right) + \beta^{2k+1}\left(\frac{\theta_1}{2k + 1} + \frac{\theta_2}{1 - \beta}\right), \quad (4.24)$$

where $0 \le \beta < 1, 0 < \theta_1 < 1$, and $0 < \theta_2 < 1$. The relations (4.20), (4.22), and (4.24) are basic for the computation of logarithms.

Putting $\beta = 2$ in (4.20), we obtain

$$\ln 2 = 1 - \frac{1}{2} + \frac{1}{3} - \frac{1}{4} + \cdots + \frac{(-1)^{n-1}}{n} + \frac{(-1)^n}{n + 1}\theta_1, \qquad (4.25)$$

where $0 < \theta_1 < 1$. While (4.25) is of considerable theoretical interest, it is not very helpful in computing $\ln 2$ accurate to, say, six decimal places because we would have to let $n = 10^6$ or more and the resulting calculation would be formidable indeed.

The relation (4.25), in contrast, is much more helpful in the computation of $\ln 2$. To get

$$\frac{1 + \beta}{1 - \beta} = 2$$

we merely need to take $\beta = \frac{1}{3}$; doing so, we get

$$\ln 2 = 2\left(\frac{1}{3} + \frac{1}{3 \cdot 3^3} + \frac{1}{5 \cdot 3^5} + \cdots + \frac{1}{(2k - 1)3^{2k-1}}\right) + \frac{1}{3^{2k+1}}\left(\frac{\theta_1}{2k + 1} + \frac{3}{2}\theta_2\right),$$
$$(4.26)$$

where $0 < \theta_1 < 1$ and $0 < \theta_2 < 1$. For example, it is easy to see that we get an accuracy of up to eight decimal places by taking $k = 9$ in (4.26).

Suppose we already calculated $\ln m$, where m is a positive integer, and we wish to calculate $\ln(m + 1)$ next. Since

$$\ln(m + 1) = \ln m + \ln\frac{m + 1}{m},$$

it is of interest to calculate $\ln\{(m + 1)/m\}$ as efficiently as possible. By setting

$$\beta = \frac{1}{2m + 1},$$

we obtain

$$\frac{1 + \beta}{1 - \beta} = \frac{m + 1}{m};$$

therefore, relation (4.25) will aid us in the computation of $\ln\{(m + 1)/m\}$ and will in fact become more convenient as m gets larger. Incidentally, note that calculating, for example, $\ln 3$ by use of the identity

$$\ln 3 = \ln 2 + \ln\tfrac{3}{2}$$

with $\beta = \tfrac{1}{5}$ in (4.25) is preferable to setting $\beta = \tfrac{1}{2}$ in (4.25) and then working with

$$\ln 3 = 2\left(\frac{1}{2} + \frac{1}{3\cdot 2^3} + \frac{1}{5\cdot 2^5} + \cdots + \frac{1}{(2k - 1)2^{2k-1}}\right) + \frac{1}{2^{2k+1}}\left(\frac{\theta_1}{2k + 1} + 2\theta_2\right).$$

J. C. Adams (in *Proceedings of the Royal Society of London*, vol. 27, 1878, p. 88 ff.) used the identities

$$\ln 2 = 7\ln\frac{10}{9} - 2\ln\frac{25}{24} + 3\ln\frac{81}{80},$$

$$\ln 3 = 11\ln\frac{10}{9} - 3\ln\frac{25}{24} + 5\ln\frac{81}{80},$$

$$\ln 5 = 16\ln\frac{10}{9} - 4\ln\frac{25}{24} + 7\ln\frac{81}{80},$$

$$\ln 7 = \frac{1}{2}\left(39\ln\frac{10}{9} - 10\ln\frac{25}{24} + 17\ln\frac{81}{80} - \ln\frac{50}{49}\right),$$

and

$$\ln 7 = 19\ln\frac{10}{9} - 4\ln\frac{25}{24} + 8\ln\frac{81}{80} + \ln\frac{126}{125}$$

to calculate $\ln 2$, $\ln 3$, $\ln 5$, and $\ln 7$ with an accuracy of up to 262 decimal places. Note that

$$\ln\frac{10}{9} = -\ln\left(1 - \frac{1}{10}\right), \quad \ln\frac{25}{24} = -\ln\left(1 - \frac{4}{100}\right), \quad \ln\frac{81}{80} = \ln\left(1 + \frac{1}{80}\right),$$

$$\ln\frac{50}{49} = -\ln\left(1 - \frac{2}{100}\right), \quad \text{and} \quad \ln\frac{126}{125} = \ln\left(1 + \frac{8}{1000}\right);$$

moreover, the numbers

$$\frac{1}{10}, \quad \frac{4}{100}, \quad \frac{1}{80}, \quad \frac{2}{100}, \quad \frac{8}{1000}$$

together with their integer powers are easily calculated to many decimal places.

Once we have $\ln 2$ and $\ln 5$ computed with high accuracy, we know $\ln 10$ with high accuracy as well because

$$\ln 10 = \ln 2 + \ln 5.$$

The relation between logarithms to the base 10 and natural logarithms is given by

$$\log_{10} N = \frac{\ln N}{\ln 10}$$

and therefore an accurate calculation of $M = \ln 10$ is of interest. We note that

$$M = 2.3025850929940\ldots \quad \text{and} \quad \frac{1}{M} = 0.4342944819032\ldots.$$

Binomial Function. Let c be any real number and put

$$\binom{c}{0} = 1 \quad \text{and} \quad \binom{c}{k} = \frac{c(c-1)(c-2)\ldots(c-n+1)}{k!}$$

for $k = 1, 2, \ldots$. For $x > -1$, let $f(t) = (1+t)^c$; then

$$\frac{1}{k!} f^{(k)}(t) = \binom{c}{k}(1+t)^{c-k}, \quad \text{that is,} \quad \frac{1}{k!} f^{(k)}(0) = \binom{c}{k},$$

for $k = 0, 1, 2, \ldots$. Taking $\alpha = 0$ and choosing for R_n Lagrange's form of the remainder (for the point x between 0 and β we put $\theta\beta$ with θ satisfying the inequality $0 < \theta < 1$), we obtain

$$(1+\beta)^c = 1 + \binom{c}{1}\beta + \binom{c}{2}\beta^2 + \cdots + \binom{c}{n}\beta^n + R_n(\beta), \qquad (4.27)$$

where

$$R_n(\beta) = \binom{c}{n+1}\beta^{n+1}(1+\theta\beta)^{c-n-1}$$

with θ dependent on n and β and satisfying the inequality $0 < \theta < 1$.

It is clear that (4.27) is an extension of the Binomial Theorem and turns into the Binomial Theorem if c is a positive integer p and $n \geq p$.

If $\beta > 0$ (the case $\beta = 0$ is trivial) and $n + 1 > c$, then

$$0 < (1+\theta\beta)^{c-n-1} < 1$$

and we can replace (4.27) by the simpler representation

$$(1 + \beta)^c = 1 + \binom{c}{1}\beta + \binom{c}{2}\beta^2 + \cdots + \binom{c}{n}\beta^n + \binom{c}{n+1}\theta_1\beta^{n+1}, \quad (4.28)$$

where c is arbitrary, $\beta > 0$, $n + 1 > c$, and $0 < \theta_1 < 1$.

Let c be fixed and not a positive integer and suppose that β is fixed and $|\beta| < 1$. We wish to convince ourselves next that the numbers

$$y_n = \binom{c}{n}\beta^n \to 0 \quad \text{as } n \to \infty. \quad (4.29)$$

Indeed,

$$\frac{y_{n+1}}{y_n} = \frac{c - n}{n + 1}\beta$$

and so

$$\frac{y_{n+1}}{y_n} \to -\beta \quad \text{as } n \to \infty, \quad (4.30)$$

that is, the absolute value of the limit is <1. Hence, if v is any fixed number satisfying $|\beta| < v < 1$, we can find a positive integer m so that

$$\left|\frac{y_{k+1}}{y_k}\right| < v \quad \text{for } k \geq m. \quad (4.31)$$

We now envisage the inequality (4.31) written out for $k = m, m + 1, \ldots, n - 1$. Multiplying all these inequalities together, we get

$$|y_n| < \frac{|y_m|}{v^m}v^n \quad \text{for } n > m.$$

But m is fixed, hence, $|y_m|/v^m$ is fixed and so $v^n \to 0$ as $n \to \infty$ implying

$$y_n \to 0 \quad \text{as } n \to \infty.$$

In the sequel we shall need to know also the stronger result that

$$z_n = ny_n \to 0 \quad \text{as } n \to \infty \quad (4.32)$$

in case $|\beta| < 1$. But

$$\frac{z_{n+1}}{z_n} = \frac{n + 1}{n}\frac{c - n}{n + 1}\beta \to -\beta$$

and we can use for the numbers z_n the same reasoning as we employed in the case of numbers y_n on the basis of the relation (4.31).

Thus, for fixed β satisfying $0 \leq \beta < 1$ we can certainly claim that the remainder term $R_n(\beta)$ in (4.27) tends to zero as n becomes arbitrarily large. What happens to $R_n(\beta)$ as $n \to \infty$ in case $-1 < \beta < 0$ is hard to make out from (4.27) or (4.28). However, using Cauchy's form of the remainder instead of Lagrange's form, it is not difficult to see that $R_n(\beta) \to 0$ as $n \to \infty$ in case β is fixed and satisfies $-1 < \beta < 0$.

Indeed, using Cauchy's form of the remainder we get the representation

$$(1 + \beta)^c = 1 + \binom{c}{1}\beta + \binom{c}{1}\beta^2 + \cdots + \binom{c}{n}\beta^n$$

$$+ (n + 1)\binom{c}{n + 1}\beta^{n+1}(1 - \omega)^n(1 + \omega\beta)^{c-n-1}, \tag{4.33}$$

where $-1 < \beta \le 0$ and $0 < \omega < 1$ (note that we have replaced x by $\omega\beta$ in Cauchy's form of the remainder). Just as in the foregoing study of the logarithmic function, we note that $-1 < \beta \le 0$ and $0 < \omega < 1$ implies

$$1 - \omega < 1 + \omega\beta$$

and therefore, for any $n \ge 0$, the number

$$\theta_2 = \left(\frac{1 - \omega}{1 + \omega\beta}\right)^n$$

satisfies

$$0 < \theta_2 \le 1.$$

We can therefore rewrite (4.33) in a somewhat simpler form as follows:

$$(1 + \beta)^c = 1 + \binom{c}{1}\beta + \binom{c}{2}\beta^2 + \cdots + \binom{c}{n}\beta^n$$

$$+ (n + 1)\binom{c}{n + 1}\beta^{n+1}\theta_2(1 + \omega\beta)^{c-1}, \tag{4.34}$$

where c is arbitrary, $-1 < \beta \le 0$, $0 < \theta_2 \le 1$, and $0 < \omega < 1$. By (4.32), we have

$$z_n = ny_n = n\binom{c}{n}\beta^n \to 0 \quad \text{as } n \to \infty$$

for fixed β satisfying $-1 < \beta \le 0$; also, for fixed β such that $-1 < \beta \le 0$ we have

$$|\theta_2(1 + \omega\beta)^{c-1}| \le 1, \quad \text{respectively} \le (1 + \beta)^{c-1},$$

depending on whether $c - 1 \ge 0$ or ≤ 0. In any event, the factors joining z_n are bounded and so $R_n(\beta) \to 0$ as $n \to \infty$ also in case the fixed β satisfies $-1 < \beta \le 0$.

To sum up: For any β satisfying $-1 < \beta < 1$ we have a representation of the form

$$(1 + \beta)^c = 1 + \binom{c}{1}\beta + \binom{c}{2}\beta^2 + \cdots + \binom{c}{n}\beta^n + R_n(\beta), \tag{4.35}$$

where $R_n(\beta) \to 0$ when we fix β and let $n \to \infty$. Moreover, if $n + 1 > c$, we have

$$R_n(\beta) = \binom{c}{n+1}\theta_1\beta^{n+1} \quad \text{for } 0 \le \beta < 1 \tag{4.36}$$

and

$$R_n(\beta) = (n+1)\binom{c}{n+1}\beta^{n+1}\theta_2(1+\omega\beta)^{c-1} \quad \text{for } -1 < \beta \le 0. \tag{4.37}$$

The numbers θ_1, θ_2, and ω are larger than 0 but do not exceed 1; nothing more definite is known about them.

We can use the foregoing result to come up with high-precision calculations. For example,

$$\sqrt{2} = 1.4\left(1 - \frac{1}{50}\right)^{-1/2}$$

$$= 1.4\left(1 + \frac{1}{2}\frac{1}{50} + \frac{3}{8}\frac{1}{50^2} + \frac{5}{16}\frac{1}{50^3} + \frac{35}{128}\frac{1}{50^4} + \frac{63}{256}\frac{1}{50^5} + \cdots\right).$$

But

$$1 + \cdots + \frac{5}{16}\frac{1}{50^3} = 1.0101525$$

$$\frac{35}{128}\frac{1}{50^4} = 0.00000004375$$

$$\frac{63}{256}\frac{1}{50^5} = \frac{0.0000000007875}{1.0101525445375}$$

However,

$$1.0101525445375 \cdot 1.4 = 1.41421356235250.$$

We note that the error in this approximation is smaller than

$$1.4 \cdot 6\binom{-1/2}{6}\frac{1}{50^6} = 1.4 \cdot 6 \cdot \frac{231}{1024}\frac{1}{50^6}$$

and the latter is smaller than $1.22 \cdot 10^{-10}$. Therefore,

$$\sqrt{2} = 1.414213562\ldots$$

with an accuracy of up to nine decimal places.

Our accuracy would be even higher if we had used the identity

$$\sqrt{2} = 1.41\left(1 - \frac{119}{20000}\right)^{-1/2}.$$

Inverse Tangent Function. Let $y = \arctan t$. We seek to express $y^{(n)}$ in terms of y. Since $t = \tan y$, we have

$$y' = \frac{1}{1 + t^2} = \cos^2 y = (\cos y)\sin\left(y + \frac{\pi}{2}\right).$$

Another differentiation yields

$$y'' = \left(-(\sin y)\sin\left(y + \frac{\pi}{2}\right) + (\cos y)\cos\left(y + \frac{\pi}{2}\right)\right)y'$$

$$= (\cos^2 y)\cos\left(2y + \frac{\pi}{2}\right) = (\cos^2 y)\sin 2\left(y + \frac{\pi}{2}\right).$$

Differentiating again, we get

$$y''' = \left(-2(\sin y)(\cos y)\sin 2\left(y + \frac{\pi}{2}\right) + 2(\cos^2 y)\cos 2\left(y + \frac{\pi}{2}\right)\right)y'$$

$$= 2(\cos^3 y)\cos\left(3y + 2\cdot\frac{\pi}{2}\right) = 2(\cos^3 y)\sin 3\left(y + \frac{\pi}{2}\right).$$

The general formula is

$$y^{(n)} = (n - 1)!(\cos^n y)\sin n\left(y + \frac{\pi}{2}\right) \tag{4.38}$$

and can be verified by induction (see also Exercise 3.18 at the end of Chapter 3).

Putting $f(t) = \arctan t$, we see that

$$f(0) = 0, \quad \frac{f'(0)}{1!} = 1, \quad \frac{f''(0)}{2!} = 0, \quad \frac{f'''(0)}{3!} = -\frac{1}{3}, \quad \frac{f^{(4)}(0)}{4!} = 0,$$

$$\frac{f^{(5)}(0)}{5!} = \frac{1}{5}, \quad \frac{f^{(6)}(0)}{6!} = 0, \quad \frac{f^{(7)}(0)}{7!} = -\frac{1}{7}, \dots.$$

Letting $\alpha = 0$, substituting for x the expression $\theta\beta$, and taking n to be the even integer $2k$, we have

$$\arctan \beta = \beta - \frac{\beta^3}{3} + \frac{\beta^5}{5} - \frac{\beta^7}{7} + \cdots + (-1)^{k-1}\frac{\beta^{2k-1}}{2k - 1} + R_n(\beta), \tag{4.39}$$

where

$$R_n(\beta) = (-1)^k\frac{\beta^{2k+1}}{2k + 1}(\cos^{2k+1}\theta\beta)\sin(2k + 1)\left(\theta\beta + \frac{\pi}{2}\right)$$

with $|\beta| \leq 1$ and $0 < \theta < 1$. It is clear that

$$|R_n(\beta)| \leq \frac{1}{n + 1} \to 0 \quad \text{as } n \to \infty.$$

The relation (4.39) was used around 1706 by John Machin to calculate the number π with an accuracy of up to 100 decimal places; W. Shanks in 1873 extended Machin's result and computed π with an accuracy of up to 707 decimal places (see *Proceedings of the Royal Society of London*, vol. 21, 1873, p. 318; corrections in vol. 22, 1874, p. 45). Machin's method makes use of the formula

$$\frac{\pi}{4} = 4 \arctan \frac{1}{5} - \arctan \frac{1}{239}$$

and goes as follows. Let $A = \arctan \frac{1}{5}$. Then

$$\tan A = \frac{1}{5}, \qquad \tan 2A = \frac{2/5}{1 - 1/25} = \frac{5}{12}, \qquad \tan 4A = \frac{10/12}{1 - 25/144} = \frac{120}{119}.$$

Since the number 120/119 is near 1, the angle $4A$ is near $\pi/4$. Putting

$$B = 4A - \frac{\pi}{4},$$

we obtain

$$\tan B = \frac{120/119 - 1}{1 + 120/119} = \frac{1}{239}, \quad \text{that is, } B = \arctan \frac{1}{239}.$$

We therefore have *Machin's Formula*

$$\pi = 16A - 4B = 16\left(\frac{1}{5} - \frac{1}{3} \cdot \frac{1}{5^3} + \frac{1}{5} \cdot \frac{1}{5^5} + \frac{1}{7} \cdot \frac{1}{5^7} + \frac{1}{9} \cdot \frac{1}{5^9} + \frac{1}{11} \cdot \frac{1}{5^{11}} + \cdots\right)$$

$$- 4\left(\frac{1}{239} - \frac{1}{3} \cdot \frac{1}{239^3} + \cdots\right).$$

Consideration of (4.39) shows that calculation of the listed terms will suffice to obtain π with an accuracy of up to seven decimal places:

$$\pi = 3.1415926\ldots.$$

REMARKS. A less effective formula than $\pi/4 = 4(\tan^{-1} \frac{1}{5}) - (\tan^{-1} \frac{1}{239})$ is the formula $\pi/4 = \tan^{-1} \frac{1}{2} + \tan^{-1} \frac{1}{3}$. Gauss found, by means of the theory of numbers, two remarkable formulas, namely,

$$\frac{\pi}{4} = 12\left(\tan^{-1} \frac{1}{18}\right) + 8\left(\tan^{-1} \frac{1}{57}\right) - 5\left(\tan^{-1} \frac{1}{239}\right),$$

$$= 12\left(\tan^{-1} \frac{1}{38}\right) + 20\left(\tan^{-1} \frac{1}{57}\right) + 7\left(\tan^{-1} \frac{1}{239}\right) + 24\left(\tan^{-1} \frac{1}{268}\right),$$

by means of which π could be calculated with great rapidity should its value be required to an accuracy beyond the one reached by W. Shanks. In recent times π was actually calculated with an accuracy of one million decimal places.

Proposition 4.12. *Suppose that f is a real-valued function and n denotes an integer ≥ 2. Let $f^{(n)}$ exist at a point c and, for $k = 1, 2, \ldots, n - 1$,*

$$f^{(k)}(c) = 0 \quad \text{and} \quad f^{(n)}(c) \neq 0.$$

Then f has a relative extremum at c if n is even. If n is even and $f^{(n)}(c) > 0$, then f has a relative minimum at c; if n is even and $f^{(n)}(c) < 0$, then f has a relative maximum at c.

PROOF. It will be sufficient to carry out the proof in the case $f^{(n)}(c) > 0$; the case $f^{(n)}(c) < 0$ is entirely analogous and can in fact be reduced to the considered case by replacing f by $-f$.

Since $f^{(n)}$ is assumed to exist at c, $f^{(n-1)}$ exists on a neighborhood of c. By Proposition 4.11, we have for sufficiently small $|h|$ and for suitable θ with $0 < \theta < 1$,

$$f(c + h) = f(c) + \frac{f'(c)}{1!} h + \frac{f''(c)}{2!} h^2 + \cdots + \frac{f^{(n-2)}(c)}{(n-2)!} h^{n-2}$$

$$+ \frac{f^{(n-1)}(c + \theta h)}{(n-1)!} h^{n-1} \tag{4.40}$$

$$= f(c) + \frac{f^{(n-1)}(c + h)}{(n-1)!} h^{n-1}.$$

But $f^{(n-1)}(c) = 0$ and $f^{(n)}(c) > 0$ by assumption and so

$$f^{(n-1)}(c + \theta h) < 0 \quad \text{for } h < 0$$
$$> 0 \quad \text{for } h > 0$$

(see Comments to Proposition 4.7). For odd n we have $h^{n-1} > 0$ and so

$$f^{(n-1)}(c + \theta h) < 0 \quad \text{for } h < 0$$
$$> 0 \quad \text{for } h > 0$$

and thus f can not have a relative extremum by (4.40). On the other hand, for even n the sign of h and the sign of h^{n-1} coincide and so, for even n,

$$f^{(n-1)}(c + \theta h) \cdot h^{n-1} > 0 \quad \text{for } h \neq 0.$$

Thus, (4.40) shows that f has a relative minimum at c. $\qquad \square$

REMARK. In the discussion following Proposition 4.10 we studied the function

$$F(0) = 0, \qquad F(t) = e^{-1/t^2} \quad \text{for } t \neq 0.$$

This function F has both a relative as well as an absolute minimum at $t = 0$; however, Proposition 4.12 is not applicable because $F^{(n)}(0) = 0$ for any integer $n \geq 0$. While Proposition 4.12 is very useful, its effectiveness is certainly less than universal.

Proposition 4.13. *If on an interval J the derivative $f^{(n+1)}$ of a function f exists and is equal to 0 for every t in J, then f is a polynomial of degree at most n (possibly the zero polynomial).*

PROOF. Let α be a point in the interior of J and θ satisfy $0 < \theta < 1$. For any point β in the interior of J we have, by Proposition 4.11,

$$f(\beta) = f(\alpha) + \frac{f'(\alpha)}{1!}(\beta - \alpha) + \frac{f''(\alpha)}{2!}(\beta - \alpha)^2 + \cdots + \frac{f^{(n)}(\alpha)}{n!}(\beta - \alpha)^n$$

$$+ \frac{f^{(n+1)}\{\alpha + \theta(\beta - \alpha)\}}{(n+1)!}(\beta - \alpha)^{n+1}$$

[we have chosen to write the point x between α and β in the form $\alpha + \theta(\beta - \alpha)$]. However, by assumption $f^{(n+1)}$ is zero on J and so

$$f(\beta) = f(\alpha) + \frac{f'(\alpha)}{1!}(\beta - \alpha) + \frac{f''(\alpha)}{2!}(\beta - \alpha)^2 + \cdots + \frac{f^{(n)}(\alpha)}{n!}(\beta - \alpha)^n. \quad (4.41)$$

This completes the proof.   □

REMARK. If f is a polynomial of degree n in the variable β and we wish to express $f(\beta)$ in terms of powers of $(\beta - \alpha)$ for some given number α, then we can use (4.41) to accomplish the given task. For example,

$$\beta^3 - 2\beta^2 + 3\beta + 5 = 11 + 7(\beta - 2) + 4(\beta - 2)^2 + (\beta - 2)^3$$

because, putting

$$f(\beta) = \beta^3 - 2\beta^2 + 3\beta + 5 \quad \text{and} \quad \alpha = 2,$$

we note that $f(2) = 11$, $f'(2) = 7$, $f''(2) = 8$, and $f'''(2) = 6$; the rest follows by completing the substitution into formula (4.41).

3. Concave Functions

Definition. Let J be an interval. A function f defined on J is said to be *concave up on J* if

$$f((1 - t)a + tb) \leq (1 - t)f(a) + tf(b) \quad (4.42)$$

whenever a and b are points of J and $0 \leq t \leq 1$. We call f *concave down on J* if $-f$ is concave up on J.

REMARKS. Putting $x = (1 - t)a + tb$, we see that inequality (4.42) is equivalent to

$$f(x) \le \frac{b-x}{b-a} f(a) + \frac{x-a}{b-a} f(b)$$

$$= f(a) + \frac{f(b) - f(a)}{b - a} (x - a),$$

(4.43)

whenever a and b are points of J and x is situated between a and b. Geometrically speaking, f is concave up on J if for any two points a and b of J the chord having endpoints $(a, f(a))$ and $(b, f(b))$ is never below the graph of f on the interval $[a, b]$. We can write in place of (4.43) the more symmetric inequality

$$(b - x)f(a) + (a - b)f(x) + (x - a)f(b) \ge 0. \qquad (4.44)$$

Proposition 4.14. *Let f be a function differentiable on an open interval J. The f is concave up on J if and only if f' is nondecreasing on J.*

PROOF. Suppose f is not concave up on J. Then there exist $a < x < b$ in J such that

$$f(x) > \frac{b-x}{b-a} f(a) + \frac{x-a}{b-a} f(b).$$

This inequality is equivalent to

$$\frac{f(x) - f(a)}{x - a} > \frac{f(b) - f(x)}{b - x}.$$

We now apply Proposition 4.5 to each of the closed intervals $[a, x]$ and $[x, b]$ and note that there exist points t and s, $a < t < x < s < b$, such that

$$f'(t) > f'(s).$$

Therefore, f' fails to be nondecreasing on J.

Conversely, if f is concave up on J, then $a < x < b$ with (a, b) in J means

$$\frac{f(x) - f(a)}{x - a} \le \frac{f(b) - f(x)}{b - x}.$$

But f' exists on J. If x tends to a or b, we get

$$f'(a) \le \frac{f(b) - f(a)}{b - a}, \qquad (4.45)$$

respectively

$$f'(b) \ge \frac{f(b) - f(a)}{b - a}, \qquad (4.46)$$

and so $f'(a) \le f'(b)$. But a and b are arbitrary points of J satisfying $a < b$ and so f' is seen to be nondecreasing on J. ☐

Proposition 4.15. *Let J be an open interval and f be a twice differentiable function on J. Then f is concave up on J if and only if $f''(x) \geq 0$ for all x in J.*

PROOF. The proposition is an immediate consequence of Proposition 4.14. □

Proposition 4.16. *Let f be differentiable on an open interval J. Then f is concave up on J if and only if the points of the tangent line at any point of J are never above the graph of f on J.*

PROOF. Let $(c, f(c))$ be a point of the graph of f on J. Then the equation of the tangent line at the point $(c, f(c))$ is

$$y = f(c) + f'(c)(x - c).$$

We assume that f is concave up and we wish to show that

$$f(x) \geq f(c) + f'(c)(x - c) \tag{4.47}$$

for any x of J. But the inequality (4.47) is equivalent to the two inequalities

$$f'(c) \leq \frac{f(x) - f(c)}{x - c} \quad \text{for } x > c \tag{4.48}$$

and

$$f'(c) \geq \frac{f(x) - f(c)}{x - c} \quad \text{for } x < c. \tag{4.49}$$

However, inequality (4.48) is merely inequality (4.45) with $a = c$ and $b = x$; inequality (4.49) is merely inequality (4.46) with $b = c$ and $a = x$. Therefore, the assumptions that f is differentiable on J and concave up on J imply that (4.48) and (4.49) hold.

Conversely, suppose that (4.47) is fulfilled or, equivalently, that (4.48) and (4.49) are satisfied. Setting $c = a$ and $x = b$, inequality (4.48) becomes inequality (4.45); setting $c = b$ and $x = a$, inequality (4.49) becomes inequality (4.46). But inequalities (4.45) and (4.46) together imply that f' is nondecreasing on J; Proposition 4.14 then ensures that f is concave up on J. □

REMARKS. Corresponding to the Propositions 4.14, 4.15, and 4.16 there are dual statements in terms of concave down functions.

Concavity is often used to decide whether an extremum is a maximum or a minimum. We consider an example.

Let $A > 0$, $B > 0$, and $p \neq 0$. Given that

$$f(x) = Ae^{px} + Be^{-px},$$

we wish to find the smallest value of f.

Since

$$f'(x) = Ape^{px} - Bpe^{-px} \quad \text{and} \quad f''(x) = p^2 f(x),$$

we can easily see that f is concave up on the entire number line as $f(x) > 0$ for any finite real number x and so $f''(x) > 0$ for any finite real number x. Now $f'(x) = 0$ gives

$$Ae^{px} = Be^{-px} \quad \text{or} \quad Ae^{2px} = B, \quad \text{implying } e^{px} = \sqrt{\frac{B}{A}} \quad \text{and} \quad e^{-px} = \sqrt{\frac{A}{B}}.$$

Therefore, the smallest value of f is

$$A\sqrt{\frac{B}{A}} + B\sqrt{\frac{A}{B}} = 2\sqrt{AB}.$$

Definition. Let f be a twice differentiable function on an open interval J. For c in J, the point $(c, f(c))$ on the graph of f is said to be a *point of inflection of* f if $f''(c) = 0$ and if

$$f''(c - h)f''(c + h) < 0$$

for all sufficiently small values of $h \neq 0$.

REMARKS. A point on a curve at which the curve changes from concave up to concave down or vice versa is a point of inflection. Since the tangent to a curve always lies opposite to the concave side of the curve (see Proposition 4.16), it follows that at a point of inflection the tangent crosses the curve.

A necessary and sufficient condition for f to have a point of inflection at $x = c$ is that $f''(c) = 0$ and $f'''(c) \neq 0$. In analogy to Proposition 4.12, if the first nonvanishing derivative at $x = c$ of order higher than the second is of odd order, then there is a point of inflection at $x = c$; if this derivative is of even order, then there is no point of inflection at $x = c$. For example, $f(x) = x^4$ has no point of inflection at $x = 0$, but $g(x) = x^5$ has a point of inflection at $x = 0$.

The function $h(x) = x^3$ has one point of inflection, namely, the point $(0, 0)$; moreover, the curve $y = x^3$ is symmetric with respect to this point because $h(-x) = -x^3$. More generally, we note that the cubic

$$y = ax^3 + bx^2 + cx + d \tag{4.50}$$

has one point of inflection, and that the curve is symmetric with respect to this point of inflection.

Indeed, $y'' = 6ax + 2b$ and so the point (A, B), where

$$A = -\frac{b}{3a} \quad \text{and} \quad B = \frac{2b^3}{27a^2} - \frac{cb}{3a} + d,$$

is the point of inflection. Using translation of axes, we set $X = x - A$ and $Y = y - B$. The substitution $x = X + A$ and $y = Y + B$ into (4.50) yields

$$Y = aX^3 + \left(c - \frac{b^2}{3a}\right)X. \tag{4.51}$$

Setting $Y = H(X)$ in (4.51), we note that $H(-X) = -H(X)$. This establishes the symmetry of the curve (4.50) with respect to the point (A, B).

4. Newton's Method for Approximating Real Roots of Functions

A technique generally called *Newton's Method* will enable us to find, to any desired degree of accuracy, the real roots of many equations of the form $f(x) = 0$. To fix ideas, we shall throughout this section suppose that the function f under discussion satisfies the following three conditions:

1. The function f and its derivatives f' and f'' are continuous on a closed interval $[a, b]$ of finite length.
2. The numbers $f(a)$ and $f(b)$ have opposite signs, that is, $f(a) \cdot f(b) < 0$.
3. The derivatives f' and f'' do not change signs on the interval $[a, b]$.

Since f is continuous on $[a, b]$ and because f changes sign on the interval $[a, b]$, the function f must have at least one root inside $[a, b]$. Since f' does not change sign on $[a, b]$, we have that f is monotonic on $[a, b]$ and so can have only one root inside $[a, b]$. The condition that f'' does not change sign on $[a, b]$ means that f is either concave up or concave down on $[a, b]$. For a polynomial f with real coefficients the situation called for by the three conditions above can always be realized; the same can be claimed for many functions f that are not polynomials, but the same can not be claimed for all functions.

If the three conditions above are satisfied for an interval $[a, b]$, the following four possibilities arise:

Case (a): $f''(x) > 0$ and $f'(x) > 0$ for all x in $[a, b]$;
Case (b): $f''(x) > 0$ and $f'(x) < 0$ for all x in $[a, b]$;
Case (c): $f''(x) < 0$ and $f'(x) > 0$ for all x in $[a, b]$;
Case (d): $f''(x) < 0$ and $f'(x) < 0$ for all x in $[a, b]$.

Figure 4.5 illustrates these four cases.

We are given a function f that satisfies conditions 1, 2, and 3 on $[a, b]$ and we let r denote the root of f inside $[a, b]$. Commencing with an endpoint of $[a, b]$, say the point b, Proposition 4.11 (using Lagrange's form of the remainder) gives

$$0 = f(r) = f(b) + f'(b) \cdot (r - b) + \tfrac{1}{2}f''(c) \cdot (r - b)^2, \qquad (4.52)$$

where $r < c < b$. Ignoring the remainder term, we can write the approximation

$$f(b) + f'(b) \cdot (r - b) \approx 0$$

from which we get the approximation

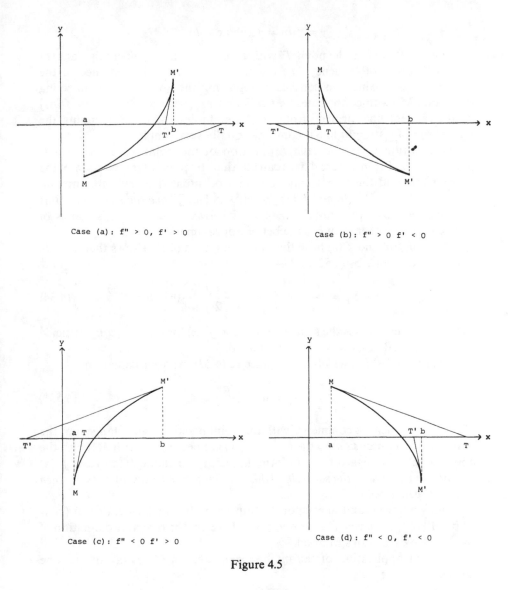

Case (a): f" > 0, f' > 0

Case (b): f" > 0 f' < 0

Case (c): f" < 0 f' > 0

Case (d): f" < 0, f' < 0

Figure 4.5

$$r \approx b - \frac{f(b)}{f'(b)}.$$

In this manner we arrive at the approximation

$$x_1 = b - \frac{f(b)}{f'(b)} \tag{4.53}$$

of the root r. Equation (4.53) can be interpreted geometrically as follows: At the point M' with coordinates $(b, f(b))$ we draw the tangent line; its equation is of the form

$$y = f(b) = f'(b) \cdot (x - b).$$

Putting $y = 0$, we find the point T' with coordinates $(x_1, 0)$ where the tangent line (with point of tangency M') intersects the x-axis. The essence of the method of approximation consists in replacing the arc $y = f(x)$ connecting the points M [with coordinates $(a, f(a))$] and M' [with coordinates $(b, f(b))$] by the tangent line formed at one of the endpoints M or M'. Finding the x-intercept of a straight line is computationally very simple.

The question as to the relative location of the point x_1 on the x-axis presents itself. From Figure 4.5 we can see that the point of intersection of the tangent line and the x-axis may very well be situated outside the interval $[a, b]$. We claim: *If $f(b)$ and $f''(x)$ for all x in $[a, b]$ have the same sign {this will be so in Cases (a) and (d)}, then x_1 is between r and b.* (This means of course that x_1 rather than b is the better approximation for r.)

Indeed, if $f(b)$ and $f'(b)$ have the same sign, then (4.53) shows that $x_1 < b$. On the other hand, by (4.52) and (4.53),

$$r - x_1 = r - b + \frac{f(b)}{f'(b)} = -\frac{1}{2}\frac{f''(c)}{f'(b)}(r - b)^2. \tag{4.54}$$

In the considered cases the signs of $f''(x)$ and $f'(x)$ for all x in $[a, b]$ coincide and so $r < x_1$. Thus, $r < x_1 < b$, as claimed.

In a similar fashion we obtain in place of (4.53) the approximation

$$x_1 = a - \frac{f(a)}{f'(a)} \tag{4.53*}$$

of the root r when we commence with the point a and draw the tangent at the point M (with abscissa a) to the arc $y = f(x)$ connecting M with M' (with the abscissa b). With regard to x_1 in formula (4.53*) we claim: *If $f(a)$ and $f''(x)$ for all x in $[a, b]$ have the same sign {this will happen in Cases (b) and (c)}, then x_1 is between a and r.*

Thus, we have found in each of the four cases, from which endpoint (be it M or M') of the arc $y = f(x)$ with $a \le x \le b$ we get the best approximation of the root r by Newton's Method.

Repeated application of the method generates in Cases (a) and (d) the decreasing sequence

$$b > x_1 > x_2 > \cdots > x_n > x_{n+1} > \cdots > r$$

and in Cases (b) and (c) the increasing sequence

$$a < x_1 < x_2 < \cdots < x_n < x_{n+1} < \cdots < c,$$

where x_{n+1} is computed from its predecessor x_n by the formula

$$x_{n+1} = x_n - \frac{f(x_n)}{f'(x_n)}. \tag{4.55}$$

[It is not difficult to show that $x_n \to r$ as $n \to \infty$. Suppose that $\{x_n\}$ is a

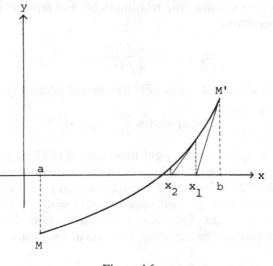

Figure 4.6

decreasing sequence. The set

$$S = \{x_n: n = 1, 2, 3, \ldots\}$$

is bounded below by r (which in turn is larger than the finite real number a). Thus, $\inf S = \beta$ exists as a finite real number (between the finite real numbers a and b) and, in fact, $x_n \to \beta$ as $n \to \infty$. But f and f' are continuous on $[a, b]$ and passage to the limit as $n \to \infty$ in (4.55) shows that

$$\frac{f(\beta)}{f'(\beta)} = 0, \quad \text{hence} \quad f(\beta) = 0 \quad \text{and} \quad \beta = r.$$

If $\{x_n\}$ is an increasing sequence we consider the supremum of the set S.]

Figure 4.6 illustrates successive applications of Newton's Method.

Now we turn to the matter of estimating the accuracy of the approximation. To estimate the deviation between r and x_n we note that, by Proposition 4.5,

$$f(x_n) = f(x_n) - f(r) = (x_n - r)f'(c),$$

where c is between x_n and r. Hence,

$$x_n - r = \frac{f(x_n)}{f'(c)}.$$

Denoting by m the smallest value of $|f'(x)|$ on the interval $[a, b]$, we see that

$$|x_n - r| \le \frac{|f(x_n)|}{m}. \tag{4.56}$$

To estimate the deviation between r and x_{n+1} in terms of the deviation between r and x_n, we consider the relation (4.54) and replace in it b by x_n and x_1 by x_{n+1}; we obtain

$$x_{n+1} - r = \frac{1}{2}\frac{f''(c)}{f'(x_n)}(x_n - r)^2.$$

Denoting by M the largest value of $|f''(x)|$ on the interval $[a, b]$, we see that

$$|x_{n+1} - r| \leq \frac{M}{2m} \cdot |x_n - r|^2. \tag{4.57}$$

Since there is a square on the right-hand side of (4.57), a rather rapid convergence of x_n to r is assured (at least beginning with some value of the index n); for example, if $M < 2m$, and if x_n approximates r with an accuracy of up to k decimal places, then x_{n+1} will approximate r with an accuracy of at least up to $2k$ decimal places. This makes Newton's Method one of the most effective techniques for the numerical solution of equations.

EXAMPLE 1. We wish to calculate the root of $x^3 - 2x - 5 = 0$ in the interval $[2, 2.1]$ with an error that is less than 10^{-10}.
  We have

$$f(x) = x^3 - 2x - 5, \qquad f(2) = -1 < 0, \qquad f(2.1) = 0.061 > 0,$$

$$f'(x) = 3x^2 - 2 > 0, \qquad f''(x) = 6x > 0 \quad \text{for } 2 \leq x \leq 2.1$$

[Case (a)]. We readily find that $m = 10$, $M < 12.6$, and

$$\frac{M}{2m} < 0.63$$

hold.
  We commence with $b = 2.1$. By (4.56) we get

$$b - r < \frac{0.061}{10} = 0.0061.$$

Using (4.57), we can determine in advance what accuracy can be expected of x_1:

$$x_1 - r < 0.63 \cdot 0.0061^2 < 0.000024.$$

Hence, we round up the number

$$x_1 = 2.1 - \frac{f(2.1)}{f'(2.1)} = 2.1 - \frac{0.061}{11.23} = 2.1 - 0.00543\ldots$$

"on the side of the root" to five decimal places: $x_1 = 2.1 - 0.00544 = 2.09456$. Since

$$f(x_1) = f(2.09456) = 0.000095078690816,$$

the error can now be determined more accurately by use of (4.56):

$$x_1 - r < \frac{0.000095}{10} < 0.00001.$$

We now pass to x_2 and use (4.57) once again to determine what accuracy can be expected of x_2:

$$x_2 - r < 0.63 \cdot 0.00001^2 < 0.000000000063.$$

Hence, the number

$$x_2 = 2.09456 - \frac{0.000095078690816}{11.1615447808} = 2.09456 - 0.000008518416\ldots,$$

which is rounded up to eleven decimal places and so comes out to

$$x_2 = 2.09456 - 0.00000851841 = 2.09455148159,$$

differs from the sought after root by less than 0.00000000007. Thus,

$$2.09455148152 < r < 2.09455148159.$$

EXAMPLE 2. The equation $2^x = 4x$ has two real roots; one root is $x = 4$ and the other is between 0 and $\frac{1}{2}$. We wish to calculate the root between 0 and $\frac{1}{2}$ with an error that is less than 10^{-5}.

For $0 \le x \le \frac{1}{2}$ we have

$$f(x) = 2^x - 4x, \qquad f'(x) = 2^x(\ln 2) - 4 < 0, \qquad f''(x) = 2^x(\ln 2)^2 > 0$$

[Case (b)]. Since $m = 4 - \sqrt{2}(\ln 2) > 3$ and $M = \sqrt{2}(\ln 2)^2 < 0.7$, we have

$$\frac{M}{2m} < 0.12.$$

We use the value $f(0.30) = 0.031144$ and estimate more accurately the error by (4.56):

$$r - x_1 < \frac{0.031144}{3} < 0.011;$$

by (4.57) we therefore obtain

$$r - x_2 < 0.12 \cdot 0.000121 < 0.000015$$

and it is seen that we are approaching the desired degree of accuracy.

In the next approximation,

$$x_2 = 0.30 - \frac{0.031144}{0.8533643\ldots - 4} = 0.30 + \frac{0.031144\ldots}{3.1466356\ldots} = 0.309897\ldots,$$

we round up "on the side of the root" to five decimal places: $x_2 = 0.30990$. Since $f(0.30990) = 0.000021\ldots > 0$, this value is still less than the root. By

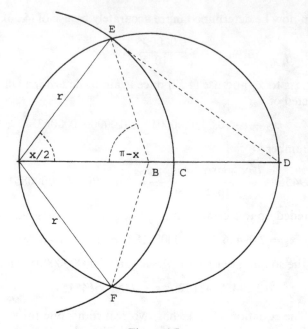

Figure 4.7

(4.56) we have in reality an error smaller than 10^{-5} because

$$r - x_2 < \frac{0.000022}{3} < 0.00001$$

and so $r = 0.30990(+0.00001)$.

EXAMPLE 3. A goat is tethered by a rope from a post on the boundary of a circular field of unit radius. What length r of rope will allow the goat access to precisely half the field?

Consider Figure 4.7. The center of the circular field of unit radius is at B, the post is at A. We denote by $x/2$ the angle $\angle BAE$ and observe that the angle $\angle AED$ is a right angle. The distance from A to E is r and the distance from A to D is 2. By similarity of triangles we have $r/2 = \cos(x/2)$, that is,

$$r = 2\left(\cos\frac{x}{2}\right).$$

The angle $\angle BDE$ equals $\pi/2 - x/2$ and so the angle $\angle BED$ equals $\pi/2 - x/2$ because the triangle $\triangle BDE$ is isosceles. Thus, the angle $\angle DBE$ equals x and the angle $\angle ABE$ equals $\pi - x$.

Consider Figure 4.8. It is easily seen that the (striped) circular segment has area $\frac{1}{2}R^2(\theta - \sin\theta)$.

From Figure 4.7 we can see that the goat will have access to a region bounded by the circular arc EAF and the circular arc FCE; this region is the

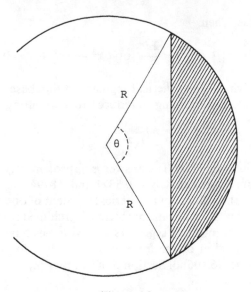

Figure 4.8

union of two circular segments and its area equals

$$\tfrac{1}{2}r^2(x - \sin x) + \tfrac{1}{2}[(2\pi - 2x) - \sin(2\pi - 2x)]$$
$$= 2(\cos^2 x/2)(x - \sin x) + \pi - x + (\sin x)(\cos x)$$
$$= (1 + \cos x)(x - \sin x) + \pi - x + (\sin x)(\cos x)$$
$$= \pi + x(\cos x) - \sin x.$$

But half of the circular field of unit radius has area $\pi/2$ and so x must satisfy

$$\sin x - x(\cos x) = \frac{\pi}{2}. \qquad (4.59)$$

Solving (4.59), we can substitute back into (4.58) to get r.

The point of this example is that a simple shepherd's problem leads to an elusive equation of the form (4.59) and does not lend itself to a simple-minded treatment.

Using a common pocket calculator we find that x satisfying (4.59) is approximately 1.9056957 radians and the corresponding r is approximately 1.1587285.

EXAMPLE 4. The volume of a spherical segment of one base is given by

$$\pi h^2\left(r - \frac{h}{3}\right),$$

where r denotes the radius of the sphere and h the height of the segment.

Suppose we divide a hemisphere of radius 1 into two equal parts by a plane parallel to the base. Then

$$2\pi x^2 \left(1 - \frac{x}{3}\right) = \frac{2\pi}{3} \quad \text{or} \quad x^3 - 3x^2 + 1 = 0,$$

where x is the height of the spherical segment of one base cut off the hemisphere of radius 1. Determining x reduces to computing the root of the equation

$$x^3 - 3x^2 + 1 = 0$$

contained in the interval $(0, 1)$. The answer is approximately $x = 0.6527$. The other two roots are approximately -0.5321 and 2.8794.

The formula for the volume of a spherical segment of one base used above is due to Archimedes; a significant problem of Archimedes was that of using a plane to cut a spere into two segments with volumes having a preassigned ratio. [For a verification of the formula $\pi h^2(r - h/3)$ for the volume of a spherical segment see (6.106) of Section 5 of Chapter 6.]

5. Arithmetic and Geometric Means

It is simple to verify that the function $f(x) = e^x - 1 - x$ with the derivative $f'(x) = e^x - 1$ has a single minimum at $x = 0$ and so

$$0 = f(0) \le f(x)$$

for any real number x. Thus,

$$e^x \ge 1 + x \quad \text{for all real numbers } x; \tag{4.60}$$

in particular,

$$e^x > 1 + x \quad \text{for } x \ne 0. \tag{4.61}$$

We shall use inequalities (4.60) and (4.61) to establish some useful inequalities.

Proposition 4.17. *Let* $a_1, a_2, \ldots, a_n$ *be positive real numbers. Then*

$$\sqrt[n]{a_1 a_2 \cdots a_n} \le \frac{a_1 + a_2 + \cdots + a_n}{n}; \tag{4.62}$$

there is equality in (4.62) if and only if all a_k, *for* $k = 1, 2, \ldots, n$, *are equal.*

PROOF. Let

$$A = \frac{a_1 + a_2 + \cdots + a_n}{n} \quad \text{and} \quad G = \sqrt[n]{a_1 a_2 \cdots a_n};$$

A is called the *arithmetic mean of* $a_1, a_2, \ldots, a_n$ and G is called the *geometric mean of* $a_1, a_2, \ldots, a_n$. For

$$x = \frac{a_k}{A} - 1, \qquad k = 1, 2, \ldots, n,$$

we get, by (4.60) and employing the notation $\exp(t) = e^t$,

$$\exp\left(\frac{a_1}{A} - 1\right) \geq \frac{a_1}{A}, \quad \exp\left(\frac{a_2}{A} - 1\right) \geq \frac{a_2}{A}, \quad \ldots, \quad \exp\left(\frac{a_n}{A} - 1\right) \geq \frac{a_n}{A}.$$

Multiplying all these inequalities together, we obtain

$$\exp\left(\frac{a_1 + a_2 + \cdots + a_n}{A} - n\right) \geq \frac{a_1 a_2 \cdots a_n}{A^n}$$

or

$$e^{n-n} \geq \frac{G^n}{A^n} \quad \text{or} \quad 1 \geq \frac{G^n}{A^n},$$

implying that

$$A \geq G.$$

Note that $A = G$ only if equality holds in all n relations. This requires that $a_k/A - 1 = 0$ in all cases, showing that $A = G$ only when all a_k are equal (to A). $\qquad\square$

REMARK. Replacing a_k by $1/a_k$ in (4.62) we get that

$$\frac{n}{\dfrac{1}{a_1} + \dfrac{1}{a_2} + \cdots + \dfrac{1}{a_n}} \leq \sqrt[n]{a_1 a_2 \cdots a_n}. \tag{4.63}$$

The term on the left-hand side of (4.63) is called the *harmonic mean of* $a_1, a_2,$ $\ldots, a_n$.

Proposition 4.18. *We have*

$$e^\pi > \pi^e. \tag{4.64}$$

PROOF. Since $\pi > e$, we have $\pi/e > 1$ and so $x = \pi/e - 1 > 0$. Thus, by (4.61),

$$\exp\left(\frac{\pi}{e} - 1\right) > 1 + \left(\frac{\pi}{e} - 1\right)$$

or

$$\frac{\exp(\pi/2)}{e} > \frac{\pi}{e} \quad \text{or} \quad e^{\pi/e} > \pi.$$

But the latter inequality is equivalent to the inequality (4.64). $\qquad\square$

REMARKS. Another way of showing the validity of the inequality (4.64) is to observe that the function $g(x) = (\ln x)/x$, defined for $x > 0$, has a single maximum at $x = e$ because $g'(x) = [1 - \ln x]/x^2$ is negative for $x > e$ and is positive for x satisfying $0 < x < e$. But $\pi > e$ and so $g(\pi) < g(e)$, and so on.

We can use a common pocket calculator and get approximations of e^π and π^e to see which is larger; however, it is more interesting to use the cutting edge of theory to establish (4.64).

Worked Examples and Comments

1. The following two statements are direct consequences of Proposition 4.17:

(i) If $a_1, a_2, \ldots, a_n$ are positive numbers satisfying the condition

$$a_1 + a_2 + \cdots + a_n = k,$$

then their product $a_1 a_2 \cdots a_n$ has maximum value, $(k/n)^n$, when $a_1 = a_2 = \cdots = a_n = k/n$.

(ii) If $a_1, a_2, \ldots, a_n$ are positive real numbers satisfying the condition

$$a_1 a_2 \cdots a_n = k,$$

then their sum $a_1 + a_2 + \cdots + a_n$ has minimum value, $n(k)^{1/n}$, when $a_1 = a_2 = \cdots = a_n = k^{1/n}$.

REMARK. For example, if a_1, a_2, a_3, and a_4 are four positive numbers whose sum is 1000, then the product $a_1 a_2 a_3 a_4$ will be a maximum if

$$a_1 = a_2 = a_3 = a_4 = 250.$$

2. The cube is the rectangular parallelepiped of maximum volume for given surface area, and of minimum surface area for given volume.

Indeed, if we denote the lengths of three adjacent edges of a rectangular parallelepiped by x, y, z, its surface area is $2(yz + zx + xy)$ and its volume is xyz. If we put $\alpha = yz, \beta = zx, \gamma = xy$, the surface area is $2(\alpha + \beta + \gamma)$ and the volume $\sqrt{\alpha\beta\gamma}$. Hence, analytically speaking, the problem is to make $\alpha\beta\gamma$ a maximum when $\alpha + \beta + \gamma$ is given, and to make $\alpha + \beta + \gamma$ a minimum when $\alpha\beta\gamma$ is given. This, by the result in the foregoing Example 1, is done in either case by making $\alpha = \beta = \gamma$, that is, $yz = zx = xy$; hence $x = y = z$.

3. The equilateral triangle has maximum area for given perimeter, and minimum perimeter for given area.

Indeed, the area is $\Delta = \sqrt{s(s-a)(s-b)(s-c)}$. Let $x = s - a$, $y = s - b$, $z = s - c$; then $x + y + z = s$ and the area is $\Delta = \sqrt{sxyz}$. Since, in the first place, s is given, we have only to make xyz a maximum subject to the condition $x + y + z = s$. By the result in Example 1, this leads to $x = y = z$.

Next, let Δ be given. Then $(x + y + z)xyz = \Delta^2$ and $s = \Delta^2/xyz$. If we put $\alpha = x^2 yz, \beta = xy^2 z, \gamma = xyz^2$, we have $\alpha + \beta + \gamma = \Delta^2$ and $s = \Delta^2/(\alpha\beta\gamma)^{1/4}$. Hence, to make s a minimum when Δ is given, we have to make $\alpha\beta\gamma$ a *maximum*, subject to the condition $\alpha + \beta + \gamma = \Delta^2$. This leads to $\alpha = \beta = \gamma$, that is,

$$x^2yz = xy^2z = xyz^2,$$

and so $x = y = z$.

REMARK. In a similar way one can prove: Of all rectangles of given perimeter, the square has the largest area; of all rectangles of given area, the square has least perimeter.

4. The fact that the arithmetic mean of n positive numbers is not less than their geometric mean (see Proposition 4.17) may also be used to show the following: For $n = 1, 2, 3, \ldots,$

$$\left(1 + \frac{1}{n}\right)^n < \left(1 + \frac{1}{n+1}\right)^{n+1} \tag{4.65}$$

and

$$\left(1 + \frac{1}{n}\right)^{n+1} > \left(1 + \frac{1}{n+1}\right)^{n+2}. \tag{4.66}$$

[Note that the inequalities (4.65) and (4.66) were already proved in Proposition 1.4.]

Indeed, consider the set of $n + 1$ numbers

$$1, \quad 1 + \frac{1}{n}, \quad 1 + \frac{1}{n}, \quad \ldots, \quad 1 + \frac{1}{n}.$$

These have an arithmetic mean of

$$1 + \frac{1}{n+1}$$

and a geometric mean of

$$\left(1 + \frac{1}{n}\right)^{n/(n+1)}.$$

Hence,

$$1 + \frac{1}{n+1} > \left(1 + \frac{1}{n}\right)^{n/(n+1)}. \tag{4.67}$$

But (4.67) is equivalent to (4.65).

Similarly, consider the set of $n + 2$ numbers

$$1, \quad \frac{n}{n+1}, \quad \frac{n}{n+1}, \quad \ldots, \quad \frac{n}{n+1}.$$

These have an arithmetic mean of

$$\frac{n+1}{n+2}$$

and a geometric mean of

$$\left(\frac{n}{n+1}\right)^{(n+1)/(n+2)}.$$

Hence,

$$\frac{n+1}{n+2} > \left(\frac{n}{n+1}\right)^{(n+1)/(n+2)}.$$

Taking reciprocals this becomes

$$1 + \frac{1}{n+1} < \left(1 + \frac{1}{n}\right)^{(n+1)/(n+2)}. \tag{4.68}$$

But (4.68) is equivalent to (4.66).

5. For any positive integer n we have

$$n\{(n+1)^{1/n} - 1\} < 1 + \frac{1}{2} + \cdots + \frac{1}{n} < n\left(1 - \frac{1}{(n+1)^{1/n}} + \frac{1}{n+1}\right). \tag{4.69}$$

The verification of inequality (4.69) will again use the fact that the arithmetic mean of n positive numbers does not exceed their geometric mean.
    First we show that

$$1 + \frac{1}{n}\left(1 + \frac{1}{2} + \cdots + \frac{1}{n}\right) > (n+1)^{1/n}.$$

But this is immediate by setting

$$a_1 = 1 + 1 = \frac{2}{1}, \quad a_2 = 1 + \frac{1}{2} = \frac{3}{2}, \quad \ldots, \quad a_n = 1 + \frac{1}{n} = \frac{n+1}{n}$$

into inequality (4.62).
    We next show that

$$\frac{n}{(n+1)^{1/n}} < n\left(1 + \frac{1}{n+1}\right) - 1 - \frac{1}{2} - \cdots - \frac{1}{n}. \tag{4.70}$$

But the right-hand side of inequality (4.70) is

$$(1 - 1) + \left(1 - \frac{1}{2}\right) + \cdots + \left(1 - \frac{1}{n}\right) + \frac{n}{n+1}$$

and (4.88) follows from inequality (4.62) if we set

$$a_1 = \frac{1}{2}, \quad a_2 = \frac{2}{3}, \quad \ldots, \quad a_n = \frac{n}{n+1}.$$

6. For any positive integer n we have

$$1 \cdot 3 \cdots (2n-1) < n^n. \tag{4.71}$$

Indeed, by inequality (4.62),

$$\frac{\{1 + 3 + \cdots + (2n-1)\}}{n} > \{1 \cdot 3 \cdots (2n-1)\}^{1/n}$$

or

$$\frac{n^2}{n} > \{1 \cdot 3 \cdots (2n-1)\}^{1/n},$$

implying (4.71).

7. Find the largest value of $(a+x)^5(a-x)^3$ for x between $-a$ and a. Here again we wish to use inequality (4.62) in obtaining the answer.

We consider the following eight positive numbers

$$a_1 = a_2 = \cdots = a_5 = \frac{a+x}{5}, \qquad a_6 = a_7 = a_8 = \frac{a-x}{3} \quad \text{for } -a < x < a.$$

Their arithmetic mean is $a/4$. Hence, by inequality (4.62),

$$\left(\frac{a+x}{5}\right)^5\left(\frac{a-x}{3}\right)^3 \le \left(\frac{a}{4}\right)^8, \tag{4.72}$$

equality arising when

$$\tfrac{1}{5}(a+x) = \tfrac{1}{3}(a-x).$$

From (4.72) we get

$$(a+x)^5(a-x)^3 \le \frac{5^5 \cdot 3^3 \cdot a^8}{4^8} \quad \text{for } -a < x < a.$$

8. Let $a_i > 0$, $b_i > 0$ $(i = 1, 2, \ldots, n)$. Then

$$\sqrt[n]{(a_1+b_1)(a_2+b_2)\cdots(a_n+b_n)} \ge \sqrt[n]{a_1 a_2 \cdots a_n} + \sqrt[n]{b_1 b_2 \cdots b_n} \tag{4.73}$$

with equality if and only if $a_1/b_1 = a_2/b_2 = \cdots = a_n/b_n$.

Indeed, by inequality (4.62),

$$\frac{\sqrt[n]{a_1 a_2 \cdots a_n} + \sqrt[n]{b_1 b_2 \cdots b_n}}{\sqrt[n]{(a_1+b_1)(a_2+b_2)\cdots(a_n+b_n)}}$$

$$= \left(\left(\frac{a_1}{a_1+b_1}\right)\cdots\left(\frac{a_n}{a_n+b_n}\right)\right)^{1/n} + \left(\left(\frac{b_1}{a_1+b_1}\right)\cdots\left(\frac{b_n}{a_n+b_n}\right)\right)^{1/n}$$

$$\le \frac{1}{n}\sum_{i=1}^{n}\frac{a_i}{a_i+b_i} + \frac{1}{n}\sum_{i=1}^{n}\frac{b_i}{a_i+b_i} = 1.$$

9. Let $w_i > 0$ for $i = 1, 2, \ldots, n$. Then

$$(1 + w_1)(1 + w_2) \cdots (1 + w_n) \geq (1 + q)^n, \qquad (4.74)$$

where $w_1 w_2 \cdots w_n = q^n$, with equality if and only if $w_1 = w_2 = \cdots = w_n$.

Indeed, inequality (4.74) follows from inequality (4.73) by setting $a_i = 1$ and $b_i = w_i$ for $i = 1, 2, \ldots, n$.

10. *Problem of Huygens.* Let $0 < a < b$. The fraction

$$u = \frac{x_1 x_2 \cdots x_k}{(a + x_1)(x_1 + x_2) \cdots (x_{k-1} + x_k)(x_k + b)}$$

assumes maximal value precisely when

$$a < x_1 < x_2 < \cdots < x_{k-1} < x_k < b$$

forms a geometric progression, that is, $x_1/a = x_2/x_1 = \cdots = x_k/x_{k-1} = b/x_k$.

Evidently, u will be maximal if and only if $1/au$ is minimal. But

$$\frac{1}{au} = (1 + w_1)(1 + w_2) \cdots (1 + w_{k-1})(1 + w_k),$$

where $w_1 = a/x_1, w_2 = x_2/x_1, \ldots, w_{k-1} = x_k/x_{k-1}, w_k = b/x_k$. By (4.74)

$$(1 + w_1)(1 + w_2) \cdots (1 + w_{k-1})(1 + w_k)$$

is minimal precisely when $w_1 = w_2 = \cdots = w_{k-1} = w_k$.

11. For $n = 2, 3, \ldots$ we have

$$n! < \left(\frac{n + 1}{2} \right)^n.$$

Indeed, let $a_k = k$. Then $\sqrt[n]{a_1 a_2 \cdots a_n} = \sqrt[n]{n!}$ and $(a_1 + a_2 + \cdots + a_n)/n = (1 + n)/2$ and the claim follows by Proposition 4.17.

REMARK. The inequality $n^{n/2} < n!$ for $n = 3, 4, 5, \ldots$ is simple to verify. Consider the equality

$$(n!)^2 = [1 \cdot n][2(n - 1)][3(n - 2)] \cdots [(n - 1)2][n \cdot 1].$$

Now, the first and the last factors in square brackets are equal and are less than the other factors in square brackets because, for $n - k > 1$ and $k > 0$ we have $(k + 1)(n - k) = k(n - k) + (n - k) > k \cdot 1 + (n - k) = n$ and so $(n!)^2 > n^n$ follows for $n = 3, 4, 5, \ldots$.

12. Let $a_1, a_2, \ldots, a_n$ be positive and put $s = a_1 + a_2 + \cdots + a_n$. Then

$$(1 + a_1)(1 + a_2) \cdots (1 + a_n) \leq 1 + s + \frac{s^2}{2!} + \frac{s^3}{3!} + \cdots + \frac{s^n}{n!}$$

with equality only in case $n = 1$.

Indeed, by Proposition 4.17,

$$(1 + a_1)(1 + a_2) \cdots (1 + a_n) \le \left(\frac{n + a_1 + a_2 + \cdots + a_n}{n} \right)^n$$

$$= \left(1 + \frac{s}{n} \right)^n$$

$$= 1 + n \left(\frac{s}{n} \right) + \frac{n(n-1)}{2} \left(\frac{s}{n} \right)^2 + \cdots + \left(\frac{s}{n} \right)^n.$$

We observe that the coefficients of s^m will be

$$\frac{n!}{m!(n-m)!} \frac{1}{n^m}.$$

But $(n - m)!n^m \ge n!$ and so

$$\frac{n!}{m!(n-m)!} \frac{1}{n^m} \le \frac{n!}{m!n!} = \frac{1}{m!},$$

proving the inequality we set out to verify. That we have equality only in case $n = 1$ follows by Proposition 4.17.

13. We have

$$1 \cdot \frac{1}{2^2} \cdot \frac{1}{3^3} \cdot \frac{1}{4^4} \cdots \frac{1}{n^n} < \left(\frac{2}{n+1} \right)^{n(n+1)/2}.$$

Indeed, the left-hand side of the inequality contains two factors 1/2, three factors 1/3 and so on, and finally n factors $1/n$; in all, there are $1 + 2 + 3 + \cdots + n = n(n+1)/2$ such factors. The geometric mean of these factors is equal to the $n(n+1)/2$ root of this product; the arithmetic mean is

$$\frac{1 \cdot 1 + 2 \cdot \frac{1}{2} + 3 \cdot \frac{1}{3} + \cdots + n \cdot \frac{1}{n}}{n(n+1)/2} = \frac{n}{n(n+1)/2} = \frac{2}{n+1}.$$

The validity of the inequality under consideration therefore follows by Proposition 4.17.

6. Miscellaneous Examples

1. Let a and b be distinct positive real numbers. We define

$$f(x) = \left(\frac{a^x + b^x}{2} \right)^{1/x} \quad \text{for } x \ne 0,$$

$$= \sqrt{ab} \quad \text{for } x = 0.$$

The function f is strictly increasing on $(-\infty, \infty)$, yet its total increase is only $|a - b|$. Specifically, we make the following claims:

(a) $f'(x) > 0$ for $x \neq 0$;
(b) $\lim_{x \to 0} f(x) = \sqrt{ab}$;
(c) $f'(0) = \frac{1}{8}\sqrt{ab}(\ln a/b)^2$;
(d) $\lim_{x \to -\infty} f(x) = \min\{a, b\}$ and $\lim_{x \to \infty} f(x) = \max\{a, b\}$.

The proof runs as follows. We verify first claim (a). For $x \neq 0$, let

$$g(x) = \ln f(x) = \frac{1}{x}\ln\left(\frac{a^x + b^x}{2}\right).$$

Then

$$f'(x) = e^{g(x)}g'(x)$$

and $f'(x)$ and $g'(x)$ have the same sign. Differentiation gives

$$g'(x) = \frac{a^x(\ln a) + b^x(\ln b)}{x(a^x + b^x)} - \frac{1}{x^2}\ln\left(\frac{a^x + b^x}{2}\right).$$

Since

$$\ln\left(\frac{a^x + b^x}{2}\right) = \frac{a^x}{a^x + b^x}\ln\left(\frac{a^x + b^x}{2}\right) + \frac{b^x}{a^x + b^x}\ln\left(\frac{a^x + b^x}{2}\right)$$

and

$$x^2 g'(x) = \frac{a^x(\ln a^x) + b^x(\ln b^x)}{a^x + b^x} - \ln\left(\frac{a^x + b^x}{2}\right),$$

we obtain

$$2x^2 g'(x) = \frac{2a^x}{a^x + b^x}\ln\left(\frac{2a^x}{a^x + b^x}\right) + \frac{2b^x}{a^x + b^x}\ln\left(\frac{2b^x}{a^x + b^x}\right).$$

We put

$$\frac{a^x - b^x}{a^x + b^x} = t; \tag{4.75}$$

since a^x and b^x are always positive, we see that $-1 < t < 1$. Moreover,

$$1 + t = \frac{2a^x}{a^x + b^x}, \qquad 1 - t = \frac{2b^x}{a^x + b^x} \tag{4.76}$$

and thus

$$2x^2 g'(x) = (1 + t)\ln(1 + t) + (1 - t)\ln(1 - t). \tag{4.77}$$

We let, for $-1 < t < 1$,

$$h(t) = (1 + t)\ln(1 + t) + (1 - t)\ln(1 - t).$$

Then

$$h'(t) = \ln(1 + t) - \ln(1 - t) = \ln\frac{1+t}{1-t}.$$

We have

$$\frac{1+t}{1-t} > 1 \quad \text{for } 0 < t < 1$$

$$< 1 \quad \text{for } -1 < t < 0$$

and so

$$\ln\frac{1+t}{1-t} > 0 \quad \text{for } 0 < t < 1$$

$$< 0 \quad \text{for } -1 < t < 0.$$

It follows that h increases on $(0, 1)$, h decreases on $(-1, 0)$, and $h(0) = 0$ is the smallest value of $h(t)$ for $-1 < t < 1$. Since $x \neq 0$ (because $1/x$ is not defined for $x = 0$) and since $a \neq b$, we see that t can not vanish. Thus, $h(t) > 0$ for $t \neq 0$, implying that $g'(x) > 0$ for $x \neq 0$ and so $f'(x) > 0$ for $x \neq 0$.

Next we verify claim (b). Let

$$y = \left(\frac{a^x + b^x}{2}\right)^{1/x} \quad \text{and so} \quad \ln y = \frac{1}{x}\ln\left(\frac{a^x + b^x}{2}\right).$$

By Proposition 4.10,

$$\lim_{x \to 0}\frac{\ln(a^x/2 + b^x/2)}{x} = \lim_{x \to 0}\frac{a^x(\ln a) + b^x(\ln b)}{a^x + b^x} = \frac{\ln a + \ln b}{2} = \ln\sqrt{ab}.$$

Thus,

$$\lim_{x \to 0}\left(\frac{a^x + b^x}{2}\right)^{1/x} = \lim_{x \to 0} f(x) = \sqrt{ab}.$$

Since $f(x)$ is differentiable for $x \neq 0$ and having defined $f(0) = \sqrt{ab}$, we see that f is continuous throughout $(-\infty, \infty)$.

We now verify claim (c). The function f is seen to satisfy the Mean Value Theorem (see Proposition 4.5) on a closed interval $[0, s]$ of finite length. Hence,

$$\frac{f(s) - f(0)}{s} = f'(x)$$

for some x between 0 and s. As before, let $g(x) = \ln f(x)$. Then

$$f'(x) = f(x)g'(x)$$

and, using the substitution (4.75), we get

$$g'(x) = \frac{(1 + t)\ln(1 + t) + (1 - t)\ln(1 - t)}{2x^2}$$

[see (4.77)]. By (4.76)

$$\frac{a^x}{b^x} = \frac{1 + t}{1 - t}$$

and so

$$x = \frac{\ln(1 + t) - \ln(1 - t)}{\ln(a/b)}.$$

Thus,

$$f'(x) = \frac{1}{2} f(x) \left(\ln \frac{a}{b} \right)^2 \frac{(1 + t)\ln(1 + t) + (1 - t)\ln(1 - t)}{[\ln(1 + t) - \ln(1 - t)]^2}.$$

But $\ln(1 + t) + \ln(1 - t) = \ln(1 - t^2)$ and so

$$f'(x) = \frac{1}{2} f(x) \left(\ln \frac{a}{b} \right)^2 \frac{(1/t^2)\ln(1 - t^2) + \dfrac{\ln(1 + t) - \ln(1 - t)}{t}}{\left(\dfrac{\ln(1 + t) - \ln(1 - t)}{t} \right)^2}.$$

As $x \to 0$, we have $t \to 0$; moreover, by Proposition 4.10,

$$\lim_{t \to 0} \frac{\ln(1 - t^2)}{-t^2} = 1 \quad \text{and} \quad \lim_{t \to 0} \frac{\ln(1 + t) - \ln(1 - t)}{2t} = 1.$$

Thus,

$$\lim_{x \to 0} f'(x) = \frac{1}{8} \sqrt{ab} \left(\ln \frac{a}{b} \right)^2.$$

Hence,

$$f'(0) = \lim_{s \to 0} \frac{f(s) - f(0)}{s} = \lim_{s \to 0} f'(x) = \lim_{x \to 0} f'(x) = \frac{1}{8} \sqrt{ab} \left(\ln \frac{a}{b} \right)^2$$

because x is between 0 and s and $\lim_{x \to 0} f'(x)$ exists.

Finally, we establish claim (d). Suppose that $a > b$ and let x take the values $1, 2, 3, \ldots$. Then

$$f(n) = \left(\frac{a^n + b^n}{2} \right)^{1/n} = a \left(\frac{1 + (b/a)^n}{2} \right)^{1/n} = a 2^{-1/n} e^{(1/n)\ln[1 + (b/a)^n]}.$$

But $(b/a)^n \to 0$ as $n \to \infty$ and so $\ln[1 + (b/a)^n] \to 0$ as $n \to \infty$. Clearly, $1/n \to 0$ as $n \to \infty$ and $2^{-1/n} \to 1$ as $n \to \infty$. Thus, $f(n) \to a$ as $n \to \infty$. But, if $n \le x < n + 1$, then $f(n) \le f(x) < f(n + 1)$ [because $f'(x) > 0$ and using Proposition 4.7] and so $f(x) \to a$ as $x \to \infty$. In an entirely similar way we show that $f(x) \to b$ as $x \to -\infty$.

REMARK. Note in particular the inequalities

$$f(-1) < f(0) < f(1);$$

$f(-1)$ is the harmonic mean, $f(0)$ is the geometric mean, and $f(1)$ is the arithmetic mean of the two distinct positive real numbers a and b.

2. We consider the problem of inscribing a right circular cylinder of maximal surface area into a right circular cone of fixed dimensions.

Let R be the radius and H the height of the cone; let r be the radius and h the height of the cylinder. Then the surface area of the cylinder is

$$S = 2\pi r^2 + 2\pi rh.$$

By similar triangles we have

$$\frac{h}{R-r} = \frac{H}{R} \quad \text{or} \quad h = \frac{R-r}{R}H$$

and so

$$S = 2\pi\left[r^2 + rH\left(1 - \frac{r}{R}\right)\right], \quad \text{where } 0 \le r \le R.$$

Differentiation with respect to r yields

$$S' = 2\pi\left(2r + H - \frac{2r}{R}H\right), \quad S'' = 4\pi\left(1 - \frac{H}{R}\right).$$

Setting $S' = 0$, we get

$$r = \frac{HR}{2(H-R)}. \tag{4.78}$$

If this value or r is to be contained in the open interval $(0, R)$, then

$$0 < \frac{HR}{2(H-R)} \quad \text{and} \quad \frac{HR}{2(H-R)} < R \tag{4.79}$$

must be satisfied. The first of these inequalities is equivalent to $H > R$. Multiplying the second inequality with the positive quantity $2(H - R)$ we get

$$R < \frac{H}{2}.$$

If the latter inequality holds, then S'' is negative; to the value of r in (4.78) corresponds the only maximum of the function S. This maximum value is easily obtained by substituting the value for r in (4.78) into the expression for S.

We assume now that the value r in (4.78) is not in the open interval $(0, R)$, that is, the inequalities in (4.79) are not satisfied. Here two possibilities can arise: either $H \le R$ or $H > R$ but $R \ge H/2$. The inequality

$$H \leq 2R \tag{4.80}$$

characterizes the two possibilities.

We rewrite the expression for S' and get

$$S' = 2\pi \left(2r + H - \frac{2rH}{R} \right) = \frac{2\pi}{R} [(2R - H)r + H(R - r)].$$

But inequality (4.80) implies $S' > 0$ for $0 < r < R$; hence S is an increasing function on $(0, R)$ and assumes its largest value for $r = R$. For $r = R$ we have $h = 0$ and the inscribed cylinder is completely flat.

3. Let (a, b) be the coordinates of a fixed point situated in the first quadrant of the x, y-plane and consider the set of all straight lines passing through the point (a, b) and intersecting both coordinate axes. Among these straight lines we wish to determine the length of the shortest segment cut off by the coordinate axes.

Let $t > 0$ and suppose that $(a + t, 0)$ is the x-intercept of a straight line passing through the point (a, b) in the first quadrant. It is easily seen that this straight line intersects the y-axis at $y = (b/t)(a + t)$ and that the square of the length of the segment cut off this straight line by the coordinate axes is

$$f(t) = (a + t)^2 + \left(\frac{b}{t}\right)^2 (a + t)^2 = (a + t)^2 \left(1 + \frac{b^2}{t^2}\right).$$

It is clear that if $f(t)$ is smallest for $t = t_0$, then $\sqrt{f(t)}$ is also smallest for $t = t_0$ and conversely. We proceed to find the smallest value $t = t_0$ of $f(t)$. Differentiation yields

$$f'(t) = 2(a + t)\left(1 + \frac{b^2}{t^2}\right) + (a + t)^2 \left(-\frac{2b^2}{t^3}\right).$$

Putting $f'(t) = 0$, we get

$$t = a^{1/3}b^{2/3}$$

and

$$f(a^{1/3}b^{2/3}) = (a^{2/3} + b^{2/3})^3.$$

A simple calculation shows that

$$f''(a^{1/3}b^{2/3}) = \frac{2}{a^{2/3}}(a^{2/3} + b^{2/3})\left[1 - 4 + \frac{3(a^{2/3} + b^{2/3})}{b^{2/3}}\right] > 0$$

because $(a^{2/3} + b^{2/3})/b^{2/3} > 1$.

We therefore see that the length of the shortest segment cut off by the coordinate axes is $(a^{2/3} + b^{2/3})^{3/2}$.

REMARKS. The foregoing result shows the validity of the following two statements:

(i) Let two corridors of width a and b, respectively, intersect at right angles. Then the length of the longest thin rod that will go horizontally around the corner is $(a^{2/3} + b^{2/3})^{3/2}$.

(ii) The length of the shortest thin beam that can be used to brace a wall, if the beam is to pass over a second wall (of negligible thickness) that is b units high and a units from the first wall is $(a^{2/3} + b^{2/3})^{3/2}$.

4. A circular sector of fixed radius R and variable central angle x (measured in radians) is to be shaped into the lateral surface of a right circular cone of height h and radius r. We wish to determine the value x for which the cone will have largest volume.

The length of the circular arc of central angle x and radius R is Rx. The circumference of the base of the cone is $2\pi r$. Thus,

$$2\pi r = Rx \quad \text{or} \quad r = \frac{Rx}{2\pi}.$$

It is clear that $h^2 + r^2 = R^2$ and so

$$h = (R^2 - r^2)^{1/2} = \left(R^2 - \frac{R^2 x^2}{4\pi^2}\right)^{1/2} = \frac{R}{2\pi}(4^2 - x^2)^{1/2}.$$

The volume of the cone is

$$V = \frac{\pi r^2 h}{3} = \frac{1}{3}\pi \frac{R^2 x^2}{4\pi^2} \cdot \frac{R}{2\pi}(4\pi^2 - x^2)^{1/2} = \frac{R^3}{24\pi^2} x^2 (4\pi^2 - x^2)^{1/2}.$$

To determine the largest value of V we only need to investigate the function

$$h(x) = 4\pi^2 x^4 - x^6 \quad \text{for } 0 < x < 2\pi.$$

But $h'(x) = 16^2 x^3 - 6x^5$ and putting $h'(x) = 0$ we get the three values

$$x_1 = 0, \quad x_2 = -2\pi\sqrt{\tfrac{2}{3}}, \quad \text{and} \quad x_3 = 2\pi\sqrt{\tfrac{2}{3}}.$$

Only x_3 is in the open interval $(0, 2\pi)$ and gives the desired answer.

5. We consider the *Law of Refraction*. Let A and B be two given points on opposite sides of the x-axis. To fix ideas, let A have coordinates $(0, a)$ and B have coordinates (c, b) with a, b, and c denoting fixed positive numbers. We wish to find the path from A to B requiring the shortest possible time if the velocity on the upper side of the x-axis is v_1 and on the lower side of the x-axis is v_2; it is assumed that v_1 and v_2 are fixed positive numbers.

It is clear that this shortest path must consist of two portions of straight lines meeting one another at a point P on the x-axis; let the point P have coordinates $(x, 0)$. See Figure 4.9. The length of the segments AP and PB are, respectively,

$$(a^2 + x^2)^{1/2} \quad \text{and} \quad [b^2 + (c - x)^2]^{1/2};$$

the time of passage along the path consisting of the line segments AP and

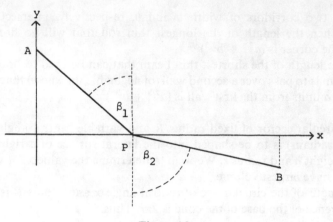

Figure 4.9

PB is

$$f(x) = \frac{1}{v_1}(a^2 + x^2)^{1/2} + \frac{1}{v_2}[b^2 + (c - x)^2]^{1/2}.$$

Differentiation gives

$$f'(x) = \frac{x}{v_1(a^2 + x^2)^{1/2}} - \frac{c - x}{v_2[b^2 + (c - x)^2]^{1/2}}$$

and

$$f''(x) = \frac{a^2}{v_1(a^2 + x^2)^{3/2}} + \frac{b^2}{v_2[b^2 + (c - x)^2]^{3/2}}.$$

Since $f''(x) > 0$ for any real number x, it follows that f' is strictly increasing on $(-\infty, \infty)$. Moreover, $f'(0) < 0$ and $f'(c) > 0$. Since f' is continuous, f' must have exactly one root between $x = 0$ and $x = c$. For this root, say $x = x_0$, we have

$$\frac{x/(a^2 + x^2)^{1/2}}{(c - x)/[b^2 + (c - x)^2]^{1/2}} = \frac{v_1}{v_2}.$$

This is the Law of Refraction due to Snell which can be put into the form

$$\frac{\sin \beta_1}{\sin \beta_2} = \frac{v_1}{v_2},$$

where β_1 is the angle of incidence and β_2 the angle of refraction. If we denote by α_1 the angle of inclination which the line segment AP forms with the x-axis and by α_2 the angle of inclination which the line segment PB forms with the x-axis, then the Law of Refraction can be written

$$\frac{\cos \alpha_1}{v_1} = \frac{\cos \alpha_2}{v_2}.$$

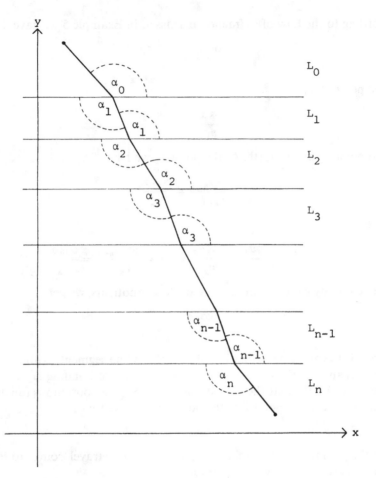

Figure 4.10

6. *The General Refraction Curve.* Suppose that the x, y-plane is partitioned into layers that run parallel to the x-axis (see Figure 4.10). Inside each layer the velocity of passage is constant. We select two points A and B situated in different layers; let A be in the layer L_0 and B in the layer L_n. Between the layers L_0 and L_n are the layers $L_1, L_2, \ldots, L_{n-1}$. The velocity of passage in the layer L_k is v_k for $k = 0, 1, 2, \ldots, n$. We are moving from the point A in layer L_0 through the consecutive layers $L_1, L_2, \ldots, L_{n-1}$ to the point B in layer L_n along a path requiring the shortest possible time. The path is made up of the line segments

$$AP_1, P_1P_2, P_2P_3, \ldots, P_{n-2}P_{n-1}, P_{n-1}P_n, P_nB.$$

The angles that these consecutive line segments make with the x-axis we denote by

$$\alpha_0, \alpha_1, \alpha_2, \ldots, \alpha_{n-2}, \alpha_{n-1}, \alpha_n.$$

According to the Law of Refraction discussed in Example 5 we have

$$\frac{\cos \alpha_0}{v_0} = \frac{\cos \alpha_1}{v_1}$$

at the point P_1 and

$$\frac{\cos \alpha_1}{v_1} = \frac{\cos \alpha_2}{v_2}$$

at the point P_2 and so forth. At the point P_n we have

$$\frac{\cos \alpha_{n-1}}{v_{n-1}} = \frac{\cos \alpha_n}{v_n}.$$

Hence,

$$\frac{\cos \alpha_0}{v_0} = \frac{\cos \alpha_1}{v_1} = \frac{\cos \alpha_2}{v_2} = \cdots = \frac{\cos \alpha_{n-1}}{v_{n-1}} = \frac{\cos \alpha_n}{v_n}.$$

If we denote by c the common value of these quotients, we get

$$\frac{\cos \alpha}{v} = c,$$

where α is the angle of inclination of any of the line segments with respect to the x-axis and v is the velocity of passage in the corresponding layer.

Suppose, finally, that the velocity of passage v is a continuous function of the ordinate y of the point P with coordinates (x, y), that is,

$$v = v(y).$$

Then the path requiring the shortest possible time to travel from A to B is the curve q characterized by the equation

$$\frac{\cos \alpha}{v} = c,$$

where α is the angle of inclination which the tangent line at the point P with coordinates (x, y) on the curve q makes with the x-axis, $v = v(y)$ is the velocity of passage at the point P, and c is a constant (see Figure 4.11). We arrive at this situation by considering layers that run parallel to the x-axis and for which the widths tend to zero.

7. *The Curve of Quickest Descent.* We assume now that the point A is on the x-axis and that the point B is below the point A but not on a vertical line going through the point A (see Figure 4.12). We assume that the y-axis is directed downward. Under the action of gravity alone an object is to travel from point A to point B; it is assumed that the object is at rest when at A and it is moving along a path from A to B such that it reaches B in the shortest possible time.

From physics we know that the gravitational force imparts a constant

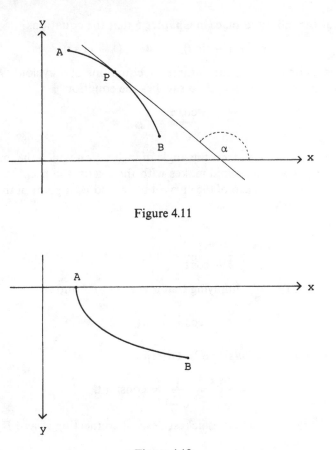

Figure 4.11

Figure 4.12

acceleration g of about 32 ft/sec² in the downward direction. But acceleration
is the rate of change of velocity v with respect to time, that is, $D_t v$. Since $v = 0$
when $t = 0$, we have $v = gt$, where t denotes time (in seconds). The displace-
ment y (measured in feet) is zero for $t = 0$ (because A is on the x-axis). Thus,
$y = gt^2/2$ and so

$$v^2 = 2gy \quad \text{or} \quad v = \sqrt{2g}\sqrt{y}.$$

Therefore, the curve connecting A and B along which an object starting from
rest at A and moving under the action of gravitational force alone reaches B
in the shortest possible time satisfies

$$\frac{\cos \alpha}{v} = \frac{\cos \alpha}{\sqrt{2g}\sqrt{y}} = c \quad \text{or} \quad \frac{\cos \alpha}{\sqrt{y}} = c_1,$$

where c and c_1 denote constants.

In Example 2 of Section 1 in Chapter 5 we shall study a curve called a
cycloid (see Figure 5.2 for a sketch of a cycloid); a cycloid is a curve traced by
a point on a circle as the circle rolls on a straight line without slipping. We

shall see at the indicated place in Chapter 5 that the equations

$$x = a(t - \sin t), \qquad y = a(1 - \cos t),$$

where a is a constant, are the parametric equations of a cycloid. What is of interest to us here is that a cycloid satisfies the equation

$$\frac{\cos \alpha}{\sqrt{y}} = c_1,$$

where c_1 is a constant and α is the angle of inclination which the tangent line at a point (x, y) on the cycloid makes with the x-axis; the point (x, y) on the cycloid must not be a cusp of the curve, but should be a point of the curve at which the derivative $D_x y$ exists.

Indeed,

$$D_x y = \frac{D_t y}{D_t x} = \frac{\sin t}{1 - \cos t} = \cot \frac{t}{2} \quad \text{and} \quad D_x y = \tan \alpha.$$

But $\tan(\pi/2 - \theta) = \cot \theta$ implying that $\alpha = \pi/2 - t/2$ and so

$$\cos \alpha = \sin \frac{t}{2}.$$

But $y = a(1 - \cos t) = 2a(\sin^2 t/2)$. Therefore,

$$\frac{\cos \alpha}{\sqrt{y}} = \frac{1}{\sqrt{2a}} = \text{constant}.$$

This shows that the curve of quickest descent connecting A and B is part of a cycloid.

REMARK. The cycloid arises in several kinematic problems involving descent of an object under gravity along a curve in a vertical plane; books on calculus of variations provide detailed treatment of kinematic problems.

8. The solution of our next problem is based on a result known as Viviani's theorem.

Viviani's Theorem. *For a point P inside an equilateral triangle* $\triangle ABC$ *the sum of the perpendiculars a, b, c from P to the sides is equal to the altitude h.*

Indeed, we have in terms of areas (see Figure 4.13)

$$\triangle ABC = \triangle PBC + \triangle PCA + \triangle PAB$$

or

$$\tfrac{1}{2}sh = \tfrac{1}{2}sa + \tfrac{1}{2}sb + \tfrac{1}{2}sc,$$

implying $h = a + b + c$. This proves Viviani's Theorem.

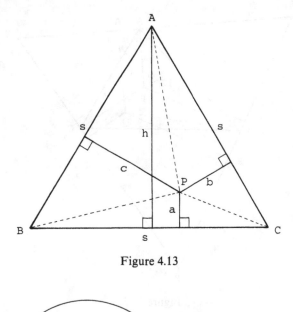

Figure 4.13

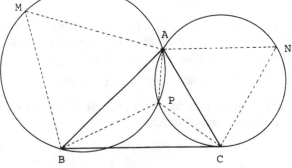

Figure 4.14

We now state the problem under consideration: We wish to determine a point P in a given triangle $\triangle ABC$ such that the sum $PA + PB + PC$ is a minimum.

It turns out that if the given triangle $\triangle ABC$ contains an angle of 120° or more the required point P is the vertex of this obtuse angle. We shall not consider this case but instead concentrate on the case in which the triangle $\triangle ABC$ has no angle as large as 120°; the required point P, called the *Fermat point of the triangle* $\triangle ABC$, is then characterized by the fact that at the point P each side of the triangle subtends an angle of 120°. Let us note that in order to find the Fermat point for a given triangle, all we need to do is draw equilateral triangles outwardly on two sides of the triangle $\triangle ABC$ and observe where their circumcircles meet (see Figure 4.14).

Consider the Fermat point P of the triangle $\triangle ABC$ in Figure 4.15; we shall

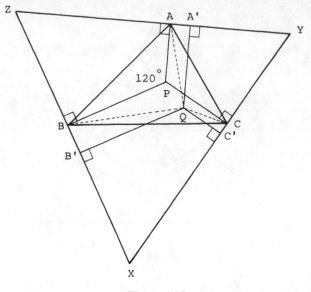

Figure 4.15

show that it solves our problem. As a first step we draw around the given triangle $\triangle ABC$ a triangle $\triangle XYZ$ by drawing perpendiculars to PA, PB, and PC. In quadrilateral $ZAPB$, then, the angle Z is

$$360° - 90° - 90° - 120° = 60°.$$

We follow a similar procedure for the angles at Y and X, making $\triangle XYZ$ equilateral. Thus, by Viviani's theorem

$$PA + PB + PC = h,$$

the altitude of $\triangle XYZ$.

For another point Q in $\triangle ABC$ we also have that the sum of the perpendiculars QA', QB', and QC' to the sides of $\triangle XYZ$ is h. But in general the hypotenuse QA of the right triangle $\triangle QAA'$ exceeds the leg QA'. Similarly, QB and QC, respectively, exceed the legs QB' and QC'. Thus, the sum

$$QA + QB + QC > QA' + QB' + QC' = h = PA + PB + PC,$$

showing P to be the solution to our problem.

We observe that at most one of the right triangles can collapse and yield the equality of a hypotenuse and leg (e.g., when Q lies on PA). Thus, at most one of QA', QB', or QC' can actually be as great as the corresponding QA, QB, or QC, giving universal validity to the inequality

$$QA + QB + QC > QA' + QB' + QC'.$$

This solves the problem under consideration.

In Figure 4.16 we indicate a second construction for the Fermat point P

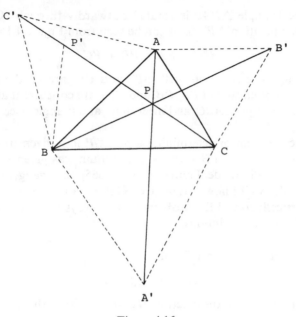

Figure 4.16

of the triangle $\triangle ABC$. Consider the triangle $\triangle ABC$ with the equilateral triangles

$$\triangle AC'B, \quad \triangle BA'C, \quad \text{and} \quad \triangle CB'A$$

erected (externally) on the three sides. After drawing the lines BB' and CC', which meet at P, we observe that a rotation through $60°$ about A takes $\triangle AC'C$ into $\triangle ABB'$. Hence, the angle $\angle C'PB = 60°$ and $C'C = BB'$. Similar reasoning shows that $A'A = CC'$. Thus,

$$AA' = BB' = CC'.$$

Moreover, since

$$\angle C'PB = 60° = \angle C'AB \quad \text{and} \quad \angle CPB' = 60° = \angle CAB',$$

the quadrangles $AC'BP$ and $CB'AP$ are cyclic; and since

$$\angle BPC = 120° \quad \text{while} \quad \angle CA'B = 60°,$$

$BA'CP$ is a third cyclic quadrangle. Therefore, the circumcircles of the triangles

$$\triangle BA'C, \quad \triangle CB'A, \quad \text{and} \quad \triangle AC'B$$

all pass through the point P. This is the Fermat point of $\triangle ABC$.

It is easy to see that

$$AA' = BB' = CC' = PA + PB + PC.$$

Indeed, if the triangle $\triangle PAB$ is rotated outwardly through $60°$ about the vertex B to the position $\triangle P'C'B$, it can be seen from Figure 4.16 that

$$C'C = PA + PB + PC.$$

Summing up: The segments AA', BB', and CC' are all the same length (equal to the minimal sum of PA, PB, and PC), are concurrent at the Fermat point P of the triangle $\triangle ABC$, and meet there at $60°$ angles (see Figure 4.16).

REMARK. The problem of determining a point P in a given triangle $\triangle ABC$ such that the sum $PA + PB + PC$ is a minimum was given by the famous number theorist Pierre de Fermat (1601–1665) to Evangelista Torricelli (1608–1647), the well-known student of Galileo and discoverer of the barometer. Torricelli solved the problem several ways and we considered the simplest of Torricelli's solutions.

9. Let $x > 0$ and $0 < \alpha < 1$. Then

$$x^{\alpha} - \alpha x \leq 1 - \alpha. \tag{4.81}$$

Indeed, differentiation of the function $f(x) = x^{\alpha} - \alpha x$ with respect to x yields

$$f'(x) = \alpha(x^{\alpha-1} - 1).$$

Clearly,

$$f'(x) > 0 \quad \text{for } 0 < x < 1,$$
$$= 0 \quad \text{for } x = 1,$$
$$< 0 \quad \text{for } x > 1;$$

hence, $f(1) \geq f(x)$ for $x > 0$ and this implies (4.81).

REMARK. Putting $x = a/b$ and setting $1 - \alpha = \beta$, we see that (4.81) gives

$$a^{\alpha}b^{\beta} \leq \alpha a + \beta b, \tag{4.82}$$

where a and b are positive numbers and α and β satisfy $0 < \alpha < 1, 0 < \beta < 1$, and $\alpha + \beta = 1$. [Note that there is strict inequality in (4.82) if and only if $a \neq b$.]

10. Let $x > 0$ and a and b be positive and $a \neq b$. Then

$$\left(\frac{a + x}{b + x}\right)^{b+x} > \left(\frac{a}{b}\right)^{b}. \tag{4.83}$$

Indeed, differentiation of the function

$$h(x) = \left(\frac{a + x}{b + x}\right)^{b+x} \quad \text{with } x > 0$$

gives

$$h'(x) = g(x)h(x),$$ (4.84)

where

$$g(x) = \frac{b - a}{a + x} + \ln\frac{a + x}{b + x}.$$

The sign of $h'(x)$ is the same as the sign of $g(x)$. Since

$$g'(x) = -\frac{(a - b)^2}{(a + x)^2(b + x)} < 0,$$

the function g is decreasing and, consequently,

$$g(x) > g(+\infty) = 0 \quad \text{for } x > 0.$$ (4.85)

By (4.85) and (4.84) we conclude that the function h is increasing and therefore (4.83) holds.

REMARK. Letting $x = a$ in (4.83) we get

$$\left(\frac{2a}{a + b}\right)^{b+a} > \left(\frac{a}{b}\right)^{b}$$

or

$$\left(\frac{a + b}{2}\right)^{a+b} < a^a b^b \quad \text{for } a \neq b,$$ (4.86)

where a and b are assumed to be positive.

11. Let a and b be positive real numbers and $a \neq b$. Then

$$a^b b^a < \left(\frac{a + b}{2}\right)^{a+b} < a^a b^b.$$ (4.87)

Indeed, in view of (4.86) it only remains to verify that

$$a^b b^a < \left(\frac{a + b}{2}\right)^{a+b}.$$ (4.88)

Putting

$$\alpha = \frac{b}{a + b} \quad \text{and} \quad \beta = \frac{a}{a + b}$$

in (4.82) and keeping in mind that $a \neq b$, we see that

$$a^b b^a < \left(\frac{2ab}{a + b}\right)^{a+b}.$$

But

$$\frac{2ab}{a+b} = \frac{2}{1/a + 1/b} < \sqrt{ab} < \frac{a+b}{2} \quad \text{for } a \neq b$$

by (4.62) and (4.63); thus,

$$\left(\frac{2ab}{a+b}\right)^{a+b} < \left(\frac{a+b}{2}\right)^{a+b}$$

and (4.88) follows.

12. The inequality

$$x(\ln x) \geq x - 1 \quad \text{for } x > 0 \tag{4.89}$$

implies

$$\sum_{i=1}^{n} p_i(\ln p_i) \geq \sum_{i=1}^{n} p_i(\ln q_i) \tag{4.90}$$

for $p_i > 0$, $q_i > 0$ $(i = 1, 2, \ldots, n)$ and

$$\sum_{i=1}^{n} p_i = \sum_{i=1}^{n} q_i.$$

Indeed, since $p_i/q_i > 0$, we get from (4.89) that

$$\frac{p_i}{q_i}\ln\frac{p_i}{q_i} \geq \frac{p_i}{q_i} - 1$$

or

$$p_i \ln\frac{p_i}{q_i} \geq p_i - q_i$$

because $q_i > 0$. Thus,

$$\sum_{i=1}^{n} p_i \ln\frac{p_i}{q_i} \geq \sum_{i=1}^{n} (p_i - q_i) = 0$$

or

$$\sum_{i=1}^{n} [p_i(\ln p_i) - p_i(\ln q_i)] \geq 0.$$

But the latter inequality implies (4.90).

REMARKS. Let a and b be positive real numbers. The inequality $a^b b^a \leq a^a b^b$ is equivalent to

$$b(\ln a) + a(\ln b) \leq a(\ln a) + b(\ln b) \tag{4.91}$$

and the inequality $[(a + b)/2]^{a+b} \leq a^a b^b$ is equivalent to

$$(a + b) \ln \frac{a + b}{2} \le a(\ln a) + b(\ln b). \tag{4.92}$$

Evidently, inequalities (4.91) and (4.92) are immediate consequences of inequality (4.90) and thus a good part of inequality (4.87) can be verified by use of inequality (4.90). Recall that inequality (4.89) is a simple consequence of inequality (1.3) in Chapter 1.

13. Let $[a, b]$ be a closed interval of finite length and f be a continuous function on $[a, b]$. Suppose that for each x between a and b we have

$$\lim_{h \to 0} \frac{f(x + h) + f(x - h) - 2f(x)}{h^2} = 0.$$

Then $f(x) = Ax + B$ for suitable numbers A and B.

Indeed, let c be any point between a and b. We put

$$Q(x) = \frac{(x - b)(x - c)}{(a - b)(a - c)} f(a) + \frac{(x - c)(x - a)}{(b - c)(b - a)} f(b) + \frac{(x - a)(x - b)}{(c - a)(c - b)} f(c).$$

Clearly, $Q(a) = f(a)$, $Q(b) = f(b)$, and $Q(c) = f(c)$. We put

$$g(x) = Q(x) - f(x) \quad \text{for } a \le x \le b.$$

Then $g(a) = g(b) = g(c) = 0$. Evidently, g is continuous on $[a, b]$. Hence, there are points S and s between a and b where g assumes its largest and its smallest values, respectively, by Proposition 3.13; if one of these points happens to be either the endpoint a or the endpoint b of the interval, we may set it equal to c and thereby will always have that S and s are situated between a and b.

But for $a < x < b$ and $h \to 0$ we have

$$\lim_{h \to 0} \frac{g(x + h) + g(x - h) - 2g(x)}{h^2} = \lim_{h \to 0} \frac{Q(x + h) + Q(x - h) - 2Q(x)}{h^2}$$

$$= \frac{2f(a)}{(a - b)(a - c)} + \frac{2f(b)}{(b - c)(b - a)} = K.$$

Thus, we also have

$$\lim_{h \to 0} \frac{g(S + h) + g(M - h) - 2g(S)}{h^2} = K$$

and

$$\lim_{h \to 0} \frac{g(s + h) + g(s - h) - 2g(s)}{h^2} = K.$$

But

$$g(S + h) \le g(S), \qquad g(S - h) \le g(S)$$

and

$$g(s + h) \geq g(s), \qquad g(s - h) \geq g(s);$$

this means that the number K has to be nonpositive and nonnegative simultaneously. Thus, $K = 0$ and so

$$f(c) = \frac{b - c}{b - a} f(a) + \frac{c - a}{b - a} f(b).$$

However, c was an arbitrary point between a and b. Moreover, our equation $g = Q - f$ remains valid if we replace c by either a or b. Hence, in the entire interval

$$f(x) = \frac{b - x}{b - a} f(a) + \frac{x - a}{b - a} f(b)$$

and so f is seen to be of the form $f(x) = Ax + B$.

14. Let w be a continuous function on the closed interval $[a, b]$ which is assumed to be of finite length. Suppose that w has a second derivative at each point of the open interval (a, b). Then

$$\frac{\dfrac{w(x) - w(a)}{x - a} - \dfrac{w(b) - w(a)}{b - a}}{x - b} = \frac{1}{2} w''(c),$$

where c is some point between a and b.

Indeed, let x be fixed with $a < x < b$. Put

$$v(t) = \begin{bmatrix} w(t) & t^2 & t & 1 \\ w(x) & x^2 & x & 1 \\ w(a) & a^2 & a & 1 \\ w(b) & b^2 & b & 1 \end{bmatrix}.$$

Then $v(a) = v(x) = v(b) = 0$. By the Mean Value Theorem (see Proposition 4.5) there exist numbers α and β such that $a < \alpha < x < \beta < b$ and $v'(\alpha) = v'(\beta) = 0$. Applying the Mean Value Theorem once again, we see that there is a number c such that $\alpha < c < \beta$ and $v''(c) = 0$. Evidently, c is between a and b. Now

$$v''(c) = \begin{bmatrix} w''(c) & 2 & 0 & 0 \\ w(x) & x^2 & x & 1 \\ w(a) & a^2 & a & 1 \\ w(b) & b^2 & b & 1 \end{bmatrix} = 0.$$

Expanding by the first row we find the desired relation.

15. *Bernoulli's Inequality.* If $x > -1$, $x \neq 0$, and n is a positive integer larger than 1, then

$$(1 + x)^n > 1 + nx. \tag{4.93}$$

Here we consider the following generalization of Bernoulli's Inequality. Let α be a fixed real number different from 0 and 1. Then $x > -1$ and $x \neq 0$ imply

$$(1 + x)^\alpha > 1 + \alpha x \quad \text{if } \alpha < 0 \text{ or } \alpha > 1 \tag{4.94}$$

and

$$(1 + x)^\alpha < 1 + \alpha x \quad \text{if } 0 < \alpha < 1. \tag{4.95}$$

Indeed, let $f(x) = (1 + x)^\alpha - (1 + \alpha x)$ for $x > -1$. Then

$$f'(x) = \alpha(1 + x)^{\alpha-1} - \alpha \quad \text{and} \quad f''(x) = \alpha(\alpha - 1)(1 + x)^{\alpha-2}.$$

Suppose $\alpha < 0$ or $\alpha > 1$. Then f is concave up on $(-1, \infty)$. Moreover, f is decreasing on $(-1, 0)$ and increasing on $(0, \infty)$. Since $f(0) = 0$, it follows that $f(x) > 0$ if $-1 < x \neq 0$. Thus, (4.94) follows. In the case where $0 < \alpha < 1$, f is concave down on $(-1, \infty)$; moreover, f is increasing on $(-1, 0)$ and decreasing on $(0, \infty)$ and $f(0) = 0$. Thus, $f(x) < 0$ if $-1 > x \neq 0$ and (4.95) is obtained, finishing the proof.

REMARK. Sometimes Bernoulli's Inequality is stated in the following form: If n is a positive integer larger than 1, then

$$(1 + x_1)(1 + x_2)\cdots(1 + x_n) > 1 + x_1 + x_2 + \cdots + x_n, \tag{4.96}$$

provided that each of the numbers $x_1, x_2, \ldots, x_n$ is different from 0 and larger than -1 and all numbers $x_1, x_2, \ldots, x_n$ have the same sign.

Indeed, from the inductive assumption

$$(1 + x_1)(1 + x_2)\cdots(1 + x_k) > 1 + x_1 + x_2 + \cdots + x_k$$

we can easily see that

$$(1 + x_1)(1 + x_2)\cdots(1 + x_k)(1 + x_{k+1})$$
$$> 1 + (x_1 + x_2 + \cdots + x_k) + x_{k+1} + (x_1 + x_2 + \cdots + x_k)x_{k+1}$$
$$> 1 + (x_1 + x_2 + \cdots + x_k + x_{k+1})$$

because $(x_1 + x_2 + \cdots + x_k)x_{k+1} > 0$. The rest of the verification of (4.96), namely, that it holds for $n = 2$, is trivial.

16. Let $y = (1 + 2x - \sqrt{1 + 4x})/2$; it is defined on $[-\frac{1}{4}, +\infty)$ and it has derivative $y' = 1 - 1/\sqrt{1 + 4x}$. Hence,

$$f(x) = \frac{1 + 2x - \sqrt{1 + 4x}}{2} \quad \text{for } x \in [0, +\infty)$$

is an increasing function and has an inverse function f^{-1}. Verify that

$$f^{-1}(x) = x + \sqrt{x} \quad \text{for } x \in [0, +\infty).$$

Indeed, interchanging x and y we have

$$x = \frac{1 + 2y - \sqrt{1 + 4y}}{2}. \tag{4.97}$$

To solve equation (4.97) for y, we observe that

$$\frac{1 + 2y - \sqrt{1 + 4y}}{2} = \left(\frac{\sqrt{1 + 4y} - 1}{2}\right)^2$$

and so

$$x + \sqrt{x} = \frac{1 + 2y - \sqrt{1 + 4y}}{2} - \frac{\sqrt{1 + 4y} - 1}{2} = y.$$

REMARK. Note that $y = (1 + 2x - \sqrt{1 + 4x})/2$ is a decreasing function on the interval $(-\frac{1}{4}, 0)$. While $y = (1 + 2x - \sqrt{1 + 4x})/2$ is defined on $[-\frac{1}{4}, +\infty)$, it is not invertible on all of $[-1/4, +\infty)$; see Proposition 2.18,

17. The polynomial

$$P(x) = 1 + \frac{x}{1} + \frac{x^2}{1 \cdot 2} + \cdots + \frac{x^n}{n!}$$

does not have multiple roots.

Indeed, a multiple root of the polynomial $P(x)$ must also be a root of its derivative

$$P'(x) = 1 + \frac{x}{1} + \cdots + \frac{x^{n-1}}{(n-1)!} = P(x) - \frac{x^n}{n!}.$$

Hence, if $P(x_0) = P'(x_0) = 0$, then $x_0 = 0$, but 0 is not a root of $P(x) = 0$.

18. Let f be a polynomial of degree n. If $f(a) > 0$, $f'(a) \geq 0, \ldots, f^{(n)}(a) \geq 0$, then all real roots of f do not exceed a.

Indeed, expanding $f(x)$ in powers of $x - a$, we get, for $x \geq a$,

$$f(x) = f(a) + \frac{f'(a)}{1!}(x - a) + \frac{f''(a)}{2!}(x - a)^2 + \cdots + \frac{f^{(n)}(a)}{n!}(x - a)^n > 0.$$

19. Let $c_k > 0$ for $k = 1, 2, \ldots, n$ and $c_1 + c_2 + \cdots + c_n = 1$. Then

$$\sum_{k=1}^{n} \left(c_k + \frac{1}{c_k}\right)^2 \geq \frac{(1 + n^2)^2}{n}. \tag{4.98}$$

Indeed, the Cauchy–Schwarz Inequality [see (1.43) in Chapter 1]

$$\left(\sum_{k=1}^{n} a_k b_k\right)^2 \leq \left(\sum_{k=1}^{n} a_k^2\right)\left(\sum_{k=1}^{n} b_k^2\right)$$

gives, putting $a_k = 1$ and $b_k = c_k + 1/c_k$,

$$\left(\sum_{k=1}^{n}\left(c_k + \frac{1}{c_k}\right)\right)^2 \le n \sum_{k=1}^{n}\left(c_k + \frac{1}{c_k}\right)^2.$$

But

$$\sum_{k=1}^{n}\left(c_k + \frac{1}{c_k}\right) = \sum_{k=1}^{n} c_k + \sum_{k=1}^{n}\frac{1}{c_k} = 1 + \sum_{k=1}^{n}\frac{1}{c_k}.$$

By (4.62) and (4.63) the arithmetic mean is larger than or equal to the harmonic mean of the numbers $1/c_1, 1/c_2, \ldots, 1/c_n$, that is,

$$\frac{n}{c_1 + c_2 + \cdots + c_n} \le \frac{1}{n}\sum_{k=1}^{n}\frac{1}{c_k} \quad \text{or} \quad n^2 \le \sum_{k=1}^{n}\frac{1}{c_k}.$$

Hence,

$$\sum_{k=1}^{n}\left(c_k + \frac{1}{c_k}\right) = 1 + \sum_{k=1}^{n}\frac{1}{c_k} \ge 1 + n^2$$

and so

$$\sum_{k=1}^{n}\left(c_k + \frac{1}{c_k}\right)^2 \ge \frac{1}{n}\left(\sum_{k=1}^{n}\left(c_k + \frac{1}{c_k}\right)\right) \ge \frac{1}{n}(1 + n^2)^2.$$

20. We are given a circle with center at O, radius r, and a tangent line AT. In Figure 4.17, AM equals arc AP and P is the intersection of the line through M and P with the line through A and O. Determine the limiting position of B as P approaches A as a limiting position.

Let $\theta = \angle POA$, $PP_1 \perp OA$, and $PP_2 \perp MA$ (see Figure 4.17). Arc $AP = r\theta = AM$, and, because the triangles $\triangle MAB$ and $\triangle MP_2P$ are similar,

$$\frac{AB}{AM} = \frac{PP_2}{P_2M}.$$

But $PP_2 = OA - OP_1 = r - r\cos\theta$ and $P_2M = AM - P_2A = AM - P_1P = r\theta - r\sin\theta$. Thus,

$$AB = AM\frac{PP_2}{P_2M} = r\frac{\theta - \theta\cos\theta}{\theta - \sin\theta}.$$

As P approaches A, $\theta \to 0$. If we let $\theta \to 0$ in our expression for AB, we get an

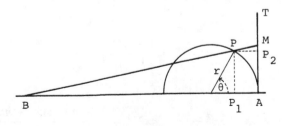

Figure 4.17

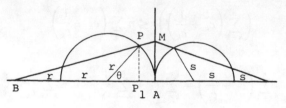

Figure 4.18

indeterminate form. Using L'Hôpital's Rules (see Proposition 4.10), we obtain

$$\lim_{P \to A} AB = r \lim_{\theta \to 0} \frac{\theta - \theta \cos \theta}{\theta - \sin \theta} = r \lim_{\theta \to 0} \frac{1 - \cos \theta + \theta \sin \theta}{1 - \cos \theta}$$

$$= r \lim_{\theta \to 0} \frac{2 \sin \theta + \theta \cos \theta}{\sin \theta} = r \lim_{\theta \to 0} \frac{3 \cos \theta - \theta \sin \theta}{\cos \theta} = 3r.$$

REMARKS. Let $AB = 3r$ and suppose $\angle AOP \le \pi/3$ in Figure 4.18. Then the line through the points B and P intersects the tangent to the circle at A at such a point M that the segment AM on the tangent and the arc AP on the circle are approximately equal. Now $PP_1 = r \sin \theta$ and $OP_1 = r \cos \theta$. We therefore get that $\theta = \angle AOP$ is approximately equal to $(3 \sin \theta)/(2 + \cos \theta)$, an approximation due to the learned Cardinal Nicolaus Cusanus of the 15th century. For $\theta = \pi/3$ the arc AP is smaller than the segment AM by only about 0.8% of the radius r. That θ is approximately equal to $(3 \sin \theta)/(2 + \cos \theta)$ is of course a consequence of the similarity of the triangles $\triangle BP_1 P$ and $\triangle BAM$. Figure 4.18 also indicates how one can measure off approximately equal arcs on two circles of different radii r and s.

21. We have, for $x \ne 1$,

$$Q_n = 1^2 + 2^2 x + 3^2 x^2 + \cdots + n^2 x^{n-1}$$

$$= \frac{1 + x - (n + 1)^2 x^n + (2n^2 + 2n - 1)x^{n+1} - n^2 x^{n+2}}{(1 - x)^3}.$$

Indeed, for $x \ne 1$, let

$$T_n = 1 + x + x^2 + x^3 + \cdots + x^n.$$

Then

$$T_n = \frac{1 - x^{n+1}}{1 - x}$$

and

$$\frac{d}{dx}(T_n) = 1 + 2x + 3x^2 + \cdots + nx^{n-1} = \frac{1 - (n + 1)x^n + nx^{n+1}}{(1 - x)^2} = S_n$$

and

$$Q_n = \frac{d}{dx}(xS_n) = \frac{1 + x - (n + 1)^2 x^n + (2n^2 + 2n - 1)x^{n+1} - n^2 x^{n+2}}{(1 - x)^3}.$$

22. We have, for $x \neq 0$,

$$\frac{1}{2}\tan\frac{x}{2} + \frac{1}{4}\tan\frac{x}{4} + \cdots + \frac{1}{2^n}\tan\frac{x}{2^n} = \frac{1}{2^n}\tan\frac{x}{2^n} - \cot x.$$

Indeed, for $x \neq 0$,

$$\sin x = 2\left(\cos\frac{x}{2}\right)\left(\sin\frac{x}{2}\right) = 2^2\left(\cos\frac{x}{2}\right)\left(\cos\frac{x}{4}\right)\left(\sin\frac{x}{4}\right)$$

$$= \cdots = 2^n\left(\cos\frac{x}{2}\right)\left(\cos\frac{x}{4}\right)\cdots\left(\cos\frac{x}{2^n}\right)\left(\sin\frac{x}{2^n}\right)$$

and so

$$\left(\cos\frac{x}{2}\right)\left(\cos\frac{x}{2^2}\right)\cdots\left(\cos\frac{x}{2^n}\right) = \frac{\sin x}{2^n(\sin x/2^n)}$$

or

$$\ln\left(\cos\frac{x}{2}\right) + \ln\left(\cos\frac{x}{2^n}\right) + \cdots + \ln\left(\cos\frac{x}{2^n}\right) = \ln(\sin x) - \ln 2^n - \ln\left(\sin\frac{x}{2^n}\right).$$

Differentiating both sides of the last equality gives the desired relation.

23. Let f be such that $f''(x)$ exists in (a, b) and $f'(a) = f'(b) = 0$. Then there is a point c satisfying $a < c < b$ such that

$$|f''(c)| \geq \frac{4}{(b - a)^2}|f(b) - f(a)|.$$

Indeed, since $f'(a) = f'(b) = 0$, we get, by Proposition 4.11,

$$f\left(\frac{a + b}{2}\right) = f(a) + \frac{f''(x_1)}{2!}\left(\frac{b - a}{2}\right)^2$$

and

$$f\left(\frac{a + b}{2}\right) = f(b) + \frac{f''(x_2)}{2!}\left(\frac{b - a}{2}\right)^2,$$

where x_1 is between a and $(a + b)/2$ and x_2 is between $(a + b)/2$ and b. But

$$|f(b) - f(a)| \leq \left|f(b) - f\left(\frac{a + b}{2}\right)\right| + \left|f\left(\frac{a + b}{2}\right) - f(a)\right|$$

$$\leq \left(\frac{b - a}{2}\right)^2\left(\frac{1}{2!}\right)[|f''(x_1)| + |f''(x_2)|].$$

Let $|f''(c)|$ be the larger of the two numbers $|f''(x_1)|$ and $|f''(x_2)|$; then

$$\tfrac{1}{2}[|f''(x_1)| + |f''(x_2)|] \le |f''(c)|$$

and the claim follows immediately.

24. Let $0 < a < b$. Then $b \le e$ implies $a^b < b^a$ and $a \ge e$ implies $a^b > b^a$. Indeed, $a^b < b^a$ if and only if $(\ln a)/a < (\ln b)/b$, which is true if

$$\frac{d}{dx}\left(\frac{\ln x}{x}\right) = \frac{1 - \ln x}{x^2} > 0$$

throughout (a, b), and this holds if $b \le e$. Similarly, $a^b > b^a$ if

$$\frac{d}{dx}\left(\frac{\ln x}{x}\right) = \frac{1 - \ln x}{x^2} < 0$$

throughout (a, b), which holds if $a \ge e$.

REMARK. What if $a < e < b$? Taking $a = 2$, $b = 3$, we have $a^b < b^a$; if $a = 2$, $b = 4$, then $a^b = b^a$; while if $a = 2$, $b = 5$, then $a^b > b^a$.

25. *The Rule of Proportional Parts.* If $a < b < c$ and the values of a function f are tabulated at the points a and c, the rule is that, approximately,

$$f(b) - f(a) = \frac{b - a}{c - a}(f(c) - f(a)).$$

In other words, we replace the arc of the curve $y = f(x)$ on the interval $[a, c]$ by the chord connecting the points $(a, f(a))$ and $(c, f(c))$.

To obtain an upper bound for the error, suppose that f is continuous on the closed interval $[a, c]$ and that f has second derivative at each point of the open interval (a, c). Applying the Mean Value Theorem (see Proposition 4.5) to the function

$$g(x) = \begin{bmatrix} f(x) & x^2 & x & 1 \\ f(a) & a^2 & a & 1 \\ f(b) & b^2 & b & 1 \\ f(c) & c^2 & c & 1 \end{bmatrix}$$

we obtain

$$f(b) - f(a) = \frac{b - a}{c - a}(f(c) - f(a)) + \frac{1}{2}(b - a)(b - c)f''(s),$$

where $a < s < c$. (The details of proof are completely analogous to the details of proof in Example 14 of this section.)

Next, we note that $b - a$ and $c - b$ are positive and that the sum

$$(b - a) + (c - b) = c - a$$

which is independent of b. But the product $(b - a)(c - b)$ is largest if

$$b - a = c - b = \frac{c - a}{2}$$

(see Number 1 of Worked Examples and Comments following Proposition 4.18). Thus,

$$|(b - a)(b - c)| \leq \frac{(c - a)^2}{4}.$$

Hence, the error involved in assuming that as x increases from a to c, the increase in $f(x)$ is proportional to the increase in x, can not be greater than $\frac{1}{8}(c - a)^2 M$, where $M = \sup\{|f''(x)|: a \leq x \leq c\}$.

26. Let f and g be continuous functions on a closed interval $[a, b]$ of finite length and both be differentiable on the open interval (a, b). If, in addition, $g'(t)$ is not zero for any t in (a, b), then

$$\frac{f(b) - f(a)}{g(b) - g(a)} = \frac{f'(x)}{g'(x)} \tag{4.99}$$

for some x in (a, b).

We have already encountered the foregoing result in part (iii) of Proposition 4.9. Our present objective is to show that formula (4.99) is an example of the application of the Mean Value Theorem (see Proposition 4.8) to a composition of functions.

Indeed, if G is a continuous function on $[\alpha, \beta]$, differentiable on (α, β), with a nonvanishing derivative in (α, β), then, since the derivative of $f[G(t)]$ is $f'[G(t)]G'(t)$, by Proposition 4.8 there is a point γ in (α, β) such that

$$\frac{f[G(\beta)] - f[G(\alpha)]}{\beta - \alpha} = f'[G(\gamma)]G'(\gamma). \tag{4.100}$$

Since $G'(t) \neq 0$ in (α, β), by Proposition 4.8 we see that $G'(t)$ is of constant sign, and so G has a differentiable inverse function g, say, such that $G'(t) = 1/g'[G(t)]$. Putting a, b, x for $G(\alpha)$, $G(\beta)$, $G(\gamma)$ so that $g(a) = \alpha$ and $g(b) = \beta$, we see that equation (4.100) becomes equation (4.99).

27. Let f and g be continuous functions on a closed interval $[\alpha, b]$ of finite length and both be differentiable n times on the open interval (α, b). Moreover, let

$$f^{(k)}(\alpha) = g^{(k)}(\alpha) = 0, \qquad 1 \leq k < n,$$

and

$$|g^{(k)}(t)| > 0 \quad \text{in } [\alpha, b], \qquad 1 \leq k \leq n.$$

Then if $\alpha \leq a < b$, we can find x in (α, b) such that

$$\frac{f(b) - f(a)}{g(b) - g(a)} = \frac{f^{(n)}(x)}{g^{(n)}(x)}.$$

Indeed, by (4.99) we can find $x_1, x_2, \ldots, x_{n-1}, x$ in turn, such that

$$\frac{f(b) - f(a)}{g(b) - g(a)} = \frac{f'(x_1)}{g'(x_1)}$$

and

$$\frac{f'(x_1)}{g'(x_1)} = \frac{f'(x_1) - f'(\alpha)}{g'(x_1) - g'(\alpha)} = \frac{f''(x_2)}{g''(x_2)} = \cdots = \frac{f^{(n-1)}(x_{n-1})}{g^{(n-1)}(x_{n-1})}$$

$$= \frac{f^{(n-1)}(x_{n-1}) - f^{(n-1)}(\alpha)}{g^{(n-1)}(x_{n-1}) - g^{(n-1)}(\alpha)} = \frac{f^{(n)}(x)}{g^{(n)}(x)},$$

which completes the proof.

28. *Formula of Huygens.* If A is the chord of any circular arc and B that of half the arc, then the length L of the arc is approximately equal to

$$\frac{8B - A}{3}.$$

Indeed, if α is the central angle and R the radius, then by the cosine law we get

$$A^2 = R^2 + R^2 - 2R^2 \cos \alpha.$$

But $2 \sin^2(\alpha/2) = 1 - \cos \alpha$ and so

$$A = 2R \sin \frac{\alpha}{2}.$$

It is clear that $B = 2R \cos(\alpha/4)$ and so

$$\frac{8B - A}{3} = \frac{R}{3} \left(16 \sin \frac{\alpha}{4} - 2 \sin \frac{\alpha}{2} \right) = R\alpha + \lambda \alpha^5,$$

where λ denotes a finite factor, and where we have replaced the sine function by its Taylor polynomial of order 5. But $L = R\alpha$.

REMARK. A better approximation for the length L of a circular arc is given by the formula

$$\tfrac{1}{45}(A + 256C - 40B), \tag{4.101}$$

where A is the chord of the circular arc, B that of half the arc, and C that of quarter of the arc.

EXERCISES TO CHAPTER 4

4.1. If f has n roots in $[a, b]$, show that $f^{(n-1)}$ has at least one root in that interval.
      [Hint: By Rolle's Theorem (see Proposition 4.3), f' has at least one root between any two of f, so that f' has at least $n - 1$ roots; hence f'' has at least $n - 2$ roots, f''' at least $n - 3$, and so on.]

4.2. If $f(a) = f(b) = 0$ and $f''(x) \neq 0$ in (a, b), show that $f(x) \neq 0$ in (a, b).
[Hint: For $a < x < b$, let

$$g(t) = f(x)(t - a)(t - b) - f(t)(x - a)(x - b).$$

Then g vanishes at $t = a$, $t = x$, and $t = b$, so that, by Rolle's Theorem, g' vanishes between a and x, and again between x and b, so that g'' vanishes (at least) once at c, say, in (a, b). But $g''(t) = 2f(x) - (x - a)(x - b)f''(t)$ and so $f(x) = \frac{1}{2}(x - a)(x - b)f''(c)$, which shows that $f(x) \neq 0$ in (a, b), since $f''(c) \neq 0$.]

4.3. If $a < b < c$, $a + b + c = 2$, and $ab + bc + ca = 1$, show that a, b, and c lie in the intervals $(0, \frac{1}{3})$, $(\frac{1}{3}, 1)$, and $(1, \frac{4}{3})$, respectively.
[Hint: The numbers a, b, and c are the roots of the cubic polynomial

$$f(t) = t^3 - 2t^2 + t - abc = t(t - 1)^2 - abc.$$

But $f'(t) = (3t - 1)(t - 1)$, and a root of f' lies between two of the roots of f, by Rolle's Theorem, and therefore $a < \frac{1}{3} < b < 1 < c$.
Since $f(c) = 0$, therefore $ab = (c - 1)^2$ so that $a > 0$, and therefore $f(0) = -abc < 0$. Since $f'(\frac{1}{3}) = 0$, $f(t) - f(\frac{1}{3})$ has the factor $(t - \frac{1}{3})^2$, and since the sum of the roots of the equation $f(t) - f(\frac{1}{3}) = 0$ is 2, the third root is $t = \frac{4}{3}$. Thus, $f(t) = (t - \frac{4}{3})(t - \frac{1}{3})^2 + f(\frac{1}{3})$. One root only of $f(t) = 0$ lies between 0 and $\frac{1}{3}$ and $f(0) < 0$, therefore $f(\frac{1}{3}) > 0$. But $f(c) = 0$ and so $(c - \frac{4}{3})(c - \frac{1}{3})^2 = -f(\frac{1}{3}) < 0$. Thus, $c < \frac{4}{3}$. Hence, $0 < a < \frac{1}{3} < b < 1 < c < \frac{4}{3}$.]

4.4. If $f(0) = a$, $f(a) = b$, $f'(0) = -1$, and if $|f''(x)| < 1/4|a|$ in the interval $[-2a, 2a]$, show that $|1 + f'(x)| < \frac{1}{2}$ in $[-2a, 2a]$ and hence that $|f(a + b)| < \frac{1}{2}|f(a)| < \frac{1}{4}|a|$.
[Hint: By the Mean Value Theorem (see Proposition 4.5), for $-2a \leq x \leq 2a$,

$$|f'(x) + 1| = |f'(x) - f'(0)| = |xf''(c)| < \frac{2|a|}{4|a|} = \frac{1}{2}.$$

Since $f(a) - a = f(a) - f(0) = af'(a)$, therefore

$$|f(a)| = |a||1 + f'(a)| < \frac{1}{2}|a|.$$

Moreover, $|a + b| = |a + f(a)| < \frac{3}{2}a$ so that $a + b$ lies in $[-2a, 2a]$ and

$$f(a + b) = f(a) + bf'(\beta), \qquad \beta \text{ in } (-2a, 2a),$$

$$= f(a)(1 + f'(\beta))$$

and so $|f(a + b)| < \frac{1}{2}|f(a)| < \frac{1}{4}|a|$.]

4.5. Let f be twice differentiable in $[a, b]$ and such that $f(a) = f(b) = 0$ and $f(c) > 0$ for $a < c < b$. Show that there is at least one value t between a and b for which $f''(t) < 0$.
[Hint: Since f'' exists, f' and f both exist and are continuous on $[a, b]$. Since c is a point between a and b, applying the Mean Value Theorem to f on the intervals $[a, c]$ and $[c, b]$, respectively, we get

$$\frac{f(c) - f(a)}{c - a} = f'(t_1), \qquad a < t_1 < c,$$

and

$$\frac{f(b) - f(c)}{b - c} = f'(t_2), \qquad c < t_2 < b.$$

But $f(a) = f(b) = 0$ and so

$$f'(t_1) = \frac{f(c)}{c - a} \quad \text{and} \quad f'(t_2) = -\frac{f(c)}{b - c} \quad \text{where } a < t_1 < c < t_2 < b.$$

Again f' is continuous and differentiable on $[t_1, t_2]$. Therefore, by the Mean Value Theorem

$$\frac{f'(t_2) - f'(t_1)}{t_2 - t_1} = f''(t), \qquad t_1 < t < t_2.$$

Substituting the values of $f'(t_2)$ and $f'(t_1)$, we get

$$f''(t) = -\frac{(b - a)f(c)}{(t_2 - t_1)(b - c)(c - a)} < 0.]$$

4.6. If h is differentiable, with positive nonzero derivative, in $[a, b]$, $h(a) = 0$, and if in the interval $[h(a), h(b)]$, $a \le f(x) \le b$, $f(0) = a$, and $f'(x) = 1/h'[f(x)]$, show that f and h are inverse functions.

[Hint: The function $h[f(x)]$ is differentiable in $[h(a), h(b)]$ with derivative $h'[f(x)]f'(x) = 1$, and so $h[f(x)] - x$ is constant; when $x = 0$, $h[f(x)] - x = 0$, and therefore $h[f(x)] = x$ for all x in $[h(a), h(b)]$.

Moreover, since $h'(x) > 0$ in $[a, b]$, therefore $h(a) \le h(x) \le h(b)$ in $[a, b]$, and so $h\{f[h(x)]\} = h(x)$ in $[a, b]$; since h is monotonic and increasing it follows that $f[h(x)] = x$ for all x in $[a, b]$.]

4.7. If $f'(a) = f'(b) = 0$ and $f'(x) \ne 0$ in (a, b), show that $f(a)$ and $f(b)$ can not be both maxima nor both minima.

[Hint: By Proposition 4.8, $f'(x)$ is of constant sign in (a, b). If $f'(x)$ is positive in (a, b), then $f(a)$ is not a maximum and $f(b)$ is not a minimum, that is, $f(a)$ and $f(b)$ are not both maximum nor both minimum. A similar argument holds if $f'(x)$ is negative in (a, b).]

4.8. If f is differentiable in $[a, b]$ and if f' vanishes at only a finite number of points in $[a, b]$, then between any two points in (a, b) where $f(x)$ is maximum there is a point where $f(x)$ is minimum, and between two minimum values, a maximum.

[Hint: If α and β lie in (a, b) and $f(\alpha)$ and $f(\beta)$ are both maximum values of $f(x)$, let $\alpha_1, \alpha_2, \ldots, \alpha_p$ be the points between α and β where $f'(x) = 0$. Since $f(\alpha)$ is a maximum, by Exercise 4.7, $f'(x)$ is negative in (α, α_1), and similarly $f'(x)$ is positive in (α_p, β). Moreover, $f'(x)$ is of constant sign in each interval (α_r, α_{r+1}). Let ρ be the least value of r for which $f'(x)$ is positive in (α_r, α_{r+1}); then $f'(x)$ is negative in $(\alpha_{\rho-1}, \alpha_\rho)$ and positive in $(\alpha_\rho, \alpha_{\rho+1})$, and therefore $f(x)$ has a minimum value at $x = \alpha_\rho$.]

4.9. Given that a and b are two real roots of the equation $f(x) = 0$, where $f(x)$ is a polynomial in x, show that there is at least one real root of the equation $f'(x) + f(x) = 0$ which lies between a and b.

[Hint: Let $g(x) = e^x f(x)$ so that $g'(x) = e^x[f'(x) + f(x)]$. Since $f(x)$ vanishes at $x = a$ and $x = b$ so does $g(x)$ and hence, by Rolle's Theorem (see Proposition

4.3), there is at least one real root of $g(x)$ between a and b. Since $e^x \neq 0$, it follows that $f'(x) + f(x)$ vanishes for at least one real value of x between a and b.]

4.10. Deduce from the result in Exercise 4.9 that if a, b, and c are real roots of the equation $f(x) = 0$ and are such that $a < b < c$, then there is a real root of the equation $f''(x) + 2f'(x) + f(x) = 0$ which lies between a and c.

4.11. Determine the range of values of A for which the equation

$$3x^4 - 8x^3 - 6x^2 + 24x + A = 0$$

has four real unequal roots.

[Hint: Let f be a polynomial. If $f(a)$ and $f(b)$ are not zero and have opposite signs, then an odd number of real roots of the equation $f(x) = 0$ lie between a and b; if $f(a)$ and $f(b)$ are not zero and have the same sign, then an even number of roots (or no root) of $f(x) = 0$ lie between a and b. Hence, by Rolle's Theorem, necessary and sufficient conditions for $f(x) = 0$ to have n unequal real roots are that $f'(x) = 0$ shall have $n - 1$ unequal real roots and, if these $n - 1$ roots are $x_1, x_2, \ldots, x_{n-1}$ in ascending order, that the signs of the succession

$$f(-\infty), \quad f(x_1), \quad f(x_2), \quad \ldots, \quad f(x_{n-1}), \quad f(\infty)$$

are alternate.

If $f(x) = 3x^4 - 8x^3 - 6x^2 + 24x + A$, then $f'(x) = 12(x + 1)(x - 1)(x - 2)$, and the roots of $f'(x) = 0$ are at $x = -1$, 1, and 2. Hence, the equation in question will have four unequal roots if the signs of $f(-\infty)$, $f(-1) = -19 + A$, $f(1) = 13 + A$, $f(2) = 8 + A$, and $f(\infty)$ alternate. But $f(x)$ is positive if $x \to \pm\infty$ and so we must require that $-19 + A < 0$, $13 + A > 0$, and $8 + A < 0$. However, these inequalities require that $-13 < A < -8$.]

4.12. Let $f(x) = 1 - x + x^2/2 - x^3/3 + \cdots + (-1)^n x^n/n$. Show that $f(x) = 0$ has one real root if n is odd and no real root if n is even.

[Hint: Clearly, $f(x) = 0$ has no negative roots. We note that $f(x) = 0$ does not have two consecutive positive roots. Suppose it had. Then $f'(x) = 0$ would have to hold for some x between two consecutive positive roots of $f(x) = 0$. But $f'(x) = -(1 + x^{n+1})/(1 + x)$.]

4.13. The functions u and v and their derivatives u' and v' are continuous in $[a, b]$, and $uv' - u'v$ never vanishes at any point of $[a, b]$. Show that between any two zeros of u lies one of v and the converse, that is, the roots of u and v separate each other.

[Hint: Let x_1 and x_2 be consecutive roots of $u(x) = 0$. Then, if $v(x) \neq 0$ when $x_1 < x < x_2$, u/v is continuous in $[a, b]$ and vanishes at x_1 and x_2; hence, $(u/v)'$ must vanish at an intermediate point: a contradiction. Nor can v vanish twice in (x_1, x_2), for then u would have a root between x_1 and x_2.]

4.14. Show that the equation

$$a_0 x^n + a_1 x^{n-1} + a_2 x^{n-2} + \cdots + a_{n-1} x + a_n$$

$$= \frac{a_0}{n+1} + \frac{a_1}{n} + \frac{a_2}{n-1} + \cdots + \frac{a_{n-1}}{2} + \frac{a_n}{1}$$

has a root between 0 and 1.

[Hint: Consider the function

$$F(x) = \frac{a_0 x^{n+1}}{n+1} + \frac{a_1 x^n}{n} + \frac{a_2 x^{n-1}}{n-1} + \cdots + \frac{a_{n-1} x^2}{2} + \frac{a_n x}{1}$$

$$- \left(\frac{a_0}{n+1} + \frac{a_1}{n} + \frac{a_2}{n-1} + \cdots + \frac{a_{n-1}}{2} + \frac{a_n}{1} \right) x.$$

Then $F(0) = F(1) = 0$. Thus, by Rolle's Theorem, there is a point x between 0 and 1 such that $F'(x) = 0$.]

4.15. Show that $(\tan x)/x > x/(\sin x)$ for $0 < x < \pi/2$.

[Hint: Since $x \sin x > 0$ for $0 < x < \pi/2$, it will be sufficient to show that $(\sin x)(\tan x) - x^2 > 0$ for $0 < x < \pi/2$ in order to have

$$\frac{(\sin x)(\tan x) - x^2}{x \sin x} > 0 \quad \text{for } 0 < x < \frac{\pi}{2}.$$

Let $f(x) = (\sin x)(\tan x) - x^2$. Then

$$f'(x) = (\cos x)(\tan x) + (\sin x)(\sec^2 x) - 2x$$

$$= \sin x + (\sin x)(\sec^2 x) - 2x$$

and

$$f''(x) = \cos x + (\cos x)(\sec^2 x) + 2(\sin x)(\sec^2 x)(\tan x) - 2$$

$$= (\sqrt{\sec x} - \sqrt{\cos x})^2 + 2(\tan^2 x)(\sec x).$$

Thus, $f''(x) > 0$ for $0 < x < \pi/2$. Since the derivative $f''(x)$ of $f'(x)$ is positive for $0 < x < \pi/2$, the function $f'(x)$ is an increasing function for $0 < x < \pi/2$. Furthermore, since $f'(0) = 0$, we have $f'(x) > 0$ for $0 < x < \pi/2$. Again, since $f'(x)$ is larger than 0 for $0 < x < \pi/2$ and $f(x) = 0$, we have $f(x) > 0$ for $0 < x < \pi/2$. Thus, it follows that $(\tan x)/x > x/(\sin x)$ for $0 < x < \pi/2$.]

4.16. Show that if $0 < w < v$, then

$$\frac{v - w}{1 + v^2} < \tan^{-1} v - \tan^{-1} w < \frac{v - w}{1 + w^2}.$$

[Hint: Apply the Mean Value Theorem (see Proposition 4.5).]

4.17. Show that if $0 < \alpha < \beta < \pi/2$, then .

$$\frac{\sin \alpha - \sin \beta}{\cos \beta - \cos \alpha} = \cot \theta$$

for some θ satisfying $\alpha < \theta < \beta$.

[Hint: Apply part (ii) of Proposition 4.9.]

4.18. If f, g, and h are functions continuous on $[a, b]$ and differentiable on (a, b), show that there is a point c situated between a and b such that

$$\begin{bmatrix} f(a) & f(b) & f'(c) \\ g(a) & g(b) & g'(c) \\ h(a) & h(b) & h'(c) \end{bmatrix} = 0.$$

[Hint: Apply Rolle's Theorem (see Proposition 4.3) to the function

$$w(x) = \begin{bmatrix} f(a) & f(b) & f(x) \\ g(a) & g(b) & g(x) \\ h(a) & h(b) & h(x) \end{bmatrix}.]$$

4.19. Show that the extreme values of

$$y = \frac{ax^2 + 2bx + c}{px^2 + 2qx + r}, \qquad pr - q^2 > 0$$

are roots of the quadratic

$$(pr - q^2)y^2 - (ar - 2bq + cp)y + (ac - b^2) = 0.$$

[Hint: We differentiate $(px^2 + 2qx + r)y = ax^2 + 2bx + c$; when $y' = 0$, $(px + q)y = ax + b$ and, from this and $(px^2 + 2qx + r)y = ax^2 + 2bx + c$, we get $(qx + r)y = bx + c$. Solving for x in

$$y = \frac{ax + b}{px + q} \quad \text{and} \quad y = \frac{bx + c}{qx + r}$$

we get

$$x = \frac{-qy + b}{py - a} \quad \text{and} \quad x = \frac{-ry + c}{qy - b},$$

respectively. Hence, elimination of x from the equations

$$y = \frac{ax + b}{px + q} \quad \text{and} \quad y = \frac{bx + c}{qx + r}$$

gives $(-qy + b)(qy - b) = (-ry + c)(py - a)$ or

$$(pr - q^2)y^2 - (ar - 2bq + cp)y + (ac - b^2) = 0.]$$

4.20. Show that the values of $y = (x^2 + 2x + 11)/(x^2 + 4x + 10)$ are confined to the interval $\frac{5}{6} \le y \le 2$.
[Hint: Apply the result in Exercise 4.19.]

4.21. Show that $f(x) = (a \cos x + b \sin x)/(c \cos x + d \sin x)$ has neither a maximum nor a minimum.
[Hint: We have $f'(x) = (bc - ad)/(c \cos x + d \sin x)^2$.]

4.22. Show that $f(x) = a \sin x + b \cos x$ has $\pm(a^2 + b^2)^{1/2}$ as extreme values.
[Hint: $f'(x) = 0$ implies $a/b = \tan x$.]

4.23. Verify that, when m and n are positive integers,

$$f(x) = (x - a)^m (x - c)^n, \qquad a < c,$$

has the extremes given in the following list, where b divides the interval $[a, c]$ in the ratio m/n:

(i) m and n even; minima at a and c, maximum at $b = (na + mc)/(m + n)$;
(ii) m and n odd; minimum at b;
(iii) m even, n odd; maximum at a, minimum at b;
(iv) m odd, n even; maximum at b, minimum at c.

4.24. When $x > 0$, show that $f(x) = (\tan^{-1} x)/(\tanh x)$ increases as x increases and that

$$\tan^{-1} x < \frac{\pi}{2}(\tanh x).$$

[Hint: $f'(x)$ has the same sign as $g(x) = (1 + x^2)^{-1}(\sinh x)(\cosh x) - \tan^{-1} x$, $g'(x)$ has the same sign as $h(x) = (1 + x^2)(\sinh x) - x\cosh x$, and $h'(x) > 0$ when $x > 0$. Moreover, $f(x) \to \pi/2$ as $x \to \infty$.]

4.25. What is wrong with the following reasoning? Let $f(x) = e^{-2x}(\cos x + 2\sin x)$ and $g(x) = e^{-x}(\cos x + \sin x)$. As $x \to \infty$, the fraction $f(x)/g(x)$ has the form 0/0. Now

$$\frac{f'(x)}{g'(x)} = \frac{-5e^{-2x}\sin x}{-2e^{-x}\sin x}.$$

Canceling $\sin x$ in f'/g' we see that $f'/g' \to 0$ as $x \to \infty$ and so we may conclude that $f/g \to 0$ as $x \to \infty$.

[Hint: Note that f'/g' has no limit as $x \to \infty$, for, as x becomes infinite along the sequence $n\pi$, f'/g' is never defined; the limit

$$\lim_{x \to \infty} \frac{f(x)}{g(x)} = \lim_{x \to \infty} e^{-x}\frac{1 + 2\tan x}{1 + \tan x}$$

is nonexistent, since $(1 + 2\tan x)/(1 + \tan x)$ is discontinuous at the points $x_n = n\pi + \pi/2$ for $n = 1, 2, 3, \ldots$.]

4.26. Verify formula (4.101).

4.27. Find constants A and B so that $(\sin 3x + A\sin 2x + B\sin x)/x^5$ may tend to a finite limit as $x \to 0$.

[Hint: Expanding the sine function in powers of x by use of Taylor's Theorem (see Proposition 4.11), we need

$$3 + 2A + B = 0, \qquad 27 + 8A + B = 0$$

to make the coefficients of x and x^3 zero. Hence, $A = -4$ and $B = 5$.]

4.28. Let $f(x) = x^2\sin(1/x)$ and $g(x) = \sin x$. Verify that $f(x)/g(x) \to 0$ as $x \to 0$ and that $f'(x)/g'(x)$ does not tend to a limit as $x \to 0$. Does this belie the truth of L'Hôpital's Rule?

4.29. Find

$$\lim_{x \to 0} \frac{x\sin(\sin x) - \sin^2 x}{x^6}.$$

[Answer: $\frac{1}{18}$.]

4.30. Find

$$\lim_{x \to 0} \frac{(\sin x)(\sin^{-1} x) - x^2}{x^6}.$$

[Answer: $\frac{1}{18}$.]

4.31. Find
$$\lim_{x\to 0}\frac{\tan(\sin x) + \tan^{-1}(\sin x) - \sin(\tan x) - \sin(\tan^{-1} x)}{x^9}.$$

[Answer: $\tfrac{1}{9}$.]

4.32. Find
$$\lim_{a\to b}\frac{a^b - b^a}{a^a - b^b}.$$

[Answer: $(1 - \ln b)/(1 + \ln b)$.]

4.33. Show that if $a > b > 0$ the minimum value of $(a - b)x/(x + a)(x + b)$ exceeds the maximum by $4\sqrt{ab}/(a - b)$.

[Hint: If $y = (a - b)x/(x + a)(x + b) = a/(x + a) - b/(x + b)$, $y = 0$ when $x = \pm\sqrt{ab}$,
$$y'' = -\frac{(\sqrt{a} - \sqrt{b})}{\sqrt{ab}(\sqrt{a} + \sqrt{b})^3} \quad \text{if } x = \sqrt{ab},$$

and
$$y'' = \frac{\sqrt{a} + \sqrt{b}}{\sqrt{ab}(\sqrt{a} - \sqrt{b})^3} \quad \text{if } x = -\sqrt{ab}.$$

Hence, $(\sqrt{a} - \sqrt{b})/(\sqrt{a} + \sqrt{b})$ is the maximum value and $(\sqrt{a} + \sqrt{b})/(\sqrt{a} - \sqrt{b})$ is the minimum value.]

4.34. If $y = (x + a)(x + b)/(x - a)(x - b)$, show that the maximum and minimum are, respectively,
$$-\left(\frac{\sqrt{a} + \sqrt{b}}{\sqrt{a} - \sqrt{b}}\right)^2 \quad \text{and} \quad -\left(\frac{\sqrt{a} - \sqrt{b}}{\sqrt{a} + \sqrt{b}}\right)^2.$$

Also show that, if x is real, $(x + a)(x + b)/(x - a)(x - b)$ can not lie between the values
$$-\left(\frac{\sqrt{a} + \sqrt{b}}{\sqrt{a} - \sqrt{b}}\right)^2 \quad \text{and} \quad -\left(\frac{\sqrt{a} - \sqrt{b}}{\sqrt{a} + \sqrt{b}}\right)^2.$$

[Hint: We have
$$\frac{(x + a)(x + b)}{(x - a)(x - b)} = 1 + \frac{2(a + b)x}{(x - a)(x - b)} = 1 + \frac{2(a + b)}{a - b}\left(\frac{a}{x - a} - \frac{b}{x - b}\right).$$

Also note that
$$\frac{x^2 + (a + b)x + ab}{x^2 - (a + b)x + ab} = y$$

or
$$(1 - y)x^2 + (a + b)(1 + y)x - ab(1 - y) = 0,$$

an equation whose roots are only real when $(a + b)^2(1 + y)^2 \geq 4ab(1 - y)^2$.]

4.35. A rancher wishes to divide a triangular field into two equal parts by a straight fence. How is this to be done so that the fence may be of the least length?

[Hint: If A is the smallest angle and b, c the adjacent sides, the distance of each end of the fence from A equals $\sqrt{bc/2}$ and the length of the fence equals $\sqrt{2bc}\,(\sin A/2)$.]

4.36. If a and b are positive, show that

$$(1 + a)\ln(1 + a) + (1 + b)\ln(1 + b) < (1 + a + b)\ln(1 + a + b),$$

and generally that if every a_r is positive,

$$\sum_{r=1}^{n}(1 + a_r)\ln(1 + a_r) < \left(1 + \sum_{r=1}^{n} a_r\right)\ln\left(1 + \sum_{r=1}^{n} a_r\right).$$

[Hint: Let $g(x) = (1 + x)\ln(1 + x)$ for $x > 0$ and suppose that $0 < a < b$. By the Mean Value Theorem (see Proposition 4.5), there is a point c such that

$$\frac{g(a + b) - g(b)}{a} = g'(c), \quad \text{where } b < c < a + b.$$

Thus,

$$(1 + a + b)\ln(1 + a + b) - (1 + b)\ln(1 + b) = a[1 + \ln(1 + c)],$$

where $0 < a < b < c < a + b$. But $a > \ln(1 + a)$ for $a > 0$ and $\ln(1 + c) > \ln(1 + a)$ for $0 < a < c$. Hence,

$$a[1 + \ln(1 + c)] = a + a[\ln(1 + c)]$$

$$> \ln(1 + a) + a[\ln(1 + a)] = (1 + a)\ln(1 + a).]$$

4.37. Let $f''(c) = f'''(c) = \cdots = f^{(n-1)}(c) = 0$, but $f^{(n)}(c) \neq 0$. Show that if n is odd, then the curve $y = f(x)$ has a point of inflection at $x = c$; if n is even, it is concave upward or downward according to whether $f^{(n)}(c) > 0$ or < 0.

4.38. Let n be a positive integer. Show that the abscissas of the points of inflection of the curve $y^n = f(x)$ are the roots of the equation

$$\frac{n - 1}{n}\{f'(x)\}^2 = f(x)f''(x).$$

4.39. Show that the points of inflection of the curve $y^2 = (x - a)^2(x - b)$ lie on the line $3x + a = 4b$.

4.40. If $f(x) = ax + 6b^2/x^2$, $b \neq 0$, show that the curve $y = e^{-f(x)}$ has at least two points of inflection.

[Hint: We have $f'(x) = a - 12b^2/x^3$, $f''(x) = 36b^2/x^4$, and therefore

$$y'' = e^{-f}(f')^2 - e^{-f}(f'') = \frac{e^{-f}}{x^6}\{(ax^3 - 6bx - 12b^2)(ax^3 + 6bx - 12b^2)\}.$$

The cubics have no common factor, since $b \neq 0$, and each cubic has either one real root or three real roots of which one must be simple, that is, each cubic has at least one nonrepeated real root, and y'' changes sign, and vanishes, as x passes through a nonrepeated real root of each cubic.]

4.41. Let n and r be positive integers larger than or equal to 3. Show that

$$(n + r)^n < n^{n+r}.$$

[Hint: The inequality in question is equivalent to

$$\frac{\ln(n + r)}{n + r} < \frac{\ln n}{n}.$$

But the function $g(x) = (\ln x)/x$, defined for $x > 0$, has a single maximum at $x = e$ because $g'(x) = (1 - \ln x)/x^2$ is negative for $x > e$ and is positive for x satisfying $0 < x < e$. Evidently, the function g is decreasing for $x > e$.]

4.42. If a plane at distance x from the center of a unit sphere cuts it into two segments, one with twice the volume of the other, show that

$$3x^3 - 9x + 2 = 0.$$

Then use Newton's Method to find x accurate to four decimal places.
[Hint: See Example 4 in Section 4 of Chapter 4. Answer: 0.2261.]

4.43. The equation $x = \tan x$ has infinitely many roots. It is easily seen that its smallest positive root is between $5\pi/4$ and $3\pi/2$. Using Newton's Method, verify that the smallest positive root of the given equation is approximately 4.4934.

4.44. The equation $x \sin x = \frac{1}{2}$ has infinitely many roots. Verify that the smallest positive root of the given equation is approximately 0.740841.

4.45. The equation $2x^3 - x^2 - 7x + 5 = 0$ has three distinct real roots. Find these three roots with an accuracy of 0.001.
[Answer: $x_1 = -1.9509$ (approximately), $x_2 = 0.756$ (approximately), and $x_3 = 1.694$ (approximately). Note that by Vieta's Theorem the sum of the three roots is equal to $\frac{1}{2}$.]

4.46. Verify that the equation $x^4 - x - 1 = 0$ has a root between the points $a = 1.22$ and $b = 1.23$. Then determine this root with an accuracy of six decimal places.
[Answer: 1.220744.]

4.47. *Problem of Viviani*. Two parallel lines are cut by a given line AB. See Figure 4.19. From the point C we draw a straight line intersecting AB. How should this line be chosen so that the sum of the areas of the two triangles $\triangle ACP$ and $\triangle QPB$ is a minimum?
[Hint: Let the lengths of AC and AB be a and b, respectively, and the lengths

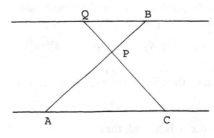

Figure 4.19

of AP and QB be x and y, respectively. Since the triangles are similar we have

$$\frac{a}{x} = \frac{y}{b-x} \quad \text{or} \quad y = a\frac{b-x}{x}.$$

If α is the angle at A, the sum of the areas is

$$\tfrac{1}{2}ax \sin \alpha + \tfrac{1}{2}(b-x)(y \sin \alpha).$$

We substitute the value of y and now the problem requires the determination of the minimum value of the function

$$f(x) = x + \frac{(b-x)^2}{x}.$$

It is easily seen that $x = b/\sqrt{2}$ gives the minimum area.]

4.48. A piece of wire k units long is cut into two parts. One part is bent into a circle, the other into a square. Show that the sum of their areas is a minimum when the wire is cut in such a way that the diameter of the circle equals the side of the square and that the sum of the areas is a maximum when the wire is left uncut and bent into a circle.

4.49. A triangle $\triangle ABC$ has a right angle at c, and the product of the sides AB and BC is constant. Show that $AC + 3BC$ is a minimum when $AC = 2BC$ and a maximum when $AC = BC$.

[Hint: Let $AB = x$ and $\angle ABC = \theta$. Then $x^2 \cos \theta$ is constant and $(AC + 3BC)^2$ is proportional to $(\sin \theta + 3 \cos \theta)^2/(\cos \theta) = \sec \theta + 6 \sin \theta + 8 \cos \theta$; the first derivative equals

$$(\sec \theta)(\tan \theta) + 6 \cos \theta - 8 \sin \theta = \frac{t^3 - 7t + 6}{(1 + t^2)^{3/2}}$$

$$= \frac{(t + 3)(t - 1)(t - 2)}{(1 + t^2)^{3/2}},$$

where $t = \tan \theta$. When $0 < t < 1$ the derivative is positive, it vanishes at $t = 1$, is negative for $1 < t < 2$, vanishes again at $t = 2$, and is positive for $t > 2$. Hence, $(AC + 3BC)^2$ is maximum when $t = 1$ and minimum when $t = 2$. Note that $f(t)^2$ and $f(t)$ are maximum and minimum together, when $f(t) > 0$, because

$$f(T)^2 - f(t)^2 = \{f(T) - f(t)\}\{f(T) + f(t)\}$$

and so $f(T)^2 - f(t)^2$ and $f(T) - f(t)$ have the same sign.]

4.50. Let $q_1, q_2, \ldots, q_n$ be positive real numbers with sum 1. Show that, for any nonnegative real numbers $a_1, a_2, \ldots, a_n$, we have

$$q_1 a_1 + q_2 a_2 + \cdots + q_n a_n \geq a_1^{q_1} \cdot a_2^{q_2} \cdots a_n^{q_n}$$

with equality only if $a_1 = a_2 = \cdots = a_n$.

[Hint: First we note that, if x is any real number, then

$$e^x \geq ex$$

with equality only if $x = 1$. Indeed, the function $f(x) = e^x - ex$ has an absolute minimum at $x = 1$.

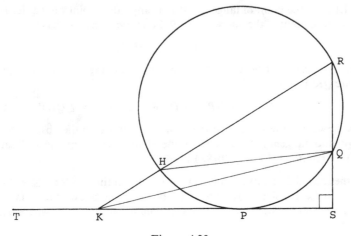

Figure 4.20

To prove the desired inequality, put $w = q_1a_1 + q_2a_2 + \cdots + q_na_n$. We now replace x in the inequality $e^x \geq ex$ by a_j/w to get

$$e^{a_j/w} \geq ea_j/w \quad \text{or} \quad e^{a_jq_j/w} \geq \left(\frac{ea_j}{w}\right)^{q_j}.$$

Multiplying these inequalities for each of the values $j = 1, 2, \ldots, n$, we get

$$e \geq \frac{e[a_1^{q_1} \cdot a_2^{q_2} \cdots a_n^{q_n}]}{w}$$

which proves the desired inequality. Evidently, there is equality here if and only if each of $a_1/w, a_2/w, \ldots, a_n/w$ equals 1, giving the condition $a_1 = a_2 = \cdots = a_n$.]

4.51. Determine the shortest distance from the point $(c,0)$ to $y^2 = 4x$, where c may be positive, negative, or zero.

[Answer: If $c \leq 2$, the point on $y^2 = 4x$ that is nearest to $(c,0)$ is the point $(0,0)$. If $c > 2$, there are two nearest points to $(c,0)$ on the curve, namely, $(c - 2, \pm 2\sqrt{c - 2})$.]

4.52. Suppose that f is a differentiable function at each point of a closed interval $[a,b]$, where $|b - a| < \infty$. If there is a number M for which $|f'(x)| < M$ for every x in $[a,b]$, show that f is uniformly continuous on $[a,b]$.
[Hint: Use the Mean Value Theorem (see Proposition 4.5).]

4.53. Let $f(x) = x(\ln x)$ if $x > 0$ and $f(0) = 0$. Show that f is uniformly continuous on the closed interval $[0,1]$, but that $f'(x)$ is not bounded on $[0,1]$. Explain why this does not contradict the result in Exercise 4.52.

4.54. A picture hangs on a wall above the level of an observer's eye. How far from the wall should the observer stand to maximize the angle of observation?
[Hint: Consider Figure 4.20, where Q and R represent the bottom and top of the picture and TS is the horizontal line at the level of the observer's eye. Next, consider the circle passing through the points Q and R, tangent to the line TS,

and let P be the point of tangency. If K is any other point on TS, let H be the intersection point of the line segment RK and the circle. We have

$$\angle QPR = \angle QHR$$

by the property that the chord QR subtends equal angles at any two points on the arc QR. Hence,

$$\angle QPR = \angle QHR = \angle QKR + \angle HQK > \angle QKR.$$

This shows that the angle of observation is maximized if the observer's eye is at the point of tangency P. The tangent line SP and the secant line SQR satisfy the distance relation $(SP)^2 = (SQ)(SR)$.]

4.55. Let the second derivative f'' of a function f be continuous for $a \leq x \leq b$ and at each point x, the sign of $f(x)$ and $f''(x)$ be the same. Show that if $f(x)$ vanishes at points c and d, where $a \leq c < d \leq b$, then it vanishes everywhere between c and d.

4.56. The normal is drawn at a variable point P of the ellipse

$$\frac{x^2}{a^2} + \frac{y^2}{b^2} = 1.$$

Show that the maximum distance of the normal from the center of the ellipse is $|a - b|$.

4.57. Verify that if in formula (4.2) we write for $f(t)$ and $g(t)$, respectively,

(i) t^2, t,
(ii) $\sin t, \cos t$,
(iii) e^t, e^{-t},

then in each case "x" is the arithmetic mean between a and b.
   If in formula (4.2) we write for $f(t)$ and $g(t)$, respectively,

$$\sqrt{t} \quad \text{and} \quad \frac{1}{\sqrt{t}},$$

then "x" is the geometric mean between a and b and if we write

$$\frac{1}{t^2} \quad \text{and} \quad \frac{1}{t},$$

then "x" is the harmonic mean between a and b.
   [The arithmetic, geometric, and harmonic means of a and b are, respectively,

$$A = \frac{a+b}{2}, \quad G = \sqrt{ab}, \quad \text{and} \quad H = \frac{2ab}{a+b}.]$$

4.58. If all the roots of the polynomial $P(x) - a$ and $P(x) - b$ are real, show that all the roots of the polynomial $P(x) - c$ are real if $a < c < b$.
   [Hint: All the roots of $P'(x)$ are clearly real; denote them by $t_1, t_2, \ldots, t_{n-1}$. Next, denote by $y_1, y_2, \ldots, y_n$ the roots of $P(x) - b$ and by $x_1, x_2, \ldots, x_n$ the roots of $P(x) - a$. Then $y_1 < t_1 < y_2 < \cdots < y_{n-1} < t_{n-1} < y_n$ and $x_1 < t_1 < x_2 < t_2 < \cdots < x_{n-1} < t_{n-1} < x_n$. It follows that intervals bounded by the points

x_1, y_i do not overlap since they lie in the nonoverlapping intervals $(-\infty, t_1)$, $(t_1, t_2), \ldots, (t_{n-1}, +\infty)$. The polynomial $P(x)$ takes on the values a and b at the endpoints of each of these intervals and passes through all intermediate values inside the interval. Hence, $P(x) - c$ vanishes n times on $(-\infty, +\infty)$.]

4.59. Let f be such that $f''(x)$ exists, $|f(x)| \leq A$, and $|f''(x)| \leq B$ for all $x > 0$, where A and B are positive constants. Show that $|f'(x)| \leq 2\sqrt{AB}$ for all $x > 0$.

[Hint: Let x and h be positive. Then, by Taylor's Theorem,

$$f(x + h) = f(x) + hf'(x) + \frac{h^2}{2!} f''(x + \theta h)$$

for some number θ satisfying $0 < \theta < 1$. Hence

$$|hf'(x)| = \left| f(x + h) - f(x) - \frac{h^2}{2!} f''(x + \theta h) \right|$$

$$\leq |f(x + h)| + |f(x)| + \frac{h^2}{2!} |f''(x + \theta h)|$$

$$\leq 2A + \frac{Bh^2}{2} \qquad \text{for any } x > 0$$

or

$$|f'(x)| \leq \frac{2A}{h} + \frac{Bh}{2} \qquad \text{for any } x > 0.$$

Since $|f'(x)|$ is independent of h and is less than or equal to $2A/h + Bh/2$ for any $h > 0$, it follows that $|f'(x)|$ is less than or equal to the least value of $2A/h + Bh/2$. But

$$\frac{2A}{h} + \frac{Bh}{2} = \left(\sqrt{\frac{2A}{h}} - \sqrt{\frac{Bh}{2}} \right)^2 + 2\sqrt{AB}$$

and so

$$2\sqrt{AB} \leq \frac{2A}{h} + \frac{Bh}{2} \qquad \text{for any } h > 0.$$

It therefore follows that $|f'(x)| \leq 2\sqrt{AB}$ for all $x > 0$.]

CHAPTER 5

Integration

1. Examples of Area Calculation

The problem of finding the area of a region in a plane bounded by a given curve has fascinated mathematicians for a long time. We shall consider a few of the celebrated examples that have come down to us from times past; the special methods of quadrature used in these examples are of a rather clever kind.

EXAMPLE 1 (The Quadrature of the Parabola by Archimedes). Consider a parabola satisfying the equation

$$y = Ax^2 + Bx + C \quad \text{with } A > 0. \tag{5.1}$$

Figure 5.1 illustrates the parabola under consideration. Let $P_1 = (x_1, y_1)$, $P_2 = (x_2, y_2)$, and $P_3 = (x_3, y_3)$ be points on this parabola such that the abscissas x_1, x_2, and x_3 form an arithmetic progression, that is,

$$x_2 - x_1 = x_3 - x_2 = h.$$

The region bounded by the parabolic arc connecting the points P_1 and P_3 and the chord from P_1 to P_3 is called the *parabolic segment between P_1 and P_3*; the chord between P_1 and P_3 is said to be the *base* of the parabolic segment between P_1 and P_3 and the point P_2 is said to be the *vertex* of the parabolic segment between P_1 and P_3. The vertex P_2 of the parabolic segment between P_1 and P_3 is that point on the parabolic arc connecting P_1 and P_3 from which the perpendicular to the base is the greatest.

Given the parabolic segment between P_1 and P_3 with vertex P_2, the parabolic arc is assumed to satisfy equation (5.1). Our problem is the determination of the area of this parabolic segment. We shall do this in two steps: We show that

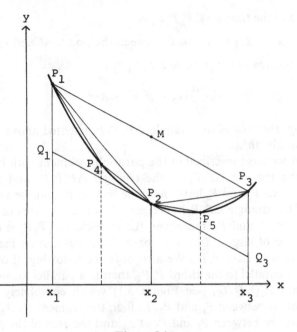

Figure 5.1

(i) the area of the triangle $\triangle P_1 P_2 P_3$ is equal to Ah^3 and
(ii) the area of the parabolic segment between P_1 and P_3, having vertex P_2, is $\frac{4}{3}$ times the area of the triangle $\triangle P_1 P_2 P_3$.

Let M denote the center of the chord from P_1 to P_3; then M has the coordinates

$$M = \left(\frac{x_1 + x_3}{2}, \frac{y_1 + y_3}{2}\right).$$

The point P_2 has the coordinates (x_2, y_2), where

$$x_2 = \frac{x_1 + x_3}{2} \quad \text{and} \quad y_2 = f\left(\frac{x_1 + x_3}{2}\right) \quad \text{with } f(x) = Ax^2 + Bx + C.$$

The distance between the points M and P_2 is

$$\frac{y_1 + y_3}{2} - f\left(\frac{x_1 + x_3}{2}\right)$$

$$= \frac{Ax_1^2 + Bx_1 + C + Ax_3^2 + Bx_3 + C}{2} - A\left(\frac{x_1 + x_3}{2}\right)^2 - B\left(\frac{x_1 + x_3}{2}\right) - C$$

$$= \frac{A}{2}(x_1^2 + x_3^2) - \frac{A}{4}(x_1 + x_3)^2$$

$$= \frac{A}{4}(x_3 - x_1)^2.$$

But the area of the triangle $\triangle P_1 P_2 P_3$ is

$$\tfrac{1}{2}(x_3 - x_1) \times (\text{distance between the points } M \text{ and } P_2).$$

Therefore, the area of the triangle $\triangle P_1 P_2 P_3$ is

$$\frac{A}{8}(x_3 - x_1)^3 = Ah^3.$$

Summing up, the area of the triangle $\triangle P_1 P_2 P_3$ situated above an interval of length $2h$ equals Ah^3.

Note that we have inscribed in the parabolic segment with base $P_1 P_3$ and vertex P_2 the triangle $\triangle P_1 P_2 P_3$; the triangle $\triangle P_1 P_2 P_3$ and the parabolic segment between P_1 and P_3 have the same base $P_1 P_3$ and the same vertex P_2. The area of the triangle $\triangle P_1 P_2 P_3$ is clearly less than the area of the parabolic sector between P_1 and P_3; moreover, the triangle $\triangle P_1 P_2 P_3$ is easily seen to be the triangle of largest area with base $P_1 P_3$ that can be inscribed in the parabolic arc with base $P_1 P_3$. We also note the following: If one constructs through P_2 a parallel to the chord $P_1 P_3$, then this parallel forms a side of the parallelogram $P_1 Q_1 Q_2 Q_3$ (see Figure 5.1) which completely contains the parabolic sector between P_1 and P_3. In fact, the segment $Q_1 Q_3$ is tangent to the parabolic arc between P_1 and P_3 at P_2 and the area of the parallelogram $P_1 Q_1 Q_3 P_3$ is exactly twice the area of the triangle $\triangle P_1 P_2 P_3$. Therefore, the area of the parabolic segment with base $P_1 P_3$ situated above an interval of length $2h$ is strictly between the numbers

$$Ah^3 \quad \text{and} \quad 2Ah^3$$

with the parabolic curve under consideration satisfying equation (5.1).

We now proceed to show that the area of the parabolic segment with base $P_1 P_3$ and vertex P_2 is $\tfrac{4}{3}$ times the area of the triangle $\triangle P_1 P_2 P_3$ with base $P_1 P_3$ and vertex P_2.

First, we inscribe in the parabolic segment with base $P_1 P_3$ and vertex P_2 the triangle $\triangle P_1 P_2 P_3$. Second, within each of the two smaller parabolic segments with bases $P_1 P_2$ and $P_2 P_3$ we inscribe the triangles $\triangle P_1 P_4 P_2$ and $\triangle P_2 P_5 P_3$ (see Figure 5.1). The parabolic segment with base $P_1 P_2$ (respectively, base $P_2 P_3$) and vertex P_4 (respectively, vertex P_5) has the same base and vertex as the triangle $\triangle P_1 P_4 P_2$ (respectively, triangle $\triangle P_2 P_5 P_3$). The abscissa x_4 is the midpoint of the interval from x_1 to x_2 and the abscissa x_5 is the midpoint of the interval from x_2 to x_3. The area of the triangle $\triangle P_1 P_4 P_2$ (respectively, triangle $\triangle P_2 P_5 P_3$) situated above an interval of length h equals

$$A\left(\frac{h}{2}\right)^3 = \frac{Ah^3}{8}$$

and hence both triangles $\triangle P_1 P_4 P_2$ and $\triangle P_2 P_5 P_3$ together have a total area of

$$\tfrac{1}{4}Ah^3.$$

Accordingly, the polygon $P_1 P_4 P_2 P_5 P_3$ has area

$$(1 + \tfrac{1}{4})Ah^3.$$

Next, we inscribe triangles within each of the four parabolic segments, the bases of which are, respectively, $P_1 P_4$, $P_4 P_2$, $P_2 P_5$, and $P_5 P_3$; the corresponding vertices are so chosen that the vertex of the parabolic segment and the vertex of its inscribed triangle coincide.

Continuing this process, at the nth stage we have a polygon with area

$$q_n = \left(1 + \frac{1}{4} + \frac{1}{4^2} + \cdots + \frac{1}{4^{n-1}}\right) Ah^3.$$

The difference between the area of the polygon and the area of the parabolic segment with base $P_1 P_3$ is nonnegative and tends to zero as n becomes arbitrarily large. The area of the parabolic segment with base $P_1 P_3$ is given by

$$\lim_{n \to \infty} q_n.$$

Putting

$$s_n = 1 + \frac{1}{4} + \frac{1}{4^2} + \cdots + \frac{1}{4^{n-1}},$$

we get

$$\frac{1}{4} s_n = \frac{1}{4} + \frac{1}{4^2} + \cdots + \frac{1}{4^n}$$

and so

$$\frac{3}{4} s_n = 1 - \frac{1}{4^n} \quad \text{or} \quad \lim_{n \to \infty} s_n = \frac{4}{3},$$

implying that

$$\lim_{n \to \infty} q_n = \frac{4}{3} Ah^3.$$

Thus, the area of the parabolic segment is $\tfrac{4}{3}$ times the area of the inscribed triangle having the same base and vertex as the parabolic segment.

We now add some remarks concerning the calculation of the area of a parabolic segment. Commencing with the triangle $\triangle P_1 P_2 P_3$, we add the two triangles $\triangle P_1 P_4 P_2$ and $\triangle P_2 P_5 P_3$, then we add four triangles, and so on, each time doubling the number of triangles in the previous step. If instead of doubling the number of triangles in the nth step, one quadruples the number of triangles, one can see that one exceeds the area of the parabolic segment under investigation. In this connection, recall that the area of the parallelogram $P_1 Q_1 Q_2 Q_3$ has twice the area of the triangle $\triangle P_1 P_2 P_3$ and contains completely the parabolic segment with base $P_1 P_3$. Hence, the area of the parabolic segment with base $P_1 P_3$ differs from the area of the triangle $\triangle P_1 P_2 P_3$ by a factor which is larger than

$$s_n = 1 + \frac{1}{4} + \frac{1}{4^2} + \cdots + \frac{1}{4^{n-2}} + \frac{1}{4^{n-1}}$$

but which is less than

$$S_n = 1 + \frac{1}{4} + \frac{1}{4^2} + \cdots + \frac{1}{4^{n-2}} + \frac{2}{4^{n-1}}.$$

Clearly, s_n increases as n gets larger; however,

$$S_{n+1} - S_n = \frac{2}{4^n} - \frac{1}{4^{n-1}} = -\frac{1}{2 \cdot 4^{n-1}}$$

and so S_n decreases as n gets larger. For $n = 2, 3, \ldots$, the closed intervals

$$J_n = [s_n, S_n],$$

where J_n has length $1/4^{n-1}$, form a nested sequence of intervals; the point $\frac{4}{3}$ is the unique point common to all intervals J_n.

COMMENT. Archimedes of Syracuse died in 212 B.C. The mathematicians Gilles Personne de Roberval and Pierre de Fermat, to whom the next two examples are due, lived in the 17th century.

EXAMPLE 2 (The Quadrature of the Cycloid by Roberval). The curve traced by a point on a circle as the circle rolls on a straight line without slipping is called a *cycloid*; the rolling circle is referred to as the *generating circle of the cycloid*.

To find simple parametric equations of the cycloid, we let the fixed line on which the circle rolls be the x-axis and we place the origin at one of the places where the tracing point comes into contact with the x-axis.

Denote the center of the rolling circle by C and its radius by a. Let the point P with the coordinates (x, y) be any position of the tracing point, and choose for a parameter the angle t through which the line segment CP has turned from its position when P was at the origin (see Figure 5.2). We assume that the generating circle is rolling to the right on top of the x-axis, that t is

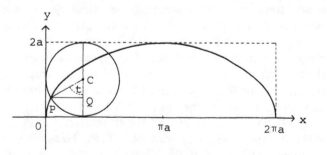

Figure 5.2

expressed in radian measure, and that the positive direction for t is clockwise. While the cycloid consists of infinitely many congruent arches, we are only interested in the arch extending from $x = 0$ to $x = 2\pi a$. Our purpose is to show that the area of the region bounded by the arch of the cycloid extending from $x = 0$ to $x = 2\pi a$ and the x-axis is three times the area of the generating circle, that is, $3\pi a^2$.

Since the generating circle rolls along the x-axis without slipping, the distance from O to N equals the length of the arc of the generating circle from P to N in the counterclockwise direction. From Figure 5.2 we can see that

$$\overline{ON} = at, \qquad \overline{QC} = a(\cot t), \qquad \overline{PQ} = a \sin t.$$

Thus,

$$x = \overline{ON} - \overline{PQ} = a(t - \sin t) \quad \text{and} \quad y = \overline{NC} - \overline{QC} = a(1 - \cos t).$$

Therefore, the parametric equations of the arch of the cycloid extending from $x = 0$ to $x = 2\pi a$ are

$$x = a(t - \sin t), \qquad y = a(1 - \cos t) \quad \text{with } 0 \le t \le 2\pi.$$

By the *companion to the cycloid* we mean the curve with the parametric equation

$$x = at, \qquad y = a(1 - \cos t) \quad \text{with } 0 \le t \le 2\pi;$$

note that the point Q is on the companion to the cycloid and that the line segment from P to Q has precisely the same length as the cross-section of the semicircle at the level $y = a(1 - \cos t)$; see Figure 5.3 for an illustration of the companion to the cycloid.

Let M be the point with coordinates $(\pi a/2, a)$. It is easy to see that the curve

$$x = at, \qquad y = a(1 - \cos t) \quad \text{with } 0 \le t \le \pi$$

is mapped onto itself by a rotation through π about the point M; note that

$$a(1 - \cos t_1) + a\left(1 - \cos\left\{\frac{\pi}{2} - t_1\right\}\right) = 2a$$

for any t_1 satisfying $0 \le t_1 \le \pi/2$. Hence, the curve

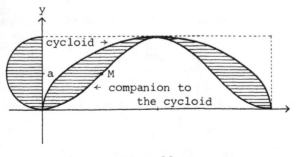

Figure 5.3

$$x = at, \qquad y = a(1 - \cos t) \quad \text{with } 0 \le t \le \pi$$

divides the rectangle with vertices $(0,0)$, $(\pi a, 0)$, $(\pi a, 2a)$, and $(0, 2a)$ into two congruent regions. Thus, the region bounded by the companion to the cycloid

$$x = at, \qquad y = a(1 - \cos t) \quad \text{with } 0 \le t \le 2\pi$$

and by the x-axis has area precisely half as large as the area of the rectangle with vertices $(0,0)$, $(2\pi a, 0)$, $(2\pi a, 2a)$, and $(0, 2a)$. Consequently, the region bounded by the companion to the cycloid and by the x-axis has area $2\pi a^2$.

On the other hand, the region bounded by the arch of the cycloid extending from $x = 0$ to $x = 2\pi a$ and the companion to the cycloid has the same area as the generating circle, namely, πa^2; the areas of two enclosed plane figures are equal provided that any system of parallel lines cuts off equal intercepts on each. Thus, the area of the region bounded by a full arch of a cycloid and the x-axis is three times the area of the generating circle.

EXAMPLE 3 (The Quadrature of $y = Ax^c$ by Fermat). Consider the region enclosed by the curve $y = Ax^c$, the x-axis, and the two straight lines $x = a$, $x = b$ with $0 < a < b$, where the coefficient A and the exponent c are arbitrary but fixed real numbers. To determine the area of the region under consideration, we insert between a and b, $n - 1$ geometric means so as to obtain the sequence

$$a, \quad a(1 + v), \quad a(1 + v)^2, \quad \ldots, \quad a(1 + v)^{n-1}, \quad b,$$

where the number v satisfies the condition $a(1 + v)^n = b$. Taking this set of numbers as the abscissas of the points of division of the interval $[a, b]$, the corresponding ordinates have the following values:

$$Aa^c, \quad Aa^c(1 + v)^c, \quad Aa^c(1 + v)^{2c}, \quad \ldots, \quad Aa^c(1 + v)^{(n-1)c}, \quad Ab^c,$$

and the area of the pth rectangle is

$$[a(1 + v)^p - a(1 + v)^{p-1}]Aa^c(1 + v)^{(p-1)c} = Aa^{c+1}v(1 + v)^{(p-1)(c+1)}.$$

Hence, the sum of the areas of all the rectangles is

$$Aa^{c+1}v[1 + (1 + v)^{c+1} + (1 + v)^{2(c+1)} + \cdots + (1 + v)^{(n-1)(c+1)}].$$

If $c + 1 \ne 0$, as we shall suppose first, the sum inside the parentheses is equal to

$$\frac{(1 + v)^{n(c+1)} - 1}{(1 + v)^{c+1} - 1};$$

or, replacing $a(1 + v)^n$ by b, the sum of the areas of all the rectangles may be written in the form

$$A(b^{c+1} - a^{c+1})\frac{v}{(1 + v)^{c+1} - 1}.$$

As v tends to zero the quotient

$$\frac{(1 + v)^{c+1} - 1}{v}$$

approaches as its limit the derivative of $(1 + v)^{c+1}$ with respect to v for $v = 0$, that is, $c + 1$; hence, the area of the region under consideration is

$$\frac{A(b^{c+1} - a^{c+1})}{c + 1}.$$

If $c = -1$, this calculation no longer applies. The sum of the areas of the inscribed rectangles is equal to nAv, and we have to find the limit of the product nv, where n and v are connected by the relation $a(1 + v)^n = b$. Hence,

$$nv = \left(\ln \frac{b}{a}\right)\frac{v}{\ln(1 + v)} = \left(\ln \frac{b}{a}\right)\frac{1}{\ln(1 + v)^{1/v}}.$$

As v tends to zero, $(1 + v)^{1/v}$ tends to e, and the product nv tends to $\ln(b/a)$. Therefore, the required area equals $A\ln(b/a)$.

2. Area of a Planar Region

Suppose that one has marked off a region of the x, y-plane with the help of a curve; it is intuitively clear that the region has an area. However, how is one to define this area arithmetically? If the region is a rectangle, the area of the region is taken to be the product of the length and the width of the rectangle. When at least part of the boundary of the region is curved, however, the definition of the area of the region is no longer obvious.

For the moment we restrict attention to particularly simple regions and consider only those bounded by the x-axis, the lines $x = a$ and $x = b$ with $a < b$, and the curve $y = f(x)$. Moreover, we assume that the function $y = f(x)$ is continuous on the closed interval $[a, b]$ and that $f(x) \geq 0$ for all x satisfying $a \leq x \leq b$. The theory of the area of general regions bounded by curves can be reduced to the case of these special regions. Let us note at once that the restriction concerning the sign of f on $[a, b]$ is not of a serious nature. Indeed, if f is a continuous function on $[a, b]$ which changes sign, we can introduce the continuous (recall the content of Proposition 2.8) non-negative functions

$$f_1(x) = \tfrac{1}{2}\{|f(x)| + f(x)\} \quad \text{and} \quad f_2(x) = \tfrac{1}{2}\{|f(x)| - f(x)\}$$

and put $f = f_1 - f_2$; see Figure 5.4.

To define the area I of a region F, we divide the interval $[a, b]$ into n subintervals with the help of the points

$$a = x_0 < x_1 < x_2 < \cdots < x_{n-1} < x_n = b \tag{5.2}$$

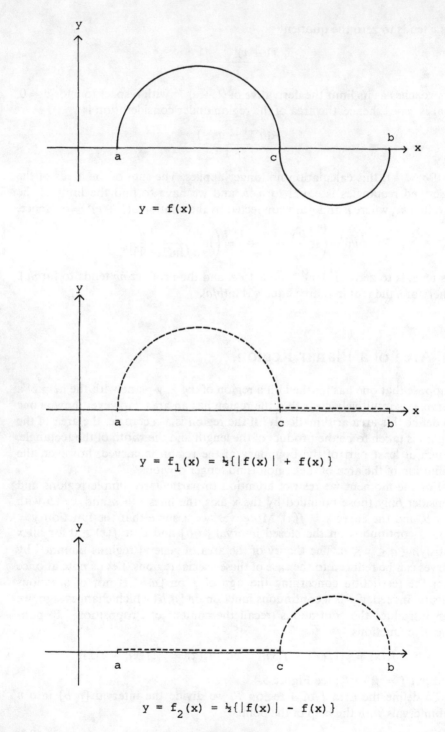

$$y = f(x)$$

$$y = f_1(x) = \tfrac{1}{2}\{\,|f(x)| + f(x)\,\}$$

$$y = f_2(x) = \tfrac{1}{2}\{\,|f(x)| - f(x)\,\}$$

Figure 5.4

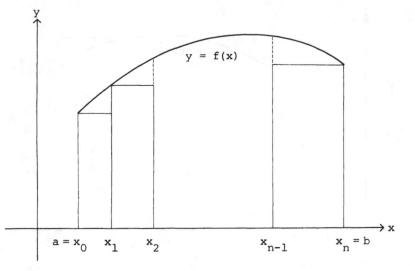

Figure 5.5

and draw lines parallel to the y-axis through the points x_k, $k = 0, 1, \ldots, n$. In this way the region F is divided into n strips. Into each of these strips we draw the largest rectangle which still fits entirely into the region (see Figure 5.5). The area of the kth of these rectangles is

$$(x_k - x_{k-1})m_k, \quad \text{where } m_k = \min\{f(x): x_{k-1} \leq x \leq x_k\}.$$

We clearly must define the area of F making it at least as large as the sum

$$s = (x_1 - x_0)m_1 + (x_2 - x_1)m_2 + \cdots + (x_n - x_{n-1})m_n = \sum_{k=1}^{n} (x_k - x_{k-1})m_k$$

of the areas of these rectangles. Now, the number s will depend on the number and the special choice of the points x_k. The set of all numbers s which can be found by different choices of the subdivision (5.2) will be denoted by $\mathscr{S}$. It seems apparent that we shall have to define I in such a way as to have numbers s in $\mathscr{S}$ arbitrarily close to I. Thus, we are compelled to set I equal to the least upper bound $\underline{I}$ of all numbers s in $\mathscr{S}$, that is, $\underline{I} = \sup \mathscr{S}$.

However, we can also obtain the definition of the area I of the region F by considering, instead of the rectangles lying inside F, those rectangles that cover F. We again draw lines parallel to the y-axis through the points x_k of the subdivision (5.2), dividing F into n strips. But now for each strip we find the smallest rectangle which still contains the strip (see Figure 5.6). The area of the region of the kth of these rectangles is

$$(x_k - x_{k-1})M_k, \quad \text{where } M_k = \max\{f(x): x_{k-1} \leq x \leq x_k\}.$$

We evidently must define the area I of F so as to have it at most as large as the sum

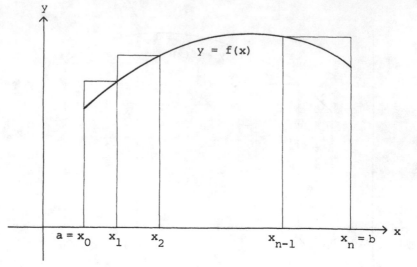

Figure 5.6

$$S = (x_1 - x_0)M_1 + (x_2 - x_1)M_2 + \cdots + (x_n - x_{n-1})M_n = \sum_{k=1}^{n} (x_k - x_{k-1})M_k$$

of the areas of all these rectangles. The set of all the numbers S which can be obtained by different choices of the subdivision (5.2) of the interval $[a, b]$ will be denoted by $\overline{\mathscr{S}}$. It again seems apparent that I must be so defined that there are numbers S in $\overline{\mathscr{S}}$ arbitrarily close to I. One is thus compelled to set I equal to the greatest lower bound $\overline{I}$ of all numbers S in $\overline{\mathscr{S}}$, that is, $\overline{I} = \inf \overline{\mathscr{S}}$.

We have thus seen that $I = \underline{I}$ and $I = \overline{I}$ must hold. However, it is not clear from the start that the numbers $\underline{I}$ and $\overline{I}$ coincide. A complete theory of area with the first goal of a correct definition of a numerical value for area of a region must verify the equation

$$\underline{I} = \overline{I}. \tag{5.3}$$

The proof of this equation is a purely arithmetic problem. The equation is a statement about continuous functions. The class of functions for which a statement corresponding to equation (5.3) holds is far greater than the set of continuous functions. In the next section we review the train of thought which led us to equation (5.3); the approach will be purely arithmetic and will avoid geometric considerations and visualizations.

3. The Riemann Integral

Definition. Let $[a, b]$ be a closed interval of finite length. By a *partition* P of $[a, b]$ we mean any finite set of points $x_0, x_1, \ldots, x_n$ such that

$$a = x_0 < x_1 < \cdots < x_n = b.$$

We write, for $1 \leq k \leq n$,

$$\Delta x_k = x_k - x_{k-1}$$

and denote by

$$\|P\| = \max\{\Delta x_k : 1 \leq k \leq n\};$$

$\|P\|$ is called the *mesh of the partition P*.

Definition. A partiton P^* of $[a, b]$ is called a *refinement of* a partition P of $[a, b]$ (in symbols, $P^* \supset P$) if every point of P is a point of P^*. Given two partitions P_1 and P_2 of $[a, b]$, we call $P^* = P_1 \cup P_2$ their *common refinement*. P^* is said to be *finer than* P if $P^* \supset P$.

REMARK. It is clear that the common refinement of two partitions of an interval is actually a refinement of each.

Definition. Let f be a bounded real-valued function on a closed interval $[a, b]$ of finite length, that is, there are two numbers, m and M, such that

$$m \leq f(x) \leq M \quad \text{for } a \leq x \leq b.$$

Corresponding to each partition P of $[a, b]$ we put

$$M_k = \sup\{f(x) : x_{k-1} \leq x \leq x_k\},$$
$$m_k = \inf\{f(x) : x_{k-1} \leq x \leq x_k\},$$

and define the *upper* and the *lower Darboux sums of f relative to P* by, respectively,

$$U(P, f) = \sum_{k=1}^{n} M_k \Delta_k \quad \text{and} \quad L(P, f) = \sum_{k=1}^{n} m_k \Delta_k;$$

finally, we put

$$\overline{\int_a^b} f(x)\, dx = \inf U(P, f), \tag{5.4}$$

$$\underline{\int_a^b} f(x)\, dx = \sup L(P, f), \tag{5.5}$$

where the infimum and the supremum are taken over all partitions P of $[a, b]$. The left-hand members of (5.4) and (5.5) are called the *upper* and the *lower Riemann integrals of f over* $[a, b]$, respectively. If the upper and the lower Riemann integrals are equal, we say that f is *Riemann integrable on* $[a, b]$ and we denote the common value of (5.4) and (5.5) by the symbol

$$\int_a^b f(x)\, dx \quad \text{or} \quad \int_a^b f(x)\, dx$$

and call it the *Riemann integral of f over* $[a, b]$. When talking about the Riemann integral of f over $[a, b]$ we *assume* that f is bounded and $[a, b]$ is of finite length.

REMARK. To see that the upper and lower Riemann integrals exist for every bounded function f on a closed interval $[a, b]$ of finite length, we observe that the numbers $L(P, f)$ and $U(P, f)$ form a bounded set. Indeed, since f is bounded,

$$m \leq f(x) \leq M \quad \text{for } a \leq x \leq b;$$

hence, for any partition P of $[a, b]$, we have

$$m(b - a) \leq L(P, f) \leq U(P, f) \leq M(b - a).$$

Proposition 5.1. *Let f be a bounded function on* $[a, b]$. *If P and P* are partitions of* $[a, b]$ *and* $P \subset P^*$, *then*

$$L(P, f) \leq L(P^*, f) \leq U(P^*, f) \leq U(P, f).$$

PROOF. The middle inequality is obvious. The verifications of the first and the third inequalities are similar and so we will only prove

$$L(P, f) \leq L(P^*, f).$$

We suppose first that P^* contains only one more point than P. Let x^* be this extra point and assume that $x_{k-1} < x^* < x_k$, where x_{k-1} and x_k are two consecutive points of P. We put

$$w_1 = \inf\{f(x): x_{k-1} \leq x \leq x^*\},$$

$$w_2 = \inf\{f(x): x^* \leq x \leq x_k\}.$$

Letting, as before,

$$m_k = \inf\{f(x): x_{k-1} \leq x \leq x_k\},$$

we see that $w_1 \geq m_k$ and $w_2 \geq m_k$. Thus,

$$L(P^*, f) - L(P, f) = w_1(x^* - x_{k-1}) + w_2(x_k - x^*) - m_k(x_k - x_{k-1})$$

$$= (w_1 - m_k)(x^* - x_{k-1}) + (w_2 - m_k)(x_k - x^*) \qquad (5.6)$$

$$\geq 0.$$

If P^* contains j points more than P, we repeat the above reasoning j times, and arrive at the inequality $L(P, f) \leq L(P^*, f)$. $\qquad\qquad \square$

Proposition 5.2. *Let f be a bounded function on* $[a, b]$ *and* P_1 *and* P_2 *be any partitions of* $[a, b]$. *Then*

$$L(P_1, f) \leq U(P_2, f).$$

PROOF. Let $P^* = P_1 \cup P_2$, the common refinement of P_1 and P_2. By Proposition 5.1 we have

$$L(P_1, f) \le L(P^*, f) \le U(P^*, f) \le U(P_2, f)$$

because $P_1 \subset P^*$ and $P_2 \subset P^*$. $\qquad\qquad\qquad\qquad\qquad\qquad\qquad$ □

Proposition 5.3. *If f is a bounded function on* $[a, b]$, *then*

$$\underline{\int}_a^b f(x)\,dx \le \overline{\int}_a^b f(x)\,dx.$$

PROOF. By Proposition 5.2, for any partitions P_1 and P_2 of $[a, b]$,

$$L(P_1, f) \le U(P_2, f).$$

By keeping P_2 fixed, and taking the supremum over all P_1, the foregoing inequality gives

$$\underline{\int}_a^b f(x)\,dx \le U(P_2, f).$$

The claim of the proposition now follows by taking the infimum over all P_2 in the foregoing inequality. $\qquad\qquad\qquad\qquad\qquad\qquad\qquad\qquad$ □

Proposition 5.4. *A bounded function f on* $[a, b]$ *is Riemann integrable over* $[a, b]$ *if and only if for any* $\varepsilon > 0$ *there exists a partition P of* $[a, b]$ *such that*

$$U(P, f) - L(P, f) < \varepsilon. \tag{5.7}$$

PROOF. Using Proposition 5.3, we see that

$$L(P, f) \le \underline{\int}_a^b f(x)\,dx \le \overline{\int}_a^b f(x)\,dx \le U(P, f),$$

where P denotes any partition of $[a, b]$. Inequality (5.7) therefore implies

$$0 < \overline{\int}_a^b f(x)\,dx - \underline{\int}_a^b f(x)\,dx < \varepsilon.$$

Thus, if (5.7) holds for any $\varepsilon > 0$, then

$$\overline{\int}_a^b f(x)\,dx = \underline{\int}_a^b f(x)\,dx,$$

implying that f is Riemann integrable over $[a, b]$.

Conversely, suppose that f is Riemann integrable over $[a, b]$ and let $\varepsilon > 0$ be given. Then there are partitions P_1 and P_2 of $[a, b]$ such that

$$U(P_2, f) - \int_a^b f(x)\,dx < \frac{\varepsilon}{2} \quad \text{and} \quad \int_a^b f(x)\,dx - L(P_1, f) < \frac{\varepsilon}{2}. \tag{5.8}$$

Let P be the common refinement of P_1 and P_2. Then, by Proposition 5.3, together with (5.8), we get

$$U(P,f) \le U(P_2,f) < \int_a^b f(x)\,dx + \frac{\varepsilon}{2} < L(P_1,f) + \varepsilon \le L(P,f) + \varepsilon;$$

thus (5.7) holds for the partition $P = P_1 \cup P_2$.   □

REMARK. In the foregoing proof we have used the following fact: If A and B are fixed real numbers such that $A \ge B$ and $A - B < \varepsilon$ for any $\varepsilon > 0$, then $A = B$. Indeed, if $A \ne B$ was possible, then $A - B = \alpha > 0$ and so $\varepsilon > \alpha$ would follows.

Proposition 5.5. *A bounded function f on $[a, b]$ is Riemann integrable over $[a, b]$ if and only if for any $\varepsilon > 0$ there exists some $\delta > 0$ such that*

$$\|P\| < \delta \quad implies \quad U(P,f) - L(P,f) < \varepsilon \tag{5.9}$$

for all partitions P of $[a, b]$.

PROOF. Proposition 5.4 shows that the condition (5.9) implies the Riemann integrability of f.

Conversely, assume that f is Riemann integrable over $[a, b]$. Let $\varepsilon > 0$ and pick a partition

$$P_0 = \{a = t_0 < t_1 < \cdots < t_m = b\}$$

of $[a, b]$ such that

$$U(P_0,f) - L(P_0,f) < \frac{\varepsilon}{2}. \tag{5.10}$$

Since f is bounded, there exists $B > 0$ such that $|f(x)| \le B$ for all x in $[a, b]$. Let $\delta = \varepsilon/8mB$ with m being the number of intervals comprising P_0.

To verify (5.9), we consider any partition

$$P = \{a = x_0 < x_1 < \cdots < x_n = b\}$$

with mesh $\|P\| < \delta$. Let $Q = P \cup P_0$. If Q has one more element than P, then a look at relation (5.6) in the proof of Proposition 5.1 reveals that

$$L(Q,f) - L(P,f) \le 2B \cdot \|P\|.$$

Since Q has at most m elements that are not in P, we see that

$$L(Q,f) - L(P,f) \le 2mB \cdot \|P\| < 2mB\delta = \frac{\varepsilon}{4}.$$

By Proposition 5.1 we have $L(P_0,f) \le L(Q,f)$ and so

$$L(P_0,f) - L(P,f) < \frac{\varepsilon}{4}.$$

Similarly, $U(P,f) - U(P_0,f) < \varepsilon/4$ and so

$$U(P,f) - L(P,f) < U(P_0,f) - L(P_0,f) + \frac{\varepsilon}{2}.$$

Now (5.10) shows that $U(P,f) - L(P,f) < \varepsilon$ and we have verified (5.9).   □

Definition. Let $[a,b]$ be a closed interval of finite length and f a bounded function on $[a,b]$. Take a partition

$$P = \{a = x_0 < x_1 < \cdots < x_n = b\}$$

of $[a,b]$. With t_k selected such that $x_{k-1} \leq t_k \leq x_k$ for $k = 1, \ldots, n$, form the sum

$$S(P,f) = \sum_{k=1}^{n} f(t_k)\Delta x_k, \quad \text{where } \Delta x_k = x_k - x_{k-1};$$

the number $S(P,f)$ is called a *Riemann sum of f associated with the partition P.* The choice of the t_k's is arbitrary apart from the restriction $x_{k-1} \leq t_k \leq x_k$ for $k = 1, \ldots, n$ and so there are infinitely many Riemann sums associated with a single function and partition. The notation

$$\lim_{\|P\| \to 0} S(P,f) = A, \tag{5.11}$$

where A is a real number and $\|P\|$ denotes the mesh of P, means that for any $\varepsilon > 0$ there is some $\delta > 0$ such that for any partition P of $[a,b]$ with $\|P\| < \delta$ and for any ·possible Riemann sum $S(P,f)$ associated with P, the inequality

$$|S(P,f) - A| < \varepsilon$$

is satisfied.

REMARKS. In analogy to Proposition 2.3 we can show that if

$$\lim_{\|P\| \to 0} S(P,f) = A \quad \text{and} \quad \lim_{\|P\| \to 0} S(P,f) = B,$$

then $A = B$; in other words, if the limit exists, it is unique.

   Equivalent to the foregoing definition of the notation (5.11) is the following: For any $\varepsilon > 0$ there is some partition P_ε of $[a,b]$ such that

$$|S(P,f) - A| < \varepsilon$$

holds for all partitions $P \supset P_\varepsilon$ of $[a,b]$, where $S(P,f)$ is any Riemann sum associated with P.

Proposition 5.6. *Let f be a bounded function on $[a,b]$. Then f is Riemann integrable over $[a,b]$ if and only if*

$$\lim_{\|P\| \to 0} S(P,f) = A.$$

Moreover,

$$\int_a^b f(x)\,dx = A.$$

PROOF. Suppose first that f is Riemann integrable over $[a, b]$. Let $\varepsilon > 0$, and let $\delta > 0$ be chosen so that condition (5.9) in Proposition 5.5 is satisfied. We verify that

$$\left| S(P, f) - \int_a^b f(x)\,dx \right| < \varepsilon \tag{5.12}$$

for every Riemann sum $S(P, f)$ associated with a partition P with $\|P\| < \delta$. Obviously, we have

$$L(P, f) \leq S(P, f) \leq U(P, f)$$

and so (5.12) follows from the inequalities

$$U(P, f) < L(P, f) + \varepsilon \leq \underline{\int_a^b} f(x)\,dx + \varepsilon = \int_a^b f(x)\,dx + \varepsilon$$

and

$$L(P, f) > U(P, f) - \varepsilon \geq \overline{\int_a^b} f(x)\,dx - \varepsilon = \int_a^b f(x)\,dx - \varepsilon.$$

This proves (5.12); hence,

$$\lim_{\|P\| \to 0} S(P, f) = \int_a^b f(x)\,dx.$$

Now suppose that $\lim_{\|P\| \to 0} S(P, f)$ exists and is equal to A. Let $\varepsilon > 0$ be given. From the definition of the notation in (5.11) we see that there exist some $\delta > 0$ such that $\|P\| < \delta$ implies

$$A - \frac{\varepsilon}{3} < S(P, f) < A + \frac{\varepsilon}{3}. \tag{5.13}$$

We choose one such partition

$$P = \{a = x_0 < x_1 < \cdots < x_n = b\}.$$

If we let the points t_k range over the intervals $[x_{k-1}, x_k]$ and take the supremum and the infimum of the numbers $S(P, f)$ so obtained, (5.13) yields

$$A - \frac{\varepsilon}{2} \leq L(P, f) \leq U(P, f) \leq A + \frac{\varepsilon}{3}.$$

Thus,

$$U(P, f) - L(P, f) \leq \frac{2\varepsilon}{3} < \varepsilon$$

and f is seen to be Riemann integrable over $[a, b]$ by Proposition 5.4. Since

$$L(P, f) \leq \int_a^b f(x) \, dx \leq U(P, f),$$

it follows that

$$\lim_{\|P\| \to 0} S(P, f) = \int_a^b f(x) \, dx.$$

This completes the proof.

Proposition 5.7. *Let σ and ω be arbitrary positive numbers and f be a bounded function on an interval $[a, b]$ of finite length. Then f is Riemann integrable over $[a, b]$ if and only if there is a mode of division of $[a, b]$ into subintervals such that the sum of the lengths of the subintervals in which the oscillation of f is greater than or equal to ω is less than σ.*

PROOF. Let $P = \{a = x_0 < x_1 < \cdots < x_n = b\}$ be a partition of $[a, b]$ and consider the sum

$$Z(P, f) = \sum_{k=1}^n \omega_k \Delta x_k,$$

where $\Delta x_k = x_k - x_{k-1}$ and $\omega_k = M_k - m_k$ with

$$M_k = \sup\{f(x) : x_{k-1} \leq x \leq x_k\}, \qquad m_k = \inf\{f(x) : x_{k-1} \leq x \leq x_k\}$$

(i.e., ω_k is the oscillation of f in the interval $[x_{k-1}, x_k]$). We let

$$\Omega = M - m,$$

where

$$M = \sup\{f(x) : a \leq x \leq b\}, \qquad m = \inf\{f(x) : a \leq x \leq b\},$$

and denote the length of the interval $[a, b]$ by K.

We now derive bounds for $Z(P, f)$; incidentally, $Z(P, f) = U(P, f) - L(P, f)$. Let δ be the sum of the lengths of the subintervals obtained by the partition P in which the oscillation of f is greater than or equal to ω. Then

$$Z(P, f) \geq \delta\omega. \tag{5.14}$$

But in these subintervals the oscillation of f is less than or equal to Ω and in the remaining subintervals (the sum of whose lengths is $K - \delta$) the oscillation of f is less than ω. Thus,

$$Z(P, f) \leq \delta\Omega + (K - \delta)\omega.$$

Since $K - \delta \leq K$, we see that

$$\delta\omega \leq Z(P, f) \leq \delta\Omega + K\Omega.$$

If f is Riemann integrable over $[a, b]$, then by Proposition 5.4 there exists a partition P such that for any preassigned positive numbers σ and ω we have

$$Z(P, f) \leq \omega\sigma. \tag{5.15}$$

From (5.14) and (5.15) it follows that $\delta\omega < \omega\sigma$, that is, $\delta < \sigma$.

Conversely, if there exists a partition P for which $\delta < \omega$, we choose

$$\omega = \frac{\varepsilon}{2K} \quad \text{and} \quad \delta = \frac{\varepsilon}{2\Omega}.$$

Then $Z(P, f) \leq \delta\Omega + K\Omega < \varepsilon/2 + \varepsilon/2 = \varepsilon$. $\square$

Proposition 5.8. Let $f_1, f_2, \ldots, f_q$ be Riemann integrable functions over $[a, b]$ and suppose that f is a function defined on $[a, b]$. Suppose there exist positive numbers $\gamma_1, \gamma_2, \ldots, \gamma_q$ such that on any subinterval J of $[a, b]$ we have

$$\omega(f, J) \leq \gamma_1 \omega(f_1, J) + \gamma_2 \omega(f_2, J) + \cdots + \gamma_q \omega(f_q, J),$$

where $\omega(f, J)$ denotes the oscillation of the function f on J and $\omega(f_j, J)$ the oscillation of f_j on J for $j = 1, \ldots, q$, then f is also Riemann integrable over $[a, b]$.

PROOF. Let P be the partition

$$P = \{a = x_0 < x_1 < \cdots < x_n = b\}$$

of $[a, b]$. Since

$$\omega(f, [x_{k-1}, x_k]) = \sup\{f(x): x_{k-1} \leq x \leq x_k\} - \inf\{f(x): x_{k-1} \leq x \leq x_k\},$$

we have

$$U(P, f) - L(P, f) = \sum_{k=1}^{n} \omega(f, [x_{k-1}, x_k])\Delta x_k$$

$$\leq \sum_{j=1}^{q} \gamma_j \sum_{k=1}^{n} \omega(f_j, [x_{k-1}, x_k])\Delta x_k$$

$$= \sum_{j=1}^{q} \gamma_j \{U(P, f_j) - L(P, f_j)\}.$$

The claim now easily follows from Proposition 5.4. Indeed, for a given $\varepsilon > 0$ we select a partition P such that

$$U(P, f_j) - L(P, f_j) < \frac{\varepsilon}{\gamma_1 + \gamma_2 + \cdots + \gamma_q}$$

holds for all $j = 1, \ldots, q$. $\square$

Proposition 5.9. Let f and g be Riemann integrable functions over $[a, b]$. Then each of the following functions is Riemann integrable over $[a, b]$:

  (i) $\alpha f + \beta g$, where α and β are any fixed real numbers;
 (ii) $|f|$;
(iii) fg;
(iv) f/g provided that $\inf\{|g(x)|: a \leq x \leq b\} > 0$.

PROOF. We apply Proposition 5.8. To obtain the verification of claims (i) and (ii) we only need to note that for any subinterval J of $[a, b]$ we have

$$\omega(\alpha f + \beta g, J) \leq |\alpha| \omega(f, J) + |\beta| \omega(g, J) \tag{5.16}$$

and

$$\omega(|f|, J) \leq \omega(f, J). \tag{5.17}$$

However, (5.16) and (5.17) can be deduced from the inequalities

$$|\alpha f(x) + \beta g(x) - \alpha f(t) - \beta g(t)| \leq |\alpha| |f(x) - f(t)| + |\beta| |g(x) - g(t)|$$

and

$$||f(x)| - |f(t)|| \leq |f(x) - f(t)|,$$

where x and t are points of the interval J. (In connection with the foregoing inequalities see Proposition 2.1 and the comments following it.)

To prove claim (iii) we also apply Proposition 5.8 and observe that for any subinterval J of $[a, b]$ we have

$$\omega(fg, J) \leq \gamma_1 \omega(f, J) + \gamma_2 \omega(g, J),$$

where

$$\gamma_1 = \sup\{|g(x)|: a \leq x \leq b\} \quad \text{and} \quad \gamma_2 = \sup\{|f(x)|: a \leq x \leq b\}$$

because, for any two points x and t in J,

$$|f(x)g(x) - f(t)g(t)| \leq |f(x)[g(x) - g(t)] + g(t)[f(x) - f(t)]|$$

$$\leq \sup\{|f(s)|: a \leq s \leq b\}|g(x) - g(t)| + \sup\{|g(s)|: a \leq s \leq b\}|f(x) - f(t)|.$$

To prove claim (iv) we only need to establish that $1/q$ is Riemann integrable; the rest follows from claim (iii). Noting that

$$\left| \frac{1}{g(x)} - \frac{1}{g(t)} \right| = \frac{1}{|g(x)| |g(t)|} |g(x) - g(t)| \leq \gamma |g(x) - g(t)|,$$

where

$$\gamma = \frac{1}{(\inf\{|g(s)|: a \leq s \leq b\})^2} \tag{5.18}$$

and x and t are any two points of a subinterval J of $[a, b]$, it follows that

$$\omega\left(\frac{1}{g}, J\right) \leq \gamma \omega(g, J), \tag{5.19}$$

where γ is given by (5.18). But (5.19) ensures that $1/g$ is Riemann integrable over $[a, b]$ via Proposition 5.8 provided that $\inf\{|g(s)|: a \leq s \leq b\} > 0$. $\qquad \square$

Proposition 5.10. *Let f and g be Riemann integrable over $[a, b]$ and let α and β be any fixed real numbers. Then*

$$\int_a^b \{\alpha f(x) + \beta g(x)\}\, dx = \alpha \int_a^b f(x)\, dx + \beta \int_a^b g(x)\, dx. \tag{5.20}$$

PROOF. Let $P = \{a = x_0 < x_1 < \cdots < x_n = b\}$ be a partition of $[a,b]$ and t_k satisfy $x_{k-1} \le t_k \le x_k$ for $k = 1, \ldots, n$. Then for any $\varepsilon > 0$ there exist $\delta_1 > 0$ and $\delta_2 > 0$ such that

$$\left| \alpha \sum_{k=1}^n f(t_k)\Delta x_k - \alpha \int_a^b f(x)\, dx \right| < \frac{\varepsilon}{2} \quad \text{whenever } \|P\| < \delta_1$$

and

$$\left| \beta \sum_{k=1}^n g(t_k)\Delta x_k - \beta \int_a^b g(x)\, dx \right| < \frac{\varepsilon}{2} \quad \text{whenever } \|P\| < \delta_2,$$

where $\Delta x_k = x_k - x_{k-1}$ for $k = 1, \ldots, n$. Thus, whenever $\|P\| < \min\{\delta_1, \delta_2\}$,

$$\left| \sum_{k=1}^n \{\alpha f(t_k) + \beta g(t_k)\}\Delta x_k - \left(\alpha \int_a^b f(x)\, dx + \beta \int_a^b g(x)\, dx \right) \right|$$

$$\le \left| \alpha \sum_{k=1}^n f(t_k)\Delta x_k - \alpha \int_a^b f(x)\, dx \right| + \left| \beta \sum_{k=1}^n g(t_k)\Delta x_k - \beta \int_a^b g(x)\, dx \right| < \varepsilon$$

showing that $\alpha f + \beta g$ is Riemann integrable over $[a,b]$ and that (5.20) is satisfied. $\qquad\square$

Proposition 5.11. *If f and g are Riemann integrable over $[a,b]$ and if $f(x) \le g(x)$ for all x in $[a,b]$, then $\int_a^b f(x)\, dx \le \int_a^b g(x)\, dx$.*

PROOF. By Proposition 5.9, $h = g - f$ is integrable over $[a,b]$. Since $h(x) \ge 0$ for all x in $[a,b]$, it is clear that $L(P,h) \ge 0$ for all partitions P of $[a,b]$ and so

$$\int_a^b h(x)\, dx = \underline{\int_a^b} h(x)\, dx \ge 0$$

Applying Proposition 5.10, we see that $\int_a^b g(x)\, dx - \int_a^b f(x)\, dx = \int_a^b h(x)\, dx \ge 0$ and the proof is complete. $\qquad\square$

Proposition 5.12. *If f is Riemann integrable over $[a,b]$, then $|f|$ is Riemann integrable over $[a,b]$ (by Proposition 5.9) and*

$$\left| \int_a^b f(x)\, dx \right| \le \int_a^b |f(x)|\, dx. \tag{5.21}$$

PROOF. Since we know already that $|f|$ is Riemann integrable over $[a,b]$, Proposition 5.11 applied to the inequality $-|f| \le f \le |f|$ shows that

$$-\int_a^b |f(x)|\, dx \le \int_a^b f(x)\, dx \le \int_a^b |f(x)|\, dx$$

and so (5.21) follows. $\qquad\square$

Proposition 5.13. *If f is Riemann integrable over* $[a, b]$ *and if* $a < c < b$, *then f is Riemann integrable over both* $[a, c]$ *and* $[c, b]$, *and*

$$\int_a^c f(x)\,dx + \int_c^b f(x)\,dx = \int_a^b f(x)\,dx.$$

PROOF. Let $\varepsilon > 0$ be given. Choose a partition P of $[a, b]$ such that

$$U(P, f) - L(P, f) < \varepsilon.$$

In view of Proposition 5.4 we may (and do) suppose that c is a point of P; say

$$P = \{a = x_0 < \cdots < x_m = c < x_{m+1} < \cdots < x_n = b\}.$$

Let

$$P_1 = \{a = x_0 < \cdots < x_m = c\} \quad \text{and} \quad P_2 = \{c = x_m < \cdots < x_n = b\}.$$

Then

$$[U(P_1, f) - L(P_1, f)] + [U(P_2, f) - L(P_2, f)] = U(P, f) - L(P, f) < \varepsilon.$$

It follows that f is Riemann integrable over both $[a, c]$ and $[c, b]$. Let

$$\int_a^c f(x)\,dx = K_1 \quad \text{and} \quad \int_a^b f(x)\,dx = K_2.$$

Clearly,

$$0 \le U(P_1, f) - K_1 < \varepsilon \quad \text{and} \quad 0 \le U(P_2, f) - K_2 < \varepsilon.$$

Adding these two inequalities, we obtain

$$0 \le U(P, f) - (K_1 + K_2) < 2\varepsilon.$$

Since a similar statement is true for $L(P, f)$, we conclude that

$$\int_a^b f(x)\,dx = K_1 + K_2$$

and the proof is complete.   □

Proposition 5.14. *Let f be a function defined on* $[a, b]$. *If* $a < c < b$ *and if f is Riemann integrable over* $[a, c]$ *and over* $[c, b]$, *then f is Riemann integrable over* $[a, b]$ *and*

$$\int_a^b f(x)\,dx = \int_a^c f(x)\,dx + \int_c^b f(x)\,dx.$$

PROOF. Since f is bounded on both $[a, c]$ and $[c, b]$, f is bounded on $[a, b]$. Let $\varepsilon > 0$. By Proposition 5.4 there exist partitions P_1 of $[a, c]$ and P_2 of $[c, b]$ such that (we decorate in this proof upper and lower sums of that it will be clear with which intervals we are dealing)

$$U_a^c(P_1, f) - L_a^c(P_1, f) < \frac{\varepsilon}{2} \quad \text{and} \quad U_c^b(P_2, f) - L_c^b(P_2, f) < \frac{\varepsilon}{2}.$$

The set $P = P_1 \cup P_2$ is a partition of $[a, b]$ and it is clear that

$$U_a^b(P, f) = U_a^c(P_1, f) + U_c^b(P_2, f) \tag{5.22}$$

with a similar identity for lower sums. It follows that

$$U_a^b(P, f) - L_a^b(P, f) < \varepsilon$$

and so f is Riemann integrable over $[a, b]$ by Proposition 5.4. Also (5.22) holds because

$$\int_a^b f(x)\,dx \leq U_a^b(P, f) = U_a^c(P_1, f) + U_c^b(P_2, f) < L_a^c(P_1, f) + L_c^b(P_2, f) + \varepsilon$$

$$\leq \int_a^c f(x)\,dx + \int_c^b f(x)\,dx + \varepsilon$$

and, similarly,

$$\int_a^b f(x)\,dx > \int_a^c f(x)\,dx + \int_c^b f(x)\,dx - \varepsilon.$$

completing the proof. □

Definition. If $b < a$, we set

$$\int_a^b f(x)\,dx = -\int_b^a f(x)\,dx$$

in case the latter integral exists. Furthermore, we put

$$\int_a^a f(x)\,dx = 0.$$

COMMENTS. In Propositions 5.13 and 5.14 we encountered the equation

$$\int_a^b f(x)\,dx = \int_a^c f(x)\,dx + \int_c^b f(x)\,dx. \tag{5.23}$$

In terms of the foregoing definition we can write (5.23) in the form

$$\int_a^b f(x)\,dx + \int_b^c f(x)\,dx + \int_c^a f(x)\,dx = 0. \tag{5.24}$$

It is easily seen that (5.24) is universally true whenever the three integrals exist, regardless of the order relation between the numbers a, b, and c. For example, if $c < b < a$, then (5.24) holds also. It can also be seen that if $a_1, a_2, \ldots, a_k$ are a finite number of points situated in an interval $[a, b]$ of finite length and if a function f is Riemann integrable over this interval, then

$$\int_{a_1}^{a_2} f(x)\,dx + \int_{a_2}^{a_3} f(x)\,dx + \cdots + \int_{a_{k-1}}^{a_k} f(x)\,dx + \int_{a_k}^{a_1} f(x)\,dx = 0,$$

whatever may be the order relation between the points $a_1, a_2, \ldots, a_k$.

If the order relation between a and b is not specified, then inequality (5.21) needs to be written in the form

$$\left| \int_a^b f(x)\,dx \right| \le \left| \int_a^b |f(x)|\,dx \right|.$$

Proposition 5.15. *Every monotonic function f on $[a,b]$ is Riemann integrable over $[a,b]$.*

PROOF. Suppose that f is increasing on $[a,b]$; if f were decreasing; we could simply consider the function $-f$. Since $f(a) \le f(x) \le f(b)$ for all x in $[a,b]$, f is clearly bounded on $[a,b]$. Let $\varepsilon > 0$ and pick a positive integer n so large that

$$\frac{f(b) - f(a)}{b - a} < n\varepsilon.$$

For the partition

$$P = \{a = x_0 < x_1 < \cdots < x_n = b\},$$

where $x_k - x_{k-1} = \Delta x_k = (b - a)/n$ for $k = 1, \ldots, n$, we have

$$U(P,f) - L(P,f) = \frac{b - a}{n} \sum_{k=1}^{n} \{f(x_k) - f(x_{k-1})\} = \frac{b - a}{n} \{f(b) - f(a)\} < \varepsilon.$$

Proposition 5.4 now implies that f is Riemann integrable over $[a,b]$. □

Proposition 5.16. *Every continuous function f on $[a,b]$ is Riemann integrable over $[a,b]$.*

PROOF. Let $\varepsilon > 0$. Since $[a,b]$ is assumed to be closed and of finite length, f is uniformly continuous on $[a,b]$ by Proposition 2.16. Hence, there exists $\delta > 0$ such that if x and y are in $[a,b]$ and $|x - y| < \delta$, then

$$|f(x) - f(y)| < \frac{\varepsilon}{b - a}. \tag{5.25}$$

Consider any partition

$$P = \{a = x_0 < x_1 < \cdots < x_n = b\}, \quad \text{where } \|P\| < \delta.$$

Since f assumes its maximum and minimum on each interval $[x_{k-1}, x_k]$ by Proposition 2.13, it follows from (5.25) that

$$M_k - m_k < \frac{\varepsilon}{b - a}$$

for each k, where $k = 1, \ldots, n$. Thus,

$$U(P,f) - L(P,f) < \sum_{k=1}^{n} \frac{\varepsilon}{b-a} \Delta x_k = \varepsilon$$

and Proposition 5.4 implies that f is Riemann integrable over $[a,b]$. □

Proposition 5.17. *Let f be Riemann integrable over $[a,b]$, let*

$$m = \inf\{f(x): a \le x \le b\}, \qquad M = \sup\{f(x): a \le x \le b\},$$

and let g be a continuous function on $[m, M]$. Then the composition function $h(x) = g[f(x)]$ is Riemann integrable over $[a, b]$.

PROOF. Let $\varepsilon > 0$ be given. By the uniform continuity of g on $[m, M]$ we can find some $\delta_1 > 0$ such that

$$|g(s) - g(t)| < \varepsilon$$

if $|s - t| < \delta_1$ and s, t being points of the closed interval $[m, M]$. Let $\delta = \min\{\delta_1, \varepsilon\}$. Corresponding to δ^2, choose a partition

$$P = \{a = x_0 < x_1 < \cdots < x_n = b\}$$

of $[a, b]$ such that

$$U(P,f) - L(P,f) < \delta^2, \tag{5.26}$$

which is possible by Proposition 5.4. As usual, let

$$m_k = \inf\{f(x): x_{k-1} \le x \le x_k\} \quad \text{and} \quad M_k = \sup\{f(x): x_{k-1} \le x \le x_k\}$$

and let

$$m_k^* = \inf\{h(x): x_{k-1} \le x \le x_k\} \quad \text{and} \quad M_k^* = \sup\{h(x): x_{k-1} \le x \le x_k\}.$$

Divide the numbers $1, 2, \ldots, n$ into two classes: $k \in A$ if $M_k - m_k < \delta$ and $k \in B$ if $M_k - m_k \ge \delta$.

If $k \in A$ and $x_{k-1} \le x \le y \le x_k$, then

$$|f(x) - f(y)| \le M_k - m_k < \delta \le \delta_1$$

and so $|g[f(x)] - g[f(y)]| < \varepsilon$, implying that $M_k^* - m_k^* \le \varepsilon$ since $h(x) = g[f(x)]$. Thus,

$$\sum_{k \in A} (M_k^* - m_k^*)\Delta x_k \le \varepsilon \sum_{k=1}^{n} \Delta x_k = \varepsilon(b - a). \tag{5.27}$$

If $k \in B$, then $M_k - m_k \ge \delta$ and we have by (5.26)

$$\delta \sum_{k \in B} \Delta x_k \le \sum_{k \in B} (M_k - m_k)\Delta x_k \le \sum_{k=1}^{n} (M_k - m_k)\Delta x_k = U(P,f) - L(P,f) < \delta^2$$

and

$$0 \le \sum_{k \in B} \Delta x_k < \delta \le \varepsilon.$$

Let $K = \sup\{g(t)\colon m \le t \le M\}$. Then $M_k^* - m_k^* \le 2K$ and

$$\sum_{k \in B} (M_k^* - m_k^*)\Delta x_k \le 2K \sum_{k \in B} \Delta x_k < 2K. \tag{5.28}$$

Thus, using (5.27) and (5.28),

$$U(P, h) - L(P, h) = \sum_{k \in A} (M_k^* - m_k^*)\Delta x_k + \sum_{k \in B} (M_k^* - m_k^*)\Delta x_k$$

$$\le \varepsilon(b - a + 2K).$$

Since ε is arbitrary while $b - a + 2K$ is fixed, we see that h is Riemann integrable over $[a, b]$ by Proposition 5.4. $\square$

REMARK. Since $f(x) = x$ is monotonic and hence Riemann integrable over $[a, b]$, by Proposition 5.15, it is clear that Proposition 5.16 is a special case of Proposition 5.17.

DISCUSSION. In Example 10 of Section 4 in Chapter 2 we considered the function f on $[0, 1]$ defined as follows: We let $f(0) = f(1) = 1$ and

$$f(x) = 0 \quad \text{for } x \text{ irrational}$$

$$= \frac{1}{q} \quad \text{for } x = \frac{p}{q},$$

where $q > 0$ and p and q are integers without common divisor. It was shown that f is discontinuous at every rational point x of $[0, 1]$ and continuous at every irrational point x of $[0, 1]$. Figure 5.7 gives an indication of the graph of f; the graph in part is reminiscent of the shape of a Christmas tree. We observe that f is Riemann integrable.

Indeed, any lower Darboux sum of f is zero. We next divide the interval $[0, 1]$ into k^3 equal parts. Since there are at most

$$1 + 2 + \cdots + (k - 1) = \frac{k(k - 1)}{2}$$

positive proper fractions with denominator $\le k$, the upper Darboux sum is

$$< \frac{k(k - 1)}{2} \frac{1}{k^3} + \frac{2}{k^3} + \frac{1}{k} \cdot 1. \tag{5.29}$$

Note that on at most $k(k - 1)/2$ subintervals, each of length $1/k^3$, in the open interval $(0, 1)$ the values of f are between $\frac{1}{2}$ and 0, on the subintervals $[0, 1/k^3]$ and $[(k^3 - 1)/k^3, 1]$ the values of f are between 1 and 0, and on the remaining subintervals the sum of whose lengths is at most 1 the values of f are between $1/k$ and 0. But the expression in (5.29) tends to zero as k becomes arbitrarily large; by Proposition 5.4 we see that f is therefore Riemann integrable over $[0, 1]$. It is simple to see that $\int_0^1 f(x)\,dx = 0$.

We also note that if $g(y) = 1$ for $0 < y \le 1$ and $g(0) = 0$, then the function

$$h(x) = g[f(x)] \quad \text{with } 0 \le x \le 1$$

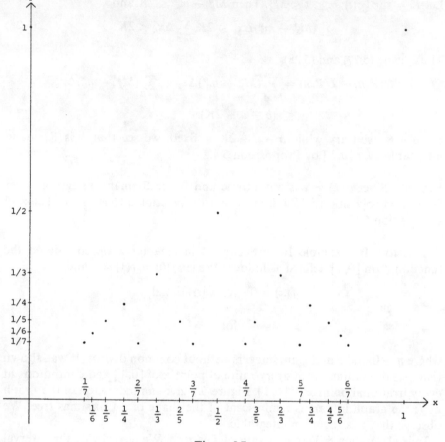

Figure 5.7

is not Riemann integrable over $[0, 1]$, because

$$h(x) = 1 \quad \text{for rational } x$$
$$= 0 \quad \text{for irrational } x;$$

however,

$$\overline{\int_0^1} h(x)\,dx = 1 \quad \text{and} \quad \underline{\int_0^1} h(x)\,dx = 0.$$

It can be seen that the function h is discontinuous everywhere on $[0, 1]$.

How badly discontinuous may a bounded function on a closed interval of finite length be and still be Riemann integrable over this interval? The answer to this question is in terms of the concept of sets having measure zero: A set of points on the x-axis is said to have *measure zero* if the sum of the lengths of intervals enclosing all the points can be made less than any given positive number ε. The integrability condition in question is: A bounded function on a closed interval of finite length is Riemann integrable if and only if the set of discontinuities of the function on this interval has measure zero. Since these matters properly belong to Lebesgue's integration theory, a significant extension of Riemann's integration theory, we shall not pursue this matter further here but refer the interested reader to books on real analysis (three such works are listed in the Bibliography at the end of this book).

In a Riemann integral $\int_a^b f(x)\,dx$ the values of f can be changed at a finite number of points without affecting either the existence of the value of the integral. To verify this, it is enough to consider the case where $f(x) = 0$ for all x in $[a, b]$ except for one point, say $x = c$. But for such a function it is clear that

$$|S(P, f)| \le |f(c)| \cdot \|P\|.$$

Since $\|P\|$ can be made arbitrarily small, it follows that $\int_a^b f(x)\,dx = 0$.

A function f on $[a, b]$ is called a *step-function* if there is a partition

$$P = \{a = x_0 < x_1 < \cdots < x_n = b\}$$

of $[a, b]$ such that f is constant on each subinterval (x_{k-1}, x_k), say $f(x) = c_k$ for x in (x_{k-1}, x_k), where $k = 1, \ldots, n$. Note that a step-function f is Riemann integrable and

$$\int_a^b f(x)\,dx = \sum_{k=1}^n c_k \Delta x_k, \quad \text{where } \Delta x_k = x_k - x_{k-1}.$$

In this connection we only need to recall Proposition 5.14 and observe that we can assign at the points of the partition P whatever (finite) values we please to the function f.

Finally, we note the following example: Let w be defined on $[0, 1]$ and

$$w(x) = (1 - x^2)^{1/2} \quad \text{when } x \text{ is rational,}$$

$$= 1 - x \quad \text{when } x \text{ is irrational.}$$

Then w is discontinuous everywhere on the open interval $(0, 1)$; moreover,

$$\underline{\int_0^1} w(x)\,dx = \frac{1}{2} \quad \text{and} \quad \overline{\int_0^1} w(x)\,dx = \frac{\pi}{4}$$

and so w is not Riemann integrable on $[0, 1]$.

4. Basic Propositions of Integral Calculus

Proposition 5.18 (Fundamental Theorem of Calculus I). *Let $[a, b]$ be a closed interval of finite length. If f is Riemann integrable over $[a, b]$ and if there exists a differentiable function F on $[a, b]$ such that $F' = f$, then*

$$\int_a^b f(x)\,dx = F(x) - F(a). \tag{5.30}$$

PROOF. Let $\varepsilon > 0$. By Proposition 5.4, there exists a partition

$$P = \{a = x_0 < x_1 < \cdots < x_n = b\}$$

of $[a, b]$ such that

$$U(P, f) - L(P, f) < \varepsilon. \tag{5.31}$$

By Proposition 3.1, F is continuous on $[a, b]$. We apply the Mean Value Theorem (see Proposition 4.5) to each interval $[x_{k-1}, x_k]$, for $k = 1, \ldots, n$, to obtain t_k in the open interval (x_{k-1}, x_k) for which

$$F(x_k) - F(x_{k-1}) = f(t_k)(x_k - x_{k-1}).$$

Hence, we have

$$F(b) - F(a) = \sum_{k=1}^{n} [F(x_k) - F(x_{k-1})] = \sum_{k=1}^{n} f(t_k)\Delta x_k,$$

where $\Delta x_k = x_k - x_{k-1}$. It follows that

$$L(P, f) \le F(b) - F(a) \le U(P, f). \tag{5.32}$$

Since $L(P, f) \le \int_a^b f(x)\, dx \le U(P, f)$, inequalities (5.31) and (5.32) imply

$$\left| \int_a^b f(x)\, dx - [F(b) - F(a)] \right| < \varepsilon.$$

Since ε is arbitrary, (5.30) holds.   □

REMARKS. The assumption that f is Riemann integrable over $[a, b]$ is an essential part of the assumptions in Proposition 5.18. For example, the function

$$F(x) = x^2 \sin \frac{1}{x^2} \quad \text{for } x \ne 0$$

and $F(0) = 0$ is differentiable on $[0, 1]$, but $F' = f$ is not Riemann integrable over $[0, 1]$ since it is unbounded.

Looking back at Example 3 in Section 1 of this chapter, we see that Proposition 5.18 produces with a minimum of effort the result

$$\int_a^b x^c \, dx = \frac{b^{c+1} - a^{c+1}}{c + 1} \quad \text{for } c \ne -1,$$

$$= \ln \frac{b}{a} \qquad \text{for } c = -1,$$

where a, b, and c are arbitrary but fixed real numbers and $0 < a < b$. Another example is

$$\int_{-1}^{1} \frac{1}{1 + x^2} \, dx = \frac{\pi}{2}.$$

Indeed, letting $F(x) = \tan^{-1} x$, we have that $F'(c) = 1/(1 + x^2)$ which is continuous, hence Riemann integrable, when $-1 \le x \le 1$. But $F(1) = \pi/4$ and $F(-1) = -\pi/4$.

Proposition 5.19 (Integration-by-Parts Formula). *Let $[a, b]$ be a closed interval of finite length. If u and v are both differentiable functions on $[a, b]$ and if u' and v' are both Riemann integrable over $[a, b]$, then*

$$\int_a^b u(x)v'(x)\,dx + \int_a^b u'(x)v(x)\,dx = u(b)v(b) - u(a)v(a). \qquad (5.33)$$

PROOF. By Proposition 3.1, u and v are continuous on $[a, b]$ and so, by Proposition 5.16, u and v are Riemann integrable over $[a, b]$. Let $g = uv$. By Proposition 3.2, $g' = uv' + u'v$. By Proposition 5.9, g' is seen to be Riemann integrable over $[a, b]$. Proposition 5.18 shows that

$$\int_a^b g'(x)\,dx = g(b) - g(a) = u(b)v(b) - u(a)v(a)$$

and so (5.33) holds.   □

EXAMPLES. Letting $u(x) = xe^x$ and $v(x) = -1/(x + 1)$, (5.33) gives

$$\int_0^1 \frac{xe^x}{(x + 1)^2}\,dx = \frac{e}{2} - 1,$$

and letting $u(x) = \tan^{-1} x$ and $v(x) = (x^2 + 1)/2$, (5.33) gives

$$\int_0^1 x(\tan^{-1} x)\,dx = \frac{\pi}{4} - \frac{1}{2}.$$

Proposition 5.20 (Fundamental Theorem of Calculus II). *Let $[a, b]$ be a closed interval of finite length. If f is a Riemann integrable function over $[a, b]$ and*

$$F(x) = \int_a^x f(t)\,dt \quad \text{for } x \in [a, b],$$

then F is continuous on $[a, b]$. If f is continuous at x_0 in the open interval (a, b), then F is differentiable at x_0 and $F'(x_0) = f(x_0)$.

PROOF. Choose $B > 0$ such that $|f(x)| \le B$ for all $x \in [a, b]$. If $x, y \in [a, b]$ and $|x - y| < \varepsilon/B$ with $x < y$, say, then

$$|F(y) - F(x)| = \left| \int_x^y f(t)\,dt \right| \le \int_x^y |f(t)|\,dt \le \int_x^y B\,dt = B(y - x) < \varepsilon.$$

We therefore see that F is (uniformly) continuous on $[a, b]$.

Next, suppose that f is continuous at $x_0 \in (a, b)$. Since

$$\frac{F(x) - F(x_0)}{x - x_0} = \frac{1}{x - x_0} \int_{x_0}^x f(t)\,dt \quad \text{for } x \ne x_0$$

and

$$f(x_0) = \frac{1}{x - x_0} \int_{x_0}^{x} f(x_0) \, dt,$$

we have

$$\frac{F(x) - F(x_0)}{x - x_0} - f(x_0) = \frac{1}{x - x_0} \int_{x_0}^{x} [f(t) - f(x_0)] \, dt. \qquad (5.34)$$

Let $\varepsilon > 0$. Since f is continuous at x_0, there exists some $\delta > 0$ such that

$$|f(t) - f(x_0)| < \varepsilon$$

if $t \in (a, b)$ and $|t - x_0| < \delta$. It follows from (5.34) that

$$\left| \frac{F(x) - F(x_0)}{x - x_0} - f(x_0) \right| < \varepsilon \qquad (5.35)$$

for $x \in (a, b)$ satisfying $|x - x_0| < \delta$; the cases $x > x_0$ and $x < x_0$ require separate arguments. From (5.35) we see that $F'(x_0) = f(x_0)$. $\qquad \square$

REMARKS. We note the following corollary to Proposition 5.20: If g is differentiable and f is continuous, then

$$\frac{d}{dx} \left(\int_{a}^{g(x)} f(t) \, dt \right) = f[g(x)] g'(x).$$

Indeed, let

$$H(x) = \int_{a}^{g(x)} f(t) \, dt$$

and observe that H is the composition of differentiable functions:

$$H(x) = F[g(x)] \quad \text{with} \quad F(x) = \int_{a}^{x} f(t) \, dt.$$

The Chain Rule (see Proposition 3.3) gives

$$H'(x) = F'[g(x)] g'(x).$$

Proposition 5.20 gives

$$F'(x) = f(x).$$

It therefore follows that

$$H'(x) = f[g(x)] g'(x),$$

which is what we had set out to show.

In the same way we can verify the following result: If g_1 and g_2 are differentiable and f is continuous, then

$$\frac{d}{dx}\left(\int_{g_1(x)}^{g_2(x)} f(t)\,dt\right) = f[g_2(x)]g_2'(x) - f[g_1(x)]g_1'(x).$$

In fact, we only need to take a number a from the domain of f and note that

$$\int_{g_1(x)}^{g_2(x)} f(t)\,dt = \int_a^{g_2(x)} f(t)\,dt - \int_a^{g_1(x)} f(t)\,dt.$$

As an illustration, observe that $H'(2) = \frac{20}{3}$ for

$$H(x) = x\left(\int_{2x}^{x^3-4} \frac{1}{1+\sqrt{t}}\,dt\right).$$

Proposition 5.21 (Change of Variable Formula). *Let u be a differentiable function on an open interval J such that u' is continuous and let I be an open interval such that $u(x) \in I$ for all $x \in J$. If f is continuous on I, then the composite function $h(x) = f[u(x)]$, for $x \in J$, is continuous on J and*

$$\int_a^b f[u(x)]u'(x)\,dx = \int_{u(a)}^{u(b)} f(u)\,du \tag{5.36}$$

for $a, b \in J$.

PROOF. The continuity of the composite function h follows from Proposition 2.9. Fix $c \in I$ and let $F(u) = \int_c^u f(t)\,dt$. Then $F'(u) = f(u)$ for all $u \in I$ by Proposition 5.20. Let $g(x) = F[u(x)]$ for $x \in J$. By the Chain Rule (see Proposition 3.3), we have $g'(x) = F'[u(x)]u'(x) = f[u(x)]u'(x)$ and so by Proposition 5.18

$$\int_a^b f[u(x)]u'(x)\,dx = \int_a^b g'(x)\,dx = g(b) - g(a) = F[u(b)] - F[u(a)]$$

$$= \int_c^{u(b)} f(t)\,dt - \int_c^{u(a)} f(t)\,dt = \int_{u(a)}^{u(b)} f(t)\,dt,$$

establishing (5.36). [Observe that $u(a)$ need not be smaller than $u(b)$, even if a is smaller than b.]   □

REMARK. Let a, b, and s be strictly positive numbers. For $f(u) = 1/u$ and $u(x) = sx$, where $x > 0$, the relation (5.36) gives

$$\int_a^b \frac{1}{x}\,dx = \int_{sa}^{sb} \frac{1}{u}\,du.$$

(Compare this with Proposition 1.1 in Chapter 1.)

DISCUSSION. Let g be a strictly monotonic and differentiable function on an open interval I with $g'(x) \neq 0$ for any $x \in I$. Then $J = g(I)$ is an open interval and the inverse function g^{-1} is differentiable on J by Proposition 3.4. We show that

$$\int_a^b g(x)\,dx + \int_{g(a)}^{g(b)} g^{-1}(u)\,du = bg(b) - ag(a) \qquad (5.37)$$

for $a, b \in I$.

Indeed, we put $f = g^{-1}$ and $u = g$ in the formula (5.36) to obtain

$$\int_a^b g^{-1}[g(x)]g'(x)\,dx = \int_{g(a)}^{g(b)} g^{-1}(u)\,du.$$

Since $g^{-1}[g(x)] = x$ for $x \in I$, we obtain

$$\int_{g(a)}^{g(b)} g^{-1}(u)\,du = \int_a^b xg'(x)\,dx.$$

Now integrate by parts using $u(x) = x$ and $v(x) = g(x)$:

$$\int_{g(a)}^{g(b)} g^{-1}(u)\,du = bg(b) - ag(a) - \int_a^b g(x)\,dx.$$

This, however, is (5.37).

Formula (5.37) can sometimes be employed to calculate the value of certain integrals which can not otherwise be determined so easily. A case in point is the integral

$$\int_0^{1/2} (\sin^{-1} x)\,dx.$$

Using (5.37), we see that

$$\int_0^{1/2} (\sin^{-1} x)\,dx + \int_0^{\pi/6} (\sin x)\,dx = \frac{1}{2}\left(\frac{\pi}{6}\right)$$

or

$$\int_0^{1/2} (\sin^{-1} x)\,dx = \frac{\pi}{12} + \frac{\sqrt{3}}{2} - 1$$

because

$$\int_0^{\pi/6} (\sin x)\,dx = -\left(\cos\frac{\pi}{6} - \cos 0\right) = 1 - \frac{\sqrt{3}}{2}.$$

Formula (5.37) is also valid if g is assumed to be only strictly monotonic and continuous on an open interval J. Indeed, then the inverse function g^{-1} exists and is a strictly monotonic and continuous function, by Proposition 2.18. Moreover, both integrals

$$\int_a^b g(x)\,dx \quad \text{and} \quad \int_{g(a)}^{g(b)} g^{-1}(u)\,du$$

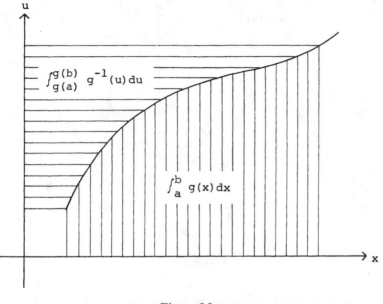

$$\int_{g(a)}^{g(b)} g^{-1}(u)\, du$$

$$\int_{a}^{b} g(x)\, dx$$

Figure 5.8

exist (by Proposition 5.15 or Proposition 5.16). These integrals can be interpreted as areas of regions; see Figure 5.8. The numbers $bg(b)$ and $ag(a)$ represent areas of rectangles. In fact, formula (5.37) is no more than might have been anticipated from geometric considerations.

Suppose now that g is a strictly increasing continuous function of x for $x \geq 0$ and $g(0) = 0$. If $A > 0$ and $B > 0$, then

$$AB \leq \int_0^A g(x)\, dx + \int_0^B g^{-1}(u)\, du. \qquad (5.38)$$

The inequality in (5.38) is called *Young's Inequality*. The validity of Young's Inequality is evident by considering Figure 5.9 and interpreting the integrals

$$\int_0^A g(x)\, dx \quad \text{and} \quad \int_0^B g^{-1}(u)\, du$$

as the areas of the regions shaded with vertical lines and horizontal lines, respectively. It is also clear that equality will hold in Young's Inequality if and only if $g(A) = B$.

From Young's Inequality we can readily obtain the following result: If $p > 1$, $A > 0$, and $B > 0$, then we have

$$AB \leq \frac{A^p}{p} + \frac{B^q}{q}, \qquad (5.39)$$

where q satisfies $1/p + 1/q = 1$; equality holds if and only if $A^p = B^q$. Indeed, let $g(x) = x^{p-1}$ and $g^{-1}(u) = u^{1/(p-1)}$. Then g and g^{-1} satisfy the conditions of

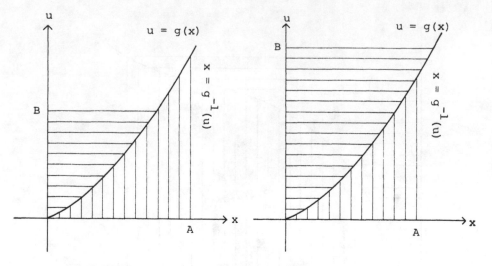

Figure 5.9

Young's Inequality (5.38). Therefore, by noting that $(p - 1)q = p$, we have

$$AB \leq \int_0^A x^{p-1}\, dx + \int_0^B u^{1/(p-1)}\, du = \frac{A^p}{p} + \frac{B^q}{q}.$$

This proves (5.39). Inequality (5.38) will be used presently to obtain an important inequality concerning integrals.

Let $p > 1$ and $q > 1$ such that $1/p + 1/q = 1$. Let u and v be Riemann integrable functions over an interval $[a, b]$ of finite length. By Propositions 5.9 and 5.17, the functions $|uv|$, $|u|^p$, and $|v|^q$ are Riemann integrable over $[a, b]$. Moreover, we have

$$\int_a^b |u(x)v(x)|\, dx \leq \left(\int_a^b |u(x)|^p\, dx \right)^{1/p} \left(\int_a^b |v(x)|^q\, dx \right)^{1/q}. \qquad (5.40)$$

Inequality (5.40) is called *Hölder's Inequality*. For $p = q = 2$, Hölder's Inequality is referred to as the *Cauchy–Schwarz Inequality*:

$$\int_a^b |u(x)v(x)|\, dx \leq \left(\int_a^b |u(x)|^2\, dx \right)^{1/2} \left(\int_a^b |v(x)|^2\, dx \right)^{1/2}. \qquad (5.41)$$

To prove (5.40) we proceed as follows. By (5.39), we have

$$\frac{|u(x)v(x)|}{AB} \leq \frac{A^{-p}|u(x)|^p}{p} + \frac{B^{-q}|v(x)|^q}{q},$$

where

$$A = \left(\int_a^b |u(x)|^p\, dx \right)^{1/p} \quad \text{and} \quad B = \left(\int_a^b |v(x)|^q\, dx \right)^{1/q}.$$

By Propositions 5.9 and 5.11, we have

$$\frac{1}{AB}\int_a^b |u(x)v(x)|\, dx \le \frac{1}{pA^p}\int_a^b |u(x)|^p\, dx + \frac{1}{qB^q}\int_a^b |v(x)|^q\, dx = \frac{1}{p}+\frac{1}{q} = 1.$$

Thus,

$$\int_a^b |u(x)v(x)|\, dx \le AB$$

and (5.40) follows.

Proposition 5.22. *Let* $[a, b]$ *be a closed interval of finite length.*

(i) *If* f *is continuous on* $[a, b]$, $f(x) \ge 0$ *for all* x *in* $[a, b]$, *and* $f(c) = k > 0$ *for some point* c *in* $[a, b]$, *then* $\int_a^b f(x)\, dx > 0$.

(ii) *If* f *and* g *are continuous on* $[a, b]$, $f(x) \le g(x)$ *for all* x *in* $[a, b]$, *and* $f(c) < g(c)$ *for some point* c *in* $[a, b]$, *then* $\int_a^b f(x)\, dx < \int_a^b g(x)\, dx$.

PROOF. We consider part (i) and assume that c is a point of the open interval (a, b). By the continuity of f, we can find an interval $(c - \sigma, c + \sigma)$ throughout which $f(x) > k/2$ (see Proposition 2.10); and then the value of the integral would be greater than σk. If c is an endpoint of $[a, b]$, say $c = a$, we can find an interval $(a, a + \sigma)$ throughout which $f(x) > k/2$; and then the value of the integral would be greater than $\sigma k/2$.

Part (ii) reduces to part (i) if we set $h = f - g$. $\qquad\qquad\square$

Proposition 5.23 (First Mean Value Theorem for Integrals). *Let* $[a, b]$ *be a closed interval of finite length. If* f *is a continuous function on* $[a, b]$, *then there exists a point* t *such that* $a < t < b$ *and*

$$\int_a^b f(x)\, dx = f(t)[b - a]. \qquad\qquad (5.42)$$

PROOF. Let m and M be the smallest and the largest values, respectively, of f on $[a, b]$ (see Proposition 2.13). By Proposition 5.16, f is integrable over $[a, b]$. If f is not constant over $[a, b]$, using part (ii) of Proposition 5.22 yields

$$m < \frac{\int_a^b f(x)\, dx}{b - a} < M.$$

But f is continuous on $[a, b]$ and hence assumes all intermediate values between m and M (see Proposition 2.12); hence, there is a point t between a and b which satisfies (5.42). $\qquad\qquad\square$

Proposition 5.24 (Generalized Form of the First Mean Value Theorem for Integrals). *Let* $[a, b]$ *be a closed interval of finite length. If* f *and* g *are continuous on* $[a, b]$ *and* g *does not change sign on* $[a, b]$, *then there exists a point* t *such that* $a < t < b$ *and*

$$\int_a^b f(x)g(x)\, dx = f(t)\int_a^b g(x)\, dx. \qquad\qquad (5.43)$$

PROOF. The details of proof are similar to the proof of Proposition 5.23. Let m and M be the smallest and the largest values, respectively, of f on $[a, b]$. To avoid the trivial case, assume that g is not identically zero on $[a, b]$. If f is not constant over $[a, b]$, then

$$m < \frac{\int_a^b f(x)g(x)\, dx}{\int_a^b g(x)\, dx} < M.$$

But f is continuous on $[a, b]$ and hence assumes all intermediate values between m and M; thus, there is a point t between a and b which satisfies (5.43).   □

REMARKS. Proposition 5.23 follows from Proposition 5.24 when we set $g(x) = x$. If f is never negative on $[a, b]$, Proposition 5.23 has a simple geometric interpretation: There is a rectangle of height $f(t)$ and length $b - a$ which has the same area as the region bounded by the curve $y = f(x)$, the lines $x = a$ and $x = b$, and the x-axis.

In our formulation of Propositions 5.23 and 5.24 it was assumed that $a < b$. A moment's reflection shows that these propositions remain valid if $b \geq a$.

Proposition 5.25 (Second Mean Value Theorem for Integrals). *Let $[a, b]$ be a closed interval of finite length. If f, g, and g' are continuous on $[a, b]$, and if g is monotonic on $[a, b]$ (equivalently, g' does not change sign on $[a, b]$), then there exists a point t such that $a < t < b$ and*

$$\int_a^b f(x)g(x)\, dx = g(a) \int_a^t f(x)\, dx + g(b) \int_t^b f(x)\, dx.$$

PROOF. Let $F(x) = \int_a^x f(s)\, ds$. By Proposition 5.19,

$$\int_a^b f(x)g(x)\, dx = F(b)g(b) - \int_a^b F(x)g'(x)\, dx;$$

by Proposition 5.24, there is some point t between a and b such that

$$\int_a^b F(x)g'(x)\, dx = F(t) \int_a^b g'(x)\, dx = F(t)[g(b) - g(a)].$$

Therefore,

$$\int_a^b f(x)g(x)\, dx = F(b)g(b) - F(t)g(b) + F(t)g(a)$$

$$= g(a)F(t) + g(b)[F(b) - F(t)]$$

$$= g(a) \int_a^t f(s)\, ds + g(b) \int_t^b f(s)\, ds$$

and the proof is complete.   □

DISCUSSION. Let $[a, b]$ be a closed interval of finite length. If u' and v' are continuous functions on $[a, b]$, then

$$\int_a^b u(x)v'(x)\,dx = u(b)v(b) - u(a)v(a) - \int_a^b u'(x)v(x)\,dx$$

by Proposition 5.19. A simpler notation for the foregoing formula is

$$\int_a^b uv'\,dx = uv \bigg|_a^b - \int_a^b u'v\,dx.$$

It is easy to see that repeated application of this formula yields the following statement:

If $u^{(n+1)}$ and $v^{(n+1)}$ are continuous functions on a closed interval $[a, b]$ of finite length, then

$$\int_a^b uv^{(n+1)}\,dx = (uv^{(n)} - u'v^{(n-1)} + \cdots + (-1)^n u^{(n)}v) \bigg|_a^b \tag{5.44}$$

$$+ (-1)^{n+1} \int_a^b u^{(n+1)}v\,dx.$$

Formula (5.44) may be used as a starting point in obtaining a Taylor-type formula with remainder (see Proposition 4.11). Let

$$v(x) = (b - x)^n.$$

Then

$$v'(x) = -n(b - x)^{n-1},$$
$$v''(x) = n(n - 1)(b - x)^{n-2},$$
$$v'''(x) = -n(n - 1)(n - 2)(b - x)^{n-3},$$
$$\vdots$$
$$v^{(n)}(x) = (-1)^n n(n - 1)(n - 2) \cdots 2 \cdot 1,$$
$$v^{(n+1)}(x) = 0;$$

Moreover, $v(b) = v'(b) = v''(b) = \cdots = v^{(n-1)}(b) = 0$. Letting $u = f$, formula (5.44) assumes the form

$$0 = (-1)^n \bigg[n!f(b) - n!f(a) - n!f'(a)(b - a)$$

$$- \frac{n!}{2!} f''(a)(b - a)^2 - \cdots - f^{(n)}(a)(b - a)^n \bigg]$$

$$+ (-1)^{n+1} \int_a^b f^{(n+1)}(x)(b - x)^n\,dx.$$

(We are assuming of course that $f^{(n+1)}$ is continuous on $[a, b]$; in Proposition 4.11 no assumption of continuity was made concerning $f^{(n+1)}$.) From this in turn we obtain

$$f(b) = f(a) + \frac{f'(a)}{1!}(b - a) + \frac{f''(a)}{2!}(b - a)^2 + \cdots$$

$$+ \frac{f^{(n)}(a)}{n!}(b - a)^n + \frac{1}{n!}\int_a^b f^{(n+1)}(x)(b - x)^n\, dx.$$

Replacing a by α, b by β, and x by t, we get

$$f(\beta) = f(\alpha) + \frac{f'(\alpha)}{1!}(\beta - \alpha) + \frac{f''(\alpha)}{2!}(\beta - \alpha)^2 + \cdots$$

$$+ \frac{f^{(n)}(\alpha)}{n!}(\beta - \alpha)^n + \frac{1}{n!}\int_\alpha^\beta f^{(n+1)}(t)(\beta - t)^n\, dt. \qquad (5.45)$$

However, $(\beta - t)^n$ does not change sign on the interval $[\alpha, \beta]$ and we may therefore apply Proposition 5.24, obtaining

$$\frac{1}{n!}\int_\alpha^\beta f^{(n+1)}(t)(\beta - t)^n\, dt = \frac{1}{n!} f^{(n+1)}(x)\int_\alpha^\beta (\beta - t)^n\, dt$$

$$= \frac{f^{(n+1)}(x)}{(n + 1)!}(\beta - \alpha)^{n+1},$$

where x is a point between α and β. But this is Lagrange's form of the remainder (see the remarks following Proposition 4.11).

FURTHER DISCUSSION. An informative example concerning the calculation of an integral is provided by the following example: Compute

$$I = \int_0^{\pi/2} \frac{\sin^\pi x}{\sin^\pi x + \cos^\pi x}\, dx.$$

We would indeed be on the wrong track, trying to determine I by use of Proposition 5.18. Instead, we proceed as follows. Letting $x = \pi/2 - t$, we obtain

$$I = \int_0^{\pi/2} \frac{\sin^\pi x}{\sin^\pi x + \cos^\pi x}\, dx = \int_0^{\pi/2} \frac{\cos^\pi t}{\cos^\pi t + \sin^\pi t}\, dt = \int_0^{\pi/2} \frac{\cos^\pi x}{\cos^\pi x + \sin^\pi x}\, dx$$

and so

$$I + I = \int_0^{\pi/2} \frac{\sin^\pi x}{\sin^\pi x + \cos^\pi x}\, dx + \int_0^{\pi/2} \frac{\cos^\pi x}{\cos^\pi x + \sin^\pi x}\, dx$$

$$= \int_0^{\pi/2} \frac{\sin^\pi x + \cos^\pi x}{\sin^\pi x + \cos^\pi x}\, dx = \int_0^{\pi/2} dx = \frac{\pi}{2}.$$

Thus, $2I = \pi/2$ or $I = \pi/4$.

In the same way we can show that, for any fixed real number r,

$$\int_0^{\pi/2} \frac{\sin^r x}{\sin^r x + \cos^r x} \, dx = \frac{\pi}{4}.$$

In Example 6 at the end of the next section we shall compute the result

$$\int_0^1 \frac{\ln(1 + x)}{1 + x^2} \, dx = \frac{\pi}{8} (\ln 2).$$

A less formidable task relating to the calculation of integrals consists in showing that

$$\int_0^b (a^2 - x^2)^{1/2} \, dx = \frac{b}{2}(a^2 - b^2)^{1/2} + \frac{a^2}{2} \arcsin \frac{b}{a} \quad \text{for } 0 \le b \le a. \quad (5.46)$$

We shall verify (5.46) by three different methods.

First method: We use integration by parts, setting $u = (a^2 - x^2)^{1/2}$ and $dv = dx$, and get

$$\int_0^b (a^2 - x^2)^{1/2} \, dx = x(a^2 - x^2) \Big|_0^b - \int_0^b \frac{-x^2}{(a^2 - x^2)^{1/2}}$$

$$= b(a^2 - b^2)^{1/2} - \int_0^b \frac{a^2 - x^2 - a^2}{(a^2 - x^2)^{1/2}} \, dx$$

$$= b(a^2 - b^2)^{1/2} - \int_0^b (a^2 - x^2)^{1/2} \, dx$$

$$+ a^2 \int_0^b \frac{1}{(a^2 - x^2)^{1/2}} \, dx,$$

implying

$$2 \int_0^b (a^2 - x^2)^{1/2} \, dx = b(a^2 - b^2)^{1/2} + a^2 \left(\arcsin \frac{b}{a} \right)$$

which is equivalent to (5.46).

Second method: Integrating by substitution, we set $x = a \sin t$ and we obtain

$$\int_0^b (a^2 - x^2)^{1/2} \, dx = a^2 \int_0^B (\cos^2 t) \, dt, \quad \text{where } B = \arcsin \frac{b}{a}.$$

But

$$\int_0^B (\cos^2 t) \, dt = \frac{1}{2} \int_0^B (1 + \cos 2t) \, dt = \frac{1}{2}(t + \frac{1}{2} \sin 2t) \Big|_0^b$$

$$= \frac{1}{2}(t + [\sin t][\cos t]) \Big|_0^b$$

$$= \frac{1}{2} \left(\left(\arcsin \frac{b}{a} \right) + \frac{b}{a} \cdot \frac{(a^2 - b^2)^{1/2}}{a} \right)$$

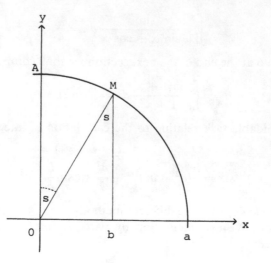

Figure 5.10

because

$$\sin B = \frac{b}{a} \quad \text{and} \quad \cos B = (1 - \sin^2 B)^{1/2} = \frac{(a^2 - b^2)^{1/2}}{a}$$

and (5.46) follows.

Third method: Knowing that the integral

$$T = \int_0^b (a^2 - x^2)^{1/2}\, dx \quad \text{for } 0 \le b \le a$$

represents the area of the region in the first quadrant bounded by the co-ordinate axes, the line $x = b$, and the circle $x^2 + y^2 = a^2$, we observe that the region in question is made up of the triangle $\triangle ObM$ and the circular sector AOM (see Figure 5.10). But the area of the triangle $\triangle ObM$ is $(b/2)(a^2 - b^2)^{1/2}$ and the area of the circular sector AOM is $\frac{1}{2}a^2 s = (a^2/2)\arc\sin(b/a)$ because $\sin s = b/a$. Thus, (5.46) follows.

ADDITIONAL COMMENTS ON INEQUALITIES RELATING TO INTEGRATION. We can use the Cauchy–Schwarz Inequality to verify that

$$\ln\frac{b}{a} < \frac{b - a}{\sqrt{ab}} \quad \text{for } b > a > 0.$$

Indeed, we merely put $u(x) = 1$ and $v(x) = 1/x$ in (5.41) and the claim follows.

An interesting inequality, due to Kantorovich, is the following: Let f be a continuous function on the interval $[0, 1]$ such that $0 < m \le f(x) \le M$ for all x in $[0, 1]$. Then

$$\left(\int_0^1 \frac{1}{f(x)}\, dx\right)\left(\int_0^1 f(x)\, dx\right) \le \frac{(m + M)^2}{4mM}. \tag{5.47}$$

[If in the foregoing we replace the unit interval $[0, 1]$ by the interval $[a, b]$, then (5.47) takes on the form

$$\left(\int_a^b \frac{1}{f(x)} \, dx \right) \left(\int_a^b f(x) \, dx \right) \leq \frac{(m + M)^2}{4mM} (b - a)^2 .]$$

Indeed, since

$$\frac{\{ f(x) - m \} \{ f(x) - M \}}{f(x)} \leq 0 \quad \text{for } 0 \leq x \leq 1,$$

we obtain by integrating $f - (m + M) + mM/f$ over $[0, 1]$

$$\int_0^1 f(x) \, dx + mM \int_0^1 \frac{1}{f(x)} \, dx \leq m + M.$$

Putting

$$u = mM \int_0^1 \frac{1}{f(x)} \, dx,$$

we get

$$\int_0^1 f(x) \, dx + u \leq m + M$$

and

$$u \int_0^1 f(x) \, dx \leq (m + M)u - u^2 \leq \frac{(m + M)^2}{4},$$

noting that $[(m + M)/2 - u]^2 \geq 0$. This establishes (5.47).

5. Numerical Integration

Let $a < b$ and $f(x) > 0$ for $a \leq x \leq b$. Then $\int_a^b f(x) \, dx$ is equal to the area of the region bounded above by the curve $y = f(x)$, below by the x-axis, and lying between the lines $x = a$ and $x = b$. We divide the interval from a to b into n subintervals, each of length

$$h = \frac{b - a}{n}$$

by the points

$$a = x_0 < x_1 < x_2 < \cdots < x_{n-1} < x_n = b$$

and put $y_k = f(x_k)$ for $k = 0, 1, 2, \ldots, n - 1, n$. For $k = 0, 1, 2, \ldots, n$, let P_k be the point with coordinate (x_k, y_k) and consider the polygonal line whose vertices are these points P_k (see Figure 5.11). Then the area of the region under the curve $y = f(x)$ may be approximated by that under the polygonal line whose vertices are P_k. The region under the polygonal line is made up of

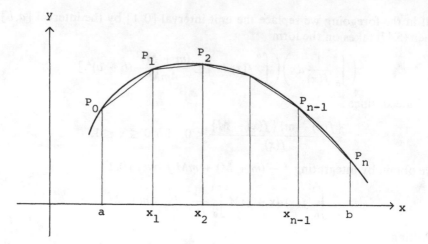

Figure 5.11

trapezoids. The area of a trapezoid is equal to half the sum of the parallel sides times the width. It follows that

$$\text{Area of first trapezoid} = \frac{h}{2}(y_0 + y_1),$$

$$\text{Area of second trapezoid} = \frac{h}{2}(y_1 + y_2),$$

$$\vdots$$

$$\text{Area of } n\text{th trapezoid} = \frac{h}{2}(y_{n-1} + y_n).$$

Adding these expressions, we get that the approximating region has area

$$T_n = h\left(\frac{y_0}{2} + y_1 + y_2 + \cdots + y_{n-1} + \frac{y_n}{2}\right).$$

The approximating formula

$$\int_a^b f(x)\, dx \approx T_n$$

is known as the *trapezoidal rule* for numerical integration.

The restriction $f(x) > 0$ for $a \le x \le b$ is not necessary for our discussion; it is easy enough to see, for example, that if (x_1, y_1) and (x_2, y_2) are connected by a straight line $y = s(x)$ and $x_2 = x_1 + h$, then

$$\int_{x_1}^{x_2} s(x)\, dx = \frac{h}{2}(y_1 + y_2).$$

In the sequel we shall establish the following *error bound for the trapezoidal rule*: If the second derivative f'' of f is continuous on the interval $[a, b]$ and if M is the largest value of $|f''(x)|$ for $a \le x \le b$, then

$$\left| T_n - \int_a^b f(x)\,dx \right| \le M \frac{(b-a)^3}{12n^2},$$

where T_n is the approximation to $\int_a^b f(x)\,dx$ given by the trapezoidal rule

$$T_n = h\left(\frac{y_0}{2} + y_1 + y_2 + \cdots + y_{n-1} + \frac{y_n}{2}\right), \qquad h = \frac{b-a}{n}.$$

Proposition 5.26. *Suppose that the second derivative f'' of a function f is continuous on a closed interval $[A, A+h]$ of finite length h. Then*

$$\int_A^{A+h} f(x)\,dx = \frac{h}{2}\{f(A) + f(A+h)\} - \frac{h^3}{12}f''(s),$$

where s is between A and $A+h$.

PROOF. We wish to determine Q such that

$$\int_A^{A+h} f(x)\,dx = \frac{h}{2}\{f(A) + f(A+h)\} + h^3 Q.$$

Let $f(x) = F'(x)$. Then

$$\int_A^{A+h} f(x)\,dx = F(A+h) - F(A)$$

and so

$$F(A+h) - F(A) - \frac{h}{2}\{f(A) + f(A+h)\} - h^3 Q = 0.$$

Let $C = A + h/2$; C is the midpoint of $[A, A+h]$. We consider the function G defined by

$$G(x) = F\left(C + \frac{x}{2}\right) - F\left(C - \frac{x}{2}\right) - \frac{x}{2}\left\{f\left(C + \frac{x}{2}\right) + f\left(C - \frac{x}{2}\right)\right\} - Qx^3,$$

where $0 \le x \le h$. But $G(0) = G(h) = 0$ and so, by Rolle's Theorem (see Proposition 4.3), we have that $G'(t) = 0$ for some t between 0 and h. But, since $F' = f$,

$$G'(x) = \tfrac{1}{2}f\left(C + \frac{x}{2}\right) + \tfrac{1}{2}f\left(C - \frac{x}{2}\right) - \tfrac{1}{2}\left\{f\left(C + \frac{x}{2}\right) + f\left(C - \frac{x}{2}\right)\right\}$$

$$- \frac{x}{4}\left\{f'\left(C + \frac{x}{2}\right) - f'\left(C - \frac{x}{2}\right)\right\} - 3Qx^2$$

$$= -\frac{x}{4}\left\{f'\left(C + \frac{x}{2}\right) - f'\left(C - \frac{x}{2}\right)\right\} - 3Qx^2.$$

Thus,

$$0 = G'(t) = -\frac{t}{4}\left\{ f'\left(C + \frac{t}{2}\right) - f'\left(C - \frac{t}{2}\right)\right\} - 3Qt^2.$$

Since $t \neq 0$,

$$Q = -\frac{1}{12}\frac{f'(C + t/2) - f'(C - t/2)}{t}.$$

Again by Rolle's Theorem,

$$\frac{f'(C + t/2) - f'(C - t/2)}{t} = f''(s)$$

for some s between the points $C - t/2$ and $C + t/2$; hence,

$$Q = -\frac{1}{12}f''(s),$$

where s is between the points $C - t/2$ and $C + t/2$ and, a fortiori, s between A and $A + h$. $\square$

Proposition 5.27. *Let $[a, b]$ be a closed interval of finite length. We divide $[a, b]$ into n subintervals, each of length $h = (b - a)/n$. Suppose that the second derivative f'' of a function f is continuous on $[a, b]$. Then*

$$\int_a^b f(x)\,dx = \frac{b - a}{2n}\left(f(a) + f(b) + 2\sum_{k=1}^{n-1} f\left(a + k\frac{b - a}{n}\right)\right) - \frac{(b - a)^3}{12n^2}f''(v),$$

where v is between a and b.

PROOF. Since $h = (b - a)/n$ and

$$\int_a^b f(x)\,dx = \int_a^{a+h} f(x)\,dx + \int_{a+h}^{a+2h} f(x)\,dx + \cdots + \int_{a+(n-1)h}^b f(x)\,dx,$$

we have

$$\int_a^b f(x)\,dx = \frac{h}{2}\{f(a) + f(a + h)\} - \frac{h^3}{12}f''(s_1)$$

$$+ \frac{h}{2}\{f(a + h) + f(a + 2h)\} - \frac{h^3}{12}f''(s_2)$$

$$+ \cdots$$

$$+ \frac{h}{2}\{f(a + [n - 1]h) + f(b)\} - \frac{h^3}{12}f''(s_n),$$

where s_1 is between a and $a + h$, s_2 is between $a + h$ and $a + 2h$, ..., s_n is between $a + (n - 1)h$ and b. But f'' is continuous on $[a, b]$ and so there is a point v between a and b such that (see Example 9 in Section 4 of Chapter 2)

$$\frac{f''(s_1) + f''(s_2) + \cdots + f''(s_n)}{n} = f''(v).$$

This completes the proof. □

REMARK. The foregoing proposition shows that

$$T_n - \int_a^b f(x)\,dx = \frac{(b - a)^3}{12n^2}f''(v),$$

where v is a point between a and b. Clearly,

$$|f''(v) \le \max\{|f''(x)|: a \le x \le b\} = M$$

for any point v between a and b. Hence,

$$\left| T_n - \int_a^b f(x)\,dx \right| \le M\frac{(b - a)^3}{12n^2}$$

and the error bound for the trapezoidal rule is established.

Lemma. *The equation of a parabola with vertical axis may be written in the form* $y = g(x)$, *where*

$$g(x) = a(x - x_1)^2 + b(x - x_1) + c.$$

If this parabola passes through the points (x_0, y_0), (x_1, y_1), *and* (x_3, y_3), *where* $x_1 = x_0 + h$ *and* $x_2 = x_1 + h$ *for some fixed, positive* h, *then*

$$\int_{x_1}^{x_2} g(x)\,dx = \frac{h}{3}(y_0 + 4y_1 + y_2).$$

PROOF. We have

$$\int_{x_1}^{x_2} g(x)\,dx = \int_{x_1-h}^{x_1+h} \{a(x - x_1)^2 + b(x - x_1) + c\}\,dx;$$

with $x = x_1 + t$, this becomes

$$\int_{-h}^{h} (at^2 + bt + c)\,dt = \frac{h}{3}(2ah^2 + 6c).$$

But

$$y_0 = g(x_0) = g(x_1 - h) = ah^2 - bh + c,$$

$$y_1 = g(x_1) = c,$$

$$y_2 = g(x^2) = g(x_1 + h) = ah^2 + bh + c.$$

It follows that

$$y_0 + y_2 = 2ah^2 + 2c \quad \text{and} \quad y_0 + 4y_1 + y_2 = 2ah^2 + 6c,$$

so that

$$\int_{x_0}^{x_2} g(x)\,dx = \frac{h}{3}(2ah^2 + 6c) = \frac{h}{3}(y_0 + 4y_1 + y_2),$$

as was to be shown.

$\square$

DISCUSSION. We again consider the definite integral $\int_a^b f(x)\,dx$. We pick an *even* integer $2n$ and divide the interval $[a, b]$ into $2n$ subintervals, each of length

$$h = \frac{b-a}{2n},$$

by the points

$$a = x_0 < x_1 < x_2 < \cdots < x_{2n-2} < x_{2n-1} < x_{2n} = b.$$

We put $y_k = f(x_k)$ for $k = 0, 1, 2, \ldots, 2n$ and let P_k denote the point with the coordinates (x_k, y_k). Now, as in Figure 5.12, we approximate the curve $y = f(x)$ for each pair of intervals, as $P_0 P_1 P_2$, by the arc of a parabola with vertical axis. It follows from the Lemma and without regard to the sign of $f(x)$ that between the x-axis and the parabolic arc

$P_0 P_1 P_2$, the area is $\dfrac{h}{3}(y_0 + 4y_1 + y_2)$,

$P_2 P_3 P_4$, the area is $\dfrac{h}{3}(y_2 + 4y_3 + y_4)$,

$$\vdots$$

$P_{n-2} P_{n-1} P_n$, the area is $\dfrac{h}{3}(y_{2n-2} + 4y_{2n-1} + y_{2n})$.

Adding these expressions, we find that the approximating area is

$$S_{2n} = \frac{h}{3}(y_0 + 4y_1 + 2y_2 + 4y_3 + 2y_4 + \cdots + 2y_{2n-2} + 4y_{2n-1} + y_{2n}).$$

The coefficients 1, 4, 2 result from the fact that each y_k with k odd is a middle ordinate once, whereas except for y_0 and y_{2n}, each y_k with k even is an end ordinate twice.

The approximation formula

$$\int_a^b f(x)\,dx = S_{2n}, \qquad h = \frac{b-a}{2n}$$

is known as *Simpson's rule* for numerical integration.

We shall establish the following *error bound for Simpson's rule*: If the fourth derivative $f^{(4)}$ of a function f is continuous on the interval $[a, b]$ and if K is

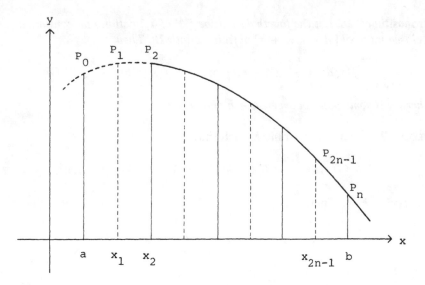

Figure 5.12

the largest value of $|f^{(4)}(x)|$ for $a \le x \le b$, then

$$\left| S_{2n} - \int_a^b f(x)\,dx \right| \le K \frac{(b-a)^5}{2880n^4},$$

where S_{2n} is the approximation to $\int_a^b f(x)\,dx$ given by Simpson's rule

$$S_{2n} = \frac{h}{3}(y_0 + 4y_1 + 2y_2 + 4y_3 + 2y_4 + \cdots + 2y_{2n-2} + 4y_{2n-1} + y_{2n})$$

and

$$h = \frac{b-a}{2n}.$$

REMARK. If f is a polynomial whose degree does not exceed three, then $K = 0$ and Simpson's rule give the exact value fo $\int_a^b f(x)\,dx$:

$$\int_a^b f(x)\,dx = \frac{b-a}{6}\left(f(a) + 4f\left(\frac{a+b}{2}\right) + f(b) \right).$$

[When the volume V of a solid is regarded as a limiting sum of thin slices of area $f(x)$, then

$$V = \int_a^b g(x)\,dx \approx \frac{b-a}{6}\left(f(a) + 4f\left(\frac{a+b}{2}\right) + f(b) \right)$$

is known as the *prismoid formula*. It is easy to see that the prismoid formula gives the volume of a cylinder, a cone, a sphere, and an ellipsoid exactly.]

Proposition 5.28. *Let the fourth derivative $f^{(4)}$ of a function f be continuous on a closed interval $[A - h, A + h]$ of finite length $2h$. Then*

$$\int_{A-h}^{A+h} f(x)\, dx = \frac{h}{3}\{f(A - h) + 4f(A) + f(A + h)\} - \frac{h^5}{90} f^{(4)}(w),$$

where w is some point between $A - h$ and $A + h$.

PROOF. We wish to determine K such that

$$\int_{A-h}^{A+h} f(x)\, dx = \frac{h}{3}\{f(A - h) + 4f(A) + f(A + h)\} + Kh^5.$$

Let $f(x) = F'(x)$. Then

$$\int_{A-h}^{A+h} f(x)\, dx = F(A + h) - F(A - h)$$

and so

$$F(A + h) - F(A - h) - \frac{h}{3}\{f(A + h) + 4f(A) + f(A - h)\} - Kh^5 = 0.$$

We consider the function

$$H(x) = F(A + x) - F(A - x) - \frac{x}{3}\{f(A + x) + 4f(A) + f(A - x)\} - Kx^5$$

for $0 \le x \le h$. But $H(0) = H(h) = 0$ and so by Rolle's Theorem (see Proposition 4.3) we have $H'(t) = 0$ for some t between 0 and h. But, since $F' = f$,

$$H'(x) = f(A + x) + f(A - x) - \tfrac{1}{3}\{f(A + x) + 4f(A) + f(A - x)\}$$

$$- \frac{x}{3}\{f'(A + x) - f'(A - x)\} - 5Kx^4$$

$$= \tfrac{2}{3}\{f(A + x) + f(A - x)\} - \tfrac{4}{3}f(A)$$

$$- \frac{x}{3}\{f'(A + x) - f'(A - x)\} - 5Kx^4.$$

Hence, $H'(0) = H'(t) = 0$ and so, by Rolle's Theorem, $H''(s) = 0$ for some s between 0 and t. But

$$H''(x) = \tfrac{2}{3}\{f'(A + x) - f'(A - x)\} - \tfrac{1}{3}\{f'(A + x) - f'(A - x)\}$$

$$- \frac{x}{3}\{f''(A + x) + f''(A - x)\} - 20Kx^3$$

$$= \tfrac{1}{3}\{f'(A + x) - f'(A - x)\} - \frac{x}{3}\{f''(A + x) + f''(A - x)\} - 20Kx^3.$$

Thus, $H''(0) = H''(s) = 0$ and so, by Rolle's Theorem, $H'''(v) = 0$ for some v

between 0 and s. But

$$H'''(x) = \tfrac{1}{3}\{f''(A+x) + f''(A-x)\} - \tfrac{1}{3}\{f''(A+x) + f''(A-x)\}$$
$$- \frac{x}{3}\{f'''(A+x) - f'''(A-x)\} - 60Kx^2.$$

It follows that

$$H'''(v) = -\frac{v}{3}\{f'''(A+v) - f'''(A-v)\} - 60Kv^2 = 0.$$

Since $v \neq 0$, we get

$$K = -\frac{1}{90}\frac{f'''(A+v) - f'''(A-v)}{2v}.$$

By Rolle's Theorem

$$\frac{f'''(A+v) - f'''(A-v)}{2v} = f^{(4)}(w)$$

for some w between $A - v$ and $A + v$. Thus,

$$K = -\frac{1}{90}f^{(4)}(w)$$

for some w between $A - v$ and $A + v$ and so, a fortiori, some w between $A - h$ and $A + h$. $\qquad\square$

REMARKS. If we divide up the interval $[a, b]$ into $2n$ subintervals, each of length

$$h = \frac{b-a}{2n},$$

by the points

$$a = x_0 < x_1 < x_2 < x_3 < x_4 < \cdots < x_{2n-2} < x_{2n-1} < x_{2n} = b,$$

then we get

$$\int_{x_0}^{x_2} f(x)\,dx = \frac{h}{3}\{f(x_0) + 4f(x_1) + f(x_2)\} - \frac{h^5}{90}f^{(4)}(w_1),$$

$$\int_{x_2}^{x_4} f(x)\,dx = \frac{h}{3}\{f(x_2) + 4f(x_3) + f(x_4)\} - \frac{h^5}{90}f^{(4)}(w_2),$$

$$\vdots$$

$$\int_{x_{2n-2}}^{x_{2n}} f(x)\,dx = \frac{h}{3}\{f(x_{2n-2}) + 4f(x_{2n-1}) + f(x_{2n})\} - \frac{h^5}{90}f^{(4)}(w_n),$$

where $w_1, w_2, \ldots, w_n$ are between a and b. But

$$\frac{f^{(4)}(w_1) + f^{(4)}(w_2) + \cdots + f^{(4)}(w_n)}{n} = f^{(4)}(w)$$

for some w between a and b because $f^{(4)}$ is continuous on $[a, b]$ (see Example 9 in Section 4 of Chapter 2). Hence,

$$\int_a^b f(x)\,dx = S_{2n} - \frac{h^5}{90}(nf^{(4)}(w)) = S_{2n} - \frac{(b-a)^5}{2880n^4} f^{(4)}(w),$$

where w is some point between a and b and

$$S_{2n} = \frac{b-a}{6}(y_0 + 4y_1 + 2y_2 + 4y_3 + 2y_4 + \cdots + 2y_{2n-2} + 4y_{2n-1} + y_{2n})$$

with $y_k = f(x_k)$ for $k = 0, 1, 2, \ldots, 2n$. It therefore follows that

$$\left| S_{2n} - \int_a^b f(x)\,dx \right| \le K \frac{(b-a)^5}{2880n^4},$$

where K is the largest value of $|f^{(4)}(x)|$ for $a \le x \le b$. This establishes the error bound for Simpson's rule.

It is clear from the foregoing proposition that if f is a fourth-degree polynomial, then the error for Simpson's rule is a computable constant. We shall look at such a situation now.

Consider a symmetric barrel that has the shape obtained by revolving a parabolic arc. Let H be the length of the barrel, R the radius of its midsection, r the radius of each end, and $\delta = R - r$. Then its volume is exactly

$$V = \pi H(\tfrac{2}{3}R^2 + \tfrac{1}{3}r^2 - \tfrac{2}{15}\delta^2),$$

according to Newton. Indeed,

$$V = \pi \int_{-H/2}^{H/2} y^2\,dx, \quad \text{where} \quad y = -\frac{4\delta}{H^2}x^2 + R.$$

Thus,

$$V = \pi \int_{-H/2}^{H/2} f(x)\,dx, \quad \text{where} \quad f(x) = \frac{16\delta^2}{H^4}x^4 - \frac{8\delta R}{H^2}x^2 + R^2.$$

But

$$f^{(4)}(x) = 24\frac{16\delta^2}{H^4}.$$

Taking $A = 0$, $h = H/2$, and multiplying by π throughout, the desired result for V follows from the equation in the foregoing proposition.

EXAMPLE 1. Letting $n = 10$ and using the trapezoidal rule, calculate

$$\ln 2 = \int_1^2 \frac{1}{x}\,dx$$

and determine the corresponding error bound.

SOLUTION. Here $f(x) = 1/x$, $[a, b] = [1, 2]$, and $h = 0.1$; hence,

$$T_{10} = \frac{1}{10}\left(\frac{1}{2} + \frac{1}{1.1} + \frac{1}{1.2} + \frac{1}{1.3} + \frac{1}{1.4} + \frac{1}{1.5} + \frac{1}{1.6} + \frac{1}{1.7} + \frac{1}{1.8} + \frac{1}{1.9} + \frac{1}{4}\right)$$

$$= 0.69377\ldots .$$

Since $0 < f''(x) = 2/x^3 \le 2$ for $1 \le x \le 2$, the error bound is

$$\frac{2}{12 \cdot 10^2} = \frac{1}{600}.$$

Incidentally, letting $n = 20$ (and hence $h = 0.05$), we get $T_{20} = 0.69330333\ldots .$ Actually, $\ln 2 = 0.69314718\ldots$ with an accuracy of eight decimal places.

EXAMPLE 2. Letting $2n = 10$ and using Simpson's rule, calculate

$$\ln 2 = \int_1^2 \frac{1}{x} dx$$

and determine the corresponding error bound.

SOLUTION. Here $f(x) = 1/x$, $[a, b] = [1, 2]$, and $h = 0.1$; hence,

$$S_{10} = \frac{1}{30}\left(1 + \frac{4}{1.1} + \frac{2}{1.2} + \frac{4}{1.3} + \frac{2}{1.4} + \frac{4}{1.5} + \frac{2}{1.6} + \frac{4}{1.7} + \frac{2}{1.8} + \frac{4}{1.9} + \frac{1}{2}\right)$$

$$= 0.693152\ldots .$$

Since $0 < f^{(4)}(x) = 24/x^5 \le 24$ for $1 \le x \le 2$, the error bound is

$$\frac{24}{2880 \cdot 5^4} = \frac{1}{75,000}.$$

Letting $2n = 20$ (and hence $h = 0.05$), we get $S_{20} = 0.69314716\ldots$ which coincides with the actual value $\ln 2 = 0.69314718\ldots$ in the first seven decimal places.

COMMENT. Comparison of Examples 1 and 2 shows that the trapezoidal rule and Simpson's rule require about the same amount of calculation; however, Simpson's rule yields a much better approximation.

EXAMPLE 3. Since

$$\frac{\pi}{4} = \int_0^1 f(x)\,dx, \quad \text{where} \quad f(x) = \frac{1}{1 + x^2},$$

the numerical calculation of this integral is of definite interest. Since

$$f'(x) = \frac{-2x}{(1 + x^2)^2}, \quad f''(x) = \frac{6x^2 - 2}{(1 + x^2)^3}, \quad f'''(x) = \frac{24x(1 - x^2)}{(1 + x^2)^4},$$

$$f^{(4)}(x) = \frac{24(5x^4 - 10x^2 + 1)}{(1 + x^2)^5}, \quad \text{and} \quad f^{(5)}(x) = \frac{-240x(3x^4 - 10x^2 + 3)}{(1 + x^2)^6}$$

we find that $|f''(x)| < 2$ and $|f^{(4)}(x)| \leq 24$ for $0 \leq x \leq 1$. A more elegant approach is to set $y = \arctan x$. Then, by (4.38) (in Chapter 4),

$$f''(x) = y''' = 2(\cos^3 y)\left[\sin 3\left(y + \frac{\pi}{2}\right)\right] = -2(\cos^3 y)(\sin 3y)$$

and

$$f^{(4)}(x) = y^{(5)} = 24(\cos^5 y)\left[\sin 5\left(y + \frac{\pi}{2}\right)\right] = 24(\cos^5 y)(\cos 5y).$$

Here the trapezoidal rule with $n = 10$ gives $T_{10} = 0.78562...$, while Simpson's rule with $2n = 4$ gives $S_4 = 0.78539...$; the actual value of $\pi/4$ with an accuracy of six decimal places is $0.785398...$.

EXAMPLE 4. Let $h(0) = 1$ and

$$h(x) = \frac{\ln(x + 1)}{x} \quad \text{for } x > 0.$$

Using Simpson's rule for the calculation of the integral of h over a suitable interval needs information concerning $h^{(4)}$. We claim that $h^{(4)}$ is a decreasing function on $[0, \infty)$ and

$$0 < h^{(4)}(x) < 4.8 = h^{(4)}(0) \quad \text{for } x > 0.$$

Indeed,

$$h'(x) = \frac{1}{(x + 1)x} - \frac{\ln(x + 1)}{x^2}, \qquad h''(x) = 2\frac{\ln(x + 1)}{x^2} - \frac{3x + 2}{(x + 1)^2 x^2},$$

$$h'''(x) = \frac{11x^2 + 15x + 6}{(x + 1)^3 x^3} - \frac{6\ln(x + 1)}{x^4},$$

$$h^{(4)}(x) = \frac{24\ln(x + 1)}{x^5} - \frac{50x^3 + 104x^2 + 84x + 24}{(x + 1)^4 x^4}$$

$$h^{(5)}(x) = \frac{274x^4 + 770x^3 + 940x^2 + 540x + 120}{(x + 1)^5 x^5} - \frac{120\ln(x + 1)}{x^5}.$$

Therefore,

$$\frac{d}{dx}(x^6 \cdot h^{(5)}(x)) = -\frac{120x^5}{(x + 1)^6}$$

and it follows that $x^6 \cdot h^{(5)}(x)$ is a decreasing function for $x > 0$ (by Proposition 4.7). Since the product $x^6 \cdot h^{(5)}(x)$ vanishes for $x = 0$, we therefore have

$$x^6 \cdot h^{(5)}(x) < 0, \qquad h^{(5)}(x) < 0$$

for $x > 0$, showing that $h^{(4)}$ is decreasing on $[0, \infty)$. In particular,

$$0 < h^{(4)}(x) < 4.8 = h^{(4)}(0) \quad \text{for } x > 0.$$

REMARK. Using suitable methods from the theory of infinite series, one can show that (see Example 1 in Section 8 of Chapter 7)

$$\int_0^1 \frac{\ln(x+1)}{x}\,dx = \frac{\pi^2}{12}.$$

Instead of actual equality, we can verify approximate equality in the foregoing relation by using Simpson's rule.

EXAMPLE 5. Let $y = e^{-x^2}$. Then

$$y' = -2xe^{-x^2}, \qquad y'' = 2(2x^2 - 1)e^{-x^2}, \qquad y''' = -4x(2x^2 - 3)e^{-x^2},$$

$$y^{(4)} = 4(4x^4 - 12x^2 + 3)e^{-x^2}, \qquad y^{(5)} = -8x(4x^4 - 20x^2 + 15)e^{-x^2}.$$

Hence, $\max\{|y^{(4)}| : 0 \le x \le 1\} = 12$. Letting $2n = 10$ and using Simpson's rule, we get

$$\int_0^1 e^{-x^2}\,dx \approx 0.746825$$

and the error bound is

$$\frac{12}{2880 \cdot 5^4} = \frac{1}{150{,}000}.$$

EXAMPLE 6. Using Simpson's rule with $2n = 20$, we get

$$\int_0^1 \frac{\ln(1+x)}{1+x^2}\,dx \approx 0.27219844.$$

Actually,

$$\int_0^1 \frac{\ln(1+x)}{1+x^2}\,dx = \frac{\pi}{8}\ln 2 = 0.2721982613\ldots.$$

Indeed, setting $x = \tan t$,

$$\int_0^1 \frac{\ln(1+x)}{1+x^2}\,dx = \int_0^{\pi/4} \ln(\cos t + \sin t)\,dt - \int_0^{\pi/4} \ln(\cos t)\,dt.$$

Substituting $t = \pi/4 - s$,

$$\int_0^{\pi/4} \ln(\cos t)\,dt = -\int_{\pi/4}^0 \ln\left[(\cos s)\left(\cos\frac{\pi}{4}\right) + (\sin s)\left(\sin\frac{\pi}{4}\right)\right]ds$$

$$= \int_0^{\pi/4} \ln\left(\frac{\cos s + \sin s}{\sqrt{2}}\right)ds$$

$$= \int_0^{\pi/4} \ln(\cos s + \sin s)\,ds - \frac{\ln 2}{2}\int_0^{\pi/4} ds$$

$$= \int_0^{\pi/4} \ln(\cos s + \sin s)\,ds - \frac{\pi}{8}\ln 2.$$

Hence,

$$\int_0^1 \frac{\ln(1 + x)}{1 + x^2} dx = \frac{\pi}{8} \ln 2.$$

EXAMPLE 7. Using Simpson's rule with $2n = 6$, we get

$$\int_0^\pi \frac{\sin x}{x} dx \approx 1.852.$$

EXAMPLE 8. We wish to calculate $\int_1^2 (1/x) dx$ with an accuracy of $5 \cdot 10^{-5}$. Into how many parts must we divide the interval $[1, 2]$ if we use (i) the trapezoidal rule and (ii) Simpson's rule?

We first consider the case (i). The error bound for the trapezoidal rule (see Proposition 5.27 and the Remark following it) is

$$\frac{M(b - a)^3}{12n^2}.$$

In our case $M = \max\{|2/x^3|: 1 \le x \le 2\} = 2$ and $b - a = 1$. We have to consider the inequality

$$\frac{2}{12n^2} < 5 \cdot 10^{-5}$$

and find the smallest positive integer n satisfying it. But

$$\frac{2}{12n^2} < 5 \cdot 10^{-5} \quad \text{if and only if} \quad \frac{100}{\sqrt{3}} < n.$$

Since $100/\sqrt{3}$ approximately equals 57.735027, we see that the number n of subdivisions of the interval $[1, 2]$ in the case of using the trapezoidal rule will be $n = 58$.

We next consider case (ii). The error bound for Simpson's rule (see the Remarks following Proposition 5.28) is

$$\frac{K(b - a)^5}{2880n^4} = \frac{K(b - a)^5}{180(2n)^4}.$$

In the case under consideration, $K = \max\{|24/x^5|: 1 \le x \le 2\} = 24$ and $b - a = 1$. Also, recall that in the case of Simpson's rule the number of subdivisions of the interval $[a, b]$ is $2n$. We therefore have to consider the inequality

$$\frac{24}{180(2n)^4} < 5 \cdot 10^{-5}$$

and find the smallest positive integer $2n$ satisfying it. But

$$\frac{24}{180(2n)^4} < 5 \cdot 10^{-5} \quad \text{if and only if} \quad \sqrt[4]{\frac{4}{15}} \cdot 10 < 2n.$$

Since $\sqrt[4]{4/15} \cdot 10$ equals approximately 7.186082, the number $2n$ of subdivisions of the interval $[1, 2]$ in case of using Simpson's rule will only be $2n = 8$. It is clear therefore that Simpson's rule is much more efficient than the trapezoidal rule in calculating $\int_1^2 (1/x)\,dx$ with an accuracy of $5 \cdot 10^{-5}$.

EXAMPLE 9. Let n be a positive integer. Then

$$\tfrac{2}{3}n^{3/2} < 1 + \sqrt{2} + \sqrt{3} + \cdots + \sqrt{n} < \frac{4n + 3}{6}\sqrt{n}.$$

Indeed, for k a positive integer and $k - 1 \leq x < k$, we have $\sqrt{k} > \sqrt{x}$ and so

$$\sqrt{k} > \int_{k-1}^{k} \sqrt{x}\,dx.$$

Adding, we get

$$1 + \sqrt{2} + \sqrt{3} + \cdots + \sqrt{n} > \int_0^n \sqrt{x}\,dx = \tfrac{2}{3}n^{3/2}.$$

Since the graph of $y = \sqrt{x}$ is concave down, it is clear that

$$\tfrac{1}{2}(\sqrt{k-1} + \sqrt{k}) < \int_{k-1}^{k} \sqrt{x}\,dx;$$

here $\tfrac{1}{2}(\sqrt{k-1} + \sqrt{k})$ is the area of the trapezoid with the vertices $(k - 1, 0)$, $(k, 0)$, $(k, \sqrt{k})$, and $(k - 1, \sqrt{k-1})$. Hence,

$$1 + \sqrt{2} + \sqrt{3} + \cdots + \sqrt{n}$$
$$= \tfrac{1}{2}(\sqrt{0} + \sqrt{1}) + \tfrac{1}{2}(\sqrt{1} + \sqrt{2}) + \tfrac{1}{2}(\sqrt{2} + \sqrt{3})$$
$$+ \tfrac{1}{2}(\sqrt{3} + \sqrt{4}) + \cdots + \tfrac{1}{2}(\sqrt{n-1} + \sqrt{n}) + \tfrac{1}{2}\sqrt{n}$$
$$< \int_0^n \sqrt{x}\,dx + \tfrac{1}{2}\sqrt{n} = \frac{4n + 3}{6}\sqrt{n}.$$

EXAMPLE 10. Let f be a differentiable function on $[a, b]$ and assume that $|f'(x)| \leq M$ for all $x \in [a, b]$. Moreover, let

$$S_n = \sum_{j=1}^{n} f(t_j)h,$$

when $h = (b - a)/n$, $x_0 = a$, $x_1 = a + h$, $x_2 = a + 2h$, ..., $x_n = a + nh = b$ and $t_j \in [x_{j-1}, x_j]$ for each $j = 1, 2, \ldots, n$. Then

$$\left| \int_a^b f(x)\,dx - S_n \right| \leq \frac{M(b - a)^2}{n}.$$

Indeed, by the First Mean Value Theorem for Integrals (see Proposition 5.23), in each interval $[x_{j-1}, x_j]$ there exists a point c_j so that

$$\int_{x_{j-1}}^{x_j} f(x)\,dx = f(c_j)(x_j - x_{j-1}) = f(c_j)h.$$

Hence,

$$\left| \int_a^b f(x)\,dx - S_n \right| = \left| \sum_{j=1}^n \int_{x_{j-1}}^{x_j} f(x)\,dx - \sum_{j=1}^n f(t_j)h \right|$$

$$= \left| \sum_{j=1}^n [f(c_j) - f(t_j)]h \right| \le \sum_{j=1}^n |f(c_j) - f(t_j)|h.$$

But, by the Mean Value Theorem of Differential Calculus (see Proposition 4.5), there exists a point e_j between c_j and t_j so that

$$|f(c_j) - f(t_j)| = |f'(e_j)| \cdot |c_j - t_j|.$$

Moreover, it is clear that $|c_j - t_j| \le |x_j - x_{j-1}| = h$ and so

$$\left| \int_a^b f(x)\,dx - S_n \right| \le \sum_{j=1}^n |f'(e_j)|h^2 \le nMh^2 = \frac{M(b-a)^2}{n}.$$

REMARK. In Example 10 we have obtained an error bound for the approximation of an integral by Riemann sums in which the interval $[a, b]$ is partitioned into subintervals of equal length. For example, the error bound for $\int_1^2 (1/x)\,dx$ with $h = 0.1$ is only $\frac{1}{10}$ (compare this with the results in Examples 1 and 2).

EXERCISES TO CHAPTER 5

5.1. Let f be a continuous function such that $0 < A \le f(x) \le B$ for $0 \le x \le 1$. Show that

$$AB \int_0^1 \frac{1}{f(x)}\,dx \le A + B - \int_0^1 f(x)\,dx.$$

[Hint: Note that

$$\frac{\{f(x) - A\}\{f(x) - B\}}{f(x)} \le 0$$

in the interval $[0, 1]$. Integrating both sides of this inequality over $[0, 1]$ gives what we want.]

5.2. Let p and q be larger than zero. Show that

$$\int_0^1 (1 - x^p)^{1/q}\,dx = \int_0^1 (1 - x^q)^{1/p}\,dx.$$

[Hint: Let f be a decreasing continuous function in $[a, b]$. Then the inverse function g exists in $[f(b), f(a)]$ and is also decreasing and continuous. Hence,

$$\int_{f(b)}^{f(a)} g(y)\,dy = \int_b^a g\{f(t)\}f'(t)\,dt = \int_b^a tf'(t)\,dt = af(a) - bf(b) + \int_a^b f(t)\,dt.$$

If additionally we have $f(a) = b$ and $f(b) = a$, then

$$\int_a^b g(t)\,dt = \int_a^b f(t)\,dt.$$

The functions $f(x) = (1 - x^q)^{1/p}$ and $g(x) = (1 - x^p)^{1/q}$ represent in $[0, 1]$ a special case of this situation.]

5.3. Explain the curiosity

$$\int_a^b \frac{1}{x \ln x} dx = 1 \bigg|_a^b + \int_a^b \frac{1}{x \ln x} dx,$$

using integration by parts: $u = 1/(\ln x)$ and $dv = (1/x) dx$.

5.4. Let $a > 1$ and $x_n = n(\sqrt[n]{a} - 1)$ for $n = 1, 2, 3, \ldots$. Show that $x_n > x_{n+1}$ for all n.
[Hint: If $x > 1$ and $p > q > 1$, then $x^{p-1} > x^{q-1}$ and so

$$\int_1^a x^{p-1} dx > \int_1^a x^{q-1} dx \quad \text{or} \quad \frac{a^p - 1}{p} > \frac{a^q - 1}{q}.$$

Putting $p = 1/n$ and $q = 1/(n + 1)$, we obtain the desired inequality.]

5.5. If

$$\int_0^1 f(xt) \, dt = 0$$

for all values of x, show that $f = 0$.
[Hint: We have

$$\int_0^1 f(xt) \, dt = \int_0^x f(u) \frac{du}{x}, \quad \text{implying} \quad \int_0^x f(u) \, du = 0 \quad \text{for all } x$$

and so

$$\frac{d}{dx} \int_0^x f(u) \, du = 0 \quad \text{or } f = 0.]$$

5.6. Find the derivative of the function

$$H(x) = \left(\int_a^x f(y) \, dy \right) \left(\int_x^b g(y) \, dy \right)$$

and hence show that

$$\int_a^b g(x) \left(\int_a^x f(y) \, dy \right) dx = \int_b^a f(x) \left(\int_b^x g(y) \, dy \right) dx.$$

[Hint: We have $H(a) = H(b) = 0$ and

$$H'(x) = f(x) \int_x^b g(y) \, dy - g(x) \int_a^x f(y) \, dy;$$

hence, integrating over the interval $[a, b]$, the desired conclusion follows.]

REMARK. In the next three Exercises we assume that f is twice differentiable.

5.7. Show that there is a point α in $(-h, h)$ such that

$$\int_{-h}^h f(x) \, dx = 2hf(0) + \frac{h^3}{3} f''(\alpha).$$

[Hint: Let $T(h) = \int_{-h}^{h} f(x)\,dx - 2hf(0)$. Then

$$T(0) = 0, \qquad T'(h) = f(h) + f(-h) - 2f(0), \qquad T'(0) = 0,$$

and, by the Mean Value Theorem (see Proposition 4.5 in Chapter 4),

$$T''(h) = f'(h) - f'(-h) = 2hf''(c_h),$$

where c_h is some point between $-h$ and h. Hence, by the result in Example 27 of Section 6 in Chapter 4, there is a point u in $(0, h)$ such that

$$\frac{T(h)}{h^3} = \frac{T''(u)}{3 \cdot 2 \cdot u} = \frac{1}{3} f''(c_u)$$

and so $T(h) = (h^3/3)f''(\alpha)$, where $\alpha = c_u$, which is a point in $(-u, u)$ and so a point in $(-h, h)$.]

5.8. Show that there is a point β in $(-h, h)$ such that

$$\int_{-h}^{h} f(x)\,dx = h\{f(h) + f(-h)\} - \frac{2h^3}{3} f''(\beta).$$

[Hint: If $T(h) = \int_{-h}^{h} f(x)\,dx - h\{f(h) + f(-h)\}$, then

$$T'(h) = -h\{f'(h) - f'(-h)\} = -2h^2 f''(c_h),$$

where c_h is between $-h$ and h, by the Mean Value Theorem (see Proposition 4.5). Hence, by the result in Example 26 of Section 6 in Chapter 4, there is a point u in $(0, h)$ such that

$$\frac{T(h)}{h^3} = \frac{T'(u)}{3u^2} = -\frac{2}{3} f''(c_u) \quad \text{with } -u < c_u < u$$

because $T(0) = 0$, hence $T(h) = -(2h^3/3)f''(\beta)$, where $\beta = c_u$.]

5.9. Show that there is a point γ in $(-h, h)$ such that

$$\int_{-h}^{h} f(x)\,dx = \frac{h}{2}\{f(h) + 2f(0) + f(-h)\} - \frac{h^3}{6} f''(\gamma).$$

[Hint: If $T(h) = \int_{-h}^{h} f(x)\,dx - (h/2)\{f(h) + 2f(0) + f(-h)\}$, then

$$T(0) = T'(0) = 0 \quad \text{and} \quad T''(h) = -\frac{h}{2}\{f''(h) + f''(-h)\}.$$

but $\{f''(h) + f''(-h)\}/2$ lies between $f''(h)$ and $f''(-h)$ and so, since f'' takes all intermediate values, equal $f''(c)$ for some c in $[-h, h]$. Thus,

$$T''(h) = -hf''(c) \quad \text{or} \quad T(h) = -\frac{h^3}{6} f''(\gamma).]$$

5.10. If f is periodic with period a and if

$$g(x) = f(x) - \frac{1}{a} \int_{0}^{a} f(x)\,dx,$$

show that $\int_{0}^{x} g(t)\,dt$ is periodic with period a.
    [Hint: We have

$$\int_0^{x+a} g(t)\,dt = \left(\int_0^a + \int_a^{a+x}\right) g(t)\,dt = \int_0^a g(t)\,dt + \int_0^x g(u)\,du, \quad t = a + u,$$

$$= \int_0^x g(u)\,du$$

because g has period a and

$$\int_0^a g(t)\,dt = \int_0^a f(t)\,dt - \frac{1}{a}\int_0^a \left(\int_0^a f(x)\,dx\right) dt = \int_0^a f(t)\,dt - \int_0^a f(t)\,dt = 0.]$$

5.11. Let f be continuous on $[a,b]$ and $\int_a^b f(x)\,dx = 0$. Show that $f(x) = 0$ for some x in (a,b).
   [Hint: See Proposition 5.23.]

REMARK. If g and h are continuous on $[a,b]$ and $\int_a^b g(x)\,dx = \int_a^b h(x)\,dx$, then $g(x) = h(x)$ for some x in (a,b). [Indeed, let $g(x) - h(x) = f(x)$ and apply the result in Exercise 5.11.]

5.12. (Theorem of Bliss) Suppose that f and g are Riemann integrable on $[a,b]$ and let $P = \{a = x_0 < x_1 < \cdots < x_n = b\}$ be a partition of $[a,b]$. Show that

$$\sum_{k=1}^{n} f(t_k)g(t_k')\Delta x_k,$$

where $x_{k-1} \le t_k \le x_k$ and $x_{k-1} \le t_k' \le x_k$, $k = 1, 2, \ldots, n$, and

$$\sum_{k=1}^{n} f(t_k)g(t_k)\Delta x_k$$

tend to the same limit as $\|P\| \to 0$.
   [Hint: Note the identity

$$\sum_{k=1}^{n} f(t_k)g(t_k')\Delta x_k = \sum_{k=1}^{n} f(t_k)g(t_k)\Delta x_k + \sum_{k=1}^{n} f(t_k)[g(t_k') - g(t_k)]\Delta x_k$$

and the inequality (since f is Riemann integrable, it is bounded)

$$\sum_{k=1}^{n} f(t_k)|g(t_k') - g(t_k)|\Delta x_k \le K\left(\sum_{k=1}^{n} |g(t_k') - g(t_k)|\Delta x_k\right).$$

Finally, let $\xi_k = t_k$ or t_k' so that

$$g(\xi_k) = \max\{g(t_k), g(t_k')\},$$

and $\eta_k = t_k$ or t_k' so that

$$g(\eta_k) = \min\{g(t_k), g(t_k')\},$$

$k = 1, 2, \ldots, n$. Then

$$\sum_{k=1}^{n} |g(t_k') - g(t_k)|\Delta x_k = \sum_{k=1}^{n} g(\xi_k)\Delta x_k - \sum_{k=1}^{n} g(\eta_k)\Delta x_k.]$$

5.13. What is wrong with

$$\int_{-1}^{1} \frac{1}{x^2}\,dx = -\frac{1}{x}\bigg|_{-1}^{1} = -2?$$

5.14. Evaluate

$$\lim_{x \to 0} \frac{x}{1 - e^{x^2}} \int_0^x e^{t^2} \, dt.$$

[Answer: -1.]

5.15. Show that if f is continuous on $[a, b]$ and $\int_a^b f(x)g(x) \, dx = 0$ for every integrable g, then $f = 0$.

   [Hint: If f is continuous on $[a, b]$ and $\int_a^b f^2(x) \, dx = 0$, then $f = 0$ on $[a, b]$, as can be seen by part (i) of Proposition 5.22.]

5.16. Let f, g, and h be Riemann integrable on $[a, b]$. Show that

$$\int_a^b |f(x) - g(x)|^2 \, dx \le 2 \int_a^b |f(x) - h(x)|^2 \, dx + 2 \int_a^b |g(x) - h(x)|^2 \, dx.$$

   [Hint: We have $f(x) - g(x) = \{f(x) - h(x)\} - \{g(x) - h(x)\}$ and so

$$|f(x) - g(x)| \le |f(x) - h(x)| + |g(x) - h(x)|$$

and

$$|f(x) - g(x)|^2 \le |f(x) - h(x)|^2 + |g(x) - h(x)|^2 + 2|f(x) - h(x)| \, |g(x) - h(x)|.$$

But $2|f(x) - h(x)| \, |g(x) - h(x)| \le |f(x) - h(x)|^2 + |g(x) - h(x)|^2.$]

5.17. If $f_1(x) = \int_0^x f(t) \, dt$ and $f_2(x) = \int_0^x f_1(t) \, dt$, show that

$$f_2(x) = \int_0^x (x - t)f(t) \, dt.$$

More generally, if $f_k(x) = \int_0^x f_{k-1}(t) \, dt$ for $k = 2, 3, \ldots$, then

$$f_k(x) = \frac{1}{(k - 1)!} \int_0^x (x - t)^{k-1} f(t) \, dt.$$

   [Hint: We integrate $\int_0^x f_1(t) \, dt$ by parts as follows: Let $u = f_1(t)$ and $dv = dt$. Then

$$\int_0^x f_1(t) \, dt = tf_1(t) \Big|_0^x - \int_0^x tf(t) \, dt = xf_1(x) - \int_0^x tf(t) \, dt$$

$$= x \int_0^x f(t) \, dt - \int_0^x tf(t) \, dt = \int_0^x (x - t)f(t) \, dt.$$

A similar integration by parts will result in the formula

$$f_n(x) = \int_0^x (x - t)f_{n-2}(t) \, dt.$$

In this equation let $u = f_{n-2}(t)$ and $dv = (x - t) \, dt$. Then

$$\int_0^x (x - t)f_{n-2}(t) \, dt = -\frac{(x - t)^2}{2} f_{n-2}(t) \Big|_0^x + \int_0^x \frac{(x - t)^2}{2} f_{n-3}(t) \, dt$$

$$= \int_0^x \frac{(x - t)^2}{2} f_{n-3}(t) \, dt.$$

Another integration by parts will give

$$f_n(x) = \int_0^x \frac{(x-t)^3}{3} f_{n-4}(t)\,dt$$

and it is clear that we can continue in this way and obtain

$$f_n(x) = \int_0^x \frac{(x-t)^{n-1}}{(n-1)!} f(t)\,dt.]$$

5.18. As an application of the result in Exercise 5.17, if m and n are positive integers, show that

$$\int_0^1 (1-x)^n x^m\,dx = \frac{m!\,n!}{(m+n+1)!}.$$

[Hint: If $f(t) = t^m$, then, in the notation of Exercise 5.17,

$$\int_0^1 (1-t)^n t^m\,dt = n! f_{n+1}(1).$$

Now

$$f_1(x) = \int_0^x t^m\,dt = \frac{x^{m+1}}{(m+1)!},$$

$$f_2(x) = \int_0^x \frac{m!\,t^{m+1}}{(m+1)!}\,dt = \frac{m!\,x^{m+2}}{(m+2)!},$$

$$\vdots$$

$$f_k(x) = \frac{m!\,x^{m+k}}{(m+k)!}$$

and so $n! f_{n+1}(1) = m!\,n!/(m+n+1)!.]$

5.19. Using Simpson's rule, compute the volume of a sphere.

5.20. Verify that if we take $2n = 10$ in the evaluation of

$$\frac{\pi}{4} = \int_0^1 \frac{1}{1+x^2}\,dx$$

by use of Simpson's rule, we obtain approximately 0.78539815 (the exact value of $\pi/4$ with an accuracy of eight decimal places is 0.78539816); the error bound in the considered case is approximately 0.000013, much larger than the actual error.

[Hint: See Example 3 following Proposition 5.28.]

5.21. Using Simpson's rule, show that the integral

$$\int_0^1 e^{-x^2}\,dx$$

approximately equals 0.7468.

5.22. Using Simpson's rule, show that the integral

$$\int_0^{\pi/2} (1 - \tfrac{1}{2}\sin^2 x)^{1/2}\, dx$$

approximately equals 1.351.

5.23. Show that $\int_0^1 P(x)\, dx = \frac{1}{18}\{5P(\alpha) + 8P(\tfrac{1}{2}) + 5P(\beta)\}$ if P is any polynomial of the fifth degree and α and β are the roots of $x^2 - x + \frac{1}{10} = 0$.

CHAPTER 6

Additional Topics in Integration

1. The Indefinite Integral

Definition. Let f be a given continuous function on an interval $[a,b]$. A function F, such that

$$F'(x) = f(x) \quad \text{with } a < x < b,$$

is called an *antiderivative* or an *indefinite integral of f*. If the latter name is used one writes

$$F(x) = \int f(x)\,dx.$$

REMARKS. Proposition 5.20 asserts that, if f is continuous on $[a,b]$, then

$$F(x) = \int_a^x f(t)\,dt \quad \text{for } x \in [a,b]$$

is an antiderivative for f on $[a,b]$. It is clear that we can add any constant to the function F and still obtain an antiderivative of f. We now show that all antiderivatives of f are obtained by adding an arbitrary constant to F.

Proposition 6.1. *Let f be a continuous function on an interval $[a,b]$. Then*

$$F(x) = \int_a^x f(t)\,dt \quad \text{for } x \in [a,b] \tag{6.1}$$

defines an antiderivative of f and every other antiderivative of f differs from this one by a constant.

PROOF. That (6.1) defines an antiderivative of f follows from Proposition 5.20. If F_1 is another antiderivative of f, then the difference $F - F_1$ is a function whose derivative is zero for all x in $[a, b]$ and by Proposition 4.6 such a function is constant.   □

REMARKS. In view of Proposition 6.1, the indefinite integral

$$\int f(x)\, dx = \int_a^x f(t)\, dt + C, \tag{6.2}$$

where C is some constant. If

$$\int f(x)\, dx = F(x), \tag{6.3}$$

we thus immediately have $F'(x) = f(x)$, but not more. If in addition to (6.3) we have

$$\int f(x)\, dx = F_1(x), \tag{6.4}$$

it does *not* follow that $F(x) = F_1(x)$ but because of the constant C in (6.2) the equations (6.3) and (6.4) yield only

$$F(x) = F_1(x) + C,$$

where C again is some constant.

   To illustrate this point further, let f and g be functions which have continuous derivatives on an interval $[a, b]$. The function

$$f(x)g(x) - \int f'(x)g(x)\, dx$$

is an antiderivative of $f(x)g'(x)$ because $f(x)g'(x) = (f(x)g(x))' - f'(x)g(x)$ by part (iii) of Proposition 3.2 and so we have the *integration by parts formula*

$$\int f(x)g'(x)\, dx = f(x)g(x) - \int f'(x)g(x)\, dx. \tag{6.5}$$

By (6.5) we get

$$\int \frac{1}{(\sin x)(\cos x)}\, dx = \int \frac{\cot x}{\cos^2 x}\, dx = \int (\cot x)(\tan x)'\, dx$$

$$= (\cot x)(\tan x) - \int (\tan x)(\cot x)'\, dx = 1 + \int \frac{\tan x}{\sin^2 x}\, dx,$$

that is,

$$\int \frac{1}{(\sin x)(\cos x)}\, dx = 1 + \int \frac{1}{(\sin x)(\cos x)}\, dx. \tag{6.6}$$

Wrongly interpreted, equation (6.6) leads to the absurdity $0 = 1$. The correct interpretation of (6.6) is: The constant 0 is, up to an additive constant, equal to the constant 1 (which is no doubt true!). This example serves only the purpose of stressing the great importance of the *constant of integration C* in the equation

$$\int f(x)\,dx = F(x) + C,$$

where F is an antiderivative of f and f is called the *integrand*. The integration by parts formula (6.5) we have already come across in Proposition 5.19 and an extension of the integration by parts formula has appeared in (5.44) of Chapter 5.

ADDITIONAL REMARKS. From Proposition 5.18 (together with Proposition 5.16) we may conclude that if f is continuous on $[a,b]$ and if F is any antiderivative of f, then

$$\int_a^b f(t)\,dt = F(b) - F(a). \qquad (6.7)$$

We shall *agree* to refer to the Riemann integral $\int_a^b f(t)\,dt$ of f on $[a,b]$ as simply the *definite integral of f on $[a,b]$* and call the numbers a and b the *lower* and *upper limits of integration*; for brevity we shall replace the term "Riemann integrable" by the phrase "*integrable.*"

Relation (6.7) shows that knowledge of the antiderivative (or the indefinite integral) suffices to determine the corresponding definite integral. We shall consider various techniques for the determination of antiderivatives of certain functions in the sequel. We recall, however, that there are definite integrals which can be evaluated explicitly without knowledge of the corresponding antiderivatives for which explicit expressions may not be available [e.g., see (5.45) in Chapter 5 and Exercises 6.3 and 6.4 at the end of this chapter].

Proposition 6.2. *The indefinite integral satisfies the following basic rules*:

(i) *If F has a continuous derivative F', then*

$$\int F'(x)\,dx = F(x) + C. \qquad (6.8)$$

(ii) *If f and g have antiderivatives, then so does $f + g$ and*

$$\int \{f(x) + g(x)\}\,dx = \int f(x)\,dx + \int g(x)\,dx. \qquad (6.9)$$

(iii) *If f has an antiderivative, and if k is a constant, then kf has an antiderivative and*

$$\int kf(x)\,dx = k \int f(x)\,dx. \qquad (6.10)$$

(iv) *If F and G have continuous derivatives F' and G', then*

$$\int \{F(x)G'(x) + F'(x)G(x)\}\, dx = F(x)G(x) + C. \tag{6.11}$$

(v) *Let G in the variable x have a continuous derivative G' in some interval (a, b). Let F in the variable u be defined in the range of G and let F have a continuous derivative F'. Then for values of x in (a, b)*

$$\int F'\{G(x)\}G'(x)\, dx = F\{G(x)\} + C. \tag{6.12}$$

PROOF. The claims follow directly from the definition and the rules of differentiation. Note, for example, that the integrand in (6.11) equals $(FG)'$ and that $FG' + F'G$ is continuous. Similarly, the integrand in (6.12) is continuous and is equal to the derivative of $F\{G(x)\}$. $\qquad\square$

REMARKS. Observe that $\int F'(x)\, dx = F(x) + C$, but $[\int F(x)\, dx]' = F(x)$.
We can obviously rewrite (6.11) in the form

$$\int F(x)G'(x)\, dx = F(x)G(x) - \int G(x)F'(x)\, dx, \tag{6.13}$$

known as the *integration by parts formula*. Repeated application of (6.13) yields the following: If $F^{(n+1)}$ and $G^{(n+1)}$ are continuous, then

$$\int F(x)G^{(n+1)}(x)\, dx = F(x)G^{(n)}(x) - F'(x)G^{(n-1)}(x) + \cdots \tag{6.14}$$
$$+ (-1)^n F^{(n)}(x)G(x) + (-1)^{n+1}\int F^{(n+1)}(x)G(x)\, dx,$$

is known as the *extended integration by parts formula*.
If we set $u = G(x)$ in (6.12), then

$$\int F'\{G(x)\}G'(x)\, dx = \int F'(u)\, du = F(u) + C = F\{G(x)\} + C; \tag{6.15}$$

when we use this procedure, we shall speak of using the method of *integration by substitution*.

2. Some Techniques of Integration

Much of the art of integration consists essentially in a repeated use of the rules contained in Proposition 6.2 together with a few basic antiderivatives. We begin by listing a number of useful integration formulas.

1. $\displaystyle\int x^n\, dx = \frac{x^{n+1}}{n + 1} + C, \qquad n \neq -1.$

2. $\displaystyle\int \frac{1}{x}\,dx = \ln|x| + C.$

3. $\displaystyle\int e^x\,dx = e^x + C.$

4. $\displaystyle\int a^x\,dx = \frac{a^x}{\ln a} + C, \qquad a \neq 1,\, a > 0.$

5. $\displaystyle\int (\sin x)\,dx = -\cos x + C.$

6. $\displaystyle\int (\cos x)\,dx = \sin x + C.$

7. $\displaystyle\int (\sec^2 x)\,dx = \tan x + C.$

8. $\displaystyle\int (\csc^2 x)\,dx = -\cot x + C.$

9. $\displaystyle\int (\sec x)(\tan x)\,dx = \sec x + C.$

10. $\displaystyle\int (\csc x)(\cot x)\,dx = -\csc x + C.$

11. $\displaystyle\int (\tan x)\,dx = \ln|\sec x| + C.$

12. $\displaystyle\int (\cot x)\,dx = \ln|\sin x| + C.$

13. $\displaystyle\int (\sec x)\,dx = \ln|\sec x + \tan x| + C.$

14. $\displaystyle\int (\csc x)\,dx = \ln|\csc x - \cot x| + C.$

15. $\displaystyle\int (\sinh x)\,dx = \cosh x + C.$

16. $\displaystyle\int (\cosh x)\,dx = \sinh x + C.$

17. $\displaystyle\int (\operatorname{sech}^2 x)\,dx = \tanh x + C.$

18. $\displaystyle\int (\operatorname{csch}^2 x)\,dx = -\coth x + C.$

19. $\displaystyle\int (\operatorname{sech} x)(\tanh x)\, dx = -\operatorname{sech} x + C.$

20. $\displaystyle\int (\operatorname{csch} x)(\coth x)\, dx = -\operatorname{csch} x + C.$

21. $\displaystyle\int \frac{1}{a^2 + x^2}\, dx = \frac{1}{a}\tan^{-1}\left(\frac{x}{a}\right) + C.$

22. $\displaystyle\int \frac{1}{(a^2 - x^2)^{1/2}}\, dx = \sin^{-1}\left(\frac{x}{a}\right) + C, \qquad |x| < a \text{ and } a > 0.$

23. $\displaystyle\int \frac{1}{x(x^2 - a^2)^{1/2}}\, dx = \sec^{-1}\left|\frac{x}{a}\right| + C, \qquad |x| > |a| \text{ and } a \neq 0.$

24. $\displaystyle\int \frac{1}{(a^2 + x^2)^{1/2}}\, dx = \sinh^{-1}\left(\frac{x}{a}\right) + C, \qquad a > 0.$

25. $\displaystyle\int \frac{1}{(x^2 - a^2)^{1/2}}\, dx = \cosh^{-1}\left(\frac{x}{a}\right) + C, \qquad x > a > 0.$

26. $\displaystyle\int \frac{1}{a^2 - x^2}\, dx = \frac{1}{a}\tanh^{-1}\left(\frac{x}{a}\right) + C, \qquad |x| < a.$

$\displaystyle\qquad\qquad\qquad = \frac{1}{a}\coth^{-1}\left(\frac{x}{a}\right) + C, \qquad |x| > a > 0.$

27. $\displaystyle\int \frac{1}{x(a^2 - x^2)^{1/2}}\, dx = -\frac{1}{a}\operatorname{sech}^{-1}\frac{|x|}{a} + C, \qquad 0 < |x| < a.$

28. $\displaystyle\int \frac{1}{x(a^2 + x^2)^{1/2}}\, dx = -\frac{1}{a}\operatorname{csch}^{-1}\frac{|x|}{a} + C, \qquad a > 0 \text{ and } x \neq 0.$

The foregoing integration formulas are immediate consequences of the corresponding differentiation formulas in Section 3 of Chapter 3. Only formulas 13 and 14 are obtained somewhat artificially by first noting that

$$\sec x = (\sec x)\frac{\sec x + \tan x}{\sec x + \tan x} \quad \text{and} \quad \csc x = (\csc x)\frac{\csc x - \cot x}{\csc x - \cot x}$$

and then making use of the relation

$$\int \frac{f'(t)}{f(t)}\, dt = \ln|f(t)| + C. \tag{6.16}$$

Relation (6.16) follows from formula 2 by putting $x = f(t)$; see also (3.8) and (3.9) in Chapter 3. Of course, formulas 11 and 12 are also arrived at via relation (6.16).

The integration of trigonometric functions will very often involve considerable use of trigonometric identities. For example, to see that

$$\int \frac{\sin nx}{\sin x}\,dx = 2 \sum_{n=1}^{n} \frac{\sin(2k-1)x}{2k-1} + C \text{ for } n = 1, 2, 3, \ldots \qquad (6.17)$$

we need to note that

$$\sin 2nx = \sum_{k=1}^{n} [\sin 2kx - \sin(2k-2)x] = 2(\sin x)\sum_{k=1}^{n} \cos(2k-1)x$$

and this evidently involves the trigonometric identity

$$\sin A - \sin B = 2\cos\frac{A+B}{2}\cdot\sin\frac{A-B}{2}.$$

In a similar fashion we obtain

$$\int \frac{\sin(2n+1)x}{\sin x}\,dx = x + 2\sum_{k=1}^{n} \frac{\sin 2kx}{2k} + C. \qquad (6.18)$$

Integrals of the form

$$\int (\sin^m x)(\cos^n x)\,dx, \qquad (6.19)$$

where m and n are constant exponents, are easy to evaluate if at least one of the exponents m, n is a positive odd integer. In this case we use the identity

$$\sin^2 x + \cos^2 x = 1$$

to rewrite the integrand either in the form $F(\cos x)\sin x$ or in the form $F(\sin x)\cos x$. In the former case, the substitution $u = \cos x$ is effective, while in the latter case, the substitution $u = \sin x$ works. For example,

$$\int \frac{\sin^3 x}{(\cos x)^{1/2}}\,dx = \int \frac{1 - \cos^2 x}{(\cos x)^{1/2}}(\sin x)\,dx = -\int \frac{1 - u^2}{u^{1/2}}\,du$$

by setting $u = \cos x$. But

$$-\int \frac{1-u^2}{u^{1/2}}\,du = \int (u^{3/2} - u^{-1/2})\,du = \tfrac{2}{5}u^{5/2} - 2u^{1/2} + C$$

$$= \tfrac{2}{5}(\cos x)^{5/2} - 2(\cos x)^{1/2} + C.$$

Integrals of the form (6.19) are also easily evaluated if both of the exponents m, n are nonnegative even integers. In this case we use the identities

$$\sin^2 x = \tfrac{1}{2}(1 - \cos 2x) \quad \text{and} \quad \cos^2 x = \tfrac{1}{2}(1 + \cos 2x),$$

which follow from the double-angle formula

$$\cos 2x = 1 - 2\sin^2 x = 2\cos^2 x - 1.$$

For example

$$\int (\cos^2 mx)\,dx = \frac{1}{2}x + \frac{1}{4m}(\sin 2mx) + C, \qquad m \neq 0, \qquad (6.20)$$

and

$$\int (\sin^2 mx)\, dx = \frac{1}{2}x - \frac{1}{4m}(\sin 2mx) + C, \qquad m \neq 0. \tag{6.21}$$

Use of the trigonometric identities

$$(\sin mx)(\cos nx) = \tfrac{1}{2}[\sin(m+n)x + \sin(m-n)x],$$
$$(\cos mx)(\cos nx) = \tfrac{1}{2}[\sin(m+n)x + \cos(m-n)x],$$

and

$$(\sin mx)(\sin nx) = \tfrac{1}{2}[\cos(m-n)x - \cos(m+n)x]$$

for $m \pm n \neq 0$ gives

$$\int (\sin mx)(\cos nx)\, dx$$

$$= -\frac{1}{2(m+n)}\cos(m+n)x - \frac{1}{2(m-n)}\cos(m-n)x + C, \tag{6.22}$$

$$\int (\cos mx)(\cos nx)\, dx$$

$$= \frac{1}{2(m+n)}\sin(m+n)x + \frac{1}{2(m-n)}\sin(m-n)x + C, \tag{6.23}$$

and

$$\int (\sin mx)(\sin nx)\, dx$$

$$= \frac{1}{2(m-n)}\sin(m-n)x - \frac{1}{2(m+n)}\sin(m+n)x + C. \tag{6.24}$$

In integrating positive integral powers or products of positive integral powers of tan, cot, sec, and csc, the identities

$$1 + \tan^2 s = \sec^2 s \quad \text{and} \quad 1 + \cot^2 s = \csc^2 s$$

can often be used to advantage. For example,

$$\int (\tan^4 x)\, dx = \int (\tan^2 x)(\sec^2 x - 1)\, dx$$

$$= \int (\tan^2 x)(\sec^2 x)\, dx - \int (\tan^2 x)\, dx$$

$$= \int (\tan^2 x)(\tan x)'\, dx - \int (\sec^2 x - 1)\, dx$$

$$= \tfrac{1}{3}\tan^3 x - \tan x + x + C.$$

However, the evaluation of the integral $\int (\sec^3 x)\, dx$ shows that we need more than just some trigonometric identities to succeed. Using integration by parts, we have [putting $F(x) = \sec x$ and $G(x) = \tan x$ in (6.13)]

$$\int (\sec^3 x)\, dx = (\sec x)(\tan x) - \int (\sec x)(\tan^2 x)\, dx$$

$$= (\sec x)(\tan x) - \int (\sec^3 x)\, dx + \int (\sec x)\, dx.$$

Now, solving for $\int (\sec^3 x)\, dx$, we obtain

$$\int (\sec^3 x)\, dx = \tfrac{1}{2}(\sec x)(\tan x) + \tfrac{1}{2}\ln |\sec x + \tan x| + C. \qquad (6.25)$$

Integration by parts can also be used in the evaluation of an integral like

$$\int (\sin mx)(\cos nx)\, dx$$

occurring in (6.22) if we do not remember the convenient trigonometric identity

$$(\sin mx)(\cos nx) = \tfrac{1}{2}\sin[(m + n)x + \sin (m - n)x].$$

The extended integration by parts formula (6.14) modified to the form

$$\int F(x)G''(x)\, dx = F(x)G'(x) - F'(x)G(x) + \int F''(x)G(x)\, dx,$$

with $F(x) = \sin mx$ and $G(x) = -(\cos nx)/n^2$, gives us an equation that we can solve for $\int (\sin mx)(\cos nx)\, dx$. Similar remarks pertain to the integrals in (6.23) and (6.24).

The extended integration by parts formula (6.14) is particularly effective in connection with integrals such as

$$\int P(x)e^{ax}\, dx, \quad \int P(x)(\sin bx)\, dx, \quad \text{and} \quad \int P(x)(\cos bx)\, dx,$$

where P is a polynomial. Putting

$$G^{(n+1)}(x) = e^{ax} \quad \text{and} \quad F(x) = P(x)$$

in (6.14), we see that

$$\int P(x)e^{ax}\, dx = e^{ax}\left(\frac{P(x)}{a} - \frac{P'(x)}{a^2} + \frac{P''(x)}{a^3} - \cdots\right) + C. \qquad (6.26)$$

Setting

$$G^{(n+1)}(x) = \sin bx \quad \text{and} \quad F(x) = P(x)$$

in (6.14) gives

$$\int P(x)(\sin bx)\,dx = (\sin bx)\left(\frac{P'(x)}{b^2} - \frac{P'''(x)}{b^4} + \cdots\right)$$
$$- (\cos bx)\left(\frac{P(x)}{b} - \frac{P''(x)}{b^3} + \cdots\right) + C. \tag{6.27}$$

In a similar way we get

$$\int P(x)(\cos bx)\,dx = (\sin bx)\left(\frac{P(x)}{b} - \frac{P''(x)}{b^3} + \cdots\right)$$
$$+ (\cos bx)\left(\frac{P'(x)}{b^2} - \frac{P'''(x)}{b^4} + \cdots\right) + C. \tag{6.28}$$

Using the integration by parts formula (6.13) we can easily see that

$$\int e^{ax}(\cos bx)\,dx = \frac{1}{a}e^{ax}(\cos bx) + \frac{b}{a}\int e^{ax}(\sin bx)\,dx \tag{6.29}$$

and

$$\int e^{ax}(\sin bx)\,dx = \frac{1}{a}e^{ax}(\sin bx) - \frac{b}{a}\int e^{ax}(\cos bx)\,dx. \tag{6.30}$$

If in (6.29) we replace the integral on the right-hand side by what (6.30) gives for that integral, we obtain

$$\int e^{ax}(\cos bx)\,dx = \frac{b\sin bx + a\cos bx}{a^2 + b^2}e^{ax} + C. \tag{6.31}$$

In a similar way we obtain

$$\int e^{ax}(\sin bx)\,dx = \frac{a\sin bx - b\cos bx}{a^2 + b^2}e^{ax} + C. \tag{6.32}$$

Another application of the integration by parts formula (6.13) is the development of certain recursion formulas. Suppose

$$J_n = \int \frac{1}{(x^2 + a^2)^n}\,dx, \qquad n = 1, 2, 3, \ldots. \tag{6.33}$$

Putting

$$F(x) = \frac{1}{(x^2 + a^2)^n} \quad \text{and} \quad G(x) = x$$

in (6.13), we get

$$J_n = \frac{x}{(x^2 + a^2)^n} + 2n\int \frac{x^2}{(x^2 + a^2)^{n+1}}\,dx.$$

But

$$\int \frac{x^2}{(x^2 + a^2)^{n+1}} dx = \int \frac{(x^2 + a^2) - a^2}{(x^2 + a^2)^{n+1}} dx = J_n - a^2 J_{n+1}$$

and so

$$J_n = \frac{x}{(x^2 + a^2)^n} + 2n J_n - 2na^2 J_{n+1}$$

or

$$J_{n+1} = \frac{1}{2na^2} \frac{x}{(x^2 + a^2)^n} + \frac{2n - 1}{2n} \frac{1}{a^2} J_n. \tag{6.34}$$

Since

$$J_1 = \frac{1}{a} \tan^{-1} \left(\frac{x}{a} \right), \tag{6.35}$$

we get

$$J_2 = \frac{1}{2a^2} \frac{x}{x^2 + a^2} + \frac{1}{2a^3} \tan^{-1} \left(\frac{x}{a} \right), \tag{6.36}$$

$$J_3 = \frac{1}{4a^2} \frac{x}{(x^2 + a^2)^2} + \frac{3}{4a^2} J_2 = \frac{1}{4a^2} \frac{x}{(x^2 + a^2)^2} + \frac{3}{8a^5} \tan^{-1} \left(\frac{x}{a} \right), \tag{6.37}$$

and so forth. In (6.35), (6.36), and (6.37) we have omitted the constants of integration on purpose.

Another approach to the evaluation of the integrals

$$J_n = \int \frac{1}{(x^2 + a^2)^n} dx, \qquad n = 1, 2, 3, \ldots \tag{6.38}$$

is by the trigonometric substitution $x = a \tan t$. We get

$$J_n = \int \frac{1}{(x^2 + a^2)^n} dx = \int \frac{a \sec^2 t}{(a^2 \sec^2 t)^n} dt = \frac{1}{a^{2n-1}} \int (\cos^{2n-2} t) \, dt.$$

For example,

$$J_1 = \frac{1}{a} \int dt = \frac{1}{a} t + C \tag{6.39}$$

and

$$J_2 = \frac{1}{a^3} \int (\cos^2 t) \, dt = \frac{1}{a^3} \frac{1}{2} \int (1 + \cos 2t) \, dt = \frac{1}{2a^3} (1 + \tfrac{1}{2} \sin 2t) + C \tag{6.40}$$

$$= \frac{1}{2a^3} [t + (\sin t)(\cos t)] + C.$$

But $\tan t = x/a$ and so $t = \tan^{-1}(x/a)$, $\sin t = x/(x^2 + a^2)^{1/2}$, and $\cos t = a/(x^2 + a^2)^{1/2}$ (see Figure 6.2). This shows that J_1 and J_2 are as given in (6.35) and (6.36) when we delete the constant of integration for J_1 and J_2.

REMARKS. By use of the trigonometric identities

$$1 - \sin^2 t = \cos^2 t, \qquad 1 + \tan^2 t = \sec^2 t, \quad \text{and} \quad \sec^2 t - 1 = \tan^2 t$$

we obtain substitutions which enable us to integrate expressions involving

$$(a^2 - x^2)^{1/2}, \qquad (a^2 + x^2)^{1/2}, \quad \text{and} \quad (x^2 - a^2)^{1/2}.$$

If $(a^2 - x^2)^{1/2}$ occurs, we substitute $x = a \sin t$; if $(a^2 + x^2)^{1/2}$ occurs, we substitute $x = a \tan t$; if $(x^2 - a^2)^{1/2}$ occurs, we substitute $x = a \sec t$. After the integration has been carried out with respect to the variable t, the answer can be written in terms of the original variable x by referring to the appropriate right triangle in Figures 6.1, 6.2, or 6.3.

Sometimes expressions involving integrals powers $a^2 - x^2$ or $a^2 + x^2$ or $x^2 - a^2$ may be most conveniently integrated by these trigonometric substitutions; relations (6.34) and (6.35) illustrate this.

We conclude with some examples.

1. To find $\int (a^2 - x^2)^{1/2} \, dx$, where $|x| \le a$ and $a > 0$, we put $x = a \sin t$. Then

$$\int (a^2 - x^2)^{1/2} \, dx = a^2 \int (\cos^2 t) \, dt = \frac{1}{2} a^2 \int (1 + \cos 2t) \, dt$$

$$= \tfrac{1}{2} a^2 (t + \tfrac{1}{2} \sin 2t) + C = \tfrac{1}{2} a^2 [t + (\sin t)(\cos t)] + C$$

$$= \frac{1}{2} a^2 \left(\sin^{-1} \left(\frac{x}{a} \right) + \frac{x}{a} \frac{(a^2 - x^2)^{1/2}}{a} \right) + C.$$

We have used Figure 6.1 in the last step for guidance.

2. To find $\int (a^2 + x^2)^{-1/2} \, dx$, where $a > 0$, we put $x = a \tan t$. Then

$$\int (a^2 + x^2)^{-1/2} \, dx = \int (\sec t) \, dt = \ln |\sec t + \tan t| + C$$

$$= \ln \left| \frac{(a^2 + x^2)^{1/2}}{a} + \frac{x}{a} \right| + C = \ln |(a^2 + x^2)^{1/2} + x| + C_1.$$

Figure 6.1. Substitution: $x = a \sin t$.

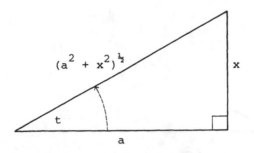

Figure 6.2. Substitution: $x = a \tan t$.

We have used Figure 6.2 when changing from the variable t back to the initial variable x. At the beginning of this section (see formula 24) the integral under consideration was evaluated in terms of the inverse of the hyperbolic sine function.

3. To find $\int (x^2 - a^2)^{-1/2}\, dx$, where $x > a > 0$, we put $x = a \sec t$. Then

$$\int (x^2 - a^2)^{-1/2}\, dx = \int (\sec t)\, dt = \ln|\sec t + \tan t| + C$$

$$= \ln\left| \frac{x}{a} + \frac{(x^2 - a^2)^{1/2}}{a} \right| + C = \ln|x + (x^2 - a^2)^{1/2}| + C_1.$$

We have used Figure 6.3 when changing from the variable t back to the initial variable x. At the beginning of this section (see formula 25) the integral under consideration was evaluated in terms of the inverse of the hyperbolic cosine function.

3. Integration of Rational Functions

Definition. A function f is said to be *rational* if it can be represented in the form

$$f(x) = \frac{P(x)}{Q(x)}, \tag{6.41}$$

where P and Q are polynomials

$$P(x) = a_m x^m + a_{m-1} x^{m-1} + \cdots + a_1 x + a_0$$

and

$$Q(x) = b_n x^n + b_{n-1} x^{n-1} + \cdots + b_1 x + b_0$$

with real coefficients $a_0, a_1, \ldots, a_m$ and $b_0, b_1, \ldots, b_n$. Also, we assume that P and Q have no common roots, in other words, that the *rational fraction* P/Q

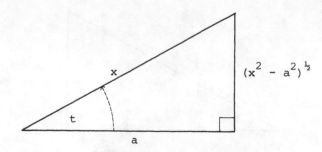

Figure 6.3. Substitution: $x = a \sec t$.

is in "canonical" form. If the degree of the numerator P is less than the degree of the denominator Q, the fraction P/Q is said to be *proper*; otherwise it is said to be *improper*.

REMARKS. If the rational fraction (6.41) is improper, then by division of polynomials we can represent P/Q in the form

$$\frac{P(x)}{Q(x)} = S(x) + \frac{R(x)}{Q(x)},$$

where S (quotient) and R (remainder) are also polynomials, but the degree of R is less than the degree of the divisor Q, that is, the rational fraction R/Q is proper. Hence, an improper rational fraction can be represented as a sum of a polynomial and a proper fraction. Since we can integrate polynomials, integration of improper rational fractions reduces to integration of proper fractions. We can therefore only consider the case when f is a proper rational fraction.

All general methods for the integration of rational functions are based on their representation in a special form convenient for integration. In these methods an important part is played by the roots of the denominator Q of the fraction. If α is a real or complex root of the polynomial Q, then Q can be divided without remainder by the binomial $x - \alpha$, that is,

$$Q(x) = (x - \alpha)Q^*(x),$$

where Q^* is also a polynomial; if $Q^*(\alpha) = 0$, we have

$$Q(x) = (x - \alpha)^2 Q^{**}(x),$$

and so forth. If

$$Q(x) = (x - \alpha)^k Q_1(x), \qquad (6.42)$$

where $k \geq 1$ and $Q_1(\alpha) \neq 0$ (i.e., α is no longer a root of the polynomial Q_1) we say that the polynomial Q has root α *of multiplicity k*.

Lemma 1. *If a real number α is a root of multiplicity $k > 0$ of the polynomial Q, we have identically*

$$\frac{P(x)}{Q(x)} = \frac{A_k}{(x-\alpha)^k} + \frac{P_1(x)}{(x-\alpha)^{k-1}Q_1(x)}, \qquad (6.43)$$

where A_k is a constant and P_1 is a polynomial.

PROOF. We note first that the polynomial Q_1 is defined in this case by equation (6.42) [so that $Q_1(\alpha) \neq 0$], the number A_k is real, all polynomials have real coefficients, and the fraction P/Q in (6.43) can be proper or improper.

Turning to the proof, the identity (6.43) is equivalent to the identity

$$P(x) - A_k Q_1(x) = (x-\alpha)P_1(x), \qquad (6.44)$$

obtained on multiplying by $Q(x)$; the identity (6.44) implies that the polynomial $P - A_k Q_1$ can be divided by the binomial $x - \alpha$. But for this to hold it is necessary and sufficient that

$$P(\alpha) - A_k Q_1(\alpha) = 0. \qquad (6.45)$$

If we therefore assume that

$$A_k = \frac{P(\alpha)}{Q_1(\alpha)}$$

[recalling that $Q_1(\alpha) \neq 0$], then equation (6.45) will be satisfied and the polynomial $P - A_k Q_1$ will be divisible by the binomial $x - \alpha$, that is, we shall have the identity (6.44) and hence also the identity (6.43). $\qquad \square$

REMARKS. If $k \geq 2$, then the rational fraction

$$\frac{P_1(x)}{(x-\alpha)^{k-1}Q_1(x)}$$

has the same form as the initial fraction P/Q; applying Lemma 1 to this fraction we get

$$\frac{P(x)}{Q(x)} = \frac{A_k}{(x-\alpha)^k} + \frac{A_{k-1}}{(x-\alpha)^{k-1}} + \frac{P_2(x)}{(x-\alpha)^{k-2}Q_1(x)}.$$

If $k \geq 3$, this process can be continued as long as the denominator or the last fraction on the right-hand side still contains the binomial $x - \alpha$ of an arbitrary positive power. Hence, we finally obtain

$$\frac{P(x)}{Q(x)} = \frac{A_k}{(x-\alpha)^k} + \frac{A_{k-1}}{(x-\alpha)^{k-1}} + \cdots + \frac{A_1}{x-\alpha} + \frac{P^*(x)}{Q_1(x)}, \qquad (6.46)$$

where $A_1, \ldots, A_k$ are real numbers and P^* is a polynomial with real coefficients.

In all these considerations we have assumed that the number α is real. If the complex number $\alpha = \beta + i\gamma$, $\gamma \neq 0$, is a root of multiplicity k of the polynomial Q (with real coefficients), then the *conjugate* complex number $\bar{\alpha} = \beta - i\gamma$ will also be a root of this polynomial of the same multiplicity k. In this case the polynomial Q is divisible by $(x - \alpha)^k$ and by $(x - \bar{\alpha})^k$ and hence also by their product; and since $(x - \alpha)(x - \bar{\alpha}) = (x - \beta)^2 + \gamma^2$, we therefore obtain

$$Q(x) = [(x - \beta)^2 + \gamma^2]^k Q_1(x), \tag{6.47}$$

where $Q_1(\alpha) \neq 0$ and $Q_1(\bar{\alpha}) \neq 0$ [the numbers β and γ and the coefficients of the polynomial Q_1 are evidently real].

Lemma 2. *If the complex number $\alpha = \beta + i\gamma$, $\gamma \neq 0$, is a root of multiplicity k of the polynomial Q, then identically*

$$\frac{P(x)}{Q(x)} = \frac{B_k x + C_k}{[(x - \beta)^2 + \gamma^2]^k} + \frac{P_1(x)}{[(x - \beta)^2 + \gamma^2]^{k-1} Q_1(x)}, \tag{6.48}$$

where B_k and C_k are constants and P_1 is a polynomial.

PROOF. The polynomial Q_1 is here defined by equation (6.47), the numbers B_k, C_k, and the coefficients of the polynomial P_1 are real numbers, and the fractions P/Q on the left-hand side of (6.48) can be proper or improper.

Turning to the proof, let

$$(x - \alpha)(x - \bar{\alpha}) = (x - \beta)^2 + \gamma^2 = q(x).$$

The identity (6.48) is equivalent to the identity

$$P(x) - (B_k x + C_k)Q_1(x) = q(x)P_1(x), \tag{6.49}$$

which because of the as yet undetermined polynomial P_1 is, in its turn, equivalent to the condition that the polynomial on the left-hand side of (6.49) is divisible by q, that is, by $x - \alpha$ and $x - \bar{\alpha}$. But for this it is necessary and sufficient that

$$P(\alpha) - (B_k \alpha + C_k)Q_1(\alpha) = P(\bar{\alpha}) - (B_k \bar{\alpha} + C_k)Q_1(\bar{\alpha}) = 0,$$

or

$$B_k \alpha + C_k = \frac{P(\alpha)}{Q_1(\alpha)} \quad \text{and} \quad B_k \bar{\alpha} + C_k = \frac{P(\bar{\alpha})}{Q_1(\bar{\alpha})}.$$

Thus, we have a system of two equations of first degree with determinant

$$\alpha - \bar{\alpha} = 2i\gamma \neq 0$$

for the evaluation of the unknowns B_k and C_k, and we can therefore determine these two numbers uniquely. It can be seen that in this case the expressions obtained for B_k and C_k depend symmetrically on α and $\bar{\alpha}$ and they are therefore real. Having B_k and C_k thus determined, we can easily find P_1 by (6.49). □

REMARKS. If $k > 1$, then, as with a real root, the last fraction on the right-hand side of (6.48) has the same form as the initial fraction on the left-hand side. We can therefore apply the same Lemma 2. Continuing this process we find, as before, that if the polynomial Q has a complex root $\alpha = \beta + i\gamma$, $\gamma \neq 0$, of multiplicity k and if the polynomial Q_1 is defined by the identity (6.47), then the following identity holds:

$$\frac{P(x)}{Q(x)} = \frac{B_k x + C_k}{\{q(x)\}^k} + \frac{B_{k-1} x + C_{k-1}}{\{q(x)\}^{k-1}} + \cdots + \frac{B_1 x + C_1}{q(x)} + \frac{P^*(x)}{Q_1(x)}, \quad (6.50)$$

where $q(x) = (x - \beta)^2 + \gamma^2$, $B_1, B_2, \ldots, B_k, C_1, C_2, \ldots, C_k$ are real numbers and P^* is a polynomial with real coefficients.

ADDITIONAL REMARKS. Concerning the identities (6.46) and (6.50) we note that if P/Q is a proper fraction, then so is P^*/Q_1. Indeed, if we assume that the variable x increases indefinitely, all terms except possibly $P^*(x)/Q_1(x)$ tend to zero; it follows from the identity that $P^*(x)/Q_1(x)$ must also tend to zero and this is possible only if P^*/Q_1 is a proper rational fraction.

Definition. A rational function is called a *partial fraction* if it is either of the form

$$\frac{A}{(x - \alpha)^u}$$

or of the form

$$\frac{Bx + C}{[(x - \beta)^2 + \gamma^2]^v};$$

here A, B, C, α, β, and γ denote fixed real number and u and v stand for positive integers.

DISCUSSION. Our goal is to show that every proper rational function P/Q can be converted to a finite sum of partial fractions. Subsequently, we shall see that partial fractions are very convenient for integration.

Suppose that Q has the real roots $\alpha_1, \alpha_2, \ldots, \alpha_r$ and the complex roots $\beta_1 \pm i\gamma_1, \beta_2 \pm i\gamma_2, \ldots, \beta_s \pm i\gamma_s$. Moreover, we assume that all these roots are distinct and that the real root α_m has multiplicity k_m with $1 \leq m \leq r$ and that the multiplicity of the pair of complex roots $\beta_n \pm i\gamma_n$ is j_n with $1 \leq n \leq s$. Then

$$Q(x) = b \prod_{m=1}^{r} (x - \alpha_m)^{k_m} \prod_{n=1}^{s} [(x - \beta_n)^2 + \gamma_n^2]^{j_n},$$

where $b \neq 0$ is a constant.

Applying formula (6.46) r times (for all r real roots α_m) we obtain the identity

$$\frac{P(x)}{Q(x)} = \frac{A_{k_1}^{(1)}}{(x - \alpha_1)^{k_1}} + \frac{A_{k_1-1}^{(1)}}{(x - \alpha_1)^{k_1-1}} + \cdots + \frac{A_1^{(1)}}{x - \alpha_1}$$

$$+ \frac{A_{k_2}^{(2)}}{(x - \alpha_2)^{k_2}} + \frac{A_{k_2-1}^{(2)}}{(x - \alpha_2)^{k_2-1}} + \cdots + \frac{A_1^{(2)}}{x - \alpha_2}$$

$$+ \cdots$$

$$+ \frac{A_{k_r}^{(r)}}{(x - \alpha_r)^{k_r}} + \frac{A_{k_r-1}^{(r)}}{(x - \alpha_r)^{k_r-1}} + \cdots + \frac{A_1^{(r)}}{x - \alpha_r} + \frac{P^*(x)}{Q^*(x)}$$

$$= \sum_{m=1}^{r} \sum_{u=1}^{k_m} \frac{A_u^{(m)}}{(x - \alpha_m)^u} + \frac{P^*(x)}{Q^*(x)},$$

where $A_u^{(m)}$ is a constant real and P^*/Q^* is a proper fraction because P/Q is and

$$Q^*(x) = b \prod_{n=1}^{s} [(x - \beta_n)^2 + \gamma_n^2]^{j_n}.$$

Applying formula (6.50) s times, we obtain

$$\frac{P^*(x)}{Q^*(x)} = \frac{B_{j_1}^{(1)}x + C_{j_1}^{(1)}}{[(x - \beta_1)^2 + \gamma_1^2]^{j_1}} + \cdots + \frac{B_1^{(1)}x + C_1^{(1)}}{(x - \beta_1)^2 + \gamma^2}$$

$$+ \frac{B_{j_2}^{(2)}x + C_{j_2}^{(2)}}{[(x - \beta_2)^2 + \gamma_2^2]^{j_2}} + \cdots + \frac{B_1^{(2)}x + C_1^{(2)}}{(x - \beta_2)^2 + \gamma_2^2}$$

$$+ \cdots$$

$$+ \frac{B_{j_s}^{(s)}x + C_{j_s}^{(s)}}{[(x - \beta_s)^2 + \gamma_s^2]^{j_s}} + \cdots + \frac{B_1^{(s)}x + C_1^{(s)}}{(x - \beta_s)^2 + \gamma_s^2} + \frac{P^{**}(x)}{Q^{**}(x)}$$

$$= \sum_{n=1}^{s} \sum_{v=1}^{j_n} \frac{B_v^{(n)}x + C_v^{(n)}}{[(x - \beta_n)^2 + \gamma_n^2]^v} + \frac{P^{**}(x)}{Q^{**}(x)},$$

where $B_v^{(n)}$ and $C_v^{(n)}$ are constant real numbers and $P^{**} = 0$ (because all roots of Q^* have been used and Q^{**} has no other roots; also, since P^*/Q^* is a proper fraction, so is P^{**}/Q^{**}). Thus, we have for the initial proper fraction P/Q the following *partial fraction expansion:*

$$\frac{P(x)}{Q(x)} = \sum_{m=1}^{r} \sum_{u=1}^{k_m} \frac{A_u^{(m)}}{(x - \alpha_m)^u} + \sum_{n=1}^{s} \sum_{v=1}^{j_n} \frac{B_v^{(n)}x + C_v^{(n)}}{[(x - \beta_n)^2 + \gamma_n^2]^v}. \qquad (6.51)$$

The foregoing shows that, given a proper rational fraction P/Q, it is possible to expand P/Q into a finite sum of partial fractions; moreover, this expansion is unique because at all stages of the successive determination of the numbers $A_u^{(m)}$, $B_v^{(n)}$, and $C_v^{(n)}$ we have found that their determination is unique. We record these facts in the following proposition.

Proposition 6.3. *Any proper rational fraction can be represented uniquely as a finite sum of partial fractions.*

COMMENTS. The method just discussed for the successive determination of coefficients of the expansion (6.51) is usually not the simplest method to use. It is generally easier and more efficient to use the so-called *method of undetermined coefficients*. We write the expansion (6.51) with *undetermined* $A_u^{(m)}$, $B_v^{(n)}$, and $C_v^{(n)}$ and, in disposing of all fractions, multiply both sides of this relation by $Q(x)$. As a result we get the given polynomial P on the left-hand side and on the right-hand side another polynomial whose coefficients, after comparison with similar terms, evidently contain the unknown numbers $A_u^{(m)}$, $B_v^{(n)}$, and $C_v^{(n)}$ and, as can readily be seen, are *linearly* dependent on these numbers.

Since the resulting equation must be an identity, the coefficients of like powers of x on the right- and left-hand sides must be equal. Comparing them with each other in pairs, we obtain a system of linear equations for the unknowns $A_u^{(m)}$, $B_v^{(n)}$, and $C_v^{(n)}$, with whose help these numbers can be determined. We know in advance that this problem has a unique solution. It can readily be seen that the number of equations of the system is equal to the number of the unknowns. Indeed, let N be the degree of the polynomial Q. On multiplying both sides of the identity (6.51) by $Q(x)$ we clearly get a polynomial of degree $N - 1$ on the right-hand side; on the left-hand side we have the polynomial P whose degree is not larger than $N - 1$, since P/Q is a proper rational fraction. Since a polynomial of degree $N - 1$ has N coefficients, a comparison of coefficients on the right- and left-hand sides gives us a system of N equations. On the other hand, the number of the $A_u^{(m)}$ ($1 \le m \le r$, $1 \le u \le k_n$) is $\sum_{m=1}^{r} k_m$; similarly, the number of $B_v^{(n)}$ is $\sum_{n=1}^{s} j_n$ and the same applies to the number of $C_v^{(n)}$. Hence, the total number of unknowns is equal to

$$\sum_{m=1}^{r} k_m + 2 \sum_{n=1}^{s} j_n;$$

but the expansion

$$Q(x) = b \prod_{m=1}^{r} (x - \alpha_m)^{k_m} \prod_{n=1}^{s} [(x - \beta_n)^2 + \gamma_n^2]^{j_n}, \qquad b \ne 0 \text{ is a constant,}$$

shows that this number is exactly equal to the degree N of the polynomial Q. Hence, the number of unknowns is, in fact, equal to the number of linear equations obtained.

EXAMPLE. As an illustration of the method of undetermined coefficients consider the partial fraction decomposition of

$$\frac{2x^2 + 2x + 13}{(x - 2)(x^2 + 1)^2}.$$

We put

$$\frac{2x^2 + 2x + 13}{(x - 2)(x^2 + 1)^2} = \frac{A}{x - 2} + \frac{Bx + C}{x^2 + 1} + \frac{Dx + E}{(x^2 + 1)^2}.$$

Clearing fractions gives

$$2x^2 + 2x + 13 = A(x^2 + 1)^2 + (Bx + C)(x^2 + 1)(x - 2) + (Dx + E)(x - 2)$$
$$= (A + B)x^4 + (-2B + C)x^3 + (2A + B - 2C + D)x^2$$
$$+ (-2B + C - 2D + E)x + (A - 2C - 2E).$$

Comparing coefficients of like powers, we get the system of equations

$$A + B = 0,$$
$$- 2B + C = 0,$$
$$2A + B - 2C + D = 2,$$
$$- 2B + C - 2D + E = 2,$$
$$A - 2C - 2E = 13.$$

If we set $x = 2$ in

$$2x^2 + 2x + 13 = A(x^2 + 1)^2 + (Bx + C)(x^2 + 1)(x - 2) + (Dx + E)(x - 2),$$

we obtain $25 = 25A$ or $A = 1$. From the system of equations we see that $B = -1$, $C = -2$, $D = -3$, and $E = -4$. Therefore,

$$\frac{2x^2 + 2x + 13}{(x - 2)(x^2 + 1)^2} = \frac{1}{x - 2} - \frac{x + 2}{x^2 + 1} - \frac{3x + 4}{(x^2 + 1)^2}. \tag{6.52}$$

DISCUSSION. We now consider the integration of partial fractions, that is, proper rational fractions of the form

$$\frac{A}{(x - \alpha)^u} \quad \text{and} \quad \frac{Bx + C}{[(x - \beta)^2 + \gamma^2]^v},$$

where A, B, C, α, β, and γ are constant real numbers and u and v are positive integers. We have

$$\int \frac{A}{x - \alpha} dx = A \ln |x - \alpha| + K \tag{6.53}$$

and

$$\int \frac{A}{(x - \alpha)^u} dx = \frac{-A}{(u - 1)(x - \alpha)^{u-1}} + K, \quad \text{for } u = 1, 2, 3, \ldots . \tag{6.54}$$

Using the substitution $x = \beta + \gamma y$ gives

$$\int \frac{Bx + C}{(x - \beta)^2 + \gamma^2} dx = \int \frac{B(\beta + \gamma y) + C}{\gamma^2(1 + y^2)} \gamma \, dy$$

$$= \frac{B}{2} \int \frac{2y}{1 + y^2} dy + \frac{B\beta + C}{\gamma} \int \frac{1}{1 + y^2} dy$$

$$= \frac{B}{2} \ln(1 + y^2) + \frac{B\beta + C}{\gamma} \tan^{-1} y + K \qquad (6.55)$$

$$= \frac{B}{2} \ln\left(1 + \left(\frac{x - \beta}{\gamma}\right)^2\right) + \frac{B\beta + C}{\gamma} \tan^{-1}\left(\frac{x - \beta}{\gamma}\right) + K.$$

Similarly, for $v = 1, 2, 3, \ldots$, we get

$$\int \frac{Bx + C}{[(x - \beta)^2 + \gamma^2]^v} dx = \int \frac{B(\beta + \gamma y) + C}{\gamma^{2v}(1 + y^2)^v} \gamma \, dy$$

$$(6.56)$$

$$= \frac{B}{2\gamma^{2v-2}} \int \frac{2y}{(1 + y^2)^v} dy + \frac{B\beta + C}{\gamma^{2v-1}} \int \frac{1}{(1 + y^2)^v} dy.$$

But

$$\int \frac{2y}{(1 + y^2)^v} dy = -\frac{1}{(v - 1)(1 + y^2)^{v-1}} + K \qquad (6.57)$$

and, setting

$$I_v = \int \frac{1}{(1 + y^2)^v} dy, \qquad (6.58)$$

we have, by (6.34),

$$I_{v+1} = \frac{2v - 1}{2v} I_v + \frac{y}{2v(1 + y^2)^v}. \qquad (6.59)$$

Hence, by knowing I_1 we can find in succession I_2, I_3, and so forth. By putting $y = (x - \beta)/\gamma$, we can change back to the initial variable x.

Summing up, we therefore see that the functions obtained by integration of partial fractions (and hence also by integration of all rational functions) can either be logarithms and arc tangents or rational functions.

EXAMPLE. By (6.52), we have

$$\int \frac{2x^2 + 2x + 13}{(x - 2)(x^2 + 1)^2} dx = \int \frac{1}{x - 2} dx - \int \frac{x + 2}{x^2 + 1} dx - \int \frac{3x + 4}{(x^2 + 1)^2} dx$$

and so

$$\int \frac{2x + 2x + 13}{(x - 2)(x^2 + 1)^2} dx = \frac{1}{2} \frac{3 - 4x}{x^2 + 1} + \frac{1}{2} \ln \frac{(x - 2)^2}{x^2 + 1} + 4 \tan^{-1} x + K.$$

FURTHER DISCUSSION. We have seen so far that the integration of rational functions for which the roots of the denominator are known does not cause much difficulty, although it sometimes can be connected with rather lengthy calculations. There is a general method, due to M. V. Ostrogradski and C. Hermite, which can often simplify and shorten these calculations. We shall consider this method next and refer to it as the *Hermite–Ostrogradski Method*.

Let us suppose again that P/Q is a proper rational fraction and

$$Q(x) = b \prod_{m=1}^{r} (x - \alpha_m)^{k_m} \prod_{n=1}^{s} [(x - \beta_n)^2 + \gamma_n^2]^{j_n}, \qquad b \neq 0. \qquad (6.60)$$

By Proposition 6.3, we can expand P/Q uniquely into a finite sum of partial fractions, that is, functions of the form

$$\frac{A}{(x - \alpha)^u} \quad \text{and} \quad \frac{Bx + C}{[(x - \beta)^2 + \gamma^2]^v},$$

where A, B, C, α, β, and γ are constant real numbers and u and v are positive integers. Integration of

$$\frac{A}{(x - \alpha)^u}$$

leads to the logarithmic function for $u = 1$ and to rational functions for $u > 1$ [see (6.53) and (6.54)]. Letting

$$I_v = \int \frac{1}{(1 + y^2)^v} \, dy,$$

relation (6.59) shows that we can represent I_v in the form

$$I_v = \lambda_v I_1 + \frac{L(y)}{(1 + y^2)^{v-1}}, \qquad (6.61)$$

where λ_v is a constant, L is a polynomial, and the fraction

$$\frac{L(y)}{(1 + y^2)^{v-1}}$$

is a proper rational fraction. Substituting (6.61) into (6.56) and returning on the right-hand side from the variable y to the initial variable x, we get

$$\int \frac{Bx + C}{[(x - \beta)^2 + \gamma^2]^v} \, dx = \frac{R(x)}{[(x - \beta)^2 + \gamma^2]^{v-1}} + \sigma_v \int \frac{1}{(x - \beta)^2 + \gamma^2} \, dx, \qquad (6.62)$$

where R is a polynomial, σ_v is a constant, and the fraction

$$\frac{R(x)}{[(x - \beta)^2 + \gamma^2]^{v-1}}$$

is a proper rational fraction. This is the situation for $v > 1$; for $v = 1$ we have

(6.55) in which there are no rational terms on the right-hand side, but only the logarithmic function and the arc tangent function.

We therefore see that if a proper rational fraction P/Q is expanded into partial fractions, those terms of the expansion in which $u > 1$ or $v > 1$ give upon integration proper rational fractions with corresponding denominators

$$(x - \alpha_m)^{u-1} \quad \text{and} \quad [(x - \beta_n)^2 + \gamma_n^2]^{v-1}.$$

On adding all these proper rational fractions we obtain another proper rational fraction,

$$\frac{P_1(x)}{Q_1(x)},$$

whose denominator evidently equals

$$Q_1(x) = \prod_{m=1}^{r} (x - \alpha_m)^{k_m-1} \prod_{n=1}^{s} [(x - \beta_n)^2 + \gamma_n^2]^{j_n-1}. \tag{6.63}$$

This is the *rational part of the integral* of the given fraction P/Q. The second *transcendental part of the integral of P/Q* will evidently consist of: (i) integrals of those terms of the expansion (6.51) in which $u = 1$ and $v = 1$, and (ii) integrals of those terms of the expansion (6.51) in which $v > 1$, as can be seen by (6.62). In both these cases the integrand belongs to one of the following types:

$$\frac{A}{x - \alpha}, \qquad \frac{Bx + C}{(x - \beta)^2 + \gamma^2};$$

the sum of these integrands will therefore be a proper rational fraction

$$\frac{P_2(x)}{Q_2(x)},$$

where

$$Q_2(x) = \prod_{m=1}^{r} (x - \alpha_m) \prod_{n=1}^{s} [(x - \beta_n)^2 + \gamma_n^2]. \tag{6.64}$$

We thus obtain the *Hermite–Ostrogradski Formula*

$$\int \frac{P(x)}{Q(x)} dx = \frac{P_1(x)}{Q_1(x)} + \int \frac{P_2(x)}{Q_2(x)} dx, \tag{6.65}$$

where the first and second terms on the right-hand side represent the rational and transcendental parts of the integral of P/Q, respectively. The polynomials Q_1 and Q_2 are determined from the formulas (6.63) and (6.64), respectively, and the fractions P_1/Q_1 and P_2/Q_2 are proper rational fractions.

The interesting feature of (6.65) is due to the fact that it can be obtained *without knowing the roots of the polynomial Q*. In fact, we can easily see that a root of multiplicity $k > 1$ of the polynomial Q is a root of multiplicity $k - 1$

of the polynomial Q' (the derivative of Q); if we therefore assume that

$$Q(x) = b \prod_{m=1}^{r} (x - \alpha_m)^{k_m} \prod_{n=1}^{s} [(x - \beta_m)^2 + \gamma_n^2]^{j_n},$$

then

$$Q'(x) = \prod_{m=1}^{r} (x - \alpha_m)^{k_m-1} \prod_{n=1}^{s} [(x - \beta_n)^2 + \gamma_n^2]^{j_n-1} \cdot R(x) = Q_1(x)R(x),$$

where the polynomials Q and R have no common root. This shows that the polynomial Q_1 is the *greatest common divisor* of the polynomial Q and its derivative Q'. By (6.60), (6.63), and (6.64), we have

$$Q(x) = bQ_1(x)Q_2(x); \tag{6.66}$$

therefore, knowing Q and Q_1, we can find the polynomial Q_2 by (6.66). Finally, to obtain P_1 and P_2, we can differentiate the expression in (6.65) and obtain

$$\frac{P(x)}{Q(x)} = \frac{Q_1(x)P_1'(x) - P_1(x)Q_1'(x)}{Q_1^2(x)} + \frac{P_2(x)}{Q_2(x)}. \tag{6.67}$$

By (6.63) each root τ of the polynomial Q_1 is a root of the polynomial Q and, by (6.64), is also a root of the polynomial Q_2. If Q_1 contains the binomial $x - \tau$ of power $k > 1$, then Q_1' contains it also, but its power is $k - 1$, and it appears in Q_2 in the first degree; therefore, the product $Q_1' Q_2$ contains $x - \tau$ of the same power k as the polynomial Q_1. And since the same also applies to any root τ of the polynomial Q_1, $Q_1' Q_2$ is divisible by Q_1, that is,

$$Q_1'(x)Q_2(x) = Q_1(x)S(x),$$

where S is a polynomial. We therefore obtain

$$\frac{Q_1(x)P_1'(x) - P_1(x)Q_1'(x)}{Q_1^2(x)} = \frac{Q_2(x)Q_1(x)P_1'(x) - Q_2(x)P_1(x)Q_1'(x)}{Q_2(x)Q_1^2(x)}$$

$$\cdot = \frac{Q_1(x)[Q_2(x)P_1'(x) - P_1(x)S(x)]}{Q_2(x)Q_1^2(x)}$$

$$= \frac{Q_2(x)P_1'(x) - P_1(x)S(x)}{Q_1(x)Q_2(x)};$$

multiplying by $Q(x) = bQ_1(x)Q_2(x)$, we get from (6.67)

$$P(x) = b[Q_2(x)P_1'(x) - P_1(x)S(x)] + bP_2(x)Q_1(x). \tag{6.68}$$

In this expansion the polynomials P, Q_1, Q_2, and S are known to us. The highest possible degrees of the polynomials P_1 and P_2 which we are trying to find is determined by the fact that the fractions P_1/Q_1 and P_2/Q_2 are proper rational fractions. Hence, the polynomials P_1 and P_2 can readily be obtained from relation (6.68) by the method of undetermined coefficients. It can easily be seen that in this case the number of unknowns coincides with the number of equations obtained and that the solution of this system is assured by the expansion (6.67).

To illustrate the theory, we look at some examples next.

EXAMPLE 1. Consider the integral

$$\int \frac{4x^5 - 1}{(x^5 + x + 1)^2}\, dx.$$

Here $Q(x) = (x^5 + x + 1)^2$ and $Q'(x) = 2(x^5 + x + 1)(5x^4 + 1)$. Since Q_1 is the largest common divisor of Q and Q', we have $Q_1(x) = x^5 + x + 1$. Since $b = 1$, $Q_2 = Q/Q_1$ and so $Q_2(x) = x^5 + x + 1$. Thus, by the Hermite–Ostrogradski Formula [see (6.65)],

$$\int \frac{4x^5 - 1}{(x^5 + x + 1)^2}\, dx = \frac{P_1(x)}{x^5 + x + 1} + \int \frac{P_2(x)}{x^5 + x + 1}\, dx$$

or

$$\frac{4x^5 - 1}{(x^5 + x + 1)^2} = \frac{(x^5 + x + 1)P_1'(x) - (5x^4 + 1)P_1(x)}{(x^5 + x + 1)^2} + \frac{P_2(x)}{x^5 + x + 1}.$$

Putting

$$P_1(x) = A_0 x^4 + A_1 x^3 + A_2 x^2 + A_3 x + A_4,$$
$$P_2(x) = B_0 x^4 + B_1 x^3 + B_2 x^2 + B_3 x + B_4,$$

we obtain, after clearing fractions,

$$4x^5 - 1 = (x^5 + x + 1)(4A_0 x^3 + 3A_1 x^2 + 2A_2 x + A_3)$$
$$- (5x^4 + 1)(A_0 x^4 + A_1 x^3 + A_2 x^2 + A_3 x + A_4)$$
$$+ (x^5 + x + 1)(B_0 x^4 + B_1 x^3 + B_2 x^2 + B_3 x + B_4).$$

Comparing coefficients of like powers of x, we find

$$A_0 = A_1 = A_2 = A_4 = 0, \qquad A_3 = -1,$$
$$B_0 = B_1 = B_2 = B_3 = B_4 = 0.$$

We therefore conclude that

$$\int \frac{4x^5 - 1}{(x^5 + x + 1)}\, dx = -\frac{x}{x^5 + x + 1} + K.$$

REMARK. In an entirely similar manner we can show that

$$\int \frac{4x^9 + 21x^6 + 2x^3 - 3x^2 - 3}{(x^7 - x + 1)^2}\, dx = -\frac{x^3 + 3}{x^7 - x + 1} + K.$$

EXAMPLE 2. Consider the integral

$$\int \frac{4x^4 + 4x^3 + 16x^2 + 12x + 8}{(x + 1)^2(x^2 + 1)^2} \, dx.$$

Here

$$Q_1(x) = Q_2(x) = (x + 1)(x^2 + 1) = x^3 + x^2 + x + 1.$$

Putting

$$P_1(x) = A_0 x^2 + A_1 x + A_2 \quad \text{and} \quad P_2(x) = B_0 x^2 + B_1 x + B_2,$$

the Hermite–Ostrogradski Formula gives

$$\frac{4x^4 + 4x^3 + 16x^2 + 12x + 8}{(x^3 + x^2 + x + 1)^2} = \left(\frac{A_0 x^2 + A_1 x + A_2}{x^3 + x^2 + x + 1}\right)' + \frac{B_0 x^2 + B_1 x + B_2}{x^3 + x^2 + x + 1}.$$

From this, in turn, we obtain

$$A_0 = -1, \quad A_1 = 1, \quad A_2 = -4, \quad B_0 = 0, \quad B_1 = 3, \quad B_2 = 3.$$

Therefore,

$$\int \frac{4x^4 + 4x^3 + 16x^2 + 12x + 8}{(x + 1)^2(x^2 + 1)^2} \, dx = -\frac{x^2 - x + 4}{x^3 + x^2 + x + 1} + 3 \tan^{-1} x + K.$$

EXAMPLE 3. Consider the integral

$$\int \frac{2x^4 - 4x^3 + 24x^2 - 40x + 20}{(x - 1)(x^2 - 2x + 2)^3} \, dx.$$

Here

$$Q_1(x) = (x^2 - 2x + 2)^2 \quad \text{and} \quad Q_2(x) = (x - 1)(x^2 - 2x + 2).$$

Thus,

$$\frac{2x^4 - 4x^3 + 24x^2 - 40x + 20}{(x - 1)(x^2 - 2x + 2)^2}$$

$$= \left(\frac{Ax^3 + Bx^2 + Cx + D}{(x^2 - 2x + 2)^2}\right)' + \frac{E}{x - 1} + \frac{Fx + G}{x^2 - 2x + 2}$$

and so

$$A = 2, \quad B = -6, \quad C = 8, \quad D = -9, \quad E = 2, \quad F = -2, \quad G = 4,$$

showing that

$$\int \frac{2x^4 - 4x^3 + 24x^2 + 12x + 8}{(x - 1)(x^2 - 2x + 2)^2} \, dx$$

$$= \frac{2x^3 - 6x^2 + 8x - 9}{(x^2 - 2x + 2)^2} + \ln \frac{(x - 1)^2}{x^2 - 2x + 2} + 2 \tan^{-1}(x - 1) + K.$$

EXAMPLE 4. Consider the integral

$$\int \frac{x^6 - x^5 + x^4 + 2x^3 + 3x^2 + 3x + 3}{(x + 1)^2(x^2 + x + 1)^3} \, dx.$$

Here

$$Q_1(x) = (x + 1)(x^2 + x + 1)^2 \quad \text{and} \quad Q_2(x) = (x + 1)(x^2 + x + 1).$$

Thus,

$$\frac{x^6 - x^5 + x^4 + 2x^3 + 3x^2 + 3x + 3}{(x + 1)^2(x^2 + x + 1)^3}$$

$$= \left(\frac{Ax^4 + Bx^3 + Cx^2 + Dx + E}{(x + 1)(x^2 + x + 1)}\right)' + \frac{Fx^2 + Gx + H}{(x + 1)(x^2 + x + 1)}$$

and so

$$A = -1, \quad B = 0, \quad C = -2, \quad D = 0, \quad E = -1, \quad F = G = H = 0,$$

showing that

$$\int \frac{x^6 - x^5 + x^4 + 2x^3 + 3x^2 + 3x + 3}{(x + 1)^2(x^2 + x + 1)^3} \, dx = -\frac{x^4 + 2x^2 + 1}{(x + 1)(x^2 + x + 1)^2} + K.$$

4. Integration by Rationalization

If a given integral can be converted by a suitable substitution into an integral of a rational function, we say that we can *integrate by rationalization*. In the foregoing section the integration of rational functions was solved in principle; in this section we consider certain integrals that can be evaluated by the method of integration by rationalization.

Definition. We call P a *polynomial in the variables* $u, v, \ldots, z$ if $P(u, v, \ldots, z)$ is a (finite, real) linear combination of expressions of the form $u^k v^j \cdots z^m$, where $k, j, \ldots, m$ are nonnegative integers. We say that R is a *rational function in the variables* $u, v, \ldots, z$ if

$$R(u, v, \ldots, z) = \frac{P(u, v, \ldots, z)}{Q(u, v, \ldots, z)},$$

where P and Q are polynomials in the variables $u, v, \ldots, z$.

Proposition 6.4. *Consider the integral*

$$\int R(\sin x, \cos x) \, dx, \tag{6.69}$$

where R is a rational function in the variables u and v with

$$u = \sin x \quad and \quad v = \cos x. \tag{6.70}$$

The integral (6.69) can be rationalized by the substitution

$$t = \tan\frac{x}{2}, \tag{6.71}$$

where $-\pi < x < \pi$.

PROOF. Since

$$\sin x = 2\sin\frac{x}{2}\cos\frac{x}{2} = 2\frac{\sin(x/2)}{\cos(x/2)}\cos^2\frac{x}{2} = \frac{2\tan(x/2)}{1 + \tan^2(x/2)},$$

$$\cos x = \cos^2\frac{x}{2} - \sin^2\frac{x}{2} = \cos^2\frac{x}{2}\left(1 - \tan^2\frac{x}{2}\right) = \frac{1 - \tan^2(x/2)}{1 + \tan^2(x/2)},$$

and

$$x = 2\tan^{-1}x,$$

we obtain

$$\sin x = \frac{2t}{1 + t^2}, \qquad \cos x = \frac{1 - t^2}{1 + t^2}, \qquad dx = \frac{2\,dt}{1 + t^2}$$

and so

$$\int R(\sin x, \cos x)\,dx = 2\int R\left(\frac{2t}{1 + t^2}, \frac{1 - t^2}{1 + t^2}\right)\frac{1}{1 + t^2}\,dt,$$

accomplishing the rationalization.   □

DISCUSSION. The substitution (6.71) at times leads to complicated expressions and it is therefore of interest to discover other substitutions resulting in less work, provided that such substitutions exist, when the rational function R satisfies some additional properties.

The situation at hand calls for some remarks concerning *even* and *odd* polynomials and rational functions first. We call a function R *even* or *odd* if

$$R(-x) = R(x) \quad \text{resp.} \quad R(-x) = -R(x)$$

holds. An *even polynomial* contains only even powers of x, hence is a polynomial with respect to x^2. An *odd polynomial* contains only odd powers of x, hence is divisible by x and thus equals the product of x and a polynomial in x^2.

Now, suppose a rational function R is *even* and let $R = P_1/P_2$, where P_1/P_2 are in reduced form, that is, P_1 and P_2 are assumed to have no common roots. Then

$$\frac{P_1(-x)}{P_2(-x)} = \frac{P_1(x)}{P_2(x)}$$

implies

$$P_2(x) = aP_2(-x), \qquad P_1(x) = aP_1(x)$$

for some constant a. Comparing in both these equations the highest powers of x, we see that a can only have the values ± 1. If it were true that $a = -1$, then P_1 and P_2 would both be odd, hence divisible by x; but we had assumed that P_1/P_2 is in reduced form. Therefore $a = 1$ and P_1 and P_2 are *even*, that is, P_1 and P_2 are polynomials in x^2.

We therefore see that an *even rational function R is a rational function in x^2* and in reduced form is representable as a quotient of even polynomials:

$$R(x) = \frac{Q_1(x^2)}{Q_2(x^2)}.$$

If, on the other hand, R is *odd*, then $R(x)/x$ is even, so that we have

$$R(x) = xR_1(x^2)$$

for some rational function R_1 in x^2.

We therefore make the following claims:

Rule I: If $R(u, v)$ is an odd function with respect to u, then the integral (6.69) can be rationalized by the substitution

$$t = \cos x.$$

Indeed, by the assumptions made and by what has been said further above,

$$R(u, v) = uR_1(u^2, v),$$

where R_1 is a rational function in the two variables u^2 and v. Therefore, the integral (6.69) assumes the form

$$\int R_1(\sin^2 x, \cos x)(\sin x)\, dx$$

and becomes via the substitution $t = \cos x$

$$-\int R_1(1 - t^2, t)\, dt.$$

Rule II: If $R(u, v)$ is an odd function with respect to v, then the integral (6.69) can be rationalized by the substitution

$$t = \sin x.$$

Indeed, in this case

$$U(u, v) = vR_1(u, v^2)$$

where R_1 is a rational function in the variables u and v^2; the integral (6.69) assumes the form

$$\int R_1(t, 1 - t^2)\, dt$$

via the substitution

$$t = \sin x.$$

Rule III: If the rational function R in the variables u and v satisfies

$$R(-u, -v) = R(u, v),$$

then the integral (6.69) can be rationalized by the substitution

$$t = \tan x.$$

Indeed, putting $u = vt$, it follows from our assumptions that

$$R(-vt, -v) = R(vt, v);$$

thus, the function $R(vt, v)$ is *even with respect to* v and therefore satisfies the relation

$$R(vt, v) = R_1(v^2, t),$$

where R_1 is a rational function in the variables v^2 and t. Putting $v = \cos x$, $t = \tan x$, we therefore obtain

$$R(\sin x, \cos x) = R_1(\cos^2 x, \tan x) = R_1\left(\frac{1}{1 + \tan^2 x}, \tan x\right)$$

and thus, with $t = \tan x$,

$$\int R(\sin x, \cos x)\, dx = \int R_1\left(\frac{1}{1 + t^2}, t\right)\frac{1}{1 + t^2}\, dt.$$

In certain cases the integrand in (6.69) can easily be decomposed into a sum in such a way that one of the foregoing Rules I, II, or III is applicable to the individual summands.

Actually, even in the most general case we can use the following decomposition of R:

$$R(u, v) = \tfrac{1}{2}[R(u, v) + R(-u, -v)] + \tfrac{1}{2}[R(u, v) - R(-u, v)]$$
$$+ \tfrac{1}{2}[R(-u, v) - R(-u, -v)].$$

To the three partial sums on the right-hand side we can apply, respectively, Rules III, I, and II. However, if all three Rules have to be brought into play, the advantage of the substitutions in these Rules over the substitution (6.71) is often insignificant in practice.

We take up some examples.

EXAMPLE 1. Let

$$I = \int \frac{1}{(\sin x)(2 + \cos x - 2\sin x)}\, dx.$$

Using the substitution $t = \tan(x/2)$, we get

$$I = \int \frac{1 + t^2}{t(t^2 - 4t + 3)} \, dt.$$

Expanding into partial fractions, we get

$$\frac{1 + t^2}{t(t - 3)(t - 1)} = \frac{A}{t} + \frac{B}{t - 3} + \frac{C}{t - 1},$$

where $A = \frac{1}{3}$, $B = \frac{5}{3}$, and $C = -1$. Hence,

$$I = \frac{1}{3} \ln |t| + \frac{5}{3} \ln |t - 3| - \ln |t - 1| + K$$

$$= \frac{1}{3} \ln \left| \tan \frac{x}{2} \right| + \frac{5}{3} \ln \left| \left(\tan \frac{x}{2} \right) - 3 \right| - \ln \left| \left(\tan \frac{x}{2} \right) - 1 \right| + K.$$

EXAMPLE 2. Let

$$J = \int \frac{(\sin^2 x)(\cos x)}{\sin x + \cos x} \, dx.$$

Here we can use Rule III and substitute $t = \tan x$. We obtain

$$J = \int \frac{t^2}{(t + 1)(t^2 + 1)^2} \, dt.$$

Expanding into partial fractions, we obtain

$$\frac{t^2}{(t + 1)(t^2 + 1)^2} = \frac{A}{t + 1} + \frac{Bt + C}{t^2 + 1} + \frac{Dt + E}{(t^2 + 1)^2},$$

where $A = \frac{1}{4}$, $B = -\frac{1}{4}$, $C = \frac{1}{4}$, $D = \frac{1}{2}$, and $E = -\frac{1}{2}$. Hence,

$$J = \frac{1}{4} \ln \left| \frac{1 + t}{(1 + t^2)^{1/2}} \right| - \frac{1}{4} \frac{1 + t}{1 + t^2} + K$$

$$= \frac{1}{4} \ln |\sin x + \cos x| - \frac{1}{4} (\cos x)(\sin x + \cos x) + K.$$

EXAMPLE 3. Let

$$H = \int \frac{1}{(\sin x)(2 \cos^2 x - 1)} \, dx.$$

Here Rule I is applicable. Putting $t = \cos x$, we obtain

$$H = -\int \frac{1}{(1 - t^2)(2t^2 - 1)} \, dt.$$

But

$$\frac{1}{(1 - t^2)(2t^2 - 1)} = \frac{1}{1 + \sqrt{2}t} + \frac{1}{1 - \sqrt{2}t} + \frac{1/2}{1 + t} + \frac{1/2}{1 - t}$$

and so

$$H = \frac{1}{\sqrt{2}} \ln \left| \frac{1 + \sqrt{2}t}{1 - \sqrt{2}t} \right| - \frac{1}{2} \ln \left| \frac{1 + t}{1 - t} \right| + K$$

$$= \frac{1}{\sqrt{2}} \ln \left| \frac{1 + \sqrt{2}(\cos x)}{1 - \sqrt{2}(\cos x)} \right| + \frac{1}{2} \ln \left| \frac{1 - \cos x}{1 + \cos x} \right| + K$$

$$= \frac{1}{\sqrt{2}} \ln \left| \frac{1 + \sqrt{2}(\cos x)}{1 - \sqrt{2}(\cos x)} \right| + \frac{1}{2} \ln \left| \tan \frac{x}{2} \right| + K.$$

EXAMPLE 4. We wish to evaluate

$$\int \frac{A \cos x + B \sin x}{C \cos x + D \sin x} \, dx,$$

where A, B, C, and D are constants.

We proceed to determine two constants λ and μ such that

$$A \cos x + B \sin x \equiv \lambda(-C \sin x + D \cos x) + \mu(C \cos x + D \sin x),$$

with the foregoing equation holding identically for all x. Equating coefficients of $\cos x$ and $\sin x$ we see that the constants λ and μ are given by the equations

$$A = D\lambda + C\mu,$$

$$B = -C\lambda + D\mu.$$

Hence,

$$\lambda = \frac{AD - BC}{D^2 + C^2} \quad \text{and} \quad \mu = \frac{AC + BD}{D^2 + C^2}.$$

Moreover,

$$\int \frac{A \cos x + B \sin x}{C \cos x + D \sin x} \, dx = \lambda \int \frac{-C \sin x + D \cos x}{C \cos x + D \sin x} \, dx + \mu \int dx$$

$$= \lambda \ln(C \cos x + D \sin x) + \mu x + K.$$

REMARK. Proposition 6.4 has a dual for hyperbolic functions. The integral

$$\int R(\sinh x, \cosh x) \, dx,$$

where R is a rational function in the variables u and v with

$$u = \sinh x \quad \text{and} \quad v = \cosh x,$$

can be rationalized by the substitution

$$t = \tanh \frac{x}{2}.$$

Indeed, using the substitution $t = \tanh(x/2)$, we have

$$\sinh x = \frac{2t}{1 - t^2}, \quad \cosh x = \frac{1 + t^2}{1 - t^2}, \quad dx = \frac{2\,dt}{1 - t^2}$$

and so

$$\int R(\sinh x, \cosh x)\,dx = 2 \int R\left(\frac{2t}{1 - t^2}, \frac{1 + t^2}{1 - t^2}\right)\frac{1}{1 - t^2}\,dt.$$

Proposition 6.5. *Let $R(x, y, z, \dots)$ be a rational function of its variables, let*

$$L(x) = \frac{ax + b}{cx + d}$$

be a fractional linear function, where a, b, c, and d are given constants and $ad - bc \neq 0$, and let α, β, $\dots$ be rational numbers. Then the integral

$$\int R(x, [L(x)]^\alpha, [L(x)]^\beta, \dots)\,dx \tag{6.72}$$

can be rationalized by the substitution

$$L(x) = t^m, \tag{6.73}$$

where m is the least common denominator of the fractions α, β, $\dots$.

PROOF. We have that αm, βm, $\dots$ are integers. Moreover,

$$x = \frac{dt^m - b}{a - ct^m}.$$

Putting $x = r(t)$, we see that the integral (6.72) becomes

$$\int R(r(t), t^{\alpha m}, t^{\beta m}, \dots)r'(t)\,dt;$$

here the integrand is obviously a rational function of t. After the integral is determined, we express t in terms of x using the substitution (6.73). □

EXAMPLE 1. Consider the integral

$$I = \int \frac{1}{\sqrt[3]{1 + x} - \sqrt[4]{1 + x}}\,dx.$$

Here $1 + x = t^{12}$ is the substitution rationalizing the integral. Hence,

$$I = 12 \int \frac{t^8}{t - 1}\,dt.$$

But

$$\frac{t^8}{t - 1} = \frac{t^8 - 1}{t - 1} + \frac{1}{t - 1} = t^7 + t^6 + t^5 + t^4 + t^3 + t^2 + t + 1 + \frac{1}{t - 1}$$

and so

$$I = 12\left(\frac{t^8}{8} + \frac{t^7}{7} + \frac{t^6}{6} + \frac{t^5}{5} + \frac{t^4}{4} + \frac{t^3}{3} + \frac{t^2}{2} + t + \ln|t - 1|\right) + K.$$

Substituting here $t = \sqrt[12]{1 + x}$ we obtain the required expression for I in terms of the initial variable x.

EXAMPLE 2. Consider the integral

$$J = \int \left(\frac{x + 1}{x - 1}\right)^{1/3} \frac{1}{x + 1} dx.$$

Here $(x + 1)/(x - 1) = t^3$ is the substitution rationalizing the integral. Thus, $x = (t^3 + 1)/(t^3 - 1)$ and

$$J = -3 \int \frac{1}{t^3 - 1} dt.$$

We therefore have

$$J = \int \left(-\frac{1}{t - 1} + \frac{t + 2}{t^2 + t + 1}\right) dt$$

$$= \frac{1}{2}\ln\left|\frac{t^2 + t + 1}{(t - 1)^2}\right| + \sqrt{3}\tan^{-1}\frac{2t + 1}{\sqrt{3}} + K$$

with $t = \sqrt[3]{(x + 1)/(x - 1)}$.

Proposition 6.6. *Suppose that a and b are arbitrary real numbers and let m, n, and p be rational numbers. Then the* binomial integral

$$I = \int x^m(a + bx^n)^p \, dx \tag{6.74}$$

can be reduced to an integral of type (6.72), that is, can be rationalized, if one of the three numbers

$$p, \qquad \frac{m + 1}{n}, \qquad \frac{m + 1}{n} + p$$

is an integer.

PROOF. We assume that $n \neq 0$, for the case $n = 0$ is evidently trivial, and we put $x^n = t$. Then

$$x = t^{1/n}, \qquad dx = \frac{1}{n}t^{1/n-1} \, dt.$$

Putting

$$q = \frac{m + 1}{n} - 1,$$

the substitution $x = \sqrt[n]{t}$ reduces the integral (6.74) to

$$I = \frac{1}{n} \int t^q (a + bt)^p \, dt. \tag{6.75}$$

Case I: Let p be an integer. Then the integrand in (6.75) depends rationally on $t^{(m+1)/n}$ and t; if

$$\frac{m+1}{n} = \frac{j}{k},$$

where j and k are integers, $k > 0$, this integrand has the form $R(t, \sqrt[k]{t})$, where R is a rational function of its arguments. This is a special case of Propositon 6.5.

Case II: Let $(m + 1)/n$ be an integer. In this case the integrand in (6.75) depends rationally on t and $(a + bt)^p$; if $p = v/w$, where v and w are integers, $w > 0$, then this integrand has the form $R(t, \sqrt[w]{a + bt})$ and we have again an integral of the type already considered in Proposition 6.5.

Case III: Let $(m + 1)/n + p$ be an integer. The integrand in (6.75) can then be written in the form $[(a + bt)/t]^p t^{q+p}$ and it therefore depends rationally on t and $(a + bt)/t$; if $p = v/w$, where v and w are integers, $w > 0$, then this integrand has the form $R(t, \sqrt[w]{(a + bt)/t})$ and we once again have an integral of the type already considered in Proposition 6.5. □

REMARKS. To evaluate an integral of the form (6.72), provided that one of the three numbers p, $(m + 1)/n$, or $(m + 1)/n + p$ is an integer, we proceed as follows:

Case I: Let p be an integer. Then, if $p > 0$, the term $(a + bx^n)^p$ is expanded by the binomial formula; but if $p < 0$, then we put $x = t^s$, where s is the least common denominator of the fractions m and n. If $p = 0$, all is trivial.

Case II: Let $(m + 1)/n$ be an integer. We put $a + bx^n = t^w$, where w is the denominator of the fraction $p = v/w$ with v and w being integers and $w > 0$.

Case III: Let $(m + 1)/n + p$ be an integer. We put $a + bx^n = t^w x^n$, where w is the denominator of the fraction $p = v/w$ with v and w being integers and $w > 0$.

It was proved by P. L. Chebyshev that an integral of the form (6.72) can not be evaluated in closed form if none of the numbers p, $(m + 1)/n$, and $(m + 1)/n + p$ is an integer. .

EXAMPLE 1. Consider the integral

$$\int x^{-1/2} (1 + x^{1/4})^{1/3} \, dx.$$

Here $m = -\frac{1}{2}$, $n = \frac{1}{4}$, $p = \frac{1}{3}$, and $(m + 1)/n = 2$. We have Case II and put

$$t = (1 + x^{1/4})^{1/3} \quad \text{or} \quad x = (t^3 - 1)^4$$

and so

$$\int x^{-1/2}(1 + x^{1/4})^{1/3}\,dx = 12\int (t^6 - t^3)\,dt = \tfrac{3}{7}t^4(4t^3 - 7) + K$$

with $t = (1 + x^{1/4})^{1/3}$.

EXAMPLE 2. Consider the integral

$$\int (1 + x^4)^{-1/4}\,dx.$$

Here $m = 0$, $n = 4$, $p = -\tfrac{1}{4}$, and $(m + 1)/n + p = 0$. We have Case III and put

$$t = \frac{(1 + x^4)^{1/4}}{x} \quad\text{or}\quad x = (t^4 - 1)^{-1/4}$$

and so

$$\int (1 + x^4)^{-1/4}\,dx = -\int \frac{t^2}{t^4 - 1}\,dt = \frac{1}{4}\int \left(\frac{1}{t + 1} - \frac{1}{t - 1}\right)dt - \frac{1}{2}\int \frac{1}{t^2 + 1}\,dt$$

$$= \frac{1}{4}\ln\left|\frac{t + 1}{t - 1}\right| - \frac{1}{2}\tan^{-1}t + K$$

with

$$t = \frac{(1 + x^4)^{1/4}}{x} = (x^{-4} + 1)^{1/4}.$$

Proposition 6.7. *Let $R(x, y)$ be a rational function of x and y and*

$$y = (ax^2 + bx + c)^{1/2}. \tag{6.76}$$

Then

$$\int R(x, y)\,dx \tag{6.77}$$

can be reduced by a rational substitution to the integral of a rational function. Specifically, the rationalizing substitutions (also called Euler *substitutions) accomplishing this are:*

(i) $(ax^2 + bx + c)^{1/2} = (x - \alpha)t$ *if* $ax^2 + bx + c = a(x - \alpha)(x - \beta)$, *that is, if the roots α and β of $ax^2 + bx + c$ are real;*

(ii) $(ax^2 + bx + c)^{1/2} = tx + \sqrt{c}$ *if $c > 0$.*

PROOF. If the roots α and β of $ax^2 + bx + c$ are real, we have

$$(ax^2 + bx + c)^{1/2} = [a(x - \alpha)(x - \beta)]^{1/2} = (x - \alpha)\left[\frac{a(x - \beta)}{x - \alpha}\right]^{1/2};$$

hence, the integrand depends rationally on x and $[a(x - \beta)/(x - \alpha)]^{1/2}$ and this brings us to a situation already considered in Proposition 6.5. Thus, rationalization can be achieved by setting

$$\frac{a(x - \beta)}{x - \alpha} = t^2 \quad \text{or} \quad x = \frac{\alpha t^2 - \beta a}{t^2 - a}.$$

The case $\alpha = \beta$ can of course be disregarded.

If the roots of $ax^2 + bx + c$ are complex, then $ax^2 + bx + c$ preserves the same sign for all values of x; we are of course assuming that the sign is always positive, for otherwise the value of the square root would be complex for all real numbers x and the problem would become void. In particular, by assuming that $x = 0$ we can see that in this case we must necessarily have $c > 0$ (the method which we are now going to describe always leads to the desired result for $c > 0$ irrespective of whether the roots of $ax^2 + bx + c$ are real or complex). Assuming that

$$\frac{(ax^2 + bx + c)^{1/2} - \sqrt{c}}{x} = t,$$

we obtain

$$ax^2 + bx + c = (tx + \sqrt{c})^2 = t^2 x^2 + 2\sqrt{c}tx + c,$$

$$ax + b = t^2 x + 2\sqrt{c}t,$$

$$x = \frac{b - 2\sqrt{c}t}{t^2 - a} = g(t),$$

$$dx = g'(t)\, dt$$

$$(ax^2 + bx + c)^{1/2} = tx + \sqrt{c} = tg(t) + \sqrt{c},$$

and therefore

$$\int R(x, (ax^2 + bx + c)^{1/2})\, dx = \int R\{g(t), tg(t) + \sqrt{c}\}g'(t)\, dt.$$

Since the function g (and therefore also its derivative g') is rational, we have accomplished the rationalization of the integral. $\qquad\square$

EXAMPLE 1. Consider the integral

$$I = \int \frac{x}{(7x - 10 - x^2)^{3/2}}\, dx.$$

Since $7x - 10 - x^2$ has the real roots $\alpha = 2$, $\beta = 5$, we use the first Euler substitution

$$(7x - 10 - x^2)^{1/2} = \sqrt{(x - 2)(5 - x)} = (x - 2)t.$$

Therefore,

$$\frac{5-x}{x-2} = t^2 \quad \text{or} \quad x = \frac{5+2t^2}{1+t^2}$$

and so

$$I = -\frac{6}{27}\int \frac{5+2t^2}{t^2}\,dt = -\frac{2}{9}\int \left(\frac{5}{t^2}+2\right)dt = -\frac{2}{9}\left(-\frac{5}{t}+2t\right)+K$$

with

$$t = \frac{(7x-10-x^2)^{1/2}}{x-2}.$$

EXAMPLE 2. Consider the integral

$$J = \int \frac{1}{x(x^2+x+1)^{1/2}}\,dx.$$

Since $c = 1 > 0$, we can apply the second Euler substitution

$$(x^2+x+1)^{1/2} = tx + 1.$$

Therefore,

$$x = \frac{-2t+1}{t^2-1}$$

and so

$$J = \int \frac{2}{2t-1}\,dt = \ln|2t-1| + K$$

with

$$t = \frac{(x^2+x+1)^{1/2}-1}{x}.$$

Lemma. *Let R be a rational function in x and y and let $y^2 = P(x)$ be a polynomial in x. Then R can be represented in the form*

$$R(x,y) = R_1(x) + R_2(x)y = R_1(x) + R_3(x)\frac{1}{y},$$

where R_1, R_2, and R_3 are rational functions in x.

PROOF. Since R is a rational function in x and y, we can consider R as a quotient of two polynomials in y whose coefficients are polynomials in x, that is, quotients of two expressions of the form

$$H_0(x) + H_1(x)y + H_2(x)y^2 + \cdots + H_n(x)y^n, \tag{6.78}$$

where $H_0, H_1, \ldots, H_n$ are polynomials in x. But $y^2 = P(x)$ is a polynomial in x, so every even power of y is a polynomial in x, and every odd power is the product of y and a polynomial in x. This means that (6.78) reduces to the form

$$P_0^*(x) + P_1^*(x)y,$$

where P_0^* and P_1^* are polynomials in x. Hence,

$$R(x, y) = \frac{P_0^*(x) + P_1^*(x)y}{Q_0^*(x) + Q_1^*(x)y}. \tag{6.79}$$

We now multiply numerator and denominator of (6.79) by $Q_0^*(x) - Q_1^*(x)y$. We get

$$(P_0^* + P_1^*y)(Q_0^* - Q_1^*y) = P_0^*Q_0^* - P_1^*Q_1^*y^2 + (P_1^*Q_0^* - P_0^*Q_1^*)y = P_0 + Q_0y$$

and

$$(Q_0^* + Q_1^*y)(Q_0^* - Q_1^*y) = (Q_0^*)^2 - (Q_1^*)^2y^2 = D,$$

where P_0, Q_0, D are polynomials in x. Finally,

$$R(x, y) = \frac{P_0(x)}{D(x)} + \frac{Q_0(x)}{D(x)}y \quad \text{or} \quad R(x, y) = R_1(x) + R_2(x)y.$$

Putting $R_3(x) = R_2(x)P(x)$ and recalling that $y^2 = P(x)$, we obtain

$$R(x, y) = R_1(x) + R_3(x)\frac{1}{y}.$$

This completes the proof. $\qquad\qquad\qquad\qquad\qquad\qquad\qquad\qquad$ □

Proposition 6.8. *If $R(x, y)$ is a rational function of x and y and if*

$$Y = ax^2 + bx + c \quad \text{and} \quad y = \sqrt{Y},$$

then, by the Lemma,

$$R(x, y) = R_1(x) + R_3(x)\frac{1}{y},$$

where R_1 and R_3 are rational functions. In the evaluation of the integral

$$\int R(x, y)\,dx = \int R_1(x)\,dx + \int R_3(x)\frac{1}{y}\,dx,$$

the integral $\int R_1(x)\,dx$ is straightforward and the integral

$$\int R_3(x)\frac{1}{y}\,dx = \int \frac{R_3(x)}{\sqrt{Y}}\,dx = \int \frac{R_3(x)}{(ax^2 + bx + c)^{1/2}}\,dx$$

reduces to integrals of the following three types:

(I) $\displaystyle\int \frac{P(x)}{(ax^2 + bx + c)^{1/2}}\, dx,$ *where P is a polynomial;*

(II) $\displaystyle\int \frac{A}{(x - \alpha)^k (ax^2 + bx + c)^{1/2}}\, dx,$ *where α is real and $k = 0, 1, 2, \ldots$;*

(III) $\displaystyle\int \frac{Bx + C}{(x^2 + px + q)^m (ax^2 + bx + c)^{1/2}}\, dx,$ *where $m = 0, 1, 2, \ldots$ and the roots of $x^2 + px + q$ are complex numbers.*

PROOF. To integrate

$$\int \frac{R_3(x)}{(ax^2 + bx + c)^{1/2}}\, dx,$$

where R_3 is a rational function of x, we first separate the integral part $P(x)$ of the rational function $R_3(x)$ and then we resolve the remaining proper fractional part of $R_3(x)$ into partial fractions. □

DISCUSSION. We now consider the evaluation of the three types of integral mentioned in Proposition 6.8.

(I). We put

$$V_m = \int \frac{x^m}{(ax^2 + bx + c)^{1/2}}\, dx = \int \frac{x^m}{\sqrt{Y}}\, dx \quad \text{for } m = 0, 1, 2, \ldots.$$

For such integrals we can easily derive a recursion formula. To this end we assume $m \geq 1$ and take the derivative

$$(x^{m-1}\sqrt{Y})' = (m - 1)x^{m-2}\sqrt{Y} + \frac{x^{m-1}Y'}{2\sqrt{Y}}$$

$$= \frac{2(m - 1)x^{m-2}(ax^2 + bx + c) + x^{m-1}(2ax + b)}{2\sqrt{Y}}$$

$$= ma\frac{x^m}{\sqrt{Y}} + \left(m - \frac{1}{2}\right)b\frac{x^{m-1}}{\sqrt{Y}} + (m - 1)c\frac{x^{m-2}}{\sqrt{Y}}$$

and integrate this identity; we obtain

$$x^{m-1}\sqrt{Y} = maV_m + (m - \tfrac{1}{2})bV_{m-1} + (m - 1)cV_{m-2}.$$

For $m = 1$ we obtain

$$V_1 = \frac{1}{a}\sqrt{Y} - \frac{b}{2a}V_0;$$

for $m = 2$ (using the expression for V_1) we get

$$V_2 = \frac{1}{4a^2}(2ax - 3b)\sqrt{Y} + \frac{1}{8a^2}(3b^2 - 4ac)V_0.$$

Continuing this process, we arrive at the general formula

$$V_m = p_{m-1}(x)\sqrt{Y} + \lambda_m V_0, \tag{6.80}$$

where $p_{m-1}(x)$ is a polynomial of degree $m - 1$ and λ_m denotes a constant. *Thus all integrals V_m can be reduced to V_0.*

If the polynomial $P(x)$ in the integral

$$\int \frac{P(x)}{\sqrt{Y}}\,dx$$

has degree n, then this integral is a linear combination of the integrals V_0, $V_1, \ldots, V_n$, and thus, on the basis of (6.80), can be expressed in the form

$$\int \frac{P(x)}{\sqrt{Y}}\,dx = Q(x)\sqrt{Y} + \lambda \int \frac{1}{\sqrt{Y}}\,dx, \tag{6.81}$$

where $Q(x)$ is a polynomial of degree $n - 1$ and λ denotes a constant.

The polynomial Q and the constant λ can be found in general by the method of undetermined coefficients. Differentiating in (6.81) and multiplying the equation so obtained by $\sqrt{Y}$, we get

$$P(x) = Q'(x)(ax^2 + bx + c) + \tfrac{1}{2}Q(x)(2ax + b) + \lambda.$$

Substituting here for $Q(x)$ a polynomial of degree $n - 1$ with undetermined coefficients, we see that on the right-hand side of the equation we will have a polynomial in x of degree n. By comparison of coefficients of like powers of x we arrive at a system of $n + 1$ linear equations; from this system we can calculate the n coefficients of Q and the constant λ. From the way the system of $n + 1$ linear equations has been derived, it is easy to see that the system is consistent and that it has a solution which is unique.

(II). The integral

$$\int \frac{1}{(x - \alpha)^k(ax^2 + bx + c)^{1/2}}\,dx$$

can be reduced to the type just considered by use of the substitution

$$x - \alpha = \frac{1}{t}.$$

Indeed, if $x - \alpha = 1/t$, then

$$dx = -\frac{1}{t^2}\,dt, \qquad ax^2 + bx + c = \frac{(a\alpha^2 + b\alpha + c)t^2 + (2a\alpha + b)t + a}{t^2}$$

so that (for simplicity we specifically assume $x > \alpha, t > 0$)

$$\int \frac{1}{(x - \alpha)^k(ax^2 + bx + c)^{1/2}}\,dx = -\int \frac{t^{k-1}}{(a\alpha^2 + b\alpha + c)t^2 + (2a\alpha + b)t + a}\,dt.$$

If $a\alpha^2 + b\alpha + c = 0$, that is, if α is a root of Y, this integral further simplifies and we obtain an integral of the type considered in Proposition 6.5.

(III). (a) We consider the integral

$$\int \frac{Bx + C}{(x^2 + px + q)^m (ax^2 + bx + c)^{1/2}} \, dx,$$

where $m = 0, 1, 2, \ldots$ and the roots of $x^2 + px + q$ are complex, and assume that the expression $ax^2 + bx + c$ differs only by the factor a from the expression $x^2 + px + q$. Then the considered integral has the form

$$\int \frac{Bx + C}{(ax^2 + bx + c)^{(2m+1)/2}} \, dx$$

which can be written

$$\frac{B}{2a} \int \frac{2ax + b}{(ax^2 + bx + c)^{(2m+1)/2}} \, dx + \left(C - \frac{Bb}{2a} \right) \int \frac{1}{(ax^2 + bx + c)^{(2m+1)/2}} \, dx.$$

$$(6.82)$$

The first integral in (6.82) can readily be calculated by the substitution $t = ax^2 + bx + c$. For the calculation of

$$\int \frac{1}{(ax^2 + bx + c)^{(2m+1)/2}} \, dx = \int \frac{1}{Y^{(2m+1)/2}} \, dx$$

we shall use the *substitution of Abel*

$$t = (\sqrt{Y})' = \frac{Y'}{2\sqrt{Y}} = \frac{ax + b/2}{(ax^2 + bx + c)^{1/2}}.$$

$$(6.83)$$

If in the foregoing equation we take the square and then multiply by $4Y$, we get

$$4t^2 Y = (Y')^2 = 4a^2 x^2 + 4abx + b^2.$$

Hence,

$$4(a - t^2)Y = 4ac - b^2$$

and so

$$Y^m = \left(\frac{4ac - b^2}{4} \right)^m \frac{1}{(a - t^2)^m}.$$

$$(6.84)$$

Next, differentiating

$$t\sqrt{Y} = ax + \frac{b}{2},$$

we obtain

$$\sqrt{Y} \, dt + t^2 \, dx = a \, dx,$$

that is,

$$\frac{dx}{\sqrt{Y}} = \frac{dt}{a - t^2}.$$

(6.85)

By (6.84) and (6.85), we have

$$\frac{dx}{Y^{(2m+1)/2}} = \left(\frac{4}{4ac - b^2}\right)^m (a - t^2)^{m-1} dt$$

and so

$$\int \frac{1}{Y^{(2m+1)/2}} dx = \left(\frac{4}{4ac - b^2}\right)^m \int (a - t^2)^{m-1} dt.$$

(6.86)

Thus, the problem is reduced to the integration of a polynomial. For example, for $m = 1$ we get

$$\int \frac{1}{(ax^2 + bx + c)^{3/2}} dx = \frac{2}{4ac - b^2} \frac{2ax + b}{(ax^2 + bx + c)^{1/2}}.$$

(b) We now consider the general case and we put

$$ax^2 + bx + c = a(x^2 + p'x + q')$$

to have symmetry in the notation; we can now assume that $x^2 + p'x + q'$ is not identical with $x^2 + px + q$. We wish to transform the variable x in such a manner that the linear terms in $x^2 + px + q$ and $x^2 + p'x + q'$ vanish simultaneously.

First we suppose that $p \neq p'$. Then we can achieve our goal with the help of the fractional linear substitution

$$x = \frac{\mu t + v}{t + 1}$$

(6.87)

if we select suitable coefficients μ and v. We obtain

$$x^2 + px + q = \frac{(\mu^2 + p\mu + q)t^2 + [2\mu v + p(\mu + v) + 2q]t + (v^2 + pv + q)}{(t + 1)^2}$$

and an analogous formula for $x^2 + p'x + q'$. The sought after coefficients can be determined with the help of the conditions

$$2\mu v + p(\mu + v) + 2q = 0, \qquad 2\mu v + p'(\mu + v) + 2q' = 0$$

or

$$\mu + v = -2\frac{q - q'}{p - p'}, \qquad \mu v = \frac{p'q - pq'}{p - p'}.$$

Hence, μ and v are the roots of

$$(p - p')z^2 + 2(q - q')z + (p'q - pq').$$

In order that these roots be real and distinct, it is (necessary and) sufficient that

$$(q - q')^2 - (p - p')(p'q - pq') > 0, \tag{6.88}$$

or, equivalently,

$$[2(q + q') - pp']^2 > 4(q - p^2)(4q' - p'^2); \tag{6.89}$$

note that for $\mu = \nu$ the substitution loses its meaning, for in that case $x = \nu$. But $4q - p^2 > 0$ (because $x^2 + px + q$ has complex roots); hence, (6.89) is certainly satisfied if simultaneously $4q' - p'^2 < 0$ holds.

It only remains to investigate the case $4q' - p'^2 > 0$. From this it follows that $q' > 0$, and since $q > 0$ as well and hence $4\sqrt{qq'} > pp'$, we obtain [because $(q + q')/2 \geq \sqrt{qq'}$]

$$[2(q + q') - pp']^2 \geq [4\sqrt{qq'} - pp']^2$$

$$= (4q - p^2)(4q' - p'^2) + 4(p\sqrt{q'} - p'\sqrt{q})^2$$

$$\geq (4q - p^2)(4q' - p'^2).$$

Here we have the sign $\geq$ twice; the equality sign can not occur in both cases simultaneously: For $q \neq q'$ there can not be equality in the first case and for $q = q'$ there can not be equality in the second case. Thus, the inequality (6.89) and with it the inequality (6.88) is established.

With the help of the substitution (6.87) we now transform the integral

$$\int \frac{Bx + C}{(x^2 + px + q)^m (ax^2 + bx + c)^{1/2}} \, dx$$

into an integral of the form

$$\int \frac{P(t)}{(t^2 + \lambda)^m (\alpha t^2 + \beta)^{1/2}} \, dt,$$

where $P(t)$ is a polynomial of degree $2m - 1$ and $\lambda > 0$. Partial fraction decomposition of the proper rational fraction (we assume $m > 1$)

$$\frac{P(t)}{(t^2 + \lambda)^m}$$

leads to a sum of integrals of the form

$$\int \frac{Mt + N}{(t^2 + \lambda)^k (\alpha t^2 + \beta)^{1/2}} \, dt \qquad (k = 1, 2, \ldots, m). \tag{6.90}$$

In the case $p = p'$, a case we have excluded so far, the first degree terms vanish even more simply, namely, by use of the substitution

$$x = t - \frac{p}{2}.$$

Using the substitution $x = t - p/2$ we get *immediately* an integral of the type appearing in (6.90).

But

$$\int \frac{Mt + N}{(t^2 + \lambda)^k(\alpha t^2 + \beta)^{1/2}} \, dt = \frac{M}{\alpha} \int \frac{\alpha t}{(t^2 + \lambda)^k(\alpha t^2 + \beta)^{1/2}} \, dt$$

$$+ N \int \frac{1}{(t^2 + \lambda)^k(\alpha t^2 + \beta)^{1/3}} \, dt. \tag{6.91}$$

The first integral on the right-hand side of (6.91) can be calculated at once by use of the substitution $s = (\alpha t^2 + \beta)^{1/2}$; to the second integral we apply Abel's substitution

$$s = \frac{\alpha t}{(\alpha t^2 + \beta)^{1/2}}.$$

By (6.86), we get

$$\frac{dt}{(\alpha t^2 + \beta)^{1/2}} = \frac{ds}{\alpha - s^2};$$

moreover, we have

$$t^2 + \lambda = \frac{(\beta - \alpha\lambda)s^2 + \lambda\alpha^2}{\alpha(\alpha - s^2)}.$$

Thus,

$$\int \frac{1}{(t^2 + \lambda)^k(\alpha t^2 + \beta)^{1/2}} \, dt = \alpha^k \int \frac{(\alpha - s^2)^{k-1}}{[(\beta - \alpha\lambda)s^2 + \lambda\alpha^2]^k} \, ds.$$

Hence, the integral under consideration has been reduced to the integral of a rational function.

REMARKS. Proposition 6.8, together with the foregoing discussion, gives us a method of calculating the integral (6.77) without the use of Euler substitutions. In practice, Euler substitutions often lead to complicated calculations.

The integral (6.77) can be brought to the form (6.69) by means of the following trigonometric substitutions:

$$x + \frac{b}{2a} = \frac{(b^2 - 4ac)^{1/2}}{2a} \sin t$$

$$(a < 0, 4ac - b^2 < 0);$$

$$= \frac{(b^2 - 4ac)^{1/2}}{2a} \cos t$$

$$x + \frac{b}{2a} = \frac{(b^2 - 4ac)^{1/2}}{2a} \sec t$$

$$(a > 0, 4ac - b^2 < 0);$$

$$= \frac{(b^2 - 4ac)^{1/2}}{2a} \csc t$$

$$x + \frac{b}{2a} = \frac{(4ac - b^2)^{1/2}}{2a} \tan t$$

$$(a > 0, 4ac - b^2 > 0).$$

$$= \frac{(4ac - b^2)^{1/2}}{2a} \cot t$$

However, this approach often leads to complicated calculations as well.

EXAMPLE 1. Let

$$I = \int \frac{x + 3}{(4x^2 + 4x - 3)^{1/2}} \, dx.$$

Using the substitution $2x + 1 = t$ or $x = (t - 1)/2$ we get

$$I = \frac{1}{4} \int \frac{t + 5}{(t^2 - 4)^{1/2}} \, dt.$$

Putting $t = 2 \sec s$, we obtain, returning to the variable t after integration,

$$I = \tfrac{1}{4}(t^2 - 4)^{1/2} + \tfrac{5}{4} \ln |t + (t^2 - 4)^{1/2}| + K$$

and so

$$I = \tfrac{1}{4}(4x^2 + 4x - 3)^{1/2} + \tfrac{5}{4} \ln |2x + 1 + (4x^2 + 4x - 3)^{1/2}| + K.$$

EXAMPLE 2. Let

$$J = \int \frac{x^3 - x - 1}{(x^2 + 2x + 2)^{1/2}} \, dx.$$

Putting $Q(x) = Ax^2 + Bx + C$, we get [see (6.81)]

$$J = (Ax^2 + Bx + C)(x^2 + 2x + 2)^{1/2} + \lambda \int \frac{1}{(x^2 + 2x + 2)^{1/2}} \, dx.$$

Differentiating this equality, we obtain

$$J' = \frac{x^3 - x - 1}{(x^2 + 2x + 2)^{1/2}}$$

$$= (2Ax + B)(x^2 + 2x + 2)^{1/2}$$

$$+ (Ax^2 + Bx + C)\frac{x + 1}{(x^2 + 2x + 2)^{1/2}} + \frac{\lambda}{(x^2 + 2x + 2)^{1/2}}.$$

Hence,

$$x^3 - x - 1 = (2Ax + B)(x^2 + 2x + 2) + (Ax^2 + Bx + C)(x + 1) + \lambda.$$

Equating coefficients of like powers of x, we get a system of equations with the solutions $A = \frac{1}{3}$, $B = -\frac{5}{6}$, $C = \frac{1}{6}$, $\lambda = \frac{1}{2}$. Thus,

$$J = (\tfrac{1}{3}x^2 - \tfrac{5}{6}x + \tfrac{1}{6})(x^2 + 2x + 2)^{1/2} + \frac{1}{2}\int \frac{1}{(x^2 + 2x + 2)^{1/2}}\,dx,$$

where

$$\int \frac{1}{(x^2 + 2x + 2)^{1/2}}\,dx = \int \frac{1}{[(x+1)^2 + 1]^{1/2}}\,d(x + 1)$$
$$= \ln|x + 1 + (x^2 + 2x + 2)^{1/2}| + K.$$

EXAMPLE 3. The integral

$$\int \frac{1}{(x - 1)^3(x^2 - 2x - 1)^{1/2}}\,dx$$

changes into the integral

$$-\int \frac{t^2}{(1 - 2t^2)^{1/2}}\,dt$$

by using the substitution $x - 1 = 1/t$ (we let $x > 1, t > 0$). The latter integral can be evaluated using the trigonometric substitution $t = \sqrt{2}\sin s$; we obtain, returning to the variable t after integration,

$$-\int \frac{t^2}{(1 - 2t^2)^{1/2}}\,dt = \frac{1}{4}t(1 - 2t^2)^{1/2} - \frac{1}{4\sqrt{2}}\sin^{-1}(\sqrt{2}t) + K.$$

Thus,

$$\int \frac{1}{(x - 1)^3(x^2 - 2x - 1)^{1/2}}\,dx = \frac{1}{4(x - 1)^2}(x^2 - 2x - 1)^{1/2}$$
$$- \frac{1}{4\sqrt{2}}\sin^{-1}\left(\frac{\sqrt{2}}{x - 1}\right) + K.$$

EXAMPLE 4. Let

$$H = \int \frac{1}{(2x^2 - x + 2)^{7/2}}\,dx.$$

Applying the substitution of Abel

$$t = \frac{4x - 1}{2(2x^2 - x + 2)^{1/2}}$$

we get, by (6.86),

$$H = \frac{64}{3375}\int (2 - t^2)^2\,dt$$

and so

$$H = \frac{64}{3375}\left(2\frac{4x-1}{(2x^2-x+2)^{1/2}} - \frac{1}{6}\frac{(4x-1)^3}{(2x^2-x+2)^{3/2}}\right.$$
$$\left. + \frac{1}{160}\frac{(4x-1)^5}{(2x-x+2)^{5/2}}\right) + K.$$

EXAMPLE 5. Consider the integral

$$\int \frac{x+3}{(x^2-x+1)(x^2+x+1)^{1/2}}\,dx.$$

Here the fractional linear substitution (6.87) gives

$$x^2 \pm x + 1 = \frac{(\mu^2 \pm \mu + 1)t^2 + [2\mu v \pm (\mu+v+2)]t + (v^2 \pm v + 1)}{(t+1)^2}.$$

The conditions

$$2\mu v + (\mu + v) + 2 = 0$$

or $\mu + v = 0$, $\mu v = -1$ are fulfilled, for example, for $\mu = 1$, $v = -1$. Hence,

$$x = \frac{t-1}{t+1}, \quad dx = \frac{2\,dt}{(t+1)^2}, \quad x+3 = \frac{4t+2}{t+1}, \quad x^2+x+1 = \frac{t^2+3}{(t+1)^2}$$

and

$$(x^2+x+1)^{1/2} = \frac{(3t^2+1)^{1/2}}{t+1},$$

if we assume $t + 1 > 0$, that is, if $x < 1$. Thus,

$$\int \frac{x+3}{(x^2-x+1)(x^2+x+1)^{1/2}}\,dx = \int \frac{8t+4}{(t^2+3)(3t^2+1)^{1/2}}\,dt.$$

The integral obtained equals the sum

$$8\int \frac{t}{(t^2+3)(3t^2+1)^{1/2}}\,dt + 4\int \frac{1}{(t^2+3)(3t^2+1)^{1/2}}\,dt.$$

The first summand can be evaluated by use of the substitution $s = (3t^2+1)^{1/2}$ and works out to

$$\sqrt{8}\tan^{-1}\left(\frac{3t^2+1}{8}\right)^{1/2} + K_1.$$

To the second summand we apply Abel's substitution

$$s = \frac{3t}{(3t^2+1)^{1/2}}$$

which transforms the integral under consideration into

$$12 \int \frac{1}{27 - 8s^2} \, ds = \frac{1}{\sqrt{6}} \ln \left| \frac{3\sqrt{3} + 2\sqrt{2}s}{3\sqrt{3} - 2\sqrt{2}s} \right| + K_2.$$

It only remains to change back to the initial variable x.

5. Some Applications of Integration

We begin with a list of formulas to give some indication of the diversity of applications of the integral to geometry.

The area A of the region R under the graph of a positive, continuous function between $x = a$ and $x = b$, $a < b$, is given by

$$A = \int_a^b f(x) \, dx. \tag{6.92}$$

When this region R is rotated about the x-axis, the volume V_x of the resulting solid is given by

$$V_x = \pi \int_a^b f^2(x) \, dx. \tag{6.93}$$

When R is rotated about the y-axis, the volume V_y of the resulting solid is given by

$$V_y = 2\pi \int_a^b x f(x) \, dx, \tag{6.94}$$

provided that either $a \geq 0$ or $b \leq 0$. Next, denote by C the graph of the function $y = f(x)$ with $a \leq x \leq b$. If f is differentiable and the derivative f' is continuous on the interval $[a, b]$, then the length L of the curve C is given by

$$L = \int_a^b \{1 + [f'(x)]^2\}^{1/2} \, dx \tag{6.95}$$

(this formula is valid even if f is not a positive function). If the curve C is rotated about the x-axis, then the surface area S_x of the resulting surface of revolution is given by

$$S_x = 2\pi \int_a^b f(x) \{1 + [f'(x)]^2\}^{1/2} \, dx. \tag{6.96}$$

If C is rotated about the y-axis, then the surface area S_y of the resulting surface of revolution is given by

$$S_y = 2\pi \int_a^b x \{1 + [f'(x)]^2\}^{1/2} \, dx, \tag{6.97}$$

provided that either $a \geq 0$ or $b \leq 0$.

Formulas (6.93) and (6.94) are only instances of a more general formula. If a solid lies between a and b on the x-axis, and its cross-sectional area at x is $A(x)$, where A is a continuous function, then its volume V is given by

$$V = \int_a^b A(x)\, dx. \tag{6.98}$$

Formulas for area and arc length can also be obtained when the region, or curve, is specified in terms of polar coordinates. Suppose that the region R is bounded by the radii vectors $\theta = \alpha$, $\theta = \beta$, and by the curve C which is the graph of $r = f(\theta)$. Then the area A of R can be obtained from

$$A = \frac{1}{2} \int_\alpha^\beta f^2(\theta)\, d\theta, \tag{6.99}$$

while the length L of C is given by

$$L = \int_\alpha^\beta \{f^2(\theta) + [f'(\theta)]^2\}^{1/2}\, d\theta. \tag{6.100}$$

There is a variant of (6.100) which is sometimes useful. If C is the graph of $\theta = g(r)$ for r between r_1 and r_2, then

$$L = \int_{r_1}^{r_2} \{r^2[g'(r)]^2 + 1\}^{1/2}\, dr. \tag{6.101}$$

More generally, if the coordinates of the curve are given by the parametric equations

$$x = x(t) \quad \text{and} \quad y = y(t),$$

if $a = x(t_1)$ and $b = x(t_2)$, then the area A of the region bounded by C, the lines $x = a$, $x = b$, and the x-axis is given by

$$A = \left| \int_{t_1}^{t_2} y(t) x'(t)\, dt \right| \tag{6.102}$$

provided that $x(t)$ and $y(t)$ have continuous derivatives, and $y(t)$ is strictly increasing, all on $[t_1, t_2]$. The length L of C between a and b is given by

$$L = \int_{t_1}^{t_2} \{[y'(t)]^2 + [x'(t)]^2\}^{1/2}\, dt, \tag{1.103}$$

provided that $x(t)$ and $y(t)$ have continuous derivatives on $[t_1, t_2]$.

Finally, if the region R is bounded by the graphs of two continuous functions f_1 and f_2 with $f_2(x) \geq f_1(x)$ for $a \leq x \leq b$ and by two straight lines $x = a$ and $x = b$, then the area A of R is given by

$$A = \int_a^b \{f_2(x) - f_1(x)\}\, dx. \tag{6.104}$$

If the boundary of the region R is a simple closed curve (one which does not intersect itself anywhere) and the boundary is given by the parametric equations $x = x(t)$ and $y = y(t)$, where $x'(t)$ and $y'(t)$ are continuous functions on $[t_1, t_2]$ with t_1 and t_2 being the values of the parameter t that correspond, respectively, to the beginning and the end of the traversal of the boundary in the positive direction of the contour (i.e., in the direction such that the region R enclosed remains on the left as we move along the boundary curve), then the area A of the region R can be evaluated by one of the three formulas:

$$A = -\int_{t_1}^{t_2} y(t)x'(t)\,dt = \int_{t_1}^{t_2} x(t)y'(t)\,dt = \frac{1}{2}\int_{t_1}^{t_2} \{x(t)y'(t) - y(t)x'(t)\}\,dt.$$

$$(6.105)$$

By way of illustration, we consider some examples next.

EXAMPLE 1. The volume of a spherical segment of one base is given by

$$\pi h^2 \left(r - \frac{h}{3} \right),$$

$$(6.106)$$

where r denotes the radius of the sphere and h the height of the segment.
Indeed, by (6.93), the volume of the solid in question equals (see Figure 6.4)

$$\pi \int_{r-h}^{r} (r^2 - x^2)\,dx = \pi \left(r^2 x - \frac{x^3}{3} \right)\Bigg|_{r-h}^{r} = \pi h^2 \left(r - \frac{h}{3} \right).$$

We have already used formula (6.106) in Example 4 of Section 4 of Chapter 4.

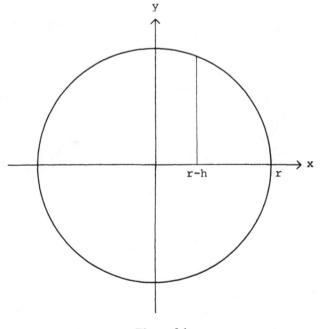

Figure 6.4

EXAMPLE 2. The surface area of a spherical segment of one base is given by

$$2\pi rh, \tag{6.107}$$

where r denotes the radius of the sphere and h the height of the segment.

Indeed, by (6.95), the surface area in question equals (see Figure 6.4)

$$2\pi \int_{r-h}^{r} r \, dx = 2\pi rx \Big|_{r-h}^{r} = 2\pi rh.$$

REMARKS. An interesting consequence of (6.107) is the following fact, already observed by Archimedes: If a sphere is inscribed in a right circular cylinder, then the surfaces of the sphere and the cylinder intercepted by a pair of planes perpendicular to the axis of the cylinder are equal in area.

Another consequence of (6.107) is the following: An observer at height H above the north pole of a sphere of radius r can see a part of the sphere having area

$$\frac{2\pi Hr^2}{H+r}. \tag{6.108}$$

Indeed, from Figure 6.5, we can see that

$$(H+r)^2 = r^2 + s^2$$

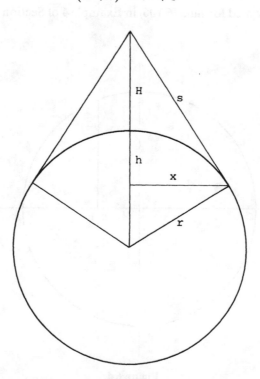

Figure 6.5

and

$$r^2 - (r - h)^2 = x^2 = s^2 - (H + h)^2.$$

Thus,

$$r^2 - (r - h)^2 = (H + r)^2 - r^2 - (H + d)^2 \quad \text{or} \quad 2hr = 2Hr - 2hH,$$

and so

$$h = \frac{Hr}{H + r}.$$

Substituting this value of h into (6.107), we obtain (6.108).

EXAMPLE 3. Given by cycloid

$$x = a(t - \sin t), \qquad y = a(1 - \cos t)$$

(see Example 2 in Section 1 of Chapter 5), we wish to find the area of the region bounded by an arc of the cycloid and the x-axis using integration.

The boundary of the region under consideration consists of an arc of the cycloid ($0 \le t \le 2\pi$) and a segment of the x-axis ($0 \le x \le 2\pi a$). We apply the formula

$$A = -\int_{t_1}^{t_2} y(t) x'(t) \, dt.$$

Since on the segment of the x-axis we have $y = 0$, it remains to compute the integral (taking into account the direction of the boundary traversal)

$$A = -\int_{2\pi}^{0} a(1 - \cos t) a(1 - \cos t) \, dt = a^2 \int_0^{2\pi} (1 - \cos t)^2 \, dt$$

$$= a^2 \int_0^{2\pi} (1 - 2\cos t + \tfrac{1}{2}(1 + \cos 2t)) \, dt = 3\pi a^2.$$

Thus, the area of the region bounded by an arc of the cycloid and the x-axis is three times the area of the generating circle.

EXAMPLE 4. To find the length of one arc of the cycloid

$$x = a(t - \sin t), \qquad y = a(1 - \cos t)$$

we use the formula (6.103) and get

$$L = \int_0^{2\pi} \{a^2 \sin^2 t + a^2(1 - \cos t)^2\}^{1/2} \, dt = 2a \int_0^{2\pi} \left(\sin \frac{t}{2}\right) dt = 8a.$$

Hence, the length of one arc of the cycloid is four times the diameter of the generating circle, a result originally found by Christopher Wren, the architect of St. Paul's Cathedral.

EXAMPLE 5. The loop of the folium of Descartes,

$$x^3 + y^3 = 3axy$$

(see Example 4 in Section 4 of Chapter 4 and Figure 6.6), has area

$$\frac{3a^2}{2}.$$

Moreover, the area of the loop is equal to the area of the region contained between the folium of Descartes and its asymptote $x + y + a = 0$.

Indeed, changing to polar coordinates, that is, letting

$$x = r\cos\theta, \qquad y = r\sin\theta,$$

and then dividing by r^2, we get the equation in polar coordinates,

$$r = \frac{3a(\sin\theta)(\cos\theta)}{\sin^3\theta + \cos^3\theta}.$$

Since the angle θ varies in the first quadrant from 0 to $\pi/2$, (6.99) gives

$$A = \frac{9a^2}{2}\int_0^{\pi/2} \frac{(\sin^2\theta)(\cos^2\theta)}{(\sin^3\theta + \cos^3\theta)^2}\,d\theta.$$

Replacing $\sin\theta$ in the integrand by $(\tan\theta)(\cos\theta)$, we obtain

$$A = \frac{9a^2}{2}\int_0^{\pi/2} \frac{\tan^2\theta}{(1 + \tan^3\theta)^2}\,d(\tan\theta)$$

$$= -\frac{9a^2}{2}\frac{1}{3}\left(\frac{1}{1 + \tan^3\theta}\right)\Bigg|_0^{\pi/2} = -\frac{3a^2}{2}\frac{\cos^3\theta}{\sin^3\theta + \cos^3\theta}\Bigg|_0^{\pi/2} = \frac{3a^2}{2}.$$

We shall now find the area of the region between the curve and its asymptote. The line (shown dashed in Figure 6.6) drawn parallel to the asymptote makes an angle $3\pi/4$ with the x-axis. We draw any line through O whose

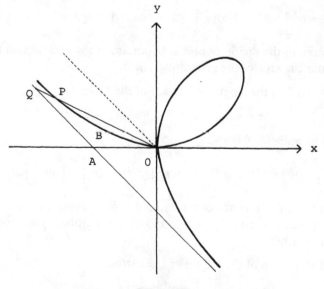

Figure 6.6

vectorial angle θ lies between $3\pi/4$ and π. Let it cut the curve and the asymptote in P and Q, respectively (see Figure 6.6).

We shall first find the area between the curve and its asymptote lying in the second quadrant. This area is the limit of the area of the curvilinear region $OBPQAO$ as the line OP starting from OA moves toward the dashed line. The area of the triangle $\triangle OAQ$ equals

$$\frac{1}{2}\int_\theta^\pi r^2\, d\theta = \frac{1}{2}\int_\theta^\pi \frac{a^2}{(\sin\theta + \cos\theta)^2}\, d\theta$$

and the area of the region bounded by the curve and the line OP equals

$$\frac{1}{2}\int_\theta^\pi \frac{9a^2(\sin^2\theta)(\cos^2\theta)}{(\sin^3\theta + \cos^3\theta)^2}\, d\theta.$$

Thus, the area of the curvilinear region $OBPQAO$ is

$$\frac{1}{2}\int_0^\pi \left(\frac{a^2}{(\sin\theta + \cos\theta)^2} - \frac{9a^2(\sin^2\theta)(\cos^2\theta)}{(\sin^3\theta + \cos^3\theta)^2}\right) d\theta$$

$$= \frac{a^2}{2}\left(-\frac{1}{1 + \tan\theta} + \frac{3}{1 + \tan^3\theta}\right)\Bigg|_0^\pi$$

because

$$\int \frac{(\sin^2\theta)(\cos^2\theta)}{(\sin^3\theta + \cos^3\theta)^2}\, d\theta = \int \frac{(\tan^2\theta)(\sec^2\theta)}{(1 + \tan^3\theta)^2}\, d\theta = -\frac{1}{3(1 + \tan^3\theta)} + K$$

and

$$\int \frac{1}{(\sin\theta + \cos\theta)^2}\, d\theta = \int \frac{\sec^2\theta}{(1 + \tan^2\theta)^2}\, d\theta = -\frac{1}{1 + \tan\theta} + K.$$

But, as $\theta \to 3\pi/4$, we have

$$\tan\theta \to -1$$

and

$$\frac{1}{1 + \tan\theta} - \frac{3}{1 + \tan^3\theta} = \frac{\tan^2\theta - \tan\theta - 2}{1 + \tan^3\theta}$$

$$= \frac{(\tan\theta + 1)(\tan\theta - 2)}{(\tan\theta + 1)(\tan^2\theta - \tan\theta + 1)}$$

$$= \frac{\tan\theta - 2}{\tan^2\theta - \tan\theta + 1} \to -1.$$

Therefore, the area between the curve and its asymptote lying in the second quadrant equals

$$\frac{a^2}{2}(2 - 1) = \frac{a^2}{2}.$$

Because of symmetry about the line $y = x$, $a^2/2$ is also the area between the curve and the asymptote lying in the fourth quadrant. Also, the area in the third quadrant, being that of a triangle, is $a^2/2$. Finally, the area between the curve and its asymptote equals

$$\frac{a^2}{2} + \frac{a^2}{2} + \frac{a^2}{2} = \frac{3a^2}{2}$$

which is the same as the area of the loop.

EXAMPLE 6. The area of the region enclosed by the ellipse

$$\frac{x^2}{a^2} + \frac{y^2}{b^2} = 1$$

is πab.

Indeed, considering the parametric representation

$$x = a(\cos t), \qquad y = b(\sin t), \qquad o \leq t \leq 2\pi,$$

we obtain

$$x(t)y'(t) - y(t)x'(t) = a(\cos t)b(\cos t) + b(\sin t)a(\sin t) = ab$$

and so, by (6.105),

$$A = \frac{1}{2}\int_0^{2\pi} \{x(t)y'(t) - y(t)x'(t)\}\, dt = \frac{1}{2}\int_0^{2\pi} ab\, dt = \pi ab.$$

EXAMPLE 7. The volume of the ellipsoid

$$\frac{x^2}{a^2} + \frac{y^2}{b^2} + \frac{z^2}{c^2} = 1$$

is $\frac{4}{3}\pi abc$.

Indeed, the section of the ellipsoid by the plane $x = $ constant is an ellipse,

$$\frac{y^2}{b^2(1 - x^2/a^2)} + \frac{z^2}{c^2(1 - x^2/a^2)} = 1,$$

with semiaxes $b(1 - x^2/a^2)^{1/2}$ and $c(1 - x^2/a^2)^{1/2}$. Hence, the area of the section, by Example 6, is

$$A(x) = \pi b\left(1 - \frac{x^2}{a^2}\right)^{1/2} c\left(1 - \frac{x^2}{a^2}\right)^{1/2} = \pi bc\left(1 - \frac{x^2}{a^2}\right)^{1/2},$$

where $-a \leq x \leq a$. By (6.98), the volume V of the ellipsoid is

$$V = \int_{-a}^{a} \pi bc\left(1 - \frac{x^2}{a^2}\right) dx = \pi bc\left(x - \frac{x^3}{3a^2}\right)\Bigg|_{-a}^{a} = \frac{4}{3}\pi abc.$$

In the particular case $a = b = c$ the ellipsoid turns into a sphere of radius a and V becomes $\frac{4}{3}\pi a^3$.

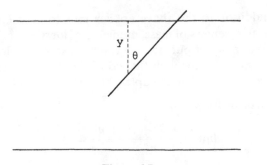

Figure 6.7

EXAMPLE 8 (Buffon's Needle Problem). A table is ruled with equidistant parallel lines a distance D apart. A needle of length L, where $L < D$, is randomly thrown on the table. What is the probability P that the needle will intersect one of the lines (the other possibility being that the needle will be completely contained in the strip between two lines)?

Let us determine the position of the needle by specifying the distance y from the middle point of the needle to the nearest parallel line, and the angle θ between the needle and the projected line of length y (see Figure 6.7). The needle will intersect a line if the hypotenuse of the right triangle in Figure 6.7 is less than $L/2$, that is, if

$$\frac{y}{\cos \theta} < \frac{L}{2} \quad \text{or} \quad y < \frac{L}{2}\cos \theta.$$

As y varies between 0 and $D/2$ and θ between 0 and $\pi/2$, it is reasonable to assume that they are independent, uniformly distributed random variables over these respective ranges. The possible cases are described by the condition $0 \le \theta \le \pi/2$, $0 \le y \le D/2$; the favorable cases by $y < (L/2)\cos \theta$. The required probability P is therefore equal to the ratio of the area of the region below the curve

$$y = \frac{L}{2}(\cos \theta) \quad \text{with } 0 \le \theta \le \pi/2$$

and the area of the rectangle $0 \le \theta \le \pi/2$, $0 \le y \le D/2$; thus,

$$P = \frac{\int_0^{\pi/2} (L/2)(\cos \theta)\, d\theta}{(\pi/2)(D/2)} = \frac{2L}{\pi D}. \tag{6.109}$$

The chief fascination with (6.109) is that from it we get

$$\pi = \frac{2L}{PD}$$

and an experimental evaluation of P leads to a statistical determination of π.

REMARKS. In addition to the above applications to geometry, integrals can also be used to find centers of mass, moments of inertia, work, fluid pressure, and so forth. We shall not pursue such topics. Instead, we look briefly at the computing of limits of sums with the aid of the definite integral in the next two examples as well as a few examples that are of general interest.

EXAMPLE 9. We have, for $p > 0$,

$$\lim_{n \to \infty} \frac{1^p + 2^p + \cdots + n^p}{n^{p+1}} = \frac{1}{p+1}.$$

Indeed, we have (see Proposition 5.6)

$$\frac{1^p + 2^p + \cdots + n^p}{n^{p+1}} = \frac{1}{n}\left(\left(\frac{1}{n}\right)^p + \left(\frac{2}{n}\right)^p + \cdots + \left(\frac{n}{n}\right)^p\right)$$

$$= \frac{1}{n}\left(f\left(\frac{1}{n}\right) + f\left(\frac{2}{n}\right) + \cdots + f\left(\frac{n}{n}\right)\right)$$

$$\to \int_0^1 f(x)\,dx \quad \text{as } n \to \infty,$$

where

$$f(x) = x^p$$

and

$$\int_0^1 x^p\,dx = \frac{1}{p+1}.$$

EXAMPLE 10. We have, as $n \to \infty$,

$$\left(g(a) \cdot g\left(a + \frac{h}{n}\right) \cdot g\left(a + \frac{2h}{n}\right) \cdots g\left(a + \frac{nh}{n}\right)\right)^{1/n} \to \exp\left(\frac{1}{h}\int_a^{a+h} \ln[g(x)]\,dx\right).$$

In particular,

$$\left(\left(1 + \frac{1}{n^2}\right)\left(1 + \frac{2^2}{n^2}\right) \cdots \left(1 + \frac{n^2}{n^2}\right)\right)^{1/n} \to 2e^{(\pi-4)/2} \quad \text{as } n \to \infty.$$

Indeed, taking logarithms,

$$\frac{1}{n}\left(\ln[g(a)] + \ln\left[g\left(a + \frac{h}{n}\right)\right] + \cdots + \ln\left[g\left(a + \frac{nh}{n}\right)\right]\right)$$

$$\to \frac{1}{h}\int_a^{a+h} \ln[g(x)]\,dx$$

as $n \to \infty$. Moreover, .

$$\int_0^1 \ln(1 + x^2)\, dx = x\{\ln(1 + x^2)\}\Big|_0^1 - \int_0^1 \frac{2x^2}{1 + x^2}\, dx$$

$$= x\{\ln(1 + x^2)\}\Big|_0^1 - 2\int_0^1 dx + 2\int_0^1 \frac{1}{1 + x^2}\, dx$$

$$= \ln 2 + \tfrac{1}{2}(\pi - 4).$$

EXAMPLE 11. We have

$$\sum_{n=1}^r \frac{1}{n} = \sum_{n=1}^r (-1)^{n+1} \frac{1}{n}\binom{r}{n}.$$

Indeed, direct evaluation of the integral

$$\int_0^1 \frac{1 - x^r}{1 - x}\, dx$$

gives the left-hand side of the given expression. Then the evaluation of the same integral after making the substitution $x = 1 - u$ gives the right-hand side of the given expression.

EXAMPLE 12. Let $E_1, E_2, \ldots, E_n$ be n intervals which are situated in the unit interval $[0, 1]$. If each point of $[0, 1]$ belongs to at least q of these intervals E_j with $j = 1, \ldots, n$, then at least one of these intervals must have length $\geq q/n$.

Indeed, for $x \in [0, 1]$, define $f_j(x) = 1$ if $x \in E_j$ and $f_j(x) = 0$ if $x \notin E_j$; then let

$$f(x) = \sum_{j=1}^n f_j(x).$$

Evidently, $f(x) \geq q$ for every x in the interval $[0, 1]$ and so

$$q \leq \int_0^1 f(x)\, dx = \int_0^1 \sum_{j=1}^n f_j(x)\, dx = \sum_{j=1}^n \int_0^1 f_j(x)\, dx = \sum_{j=1}^n |E_j|,$$

where $|E_j|$ denotes the length of the interval E_j. It is clear that not every summand in the last sum can be less than q/n, for if it were then we would have $q < n(q/n)$.

EXAMPLE 13. Let f be integrable over every interval of finite length and

$$f(x + y) = f(x) + f(y) \tag{6.110}$$

for any real numbers x and y. Then $f(x) = cx$, where $c = f(1)$.

Indeed, integrating $f(y) = f(u + y) - f(u)$ with respect to u over the interval $[0, x]$, we see that

$$xf(y) = \int_0^x f(u + y)\, du - \int_0^x f(u)\, du.$$

Putting $u + y = s$, we see that

$$\int_0^x f(u + y)\, du = \int_y^{x+y} f(s)\, ds = \int_0^{x+y} f(s)\, ds - \int_0^y f(s)\, ds.$$

Therefore,

$$xf(y) = \int_0^{x+y} f(u)\, du - \int_0^x f(u)\, du - \int_0^y f(u)\, du. \qquad (6.111)$$

Since the right-hand side of (6.111) is invariant under the interchange of x and y, it follows that $xf(y) = yf(x)$. Thus, for $x \neq 0$, $x^{-1}f(x) = c$, a constant; hence, $f(x) = cx$. Since in (6.110) $f(0) = 0$, $f(x) = cx$ also holds for $x = 0$. Taking $x = 1$ in $f(x) = cx$, we obtain $c = f(1)$.

REMARK. If f satisfies (6.110), then f being continuous at a single point implies that f is continuous everywhere. Indeed,

$$|f(x + h) - f(x)| = |f(h)| = |f(y + h) - f(y)|.$$

EXAMPLE 14. An arithmetic progression and a geometric progression each have n terms and also have the same first term a and the same last term b. Their sums are s_1 and s_2, respectively. Then

$$\lim_{n \to \infty} \frac{s_1}{s_2} = \frac{1}{2} \frac{b + a}{b - a} \ln \frac{b}{a}.$$

Indeed, the clue here is to note that the exponential function takes an arithmetic progression into a geometric progression. Actually,

$$\lim_{n \to \infty} \frac{s_1}{s_2} = \frac{\dfrac{1}{b - a} \displaystyle\int_a^b x\, dx}{\dfrac{1}{\ln b - \ln a} \displaystyle\int_{\ln a}^{\ln b} e^x\, dx}$$

and working out these integrals, we get the desired result.

REMARKS. By Proposition 5.6,

$$\frac{1}{B - A} \int_A^B f(x)\, dx = \lim_{n \to \infty} \frac{1}{n} \sum_{k=1}^n f\left(A + (B - A)\frac{k}{n}\right),$$

where f is integrable on the interval $[A, B]$. Since

$$\lim_{n \to \infty} \frac{1}{n} f(A) = 0,$$

we may also write

$$\frac{1}{B - A} \int_A^B f(x)\, dx = \lim_{n \to \infty} \frac{1}{n} \sum_{k=0}^n f\left(A + (B - A)\frac{k}{n}\right).$$

EXAMPLE 15. The number π is irrational, that is, supposing $\pi = a/b$, where a and b are integers, leads to a contradiction.

Indeed, let

$$f(x) = \frac{x^n(a - bx)^n}{n!},$$

where n is an integer (to be specified more precisely later). Thus $f(x) = f(\pi - x)$. We put

$$F(x) = f(x) - f''(x) + f^{(4)}(x) + \cdots + (-1)^n f^{(2n)}(x)$$

and observe that

$$F(x) + F''(x) = f(x)$$

since $f^{(k)}(x) = 0$ for $k > 2n$. Moreover, $f^{(k)}(0)$ is an integer for all k. In fact, this is obvious for $k > 2n$ and for $k < n$ since then the derivative is zero. For other values of k the derivative is the product of $k!/n!$, which is an integer, and the coefficient of x^k in $x^n(a - bx)^n$, which is also an integer. Since $f(x) = f(\pi - x)$, it follows that $f^{(k)}(\pi)$ is an integer as well for all k, so that both $F(0)$ and $F(\pi)$ are integers.

Now, since $F(x) + F''(x) = f(x)$,

$$\frac{d}{dx}(F'(x)\sin x - F(x)\cos x) = [F''(x) + F(x)]\sin x = f(x)\sin x$$

and so

$$\int_0^\pi f(x)(\sin x)\,dx = (F'(x)\sin x - F(x)\cos x)\Big|_0^\pi = F(\pi) + F(0),$$

implying that the integral here is an integer, say N. But, if $0 < x < \pi$, then

$$0 < f(x)\sin x \le f(x) < \frac{\pi^n}{n!}a^n.$$

Hence, by Proposition 5.22,

$$0 < N < \frac{\pi^n a^n}{n!}\pi. \tag{6.112}$$

But

$$\lim_{n\to\infty} \frac{(\pi a)^n}{n!} = 0$$

(e.g., see Number 11 of the worked examples at the end of Section 1 in Chapter 7), so that the right-hand term in inequality (6.112) is less than 1 for sufficiently large n, and thus $0 < N < 1$. Since N is an integer, this is clearly impossible, and our assumption that π is rational has led us to a contradiction. Thus, π is irrational.

EXAMPLE 16. If $t \geq 1$ and $s \geq 0$, then $ts \leq t(\ln t) - t + e^s$.

Indeed, let $f(x) = \ln(x + 1)$, $A = t - 1$, and $B = s$ in Young's Inequality [see relation (5.38) in Chapter 5].

EXAMPLE 17. Find the position of a normal chord of a parabola such that it cuts off from the parabola a segment of minimum area. (Note that a *normal* to a curve is a straight line which is perpendicular to a tangent at the point of tangency.)

To solve the problem we choose coordinates so that the equation of the parabola is $4ay = x^2$, $a > 0$. The chord connecting the point $P(2as, as^2)$ to the point $Q(2at, at^2)$ has the equation

$$y = \tfrac{1}{2}(t + s)x - ast \tag{6.113}$$

and the tangent line at $(2at, at^2)$ has slope t. Hence, the line (6.113) will be normal to the parabola at Q if and only if $\tfrac{1}{2}t(t + s) = -1$ which may be written

$$s = -\frac{2}{t} - t. \tag{6.114}$$

We see therefore that s and t have opposite signs. Take $s < 0$ and $t > 0$. Then the area cut off by the chord is

$$\int_{2as}^{2at} \left[\frac{1}{2}(t + s)x - ast - \frac{1}{4a}x^2 \right] dx = \tfrac{1}{3}a^2(t - s)^3.$$

This area will be minimal when $t - s$ is minimal. But, by (6.114),

$$t - s = 2t + \frac{2}{t} = 2\left(\sqrt{t} - \frac{1}{\sqrt{t}} \right)^2 + 4 \geq 4.$$

Equality is attained only when $\sqrt{t} = 1$ and hence $t = 1$.

Thus, of all normals to the parabola at points to the right of the axis the normal at $(2a, a)$ cuts off the largest area. The area cut off is $64a^2/3$. By symmetry, the normal at $(-2a, a)$ cuts off the least area among normals at the point to the left of the axis. The critical normals can be characterized as those which meet the axis at an angle of $\pi/4$.

EXERCISES TO CHAPTER 6

6.1. A regular polygon of n sides rolls on a straight line. Show that the length of the path described by a vertex of the polygon in a complete revolution is

$$4a \sum_{r=1}^{n-1} \frac{\pi}{n} \sin \frac{r\pi}{n},$$

where a is the circumradius of the polygon. Deduce that if a circle of radius a rolls on a line, in a complete revolution, the length of the path of a point on the circumference, that is, the length of an arch of a cycloid having a generating circle of radius a, is $8a$.

In a similar way establish that the area bounded by an arch of the cycloid and the base line is three times the area of the generating circle.

[Hint: Let the vertices be numbered in order from 1 to n. Initially, the side $n1$ rests on the line, then the side 12, then 23, and so on. The polygon turns about the vertices 1, 2, 3, ... in turn, turning through an angle $2\pi/n$ about each vertex. When the polygon turns about the vertex r, the vertex n describes an arc of a circle of angle $2\pi/n$ and of radius rn. The diagonal rn subtends an angle $2r\pi/n$ at the center, and so the length of the diagonal is $2a\sin(r\pi/n)$. Hence, the length of the path of the vertex n, in one revolution of the polygon, is

$$4a \sum_{r=1}^{n-1} \frac{\pi}{n} \sin \frac{r\pi}{n}.$$

But

$$\sum_{r=1}^{n-1} \frac{\pi}{n} \sin \frac{r\pi}{n} \to \int_0^\pi (\sin x)\, dx = 2 \quad \text{as } n \to \infty.$$

Thus, the length of an arch of the cycloid, the locus of a point on the circumference of a rolling circle of radius a, is $8a$.

The area bounded by the path of the vertex n, in one revolution, and the line on which the polygon rolls, is the sum of the areas of the $n-1$ sectors with centers 1, 2, 3, ..., $n-1$, angle $2\pi/n$, and radii $2a\sin(r\pi/n)$, $r = 1, 2, 3, ..., n-1$, together with the sum of the areas of triangles $12n, 23n, 34n, ..., (n-2)(n-1)n$, that is

$$\sum_{r=1}^{n-1} \frac{1}{2} \frac{2\pi}{n} 4a^2 \sin^2 \frac{r\pi}{n} + \frac{n}{2} a^2 \sin \frac{2\pi}{n} = 4a^2 \sum_{r=1}^{n-1} \frac{\pi}{n} \sin^2 \frac{r\pi}{n} + \frac{n}{2} a^2 \sin \frac{2\pi}{n}.$$

But, as $n \to \infty$,

$$\sum_{r=1}^{n-1} \frac{\pi}{n} \sin^2 \frac{r\pi}{n} \to \int_0^\pi (\sin^2 x)\, dx = \tfrac{1}{2}\pi$$

and

$$\frac{na^2}{2} \sin \frac{2\pi}{n} = \pi a^2 \frac{\sin(2\pi/n)}{2\pi/n} \to \pi a^2$$

and so the area bounded by an arch of the cycloid and the base line is $2\pi a^2 + \pi a^2 = 3\pi a^2$.]

6.2. The centers of two spheres of radii a and b are at a distance c apart, where $c > a + b$. Where must a point source of light be placed on the line of centers between the two spheres so as to illuminate the greatest total surface?

[Hint: Use formula (6.108). Answer: Divide line of centers in the ratio $a^{3/2}$ to $b^{3/2}$.]

6.3. Show that

$$\int_0^\pi \frac{x \sin x}{1 + \cos^2 x}\, dx = \frac{\pi^2}{4}.$$

[Hint: It is easy to see that $\int_0^a f(x)\, dx = \int_0^a f(a - x)\, dx$. Hence,

$$\int_0^\pi \frac{x \sin x}{1 + \cos^2 x} dx = \int_0^\pi \frac{(\pi - x)\sin(\pi - x)}{1 + \cos^2(\pi - x)} dx = \int_0^\pi \frac{(\pi - x)\sin x}{1 + \cos^2 x}$$

and so we obtain

$$2 \int_0^\pi \frac{x \sin x}{1 + \cos^2 x} dx = \pi \int_0^\pi \frac{\sin x}{1 + \cos^2 x} dx = -\pi \tan^{-1}(\cos x)\Big|_0^\pi = \frac{\pi^2}{2}.]$$

6.4. Show that

$$\int_0^{\pi/2} \frac{\sin^2 x}{\sin x + \cos x} dx = \frac{1}{\sqrt{2}} \ln(1 + \sqrt{2}).$$

[Hint: See the hint to Exercise 6.3. Let

$$f(x) = \frac{\sin^2 x}{\sin x + \cos x}.$$

Then

$$f(\tfrac{1}{2}\pi - x) = \frac{\cos^2 x}{\cos x + \sin x}$$

and so

$$\int_0^{\pi/2} \frac{\sin^2 x}{\sin x + \cos x} dx = \int_0^{\pi/2} \frac{\cos^2 x}{\cos x + \sin x} dx = I$$

Hence,

$$2I = \int_0^{\pi/2} \frac{1}{\cos x + \sin x} dx.$$

Putting $\tan(x/2) = t$, we get (see the proof of Proposition 6.4)

$$2I = \int_0^{\pi/2} \frac{1}{\cos x + \sin x} dx = 2 \int_0^1 \frac{1}{2 - (t - 1)^2} dt$$

$$= 2 \frac{1}{2\sqrt{2}} \int_0^1 \left(\frac{1}{\sqrt{2} - (t - 1)} + \frac{1}{\sqrt{2} + (t - 1)} \right) dt$$

$$= \frac{1}{\sqrt{2}} \ln \left| \frac{\sqrt{2} + (t - 1)}{\sqrt{2} - (t - 1)} \right| \Big|_0^1 = \frac{2}{\sqrt{2}} \ln(\sqrt{2} + 1)$$

and so $I = (1/\sqrt{2})\ln(1 + \sqrt{2}).]$

6.5. Evaluate $\int [\sin(x - a)\sin(x - b)]^{-1} dx$.
   [Hint: Since

$$\sin(b - a) = \sin[(x - a) - (x - b)]$$

$$= \sin(x - a)\cos(x - b) - \cos(x - a)\sin(x - b),$$

we have

$$\frac{\sin(b - a)}{\sin(x - a)\sin(x - b)} = \cot(x - b) - \cot(x - a).$$

Therefore, the integral in question equals

$$\frac{1}{\sin(a-b)} \ln \left| \frac{\sin(x-a)}{\sin(x-b)} \right| + c.]$$

6.6. Reduce the evaluation of

$$\int \frac{A \cos x + B \sin x + C}{D \cos x + E \sin x + F} \, dx$$

to the evaluation of

$$\int \frac{1}{D \cos x + E \sin x + F} \, dx.$$

[Hint: We determine three constants λ, μ, and v such that

$$A \cos x + B \sin x + C = \lambda(D \cos x + E \sin x + F) + \mu(-D \sin x + E \cos x) + v.$$

They are given by the equations

$$A = D\lambda + E\mu, \qquad B = E\lambda - D\mu, \qquad C = F\lambda + v.$$

With these values of λ, μ, and v, we have

$$\int \frac{A \cos x + B \sin x + C}{D \cos x + E \sin x + F} \, dx$$

$$= \int \lambda \, dx + \mu \int \frac{-D \sin x + E \cos x}{D \cos x + E \sin x + F} \, dx + v \int \frac{1}{D \cos x + E \sin x + F} \, dx$$

$$= \lambda x + \mu \ln |D \cos x + E \sin x + F| + v \int \frac{1}{D \cos x + E \sin x + F} \, dx.]$$

6.7. For $n = 1, 2, 3, \dots$, let

$$S_n = \frac{1}{n\sqrt{n}} (\sqrt{1} + \sqrt{2} + \cdots + \sqrt{n}).$$

Find S_n as $n \to \infty$.

[Hint: We have

$$S_n = \frac{1}{n} \left(\frac{\sqrt{1}}{n} + \frac{\sqrt{2}}{n} + \cdots + \frac{\sqrt{n}}{n} \right) \to \int_0^1 \sqrt{x} \, dx = \frac{2}{3} \quad \text{as } n \to \infty.]$$

6.8. For $n = 1, 2, 3, \dots$, let

$$S_n = \frac{1}{n} + \frac{1}{(n^2 - 1)^{1/2}} + \frac{1}{(n^2 - 2^2)^{1/2}} + \cdots + \frac{1}{(n^2 - \{n-1\}^2)^{1/2}}.$$

Find S_n as $n \to \infty$.

[Hint: We have, putting $h = 1/n$,

$$S_n = h \left(1 + \frac{1}{(1 - h^2)^{1/2}} + \frac{1}{(1 - \{2h\}^2)^{1/2}} + \cdots + \frac{1}{(1 - \{n-1\}^2 h^2)^{1/2}} \right)$$

$$\to \int_0^1 \frac{1}{(1 - x^2)^{1/2}} \, dx = \frac{\pi}{2} \quad \text{as } n \to \infty.]$$

6.9. Using partial fraction decomposition, evaluate

$$\int \frac{x^2 + 5x + 41}{(x + 3)(x - 1)(2x - 1)}\,dx.$$

[Answer: $\frac{5}{4}\ln|x + 3| + \frac{47}{4}\ln|x - 1| - \frac{25}{2}\ln|2x - 1| + c.$]

6.10. Evaluate

$$\int \frac{1}{x^3(x - 1)^2(x + 1)}\,dx.$$

[Answer:

$$2\ln|x| - \frac{1}{x} - \frac{1}{2x^2} - \frac{7}{4}\ln|x - 1| - \frac{1}{2(x - 1)} - \frac{1}{4}\ln|x + 1| + c.$$]

6.11. Evaluate

$$\int \frac{1}{(x - 1)^2(x^2 + 4)}\,dx.$$

[Answer:

$$-\frac{2}{25}\ln|x - 1| - \frac{1}{5(x - 1)} + \frac{1}{25}\ln(x^2 + 4) - \frac{3}{50}\tan^{-1}\frac{x}{2} + c.$$]

6.12. Evaluate

$$\int \frac{(x^2 + 1)(x^2 + 2)}{(x^2 + 3)(x^2 + 4)}\,dx.$$

[Hint: Put $x^2 = y$. Answer:

$$x + \frac{2}{\sqrt{3}}\tan^{-1}\frac{2}{\sqrt{3}} - 3\tan^{-1}\frac{x}{2} + c.$$]

6.13. Evaluate

$$\int \frac{1 + x^2}{1 + x^4}\,dx.$$

[Hint: Put $x - 1/x = y$. Answer:

$$\frac{1}{\sqrt{2}}\tan^{-1}\frac{x^2 - 1}{\sqrt{2x}} + c.$$]

6.14. The integration of $(\sin^p x)(\cos^q x)$, when $p + q$ is a negative even integer, is easily accomplished by use of the substitution $\tan x = t$. Indeed, we have

$$\sin x = \frac{t}{(1 + t^2)^{1/2}} \quad \text{and} \quad \cos x = \frac{1}{(1 + t^2)^{1/2}}$$

and so, putting $p + q = -2n$,

$$\int (\sin^p x)(\cos^q x)\,dx = \int t^p(1 + t^2)^{n-1}\,dt.$$

Evaluate $\int (\cot t)^{1/2}(\sec^4 x)\,dx$.
[Answer: $2(\tan x)^{1/2} + \frac{2}{5}(\tan^5 x)^{1/2} + c.$]

6.15. Evaluate

$$\int_{-\pi/2}^{\pi/2} \frac{1}{5 + 7 \cos x + \sin x} \, dx.$$

[Hint: Substitute $\tan(x/2) = t$. Answer: $\frac{1}{3} \ln 6$.]

6.16. Evaluate

$$\int \frac{1}{a^2 \sin^2 x + b^2 \cos^2 x} \, dx = \int \frac{\sec^2 x}{b^2 + a^2 \tan^2 x} \, dx.$$

[Hint: Substitute $\tan x = t$.

$$\text{Answer: } \frac{1}{ab} \tan^{-1}\left(\frac{a \tan x}{b}\right) + c.]$$

6.17. Evaluate

$$\int \frac{\tan x}{a^2 + b^2 \tan^2 x} \, dx.$$

[Hint: Substitute $\tan^2 x = t$. Answer:

$$-\frac{1}{2(a^2 - b^2)} \ln(a^2 \cos^2 x + b^2 \sin^2 x) + c.]$$

6.18. Evaluate

$$\int \frac{1}{(1 + x)^{1/2} + (1 + x)^{1/3}} \, dx.$$

[Hint: Substitute $1 + x = t^6$. Answer:

$$2(1 + x)^{1/2} - 3(1 + x)^{1/3} + 6(1 + x)^{1/6} - 6 \ln|1 + (1 + x)^{1/6}| + c.]$$

6.19. Evaluate

$$\int \left(\frac{1 - x}{1 + x}\right)^{1/2} \frac{1}{x} \, dx.$$

[Hint: Substitute $(1 - x)/(1 + x) = y^2$. Answer:

$$\ln\left|\frac{(1 + x)^{1/2} - (1 - x)^{1/2}}{(1 + x)^{1/2} + (1 - x)^{1/2}}\right| + 2 \tan^{-1}\left(\frac{1 - x}{1 + x}\right)^{1/2} + c.]$$

6.20. Using the Hermite–Ostrogradski Formula (6.65), evaluate

$$\int \frac{1}{(x^3 - 1)^2} \, dx.$$

[Hint: We have

$$\int \frac{1}{(x^3 - 1)^2} \, dx = \frac{Ax^2 + Bx + C}{x^3 - 1} + \int \frac{Dx^2 + Ex + F}{x^3 - 1} \, dx,$$

where $A = 0$, $B = -\frac{1}{3}$, $C = D = E = 0$, and $F = -\frac{2}{3}$. Answer:

$$-\frac{x}{3(x^3 - 1)} + \frac{1}{9} \ln \frac{x^2 + x + 1}{(x - 1)^2} - \frac{1}{\sqrt{3}} \tan^{-1} \frac{2x + 1}{\sqrt{3}} + c.]$$

6.21. Evaluate

$$\int \frac{1}{(x^2 + 1)^4} dx.$$

[Answer:

$$\frac{15x^5 + 40x^3 + 33x}{48(1 + x^2)^3} + \frac{15}{48} \tan^{-1} x + c.]$$

6.22. Evaluate

$$\int \frac{x^4 - 2x^2 + 2}{(x^2 - 2x + 2)} dx.$$

[Answer:

$$x - \frac{x - 1}{x^2 - 2x + 2} + 2\ln(x^2 - 2x + 2) + 3\tan^{-1}(x - 1) + c.]$$

6.23. Evaluate

$$\int \frac{1}{(x + 1)^2(x^2 + 1)} dx.$$

[Answer:

$$\frac{-x^2 + x}{4(x + 1)(x^2 + 1)} + \frac{1}{2}\ln|x + 1| - \frac{1}{4}\ln(x^2 + 1) + \frac{1}{4}\tan^{-1} x + c.]$$

6.24. Using the result in Proposition 6.5, evaluate

$$\int \frac{x + x^{2/3} + x^{1/6}}{x(1 + x^{1/3})} dx.$$

[Hint: Use the substitution $x = t^6$. Answer: $\frac{3}{2}x^{2/3} + 6\tan^{-1} x^{1/6} + c.]$

6.25. Evaluate

$$\int \frac{(2x - 3)^{1/2}}{(2x - 3)^{1/3} + 1} dx.$$

[Hint: Use the substitution $2x - 3 = t^6$. Answer:

$$\tfrac{3}{7}(2x - 3)^{7/6} - \tfrac{3}{5}(2x - 3)^{5/6} + (2x - 3)^{1/2} - 3(2x - 3)^{1/6} + \tan^{-1}(2x - 3)^{1/6} + c.]$$

6.26. Evaluate

$$\int \frac{2}{(2 - x)^2} \left(\frac{2 - x}{2 + x}\right)^{1/3} dx.$$

[Hint: Use the substitution $(2 - x)/(2 + x) = t^3$. Answer:

$$\frac{3}{4}\left(\frac{2 - x}{2 + x}\right)^{2/3} + c.]$$

6.27. Evaluate

$$\int \frac{1}{[(x - 1)^3(x + 2)^5]^{1/4}} dx.$$

[Hint: Use the substitution $(x + 2)/(x - 1) = t^4$. Answer:

$$\frac{4}{3}\left(\frac{x-1}{x+2}\right)^{1/4} + c.]$$

6.28. Using the result in Proposition 6.6, evaluate

$$\int x^{-2/3}(1 + x^{1/3})^{1/2}\,dx.$$

[Hint: Use the substitution $1 + x^{1/3} = t^2$. Answer: $2(1 + x^{1/3})^{3/2} + c.$]

6.29 Evaluate

$$\int x^{-11}(1 + x^4)^{-1/2}\,dx.$$

[Hint: We use the substitution $1 + x^4 = x^4t^2$. Answer:

$$-\frac{1}{10x^{10}}(1 + x^4)^{5/2} + \frac{1}{3x^6}(1 + x^4)^{3/2} - \frac{1}{2x^2}(1 + x^4)^{1/2} + c.]$$

6.30. Using the result in Proposition 6.7, evaluate

$$\int \frac{1}{x + (x^2 - x + 1)^{1/2}}\,dx.$$

[Hint: We use the substitution $(x^2 - x + 1)^{1/2} = tx + 1$ and so the given integral becomes

$$\int \frac{2 + 2t + 2t^2}{(2 + t)(1 - t)(1 + t)^2}\,dx$$

$$= 2\ln|2 + t| - \tfrac{1}{2}\ln|1 - x| - \tfrac{3}{2}\ln|1 + x| - \frac{1}{1+t} + c.$$

We then set

$$t = \frac{(x^2 - x + 1)^{1/2} - 1}{x}.]$$

6.31. Using relation (6.81), verify that

$$\int \frac{30x^5 + 30x^4 + 12x^3 + 21x^2 - 15x - 1}{(4 + 2x + 3x^2)^{1/2}}\,dx$$

$$= (2x^4 + x^3 - 3x^2 + 4x - 1)(4 + 2x + 3x^2)^{1/2} - 16\int \frac{1}{(4 + 2x + 3x^2)^{1/2}}\,dx.$$

Moreover, verify that

$$\int \frac{1}{(3x^2 + 2x + 4)^{1/2}}\,dx = \frac{1}{\sqrt{3}}\ln|1 + 3x + (9x^2 + 6x + 12)^{1/2}| + c.$$

6.32. Evaluate the integral

$$I = \int \frac{x + 4}{(x - 1)(x + 2)^2(x^2 + x + 1)^{1/2}}\,dx.$$

[Hint: Noting that

$$\frac{x+4}{(x-1)(x+2)^2} = \frac{A}{x-1} + \frac{B}{(x+2)^2} + \frac{C}{x+2},$$

where $A = \frac{5}{9}$, $B = -\frac{2}{3}$, and $C = -\frac{5}{9}$, we obtain that

$$I = \frac{5}{9} \int \frac{1}{(x-1)(x^2+x+1)^{1/2}} \, dx - \frac{2}{3} \int \frac{1}{(x+2)^2(x^2+x+1)^{1/2}} \, dx$$

$$- \frac{5}{9} \int \frac{1}{(x+2)(x^2+x+1)^{1/2}} \, dx.$$

The first integral is calculated by the substitution $x - 1 = 1/t$, the second and the third by the substitution $x + 2 = 1/t$.]

6.33. Evaluate

$$\int \frac{1}{(x^2+2)(2x^2-2x+5)^{1/2}} \, dx.$$

[Answer:

$$\frac{1}{6\sqrt{2}} \ln \left| \frac{[2(2x^2-2x+5)]^{1/2} - (x+1)}{[2(2x^2-2x+5)]^{1/2} + (x+1)} \right| - \frac{1}{3} \tan^{-1} \frac{(2x^2-2x+5)^{1/2}}{x+1} + c.]$$

6.34. Evaluate, using the result in Proposition 6.7,

$$\int \frac{1}{[1 + \{x(1+x)\}^{1/2}]^2} \, dx.$$

[Answer:

$$\frac{2(3-4z)}{5(1-z-z^2)} + \frac{2}{5\sqrt{5}} \ln \left| \frac{\sqrt{5}+1+2z}{\sqrt{5}-1-2z} \right| + c, \quad \text{where } z = -x + \{x(1+x)\}^{1/2}.]$$

CHAPTER 7

Infinite Series

1. Numerical Sequences

Definition. A *sequence* of real numbers $x_1, x_2, \ldots, x_n, \ldots$ is a function that assigns to each positive integer n a number x_n. The number x_n is called the nth *term* of the sequence. Sometimes the notation $\{x_n\}_{n=1}^{\infty}$ or, more simply, $\{x_n\}$ is used as an abbreviation of the sequence $x_1, x_2, \ldots, x_n, \ldots$.

Definition. A sequence $\{x_n\}$ of real numbers is called a *null sequence* if for any $\varepsilon > 0$ there exists a positive integer n_0 such that

$$|x_n| < \varepsilon \quad \text{for any } n \geq n_0. \tag{7.1}$$

In place of the expression "for any $n \geq n_0$" we shall often use the phrase "for all sufficiently large n."

REMARKS. The defining condition means that given an arbitrarily small positive number ε, we can always find an n_0 (in general dependent on ε) such that (7.1) is satisfied. The property that $\{x_n\}$ is a null sequence may also be expressed by saying that for any $\varepsilon > 0$ we have

$$|x_n| < K\varepsilon \quad \text{for all sufficiently large } n,$$

where K is any fixed (i.e., independent of ε) positive real number.

It is clear that $\{1/n\}_{n=1}^{\infty}$ is a null sequence because $1/n \leq 1/n_0$ for all $n \geq n_0 > 0$.

Definition. A sequence $\{x_n\}$ is said to be *bounded* if there is a real number M such that

$$|x_n| < M \quad \text{for all } n.$$

Proposition 7.1. *Any null sequence $\{x_n\}$ is bounded.*

PROOF. Since $\{x_n\}$ is a null sequence, we have $|x_n| \le 1$ for all $n \ge n_0$. Therefore,

$$|x_n| \le 1 + \max\{|x_n| : n < n_0\}$$

and the claim is established. $\square$

Proposition 7.2. *The sum, difference, and product of null sequences are again null sequences.*

PROOF. Let $\{x_n\}$ and $\{y_n\}$ be null sequences. Now $|x_n| < \varepsilon$, $|y_n| < \varepsilon$ for all sufficiently large n implies

$$|x_n \pm y_n| \le |x_n| + |y_n| < 2\varepsilon \quad \text{for all sufficiently large } n,$$

and, if $|x_n| < M$ for all n, then

$$|x_n y_n| = |x_n||y_n| < M\varepsilon \quad \text{for all sufficiently large } n, \tag{7.2}$$

completing the proof. $\square$

REMARK. Relation (7.2) shows that the product of a bounded sequence and a null sequence is a null sequence.

Definition. A real number a is said to be the *limit* of a sequence $\{x_n\}$ (in symbols: $a = \lim_{n \to \infty} x_n$) if $\{x_n - a\}$ is a null sequence, that is, if for any $\varepsilon > 0$ we have

$$|x_n - a| < \varepsilon \quad \text{for all sufficiently large } n.$$

We also say that x_n *tends to* or *converges to* a as n *becomes arbitrarily large* and we write $x_n \to a$ (as $n \to \infty$). A sequence of constant terms c tends to c.

Proposition 7.3. *Every sequence has at most one limit.*

PROOF. Let a and b be limits of $\{x_n\}$. Suppose that $\varepsilon > 0$ is arbitrary. There exists a positive integer n such that

$$|x_n - a| < \varepsilon \quad \text{and} \quad |x_n - b| < \varepsilon \tag{7.3}$$

[note that (7.3) holds for all sufficiently large n] and so

$$|a - b| = |(x_n - b) - (x_n - a)| \le |x_n - a| + |x_n - b| < 2\varepsilon. \tag{7.4}$$

Since $a - b$ is fixed and 2ε is an arbitrarily small positive number, (7.4) implies that $a - b = 0$. $\square$

Definition. A sequence is said to be *convergent* if it has a (finite real number as) limit. A sequence that is not convergent is said to be *divergent*.

REMARKS. Altering a finite number of terms of a sequence or the deletion or adjunction of a finite number of terms does not affect the convergence or divergence of a sequence and, in case of convergence, it also does not affect the limit.

Definition. Given a sequence $\{x_n\}$, consider a sequence $\{n_k\}_{k=1}^{\infty}$ of positive integers, such that $n_1 < n_2 < n_3 < \cdots$. Then the sequence

$$x_{n_1}, \quad x_{n_2}, \quad x_{n_3}, \ldots$$

is called a *subsequence of* $\{x_n\}$.

REMARK. A sequence $\{x_n\}$ converges to a if and only if every subsequence of $\{x_n\}$ converges to a.

Proposition 7.4. *Any convergent sequence is bounded.*

PROOF. Let $\{x_n\}$ converge to a. Since

$$|x_n| \le |x_n - a| + |a|,$$

the claim follows because $\{x_n - a\}$ is a null sequence and hence bounded (by Proposition 7.1). $\square$

Proposition 7.5. *The sum, difference, and product of convergent sequences* $\{x_n\}$ *and* $\{y_n\}$ *are convergent. Moreover, if* $x_n \to a$ *and* $y_n \to b$ *as* $n \to \infty$ *and* c *is a fixed real number, then*

$$x_n \pm y_n \to a \pm b, \quad x_n y_n \to ab, \quad and \quad cx_n \to ca.$$

If all y_n *and the limit* b *are different from zero, then the quotient* x_n/y_n *converges and we have*

$$\frac{x_n}{y_n} \to \frac{a}{b}.$$

PROOF. The statements concerning sum and difference reduce to the corresponding statements about null sequences:

$$(x_n \pm y_n) - (a \pm b) = (x_n - a) \pm (y_n - b) \to 0;$$

the statement concerning the product follows from the Remark to Proposition 7.2 and from Proposition 7.4:

$$x_n y_n - ab = (x_n - a)y_n + a(y_n - b) \to 0.$$

Moreover,

$$cx_n - ca = c(x_n - a) \to 0.$$

Having already considered the product, the statement concerning the quotient will follow if we can show that $1/y_n - 1/b \to 0$. But for all sufficiently large n we have $|y_n - b| < |b|/2$ and so

$$|y_n| \geq |b| - |y_n - b| > \frac{|b|}{2}, \qquad \frac{1}{|y_n|} < \frac{2}{|b|}.$$

This means that the sequence $\{1/y_n\}$ is bounded. Using the Remark to Proposition 7.2 and using Proposition 7.4 we conclude

$$\frac{1}{y_n} - \frac{1}{b} = -\frac{1}{by_n}(y_n - b) \to 0$$

and the proof is finished. $\qquad\qquad\square$

EXAMPLE. Let $a_0, a_1, \ldots, a_p$ with $a_p \neq 0$ be given real numbers and

$$x_n = \sum_{k=0}^{p} a_k n^k \quad \text{and} \quad x_n^* = \frac{x_n}{n^p} = \sum_{k=0}^{p} \frac{a_k}{n^{p-k}} \quad \text{for } n = 1, 2, 3, \ldots.$$

Since $1/n \to 0$ as $n \to \infty$, Proposition 7.5 gives $x_n^* \to a_p$ as $n \to \infty$. Similarly, let $b_0, b_1, \ldots, b_q$ with $b_q \neq 0$ be given real numbers and

$$y_n = \sum_{k=0}^{q} b_k n^q \quad \text{and} \quad y_n^* = \frac{y_n}{n^q} = \sum_{k=0}^{q} \frac{b_k}{n^{q-k}} \quad \text{for } n = 1, 2, 3, \ldots.$$

Since $y_n^* \to b_q \neq 0$ as $n \to \infty$, we have $y_n \neq 0$ for all sufficiently large n, say, for all $n \geq n_1 \geq 1$. We consider the sequence $\{x_n/y_n\}$ for $n \geq n_1$. Since

$$\frac{x_n^*}{y_n^*} \to \frac{a_p}{b_q} \quad \text{as} \quad n \to \infty,$$

the sequence

$$\frac{x_n}{y_n} = n^{p-q} \left(\frac{x_n^*}{y_n^*} \right)$$

tends to 0 for $q > p$ and to a_p/b_p for $q = p$ as $n \to \infty$; for $q < p$ we apparently have an unbounded (hence divergent) sequence.

Proposition 7.6. If $x_n \to a$, then $|x_n| \to |a|$ as $n \to \infty$. For $a = 0$ the converse holds as well: If $|x_n| \to 0$, then $x_n \to 0$ as $n \to \infty$.

PROOF. The first statement follows directly from

$$||x_n| - |a|| \leq |x_n - a|$$

and the proof of the second statement is trivial. $\qquad\qquad\square$

Proposition 7.7. If $x_n \to a$, $y_n \to b$ as $n \to \infty$ and $x_n \leq y_n$ for all sufficiently large n, then $a \leq b$. In particular, if $x_n \to a$, $|x_n| \leq c$ for all sufficiently large n, then $|a| \leq c$.

REMARK. The proof of Proposition 7.7 is trivial.

Definition. A sequence $\{x_n\}$ of real numbers is said to be *nondecreasing* if $x_n \leq x_{n+1}$ for all n and $\{x_n\}$ is said to be *nonincreasing* if $x_n \geq x_{n+1}$ for all n. Observe that if $\{x_n\}$ is nondecreasing then $x_n \leq x_m$ whenever $n < m$ and if $\{x_n\}$ is nonincreasing then $x_n \geq x_m$ whenever $n < m$. A sequence that is nondecreasing or nonincreasing is called a *monotonic sequence*.

Proposition 7.8. *Any bounded monotonic sequence converges.*

PROOF. It will be enough to consider the case of a nondecreasing sequence; if $\{x_n\}$ is nonincreasing, then $\{-x_n\}$ is nondecreasing.

   Let $\{x_n\}$ be a bounded nondecreasing sequence; suppose that S denotes the set of terms $\{x_n: n = 1, 2, 3, \ldots\}$ and let $M^* = \sup S$. Then

$$x_n \leq M^* \quad \text{for all } n. \tag{7.5}$$

By the Remark following the definition of supremum in Section 2 of Chapter 2, for any $\varepsilon > 0$ there exists a positive integer n_0 such that

$$x_{n_0} > M^* - \varepsilon.$$

Since the sequence $\{x_n\}$ is nondecreasing, it follows that

$$x_n > M^* - \varepsilon \quad \text{for all } n \geq n_0. \tag{7.6}$$

By (7.5) and (7.6) we have

$$|x_n - M^*| < \varepsilon \quad \text{for all sufficiently large } n$$

and the proof is finished. $\square$

REMARK. Note that the Axiom of Completeness (see Section 2 in Chapter 2) in an essential ingredient in the proof of Proposition 7.8.

Definition, A real number a is called an *accumulation point of the sequence* $\{x_n\}$ if for any $\varepsilon > 0$ we have

$$|x_n - a| < \varepsilon \quad \text{for infinitely many } n. \tag{7.7}$$

REMARK. Every accumulation point of a subsequence of $\{x_n\}$ is also an accumulation point of the sequence $\{x_n\}$. The limit of a convergent sequence $\{x_n\}$ is an accumulation point of $\{x_n\}$, in fact, the only one (as the next proposition will show).

Proposition 7.9. *If a is an accumulation point of the sequence $\{x_n\}_{n=1}^{\infty}$, then there is a subsequence $\{x_{n_k}\}_{k=1}^{\infty}$ which converges to a.*

PROOF. Using induction, we shall construct a subsequence such that

$$|x_{n_k} - a| < \frac{1}{k} \tag{7.8}$$

for all $k = 1, 2, 3, \ldots$. Evidently, we will then have $x_{n_k} \to a$ as $k \to \infty$.

We begin by choosing n_1 such that (7.8) is satisfied for $k = 1$. The condition (7.7) allows us to find such a positive integer n_1. Then we choose n_2 such that $n_2 > n_1$ and that (7.8) is satisfied for $k = 2$. In general, we pick n_k such that n_k is larger than its predecessor n_{k-1} and which satisfies (7.8); the condition (7.7) with $\varepsilon = 1/k$ guarantees the existence of such a number n_k. $\square$

Proposition 7.10 (Theorem of Bolzano and Weierstrass). *Every bounded sequence of real numbers has at least one accumulation point.*

PROOF. Let $\{x_n\}$ be a given bounded sequence and let us assume that

$$M_0 \le x_n \le N_0 \quad \text{for all } n,$$

where $M_0 < N_0$. Commencing with the closed interval $[M_0, N_0]$, we construct by continued bisection a nested sequence of intervals

$$[M_0, N_0] \supset [M_1, N_1] \supset [M_2, N_2] \supset \cdots \supset [M_k, N_k] \supset \cdots$$

such that each of these intervals contains an infinite number of terms of the sequence $\{x_n\}$, that is, for each k, where $k = 0, 1, 2, \ldots$, we have

$$M_k \le x_n \le N_k \quad \text{for infinitely many } n.$$

The possibility of such a construction is clear; if an interval J contains an infinite number of terms of a sequence, then at least one half of the interval J must contain an infinite number of terms of that sequence. It is clear that $\{M_k\}$ is a nondecreasing sequence and $\{N_k\}$ is a nonincreasing sequence; moreover, both sequences are bounded because all terms of both sequences belong to the interval $[M_0, N_0]$. It is also clear that

$$M_k \to a, \quad N_k \to a \quad \text{as} \quad k \to \infty,$$

where a is the unique point common to all closed intervals of the nested sequence $\{[M_k, N_k]\}_{k=0}^{\infty}$ (see Nested Interval Principle in Section 1 of Chapter 1). Hence, for each $\varepsilon > 0$ there exists k such that

$$a - \varepsilon < M_k < N_k < a + \varepsilon$$

and so

$$a - \varepsilon < x_n < a + \varepsilon \quad \text{for infinitely many } n,$$

implying that a is an accumulation point of $\{x_n\}$. $\square$

Proposition 7.11. *Suppose that the set A of all accumulation points of a bounded sequence $\{x_n\}$ is nonempty. Then A contains both $\sup A$ and $\inf A$.*

PROOF. Let $\alpha = \sup A$. Then for every $\varepsilon > 0$ we have that (i) at least one element a of A satisfies the inequality $a > \alpha - \varepsilon$ and (ii) every element a of A satisfies the inequality $a < \alpha - \varepsilon$.

Thus, by (i) there is some $a \in A$ with $a > \alpha - \varepsilon$. Since $a \in A$ means that a is an accumulation point of $\{x_n\}$ and because $\alpha - \varepsilon$ is a point to the left of a on the number line, we have that

$$x_n > \alpha - \varepsilon \quad \text{for infinitely many } n.$$

Since $\alpha + \varepsilon$ is to the right of every point a of A by (ii), the inequality $x_n \geq \alpha + \varepsilon$ can hold at most for a finite number of indices n and so

$$x_n < \alpha + \varepsilon \quad \text{for all sufficiently large } n.$$

Thus, $\alpha \in A$. The dual statement concerning $\inf A$ can be proved along similar lines of reasoning. $\qquad\qquad\square$

REMARK. Proposition 7.11 shows that the set A of all accumulation points of a bounded sequence has a maximum and a minimum.

Definition. The real numbers $\sup A$ and $\inf A$ occurring in Proposition 7.11 are called the *limit superior* and the *limit inferior of the sequence* $\{x_n\}$, respectively, and are denoted by

$$\varlimsup_{n \to \infty} x_n \quad \text{and} \quad \varliminf_{n \to \infty} x_n.$$

REMARKS. The numbers $\bar{a} = \varlimsup x_n$ and $\underline{a} = \varliminf x_n$ are characterized by the following properties: For any $\varepsilon > 0$

$$x_n > \bar{a} - \varepsilon \quad \text{for infinitely many } n \text{ and}$$

$$x_n > \bar{a} + \varepsilon \quad \text{for all sufficiently large } n,$$

$$x_n < \underline{a} + \varepsilon \quad \text{for infinitely many } n \text{ and}$$

$$x_n > \underline{a} - \varepsilon \quad \text{for all sufficiently large } n.$$

For a bounded sequence $\{x_n\}$ the $\varlimsup x_n$ and $\varliminf x_n$ exist and $\varliminf x_n \leq \varlimsup x_n$; equality occurs precisely when the sequence converges (and then we have $\lim x_n = \varliminf x_n = \varlimsup x_n$). It is also easy to see that $\varliminf x_n = -\varlimsup(-x_n)$.

For an *unbounded* sequence we define $\varlimsup x_n = \infty$ if $\{x_n\}$ is not bounded above and we define $\varliminf x_n = -\infty$ if $\{x_n\}$ is not bounded below; moreover, if $x_n \to \infty$ as $n \to \infty$, we put $\lim x_n = \varliminf x_n = \varlimsup x_n = \infty$ and if $x_n \to -\infty$ as $n \to \infty$, we put $\lim x_n = \varliminf x_n = \varlimsup x_n = -\infty$.

Definition. A sequence $\{x_n\}$ of real numbers is said to be a *Cauchy sequence* if for any $\varepsilon > 0$ there exists a positive integer n_0 such that

$$|x_n - x_m| < \varepsilon \quad \text{for all } n \geq n_0 \text{ and all } m \geq n_0. \qquad (7.9)$$

REMARKS. In (7.9) we may replace ε in the inequality by $M\varepsilon$, where M denotes a fixed but arbitrary positive real number independent of ε.

Without loss of generality we may suppose that $m > n$ and reformulate (7.9) as follows:

$$|x_{n+k} - x_n| < \varepsilon \quad \text{for all } n \geq n_0 \text{ and all positive integers } k.$$

Proposition 7.12. *A sequence $\{x_n\}$ of real numbers is convergent if and only if it is a Cauchy sequence.*

PROOF. Suppose that $\{x_n\}$ is convergent and let $x_n \to a$ as $n \to \infty$. Then, given any $\varepsilon > 0$, we have

$$|x_n - a| < \varepsilon, \quad |x_m - a| < \varepsilon \quad \text{for all sufficiently large } m \text{ and } n.$$

Thus,

$$|x_n - x_m| = |(x_n - a) - (x_m - a)| \leq |x_n - a| + |x_m - a| < 2\varepsilon$$

for all sufficiently large m and n, showing that $\{x_n\}$ is a Cauchy sequence.

Conversely, let $\{x_n\}$ be a Cauchy sequence, that is, given any $\varepsilon > 0$, then there exists a positive integer n_0 such that condition (7.9) is satisfied. It follows in particular that

$$|x_n - x_{n_0}| < \varepsilon \quad \text{for all } n \geq n_0$$

and so

$$|x_n| < |x_{n_0}| + \varepsilon \quad \text{for all } n \geq n_0.$$

Therefore, the larger of the two numbers

$$|x_{n_0}| + \varepsilon \quad \text{and} \quad \max\{|x_n|: n < n_0\}$$

is a bound for the $|x_n|$ and, by Proposition 7.10, the sequence $\{x_n\}$ has an accumulation point a. Therefore, for infinitely many m we have

$$|x_m - a| < \varepsilon;$$

we choose *one* such $m \geq n_0$. Then for all $n \geq n_0$ we have

$$|x_n - a| \leq |x_n - x_m| + |x_m - a| < 2\varepsilon.$$

This means that a is the limit of $\{x_n\}$ and the proof is finished.   □

Proposition 7.13. *Let $\{x_n\}$ be a convergent sequence with limit a. Then*

$$y_n = \frac{x_1 + x_2 + \cdots + x_n}{n} \to a \quad \text{as } n \to \infty.$$

PROOF. Let $x_n = v_n + a$. We must show that

$$\frac{v_1 + v_2 + \cdots + v_n}{n} \to 0 \quad \text{as } n \to \infty$$

if $\{v_n\}$ is a null sequence. Now, for $n > m$,

$$\frac{v_1 + v_2 + \cdots + v_n}{n} = \frac{v_1 + v_2 + \cdots + v_m}{n} + \frac{v_{m+1} + v_{m+2} + \cdots + v_n}{n},$$

so that

$$\left| \frac{v_1 + v_2 + \cdots + v_n}{n} \right| \leq \frac{|v_1 + v_2 + \cdots + v_m|}{n} + \frac{|v_{m+1}| + |v_{m+2}| + \cdots + |v_n|}{n}.$$

Given $\varepsilon > 0$, we can choose m so that $|v_n| < \varepsilon/2$ for all $n > m$. For all such n we have

$$\left| \frac{v_1 + v_2 + \cdots + v_n}{n} \right| \leq \frac{|v_1 + v_2 + \cdots + v_m|}{n} + \frac{\varepsilon}{2} \cdot \frac{n - m}{n}.$$

Since $|v_1 + v_2 + \cdots + v_m|$ is a fixed real number, we can pick a positive integer $n_0 > m$ such that for all $n > n_0$ we have

$$\frac{|v_1 + v_2 + \cdots + v_m|}{n} < \frac{\varepsilon}{2}.$$

Then

$$\left| \frac{v_1 + v_2 + \cdots + v_n}{n} \right| < \varepsilon \quad \text{for all } n > n_0.$$

This completes the proof.   □

Proposition 7.14. *Let $\{u_n\}$ be a convergent sequence with limit b and assume that each u_n and b are positive. Then*

$$w_n = \sqrt[n]{u_1 u_2 \cdots u_n} \to b \quad \text{as } n \to \infty.$$

PROOF. Since the natural logarithm ln is a continuous function, we get that $u_n \to b$ implies (see Proposition 2.4 and the definition of continuity)

$$x_n = \ln u_n \to a = \ln b \quad \text{as } n \to \infty.$$

By Proposition 7.12,

$$y_n = \frac{x_1 + x_2 + \cdots + x_n}{n} = \ln \sqrt[n]{u_1 u_2 \cdots u_n} = \ln w_n \to \ln b \quad \text{as } n \to \infty.$$

But the exponential function is continuous and so

$$\ln w_n \to \ln b \quad \text{implies} \quad w_n \to b \quad \text{as } n \to \infty.$$

This completes the proof.   □

APPLICATIONS. Applying Proposition 7.14 to the sequence

$$u_1, \quad \frac{u_2}{u_1}, \quad \ldots, \quad \frac{u_n}{u_{n-1}}, \quad \frac{u_{n+1}}{u_n}, \quad \ldots,$$

where $\{u_{n+1}/u_n\}$ is assumed to be convergent, we obtain

$$\lim_{n \to \infty} \sqrt[n]{u_n} = \lim_{n \to \infty} \frac{u_{n+1}}{u_n}$$

(of course $\lim_{n \to \infty} u_n/u_{n-1} = \lim_{n \to \infty} u_{n+1}/u_n$). Putting $u_n = n!/n^n$, we get

$$\frac{u_{n+1}}{u_n} = \frac{(n+1)!}{(n+1)^{n+1}} \cdot \frac{n^n}{n!} = \frac{1}{(1+1/n)^n} \to \frac{1}{e} \quad \text{as } n \to \infty$$

and so

$$\lim_{n\to\infty} \frac{\sqrt[n]{n!}}{n} = \frac{1}{e}.$$

If we put $u_n = (n+1)(n+2)\cdots(n+n)/n^n$, we obtain $u_{n+1}/u_n \to 4/e$ as $n \to \infty$ and so

$$\lim_{n\to\infty} \frac{1}{n} \sqrt[n]{(n+1)(n+2)\cdots(n+n)} = \frac{4}{e}.$$

By taking $u_n = n$, we obtain $u_{n+1}/u_n \to 1$ as $n \to \infty$ and so $\sqrt[n]{n} \to 1$ as $n \to \infty$ (a result we already know from the Lemma preceding Proposition 1.8).

Proposition 7.15. *If $\{x_n\}$ and $\{y_n\}$ converge to X and Y, respectively, then $\{z_n\}$, where*

$$z_n = \frac{x_1 y_n + x_2 y_{n-1} + \cdots + x_n y_1}{n},$$

converges to XY.

PROOF. Let $x_n = X + a_n$. Then $\{a_n\}$ is a null sequence. Since $\{y_n\}$ is convergent, it is bounded (see Proposition 7.4) and so there exists a number K such that

$$|y_n| < K \quad \text{for all } n.$$

We have

$$z_n = X \frac{y_1 + y_2 + \cdots + y_n}{n} + \frac{a_1 y_n + a_2 y_{n-1} + \cdots + a_n y_1}{n}.$$

By Proposition 7.13,

$$\frac{y_1 + y_2 + \cdots + y_n}{n} \to y$$

and

$$\left| \frac{a_1 y_n + a_2 y_{n-1} + \cdots + a_n y_1}{n} \right| \le K \frac{a_1 + a_2 + \cdots + a_n}{n} \to 0$$

as $n \to \infty$ and so z_n converges to XY. $\qquad \square$

Worked Examples

1. (i) Let

$$\frac{a_1}{b_1}, \quad \frac{a_2}{b_2}, \quad \ldots, \quad \frac{a_n}{b_n}$$

be n fractions with $b_i > 0$ for $i = 1, 2, \ldots, n$. Show that the fraction

$$\frac{a_1 + a_2 + \cdots + a_n}{b_1 + b_2 + \cdots + b_n}$$

is contained between the largest and the smallest of these fractions.

(ii) Use part (i) to show that if $\{a_n/b_n\}$ is a monotonic sequence with $b_n > 0$ for $n = 1, 2, 3, \ldots$, then $\{c_n\}$, where

$$c_n = \frac{a_1 + a_2 + \cdots + a_n}{b_1 + b_2 + \cdots + b_n},$$

is a monotonic sequence also.

SOLUTION. To verify the claim in part (i), let m and M denote the smallest and the largest of the fractions, respectively. Then

$$m \le \frac{a_i}{b_i} \le M \quad \text{or} \quad mb_i \le a_i \le Mb_i \quad \text{for } i = 1, 2, \ldots, n.$$

Summing these inequalities, we find

$$m \sum_{i=1}^{n} b_i \le \sum_{i=1}^{n} a_i \le M \sum_{i=1}^{n} b_i$$

or

$$m \le \frac{\displaystyle\sum_{i=1}^{n} a_i}{\displaystyle\sum_{i=1}^{n} b_i} \le M.$$

To verify part (ii), let us assume that $\{a_n/b_n\}$ is nondecreasing:

$$\frac{a_1}{b_1} \le \frac{a_2}{b_2} \le \cdots \le \frac{a_n}{b_n} \le \frac{a_{n+1}}{b_{n+1}} \le \cdots.$$

By part (i),

$$\frac{a_1}{b_1} \le \frac{a_1 + a_2 + \cdots + a_n}{b_1 + a_2 + \cdots + a_n} \le \frac{a_n}{a_n}$$

and

$$\frac{a_1 + a_2 + \cdots + a_n}{b_1 + b_2 + \cdots b_n} \le \frac{a_1 + a_2 + \cdots + a_n + a_{n+1}}{b_1 + b_2 + \cdots + b_n + b_{n+1}} \le \frac{a_{n+1}}{b_{n+1}}.$$

It follows in particular that $\{c_n\}$ is nondecreasing. The case of nonincreasing sequences is handled in an entirely similar manner.

REMARKS. From the foregoing we can easily deduce the following: If $\{a_n\}$ is a monotonic sequence, then $\{c_n\}$, where

$$c_n = \frac{a_1 + a_2 + \cdots + a_n}{n},$$

is a monotonic sequence as well. Indeed, we only have to let $b_n = 1$ for $n = 1$, $2, 3, \ldots$ in part (ii).

Similarly, if $\{a_n\}$ *is a monotonic sequence and* $a_n > 0$ for $n = 1, 2, 3, \ldots$, then $\{c_n\}$, where

$$c_n = \sqrt[n]{a_1 a_2 \cdots a_n},$$

is a monotonic sequence as well. In this case we only need to note that

$$\ln c_n = \frac{\ln a_1 + \ln a_2 + \cdots + \ln a_n}{n}$$

and that the logarithmic function preserves the direction of the monotonicity.

2. Show that the sequence $\{x_n\}$, where

$$x_n = \frac{1}{n+1} + \frac{1}{n+2} + \cdots + \frac{1}{n+n},$$

is convergent and find its limit.

SOLUTION. Since

$$x_{n+1} - x_n = \frac{1}{2n+1} + \frac{1}{2n+2} - \frac{1}{n} < \frac{1}{2n} + \frac{1}{2n} - \frac{1}{n} = 0,$$

the sequence in question is seen to be decreasing; moreover, $\frac{1}{2}$ is a lower bound of the sequence because

$$\frac{1}{n+1} + \frac{1}{n+2} + \cdots + \frac{1}{n+n} > \frac{1}{2n} + \frac{1}{2n} + \cdots + \frac{1}{2n} = \frac{1}{2}$$

and 1 is an upper bound of the sequence because

$$\frac{1}{n+1} + \frac{1}{n+2} + \cdots + \frac{1}{n+n} < \frac{1}{n+1} + \frac{1}{n+1} + \cdots + \frac{1}{n+1} = \frac{n}{n+1} < 1.$$

By Proposition 7.8 the sequence $\{x_n\}$ is convergent.

An an application of Proposition 1.6 we have already shown that

$$\frac{1}{n+1} + \frac{1}{n+2} + \cdots + \frac{1}{n+n} \to \ln 2 \quad \text{as } n \to \infty.$$

Here is another way of establishing this result: Dividing the interval $[0, 1]$ into n subintervals of equal length and considering the sum of approximating rectangles under the curve $y = 1/(1 + x)$ lead to

$$\frac{1}{n+1} + \frac{1}{n+2} + \cdots + \frac{1}{n+n}$$

$$= \frac{1}{n}\left(\frac{1}{1 + 1/n} + \frac{1}{1 + 2/n} + \cdots + \frac{1}{1 + n/n}\right) \to \int_0^1 \frac{1}{1+x}\,dx = \ln 2 \quad \text{as } n \to \infty.$$

3. Let $a_1 > b_1 > 0$ be given. We form the numbers

$$a_2 = \frac{a_1 + b_1}{2} \quad \text{and} \quad b_2 = \sqrt{a_1 b_1},$$

$$a_3 = \frac{a_2 + b_2}{2} \quad \text{and} \quad b_3 = \sqrt{a_2 b_2},$$

$$\vdots$$

$$a_{n+1} = \frac{a_n + b_n}{2} \quad \text{and} \quad b_{n+1} = \sqrt{a_n b_n}$$

$$\vdots$$

Show that the sequences $\{a_n\}$ and $\{b_n\}$ tend to a common limit $L(a_1, b_1)$ and verify that

$$L(a_1, b_1) = \frac{\pi}{2G},$$

where

$$G = \int_0^{\pi/2} \frac{1}{(a_1^2 \cos^2 x + b_1^2 \sin^2 x)^{1/2}} dx.$$

SOLUTION. We observe that $a_1 > a_2 > b_2 > b_1$ and that in general

$$a_1 > a_2 > \cdots > a_n > a_{n+1} > b_{n+1} > b_n > \cdots > b_2 > b_1$$

and hence $\{a_n\}$ is a decreasing and bounded sequence and $\{b_n\}$ an increasing and bounded sequence because the a_n's and the b_n's are, in fact, the consecutive arithmetic and geometric means of the initially given numbers a_1 and b_1 with $a_1 > b_1 > 0$. Indeed, it is evident that $a_1 > a_2$ and $b_2 > b_1$ (because $a_1 > b_1 > 0$). To see that $a_2 > b_2$, we note that

$$\frac{a_1 + b_1}{2} - \sqrt{a_1 b_1} = \frac{(\sqrt{a_1} - \sqrt{b_1})^2}{2} > 0 \quad \text{for } a_1 \neq b_1.$$

In the same way we can show that

$$a_n > a_{n+1} > b_{n+1} > b_n.$$

Moreover, it is easy to see that

$$a_1 > a_n > b_n > b_1.$$

By Proposition 7.8 the sequences $\{a_n\}$ and $\{b_n\}$ are therefore convergent; let

$$\alpha = \lim_{n \to \infty} a_n \quad \text{and} \quad \beta = \lim_{n \to \infty} b_n.$$

But

$$a_{n+1} = \frac{a_n + b_n}{2}$$

and so, for $n \to \infty$, we get

$$\alpha = \frac{\alpha + \beta}{2};$$

hence, $\alpha = \beta$. We denote this common limit by $L = L(a_1, b_1)$.

We now assume that $a > b > 0$ and let

$$A = \frac{a + b}{2} \quad \text{and} \quad B = \sqrt{ab}$$

and we show that

$$\int_0^{\pi/2} \frac{dx}{(a^2 \cos^2 x + b^2 \sin^2 x)^{1/2}} = \int_0^{\pi/2} \frac{dt}{(A^2 \cos^2 t + B^2 \sin^2 t)^{1/2}}. \quad (7.10)$$

By repeated applications of (7.10) we get

$$G = \int_0^{\pi/2} \frac{dx}{(a_n^2 \cos^2 x + b_n^2 \sin^2 x)^{1/2}} \quad \text{for } n = 1, 2, 3, \ldots,$$

where a_n and b_n are defined by the recursion formula

$$a_n = \frac{a_{n-1} + b_{n-1}}{2}, \qquad b_n = \sqrt{a_{n-1} b_{n-1}}.$$

As we know already, these two sequences converge to the common value $L = L(a_1, b_1)$. It is easy to see that

$$\frac{\pi}{2a_n} < G < \frac{\pi}{2b_n}$$

and passage to the limit as $n \to \infty$ gives

$$G = \frac{\pi}{2L(a_1, b_1)} \quad \text{or} \quad L(a_1, b_1) = \frac{\pi}{2G}.$$

We return to the verification of the transformation (7.10) and put

$$\sin x = \frac{2a \sin t}{(a + b) + (a - b)\sin^2 t}.$$

As t changes from 0 to $\pi/2$, x grows from 0 to $\pi/2$. Differentiation gives

$$\cos x \, dx = 2a \frac{(a + b) - (a - b)\sin^2 t}{[(a + b) + (a - b)\sin^2 t]^2} \cos t \, dt.$$

But

$$\cos x = \frac{[(a + b)^2 - (a - b)^2 \sin^2 t]^{1/2}}{(a + b) + (a - b)\sin^2 t} \cos t,$$

and thus

$$dx = 2a \frac{(a + b) - (a - b)\sin^2 t}{(a + b) + (a - b)\sin^2 t} \cdot \frac{dt}{[(a + b)^2 - (a - b)^2 \sin^2 t]^{1/2}}.$$

On the other hand,

$$(a^2 \cos^2 x + b^2 \sin^2 x)^{1/2} = a \frac{(a+b) - (a-b)\sin^2 t}{(a+b) + (a-b)\sin^2 t}$$

and thus

$$\frac{dx}{(a^2 \cos^2 x + b^2 \sin^2 x)^{1/2}} = \frac{dt}{\{[(a+b)/2]^2 \cos^2 t + ab \sin^2 t\}^{1/2}},$$

implying (7.10).

REMARKS. We can use the formula

$$G = \frac{\pi}{2L(a_1, b_1)}$$

in the approximate calculation of certain types of integral. For example,

$$G = \int_0^{\pi/2} \frac{dx}{(1 + \cos^2 x)^{1/2}} = \int_0^{\pi/2} \frac{dx}{(2\cos^2 x + \sin^2 x)^{1/2}}$$

and we take $a_1 = \sqrt{2}$ and $b_1 = 1$. The sequences $\{a_n\}$ and $\{b_n\}$ in this case are rapidly converging to $L = L(a_1, b_1)$ and both a_4 and b_4 are seen to be approximately equal to 1.198154. Taking L approximately equal to 1.198154 we obtain

$$G = \frac{\pi}{2L} = 1.3110138 \quad \text{(approximately)}.$$

Formula (7.10) is due to Carl Friedrich Gauss (1777–1855).

4. Let $a_1, a_2, \ldots, a_p$ be p positive real numbers and suppose that A denotes the largest of these numbers. Show that

$$\lim_{n \to \infty} (a_1^n + a_2^n + \cdots + a_p^n)^{1/n} = A.$$

SOLUTION. Since

$$A \le (a_1^n + a_2^n + \cdots + a_p^n)^{1/n} \le A \cdot \sqrt[n]{p}$$

and $\sqrt[n]{p} \to 1$ as $n \to \infty$ by the Lemma preceding Proposition 1.10, the claim follows.

5. Let

$$z_n = \frac{1}{(n^2 + 1)^{1/2}} + \frac{1}{(n^2 + 2)^{1/2}} + \cdots + \frac{1}{(n^2 + n)^{1/2}}.$$

Find $\lim_{n \to \infty} z_n$.

SOLUTION. Let

$$x_n = \frac{n}{(n^2 + n)^{1/2}} \quad \text{and} \quad y_n = \frac{1}{(n^2 + 1)^{1/2}}.$$

It is clear that each summand in the expression for z_n is larger than the first and smaller than the last; hence,

$$x_n = \frac{n}{(n^2 + n)^{1/2}} < z_n < \frac{n}{(n^2 + 1)^{1/2}} = y_n.$$

But $x_n \to 1$ and $y_n \to 1$ as $n \to \infty$ and so $z_n \to 1$ as $n \to \infty$.

6. Let

$$a_n = \ln\left(1 + \frac{1}{n^2}\right) + \ln\left(1 + \frac{2}{n^2}\right) + \cdots + \ln\left(1 + \frac{n}{n^2}\right).$$

Find $\lim_{n \to \infty} a_n$.

SOLUTION. Since $x/(1 + x) < \ln(1 + x) < x$ for $x > 0$, we get by setting $x = k/n^2$ with $k \leq n < n^2$:

$$\frac{k}{n^2 + n} \leq \frac{k}{n^2 + k} < \ln\left(1 + \frac{k}{n^2}\right) < \frac{k}{n^2}.$$

Therefore,

$$\sum_{k=1}^{n} \frac{k}{n^2 + n} < \sum_{k=1}^{n} \ln\left(1 + \frac{k}{n^2}\right) < \sum_{k=1}^{n} \frac{k}{n^2}$$

and so

$$\frac{1 + 2 + \cdots + n}{n^2 + n} < a_n < \frac{1 + 2 + \cdots + n}{n^2}$$

or

$$\frac{n(n + 1)}{2(n^2 + n)} < a_n < \frac{n(n + 1)}{2n^2},$$

implying

$$\frac{1}{2} < a_n < \frac{1}{2}\left(1 + \frac{1}{n}\right).$$

It follows that $a_n \to \frac{1}{2}$ as $n \to \infty$.

7. Let $x_1 = 1$ and

$$x_{n+1} = \left(1 - \frac{1}{(n + 1)^2}\right)x_n \quad \text{for } n = 1, 2, 3, \ldots.$$

Find $\lim_{n \to \infty} x_n$.

SOLUTION. It is clear that $0 < x_n < x_{n+1} \leq 1$ and so $\lim_{n \to \infty} x_n$ exists by Proposition 7.8. Since

$$x_n = \frac{n+1}{2n},$$

we see that $x_n \to \frac{1}{2}$ as $n \to \infty$.

8. Let $a_0 = 1$ and

$$a_{n+1} = \frac{1}{1 + a_n} \quad \text{for } n = 0, 1, 2, \ldots.$$

Find $\lim_{n \to \infty} a_n$.

SOLUTION. First we verify that $\{a_n\}$ is convergent. Evidently $a_n \leq 1$ for all n and thus $a_n \geq \frac{1}{2}$; therefore, for arbitrary $n, k \geq 0$,

$$a_{n+1+k} - a_{n+1} = \frac{1}{1 + a_{n+k}} - \frac{1}{1 + a_n} = -\frac{a_{n+k} - a_n}{(1 + a_{n+k})(1 + a_n)},$$

$$|a_{n+1+k} - a_{n+1}| = \frac{|a_{n+k} - a_n|}{(1 + a_{n+k})(1 + a_n)} \leq \frac{1}{(1 + 1/2)^2}|a_{n+k} - a_n| = \frac{4}{9}|a_{n+k} - a_n|.$$

Thus,

$$|a_{n+k} - a_n| \leq (\tfrac{4}{9})^n |a_k - a_0| \leq 2 \cdot (\tfrac{4}{9})^n.$$

Since $(\tfrac{4}{9})^n \to 0$ as $n \to \infty$, we see that $\{a_n\}$ is a Cauchy sequence; by Proposition 7.12 the sequence $\{a_n\}$ is convergent. Passage to the limit as $n \to \infty$ in

$$a_{n+1} = \frac{1}{1 + a_n}$$

gives $a = 1/(1 + a)$, that is, $a^2 + a = 1$, where $a = \lim_{n \to \infty} a_n$.

9. Let

$$a_{n+1} = \frac{a_n + b_n}{2} \quad \text{and} \quad b_{n+1} = \sqrt{a_{n+1} b_n},$$

where $a_1 = \cos t$ with $-\pi/2 < t < \pi/2$ and $b_1 = 1$. Show that $\{a_n\}$ and $\{b_n\}$ converge to the same limit $L = (\sin t)/t$.

SOLUTION. First we show that, for $t \neq 0$,

$$\lim_{n \to \infty} \left(\cos \frac{t}{2} \right) \left(\cos \frac{t}{2^2} \right) \left(\cos \frac{t}{2^3} \right) \cdots \left(\cos \frac{t}{2^n} \right) = \frac{\sin t}{t}. \qquad (7.11)$$

Indeed,

$$\sin t = 2 \left(\cos \frac{t}{2} \right) \left(\sin \frac{t}{2} \right) = 2^2 \left(\cos \frac{t}{2} \right) \left(\cos \frac{t}{2^2} \right) \left(\sin \frac{t}{2^2} \right) = \cdots$$

$$= 2^n \left(\cos \frac{t}{2} \right) \left(\cos \frac{t}{2^2} \right) \cdots \left(\cos \frac{t}{2^n} \right) \left(\sin \frac{t}{2^n} \right)$$

and so

$$\frac{\sin t}{2^n \sin(t/2^n)} = \frac{\sin t}{t} \frac{1/2^n}{\sin(t/2^n)}$$

holds. But $x_n = t/2^n \to 0$ as $n \to \infty$ and $(\sin x)/x \to 1$ as $x \to 0$.
Returning to the problem at hand, we note that

$$a_2 = \frac{\cos t + 1}{2} = \cos^2 \frac{t}{2},$$

$$a_3 = \left(\cos \frac{t}{2}\right)\left(\cos^2 \frac{t}{2^2}\right),$$

$$a_4 = \left(\cos \frac{t}{2}\right)\left(\cos \frac{t}{2^2}\right)\left(\cos^2 \frac{t}{2^3}\right),$$

$$a_5 = \left(\cos \frac{t}{2}\right)\left(\cos \frac{t}{2^2}\right)\left(\cos \frac{t}{2^3}\right)\left(\cos^2 \frac{t}{2^4}\right),$$

$$b_2 = \cos \frac{t}{2},$$

$$b_3 = \left(\cos \frac{t}{2}\right)\left(\cos \frac{t}{2^2}\right),$$

$$b_4 = \left(\cos \frac{t}{2}\right)\left(\cos \frac{t}{2^2}\right)\left(\cos \frac{t}{2^3}\right),$$

$$b_5 = \left(\cos \frac{t}{2}\right)\left(\cos \frac{t}{2^2}\right)\left(\cos \frac{t}{2^3}\right)\left(\cos \frac{t}{2^4}\right),$$

and so forth. But $\cos x \to 1$ as $x \to 0$ and all is clear in view of (7.11).

10. Following the proof of Proposition 7.14 we showed that

$$\lim_{n \to \infty} \frac{\sqrt[n]{n!}}{n} = \frac{1}{e} \quad \text{and} \quad \lim_{n \to \infty} \frac{1}{n} \sqrt[n]{(n+1)(n+2)\cdots(n+n)} = \frac{4}{e}.$$

Verify these results by integration theory.

SOLUTION. We have, as $n \to \infty$,

$$\ln\left(\frac{n!}{n^n}\right)^{1/n} = \frac{1}{n}\left(\ln\frac{1}{n} + \ln\frac{2}{n} + \cdots + \ln\frac{n}{n}\right)$$

$$\to \int_0^1 (\ln x)\, dx = [x(\ln x) - x]\Big|_0^1 = -1$$

and

$$\ln\left(\frac{(n+1)(n+2)\cdots(n+n)}{n^n}\right)$$

$$= \frac{1}{n}\left(\ln\left(1+\frac{1}{n}\right) + \ln\left(1+\frac{2}{n}\right) + \cdots + \ln\left(1+\frac{n}{n}\right)\right)$$

$$\rightarrow \int_0^1 \ln(1+x)\,dx = \left[(1+x)\ln(1+x) - x\right]\Big|_0^1 = (\ln 4) - 1.$$

11. Let f be a (finite, real) linear combination of functions of the form $h(t) = t^r$, where r is a fixed real number; in particular, f can be any polynomial. Let $[a, b]$ be an interval of finite length with $0 < a < b$. Put

$$p_1 = ad, \quad p_2 = ad^2, \quad \ldots, \quad p_{n-1} = ad^{n-1},$$

where $d = \sqrt[n]{b/a}$, and

$$q_1 = a + s, \quad q_2 = a + 2s, \quad \ldots, \quad q_{n-1} = a + (n-1)s,$$

where $s = (b - a)/n$. Denote by A the arithmetic mean of the values

$$f(a), \quad f(p_1), \quad \ldots, \quad f(p_{n-1}), \quad f(b)$$

any by G the arithmetic mean of the values

$$\frac{f(a)}{a}, \quad \frac{f(q_1)}{q_1}, \quad \ldots, \quad \frac{f(q_{n-1})}{q_{n-1}}, \quad \frac{f(b)}{b}.$$

Then, as $n \rightarrow \infty$, the fraction G/A tends to a limit which is *independent of f*; this limit is $\ln(b/a)/(b-a)$.

SOLUTION. Suppose first that $f(t) = t^r$, where r is a fixed real number. Then

$$G = \frac{1}{n+1}\left(a^{r-1} + q_1^{r-1} + \cdots + q_{n-1}^{r-1} + b^{r-1}\right)$$

and so

$$\lim_{n\to\infty} G = \frac{\ln(b/a)}{b-a} \quad \text{if } r = 0,$$

$$= \frac{b^r - a^r}{(b-a)r} \quad \text{if } r \neq 0.$$

Indeed, as $n \rightarrow \infty$,

$$\frac{1}{n+1}\left(a^{r-1} + q_1^{r-1} + \cdots + q_{n-1}^{r-1} + b^{r-1}\right) \rightarrow \frac{1}{b-a}\int_a^b t^{r-1}\,dt.$$

Moreover,

$$A = \frac{1}{n+1}\left(a^r + (ad)^r + (ad^2)^r + \cdots + (ad^{n-1})^r + b^r\right)$$

$$= \frac{a^r}{n+1}\left(1 + d^r + d^{2r} + \cdots + d^{(n-1)r} + d^{nr}\right)$$

and so

$$\lim_{n\to\infty} A = 1 \qquad \text{if } r \neq 0,$$

$$= \frac{b^r - a^r}{r\ln(b/a)} \qquad \text{if } r \neq 0.$$

This is trivial if $r = 0$; if $r \neq 0$, we have, as $n \to \infty$,

$$\frac{a^r}{n+1}\left(1 + d^r + d^{2r} + \cdots + d^{(n-1)r} + d^{nr}\right) \to \frac{a^r}{r}\int_0^r \left(\frac{b}{a}\right)^t dt.$$

Thus, G/A tends to $\ln(b/a)/(b-a)$ as $n \to \infty$ for $f(t) = t^r$.

If we take $f(t) = Kt^r$, where K is a constant, the fraction G/A is clearly not affected; the factor K introduced in the numerator and denominator in G/A cancels out. Finally, to see that the claim is valid when f is a linear combination of functions of the form $h(t) = t^r$, we observe: In a finite sequence of equal ratios, the sum of the numerators divided by the sum of the denominators gives a ratio equal to each of the given ratios in the sequence.

12. For any fixed real number a, we have $a^n/n! \to 0$ as $n \to \infty$.

SOLUTION. Assume that $a > 0$ and let the integer k be such that $a < k + 1$. For $n > k$ we have

$$\frac{a^n}{n!} = \frac{a^k}{k!} \cdot \frac{a}{k+1} \cdot \frac{a}{k+2} \cdots \frac{a}{n}.$$

Thus,

$$\frac{a^n}{n!} \leq \frac{a^k}{k!} \cdot \left(\frac{a}{k+1}\right)^{n-k}.$$

But $a^k/k!$ is fixed and $[a/(k+1)]^{n-k} \to 0$ as $n \to \infty$.

13. Let

$$x_n = n\left(1 - \frac{\ln n}{n}\right)^n \qquad \text{for } n = 2, 3, \ldots.$$

Find $\lim_{n\to\infty} x_n$.

SOLUTION. We put $v = n/(\ln n)$. Then

$$\ln x_n = \left(1 + v\ln\left(1 - \frac{1}{v}\right)\right)(\ln n).$$

But

$$-\ln\left(1 - \frac{1}{v}\right) = -\ln\frac{v-1}{v} = \ln\frac{v}{v-1} = \ln\left(1 + \frac{1}{v-1}\right)$$

and [recall inequality (1.3)]

$$\frac{1}{v} < -\ln\left(1 - \frac{1}{v}\right) = \ln\left(1 + \frac{1}{v-1}\right) < \frac{1}{v-1}.$$

Moreover,

$$\ln x_n > \left(\ln\left(1 - \frac{1}{v}\right)\right)(\ln n).$$

Therefore,

$$0 < -\ln x_n < (\ln n)\left(-\ln\left(1 - \frac{1}{v}\right)\right)$$

or

$$0 < -\ln x_n < \frac{\ln n}{v-1}.$$

Hence,

$$0 < -\ln x_n < \frac{\ln n}{v-1} = \frac{(\ln n)^2}{\ln n} \cdot \frac{n}{n} \cdot \frac{1}{v-1}.$$

or

$$0 < -\ln x_n < \frac{v}{v-1} \cdot \frac{(\ln n)^2}{n}.$$

Passing to the limit (if $n \to \infty$, then $v \to \infty$) we get

$$0 \le -\lim_{n \to \infty} \ln x_n \le \lim_{v \to \infty} \frac{v}{v-1} \cdot \lim_{n \to \infty} \frac{(\ln n)^2}{n} = 1 \cdot 0 = 0$$

and so $x_n \to 1$ as $n \to \infty$.

14. Let $\{a_n\}$ be a sequence such that $(2 - a_n)a_{n+1} = 1$. Show that $\lim_{n \to \infty} a_n = 1$.

SOLUTION. If $a_n = 1$ for some n, then $a_n = 1$ for all n. Otherwise, let

$$b_{n+1} = \frac{1}{1 - a_{n+1}} = \frac{1}{1 - \dfrac{1}{2 - a_n}} = \frac{2 - a_n}{1 - a_n} = b_n + 1.$$

Hence,

$$b_n = b_1 + n - 1 \quad \text{and} \quad a_n = 1 - \frac{1}{b_n} = \frac{b_1 + n - 2}{b_1 + n - 1}.$$

Thus, $a_n \to 1$ as $n \to \infty$.

15. For each positive integer n, put

$$a_n = \left(1 + \frac{1}{n}\right)^n, \quad b_n = \left(1 + \frac{1}{n}\right)^{n+1}, \quad \text{and} \quad c_n = \frac{2a_n b_n}{a_n + b_n}.$$

Show that $c_1 < c_2 < c_3 < \cdots < c_n < \cdots$.

SOLUTION. We find $c_n = 2(n + 1)^{n+1} n^{-n} (2n + 1)^{-1}$. Let

$$g(x) = \ln 2 + (x + 1)\ln(x + 1) - x(\ln x) - \ln(2x + 1).$$

Then

$$g'(x) = \ln(x + 1) - \ln x - \frac{2}{2x + 1}$$

and

$$g''(x) = \frac{1}{x + 1} - \frac{1}{x} + \frac{4}{(2x + 1)^2} = -\frac{1}{x(x + 1)(2x + 1)^2} < 0$$

for $0 < x < \infty$. Hence, g' decreases on $(0, \infty)$. Since

$$\lim_{x \to \infty} g'(x) = \lim_{x \to \infty} \ln \frac{x + 1}{x} - \lim_{x \to \infty} \frac{2}{2x + 1} = 0,$$

if follows that g' is positive on $(0, \infty)$. Thus, g increases on $(0, \infty)$, so

$$c_n = e^{g(n)}$$

is strictly increasing for positive integers n.

REMARK. Since $a_n \to e$ and $b_n \to e$ as $n \to \infty$, it follows by Proposition 7.5 that $c_n \to e$ as $n \to \infty$. We also note that c_n is the harmonic mean of a_n and b_n.

16. Let

$$J_n = \int_0^h \frac{\sin nx}{x} \, dx,$$

where n denotes a positive integer and $h > 0$.

The symbol $(\sin nx)/x$ is meaningless for $x = 0$. By defining $(\sin nx)/x$ to be equal to n for $x = 0$, we see that $(\sin nx)/x$ is continuous for all real numbers x.

We put $nx = u$ and see that

$$J_n = \int_0^{nh} \frac{\sin u}{u} \, du;$$

thus, for n and m being positive integers and $n > m$,

$$J_n - J_m = \int_{mh}^{nh} \frac{\sin u}{u} \, du.$$

We wish to verify that $\{J_n\}$ is a Cauchy sequence and hence is convergent by Proposition 7.12.

Indeed, putting $a = mh$, $b = nh$, we have $0 < a < b$ and therefore

$$\int_a^b \frac{\sin u}{u} \, du = \left(\frac{-\cos u}{u} \right) \Big|_a^b - \int_a^b \frac{\cos u}{u^2} \, du.$$

Thus,

$$\left| \int_a^b \frac{\sin u}{u} \, du \right| \le \frac{1}{a} + \frac{1}{b} + \int_a^b \frac{1}{u^2} \, du = \frac{2}{a}$$

and so

$$\left| \int_a^b \frac{\sin u}{u} \, du \right| < \varepsilon \quad \text{whenever } a > \frac{2}{\varepsilon}.$$

Consequently, if $\varepsilon > 0$ is given, we can select m such that $2/mh < \varepsilon$. We then have for $n > m$

$$|J_n - J_m| < \frac{2}{mh} < \varepsilon$$

and the sequence $\{J_n\}$ is seen to be Cauchy.

17. Let $[a, b]$ be a closed interval of finite length, $p = 1, 2, \ldots$, and

$$A_p = \int_a^b f(x)(\sin px) \, dx \quad \text{and} \quad B_p = \int_a^b f(x)(\cos px) \, dx,$$

where f is a Riemann integrable function over $[a, b]$. Then $A_p \to 0$ and $B_p \to 0$ as $p \to \infty$.

Indeed, for any interval $[\alpha, \beta]$ of finite length we have

$$\left| \int_\alpha^\beta (\sin px) \, dx \right| = \left| \frac{\cos p\alpha - \cos p\beta}{p} \right| \le \frac{2}{p}$$

because $|\cos t| \le 1$ for all t. Let $|f(x)| \le M$ on $[a, b]$ and ε be the usual arbitrary positive number. There is a partition P of $[a, b]$, say

$$P = \{a = x_0 < x_1 < \cdots < x_n = b\},$$

such that $U(P, f) - L(P, f) < \varepsilon/2$ (see Proposition 5.4). Thus,

$$\left| \int_a^b f(x)(\sin px)\,dx \right| = \left| \sum_{k=1}^{n} \int_a^b [f(x_k) + f(x) - f(x_k)](\sin px)\,dx \right|$$

$$\leq \sum_{k=1}^{n} \left(|f(x_k)| \left| \int_{x_{k-1}}^{x_k} (\sin px)\,dx \right| \right)$$

$$+ \int_{x_{k-1}}^{x_k} |f(x) - f(x_k)||\sin px|\,dx$$

$$< \frac{2nM}{p} + [U(P,f) - L(P,f)] < \frac{2nM}{p} + \frac{\varepsilon}{2} < \varepsilon$$

when $p \geq 4nM/\varepsilon$. Hence, $A_p = \int_a^b f(x)(\sin px)\,dx$ tends to 0 as $p \to \infty$. In the same way, $B_p = \int_a^b f(x)(\cos px)\,dx$ tends to 0 as $p \to \infty$.

18. Returning to the integral

$$J_n = \int_0^h \frac{\sin nx}{x}\,dx$$

of Example 16, we wish to calculate the limit of the sequence $\{J_n\}$ as $n \to \infty$.
  By the result in Example 17,

$$\lim_{n \to \infty} \int_h^{h'} \frac{\sin nx}{x}\,dx = 0$$

whenever h and h' are two numbers larger than 0. This shows that

$$\lim_{n \to \infty} \int_0^h \frac{\sin nx}{x}\,dx$$

is entirely *independent* of what value we choose for the positive number h. We may therefore restrict ourselves to consideration of

$$\lim_{n \to \infty} \int_0^{\pi/2} \frac{\sin nx}{x}\,dx.$$

  For $0 < x \leq \pi/2$ we have that

$$\frac{1}{x} - \frac{1}{\sin x} = \frac{\sin x - x}{x(\sin x)}$$

is a continuous function. Using L'Hôpital's Rules (see Proposition 4.10) twice, we see that

$$\frac{\sin x - x}{x(\sin x)} \to 0 \quad \text{as } x \to 0.$$

By defining the meaningless expression $1/x - 1/\sin x$ to be 0 for $x = 0$, we see that

$$\frac{1}{x} - \frac{1}{\sin x} = \frac{\sin x - x}{x \sin x}$$

is continuous for $0 \leq x \leq \pi/2$. Thus, by the result in Example 17,

$$\lim_{n \to \infty} \int_0^{\pi/2} \left(\frac{1}{x} - \frac{1}{\sin x} \right) (\sin nx) \, dx = 0,$$

implying that

$$\lim_{n \to \infty} \int_0^{\pi/2} \frac{\sin nx}{\sin x} \, dx = \lim_{n \to \infty} \int_0^{\pi/2} \frac{\sin nx}{x} \, dx.$$

To find the limit of a convergent sequence, it is enough to consider a subsequence. We shall consider the subsequence corresponding to the odd indices.

We commence with the observation that

$$\cos(p - 1)t - 2 \cos pt + \cos(p + 1)t = 2(\cos t - 1)(\cos pt).$$

Putting $u_p = \cos(p - 1)t - \cos pt$, we can write this identity in the form

$$u_{p+1} - u_p = 2(1 - \cos t)(\cos pt).$$

We write down the following chain of equations:

$$u_1 = 1 - \cos t,$$

$$u_2 - u_1 = 2(1 - \cos t)(\cos t),$$

$$u_3 - u_2 = 2(1 - \cos t)(\cos 2t),$$

$$\vdots$$

$$u_{p+1} - u_p = 2(1 - \cos t)(\cos pt).$$

From this we get

$$\tfrac{1}{2}u_{p+1} = (\tfrac{1}{2} + \cos t + \cos 2t + \cdots + \cos pt)(1 - \cos t);$$

hence, in case that $\cos t \neq 1$,

$$\tfrac{1}{2} + \cos t + \cos 2t + \cdots + \cos pt = \frac{\cos pt - \cos(p + 1)t}{2(1 - \cos t)}.$$

Since

$$\cos(p + 1)t = \cos[(p + \tfrac{1}{2})t + \tfrac{1}{2}t] \quad \text{and} \quad \cos pt = \cos[(p + \tfrac{1}{2})t - \tfrac{1}{2}t],$$

we see that

$$\cos pt - \cos(p + 1)t = [2 \sin(p + \tfrac{1}{2})t](\cos \tfrac{1}{2}t);$$

but

$$1 - \cos t = 2 \sin^2 \tfrac{1}{2}t$$

and so

$$\tfrac{1}{2} + \cos t + \cos 2t + \cdots + \cos pt = \frac{\sin(p + \tfrac{1}{2})t}{2 \sin \tfrac{1}{2}t}.$$

Restricting t to the interval $[0, \pi]$, then $\sin \tfrac{1}{2}t$ is only zero for $t = 0$. If we define

$$\frac{\sin(p + \tfrac{1}{2})t}{2 \sin \tfrac{1}{2}t}$$

to be equal to $p + \tfrac{1}{2}$ for $t = 0$, then the formula

$$\tfrac{1}{2} + \cos t + \cos 2t + \cdots + \cos pt = \frac{\sin(p + \tfrac{1}{2})t}{2 \sin \tfrac{1}{2}t}$$

is also valid for $t = 0$.

In the integral

$$\int_0^{\pi/2} \frac{\sin(2p + 1)x}{\sin x}\, dx$$

we may therefore write

$$\frac{\sin(2p + 1)x}{\sin x} = 2(\tfrac{1}{2} + \cos 2x + \cos 4x + \cdots + \cos 2px).$$

Hence, we get

$$\int_0^{\pi/2} \frac{\sin(2p + 1)x}{\sin x}\, dx = \int_0^{\pi/2} dx + \sum_{k=1}^{n} 2 \int_0^{\pi/2} (\cos 2kx)\, dx.$$

But

$$2 \int_0^{\pi/2} (\cos 2kx)\, dx = \frac{1}{k}(\sin 2kx) \Big|_0^{\pi/2} = 0$$

and so

$$\int_0^{\pi/2} \frac{\sin(2p + 1)x}{\sin x}\, dx = \frac{\pi}{2}.$$

From this we can conclude that

$$\lim_{n \to \infty} \int_0^{h} \frac{\sin nx}{x}\, dx = \frac{\pi}{2} \quad \text{for } h > 0.$$

19. The decimal fraction

$$\alpha = 0.12345678910111213 14\ldots$$

(the positive integers written consecutively) represents an irrational number. Denoting by $[t]$ the integer part of t, consider the sequence

$$10^n \alpha - [10^n \alpha] \quad \text{for } n = 0, 1, 2, 3, \ldots.$$

Note that we obtain $10^n\alpha - [10^n\alpha]$ by shifting the decimal point in α by n places to the right and deleting all digits to the left of the decimal point. Let

$$\beta = 0. \, \beta_1\beta_2 \ldots \beta_k$$

be a finite decimal fraction. If we choose n such that $10^n\alpha - [10^n\alpha]$ begins with the digits $\beta_1, \beta_2, \ldots, \beta_k$ and is followed by r zeros, then

$$|10^n\alpha - [10^n\alpha] - \beta| < \frac{1}{10^{k+r}}.$$

This shows that any rational number β between 0 and 1 is a point of accumulation of the sequence $10^n\alpha - [10^n\alpha]$ for $n = 0, 1, 2, 3, \ldots$; since any real number between 0 and 1 can be obtained as a limit of a sequence of rational numbers between 0 and 1, it follows ultimately that any number in the interval $[0, 1]$ is an accumulation point of the sequence $10^n\alpha - [10^n\alpha]$ for $n = 0, 1, 2, 3, \ldots$.

20. Let A, B, C, and D be distinct real numbers and $\{x_n\}$ be a null sequence. Then the sequence $\{u_n\}$ defined by

$$u_{4n-3} = A + x_n, \qquad u_{4n-2} = B + x_n, \qquad u_{4n-1} = C + x_n, \qquad u_{4n} = D + x_n,$$

for $n = 1, 2, 3, \ldots$ has $\{A, B, C, D\}$ as its set of accumulation points.

21. For every integer $p \geq 2$ there are $p - 1$ numbers of the form $1/k + 1/m$ for which the sum of the positive integers k and m equals p. For $p = 2, 3, 4, \ldots$, consider these numbers enumerated. Then we obtain the sequence

$$2, \tfrac{3}{2}, \tfrac{3}{2}, \tfrac{4}{1}, 1, \tfrac{4}{3}, \tfrac{5}{4}, \tfrac{5}{6}, \tfrac{5}{6}, \tfrac{5}{4}, \ldots.$$

If in $1/k + 1/m$ we keep m fixed and let $k \to \infty$, we see that $1/m$ is an accumulation point for $m = 1, 2, 3, \ldots$. Thus, $1, \tfrac{1}{2}, \tfrac{1}{3}, \ldots$ are accumulation points of the sequence under consideration. Similarly, keeping k fixed and letting $m \to \infty$, we get that $1, \tfrac{1}{2}, \tfrac{1}{3}, \ldots$ are accumulation points of the sequence. Now 0 is an accumulation point of $1, \tfrac{1}{2}, \tfrac{1}{3}, \ldots$ and so the sequence under consideration has the accumulation points

$$0, 1, \tfrac{1}{2}, \tfrac{1}{3}, \tfrac{1}{4}, \ldots$$

and no other accumulation points.

22. Let $[t]$ denote the integer part of t. The sequence

$$n!e - [n!e] \quad \text{for } n = 1, 2, 3, \ldots$$

has 0 as its only point of accumulation (hence $\lim_{n \to \infty}(n!e - [n!e]) = 0$). Indeed, by Taylor's Theorem (see Proposition 4.11) and especially (4.13),

$$e = 1 + \frac{1}{1!} + \frac{1}{2!} + \cdots + \frac{1}{n!} + \frac{e^\theta}{(n+1)!} \quad \text{with } 0 < \theta < 1$$

and so

$$n!e = n! + \frac{n!}{1!} + \frac{n!}{2!} + \cdots + 1 + \frac{e^\theta}{n+1}.$$

For $n \geq 2$, we have $e^\theta/(n+1) < e/(n+1) < 1$; hence,

$$n!e - [n!e] < \frac{e^\theta}{n+1} < \frac{e}{n+1}.$$

2. The Formulas of Wallis and Stirling

Proposition 7.16 (Formula of Wallis). *We have*

$$\frac{\pi}{2} = \lim_{n \to \infty} \left(\frac{2^2 \cdot 4^2 \cdots (2n)^2}{1^2 \cdot 3^2 \cdots (2n-1)^2} \cdot \frac{1}{2n} \right). \tag{7.12}$$

PROOF. Let k be a positive integer larger than or equal to 2. Using integration by parts, we get

$$\int_0^{\pi/2} (\sin^k x) \, dx = \int_0^{\pi/2} (\sin^{k-1} x)(\sin x) \, dx$$

$$= \left[-(\sin^{k-1} x)(\cos x) \right] \Big|_0^{\pi/2} + (k-1) \int_0^{\pi/2} (\sin^{k-2} x)(\cos^2 x) \, dx$$

$$= (k-1) \int_0^{\pi/2} (\sin^{k-2} x) \, dx - (k-1) \int_0^{\pi/2} (\sin^k x) \, dx,$$

implying that

$$\int_0^{\pi/2} (\sin^k x) \, dx = \frac{k-1}{k} \int_0^{\pi/2} (\sin^{k-2} x) \, dx. \tag{7.13}$$

If $k = 2n$ is an even number, then (7.13) gives

$$\int_0^{\pi/2} (\sin^{2n} x) \, dx = \frac{1 \cdot 3 \cdots (2n-1)}{2 \cdot 4 \cdots (2n)} \cdot \frac{\pi}{2}.$$

If $k = 2n + 1$ is an odd number, we deduce from (7.13) that

$$\int_0^{\pi/2} (\sin^{2n+1} x) \, dx = \frac{2 \cdot 4 \cdots (2n)}{3 \cdot 5 \cdots (2n+1)}$$

because

$$\int_0^{\pi/2} (\sin x) \, dx = 1.$$

If $0 \leq x \leq \pi/2$ and n is a positive integer, then

$$\sin^{2n+1} x \leq \sin^{2n} x \leq \sin^{2n-1} x.$$

By Proposition 5.11 and in view of the foregoing results we therefore see that for $n \geq 2$

$$\frac{2 \cdot 4 \cdots (2n)}{3 \cdot 5 \cdots (2n+1)} \leq \frac{1 \cdot 3 \cdots (2n-1)}{2 \cdot 4 \cdots (2n)} \cdot \frac{\pi}{2} \leq \frac{2 \cdot 4 \cdots (2n-2)}{3 \cdot 5 \cdots (2n-1)}$$

holds, implying that

$$\frac{2^2 \cdot 4^2 \cdots (2n)^2}{1^2 \cdot 3^2 \cdots (2n-1)^2} \cdot \frac{1}{2n+1} \leq \frac{\pi}{2} \leq \frac{2^2 \cdot 4^2 \cdots (2n)^2}{1^2 \cdot 3^2 \cdots (2n-1)^2} \cdot \frac{1}{2n}. \qquad (7.14)$$

Evidently, (7.14) is true for $n \geq 1$ and for each such n there exists a real number θ_n satisfying $0 \leq \theta_n \leq 1$ such that

$$\frac{\pi}{2} = \frac{2^2 \cdot 4^2 \cdots (2n)^2}{1^2 \cdot 3^2 \cdots (2n-1)^2} \cdot \frac{1}{2n + \theta_n}. \qquad (7.15)$$

Replacing the fraction

$$\frac{1}{2n + \theta_n} \quad \text{by} \quad \frac{1}{2n} \cdot \frac{2n}{2n + \theta_n}$$

in (7.15) and noting that

$$\frac{2n}{2n + \theta_n} \to 1 \quad \text{as } n \to \infty,$$

we see that (7.15) implies (7.12). $\qquad \square$

REMARKS. Formula (7.12) easily yields

$$\lim_{n \to \infty} \left(\frac{1 \cdot 3 \cdots (2n-1)}{2 \cdot 4 \cdots (2n)} \cdot \sqrt{2n} \right) = \left(\frac{2}{\pi} \right)^{1/2}. \qquad (7.16)$$

Another consequence of (7.12) is

$$\lim_{n \to \infty} \left(\left(1 - \frac{1}{2^2} \right) \left(1 - \frac{1}{4^2} \right) \cdots \left(1 - \frac{1}{(2n)^2} \right) \right) = \frac{\pi}{2}. \qquad (7.17)$$

To see that (7.17) is valid, observe that

$$\frac{2}{\pi} = \lim_{n \to \infty} \left(\frac{1 \cdot 3 \cdot 3 \cdot 5 \cdots (2n-1)(2n-1)}{2 \cdot 2 \cdot 4 \cdot 4 \cdots (2n-2)(2n)} \right)$$

by (7.12). But

$$v_n = \frac{1 \cdot 3 \cdot 3 \cdot 5 \cdots (2n-1)(2n-1)}{2 \cdot 2 \cdot 4 \cdot 4 \cdots (2n-2)(2n)} \quad \text{and} \quad w_n = v_n \cdot \frac{2n+1}{2n}$$

tend to the same limit $2/\pi$ as $n \to \infty$ and

$$w_n = \left(1 - \frac{1}{2^2} \right) \left(1 - \frac{1}{4^2} \right) \cdots \left(1 - \frac{1}{(2n)^2} \right).$$

Given the sequence $s_1 = 1$ and $s_{n+1} = [1 - 1/(2n)^2] \cdot s_n$ for $n = 1, 2, 3, \ldots$, it takes little effort to show that $\{s_n\}$ is bounded and monotonic and thus convergent by Proposition 7.8. However, finding the limit of $\{s_n\}$ is not so easy and we need relation (7.17).

Proposition 7.17 (Stirling's Formula). *For any positive integer n we have*

$$n! = n^n e^{-n} \sqrt{2\pi n} \cdot (e)^{\theta_n/4n}, \quad \text{where } 0 \leq \theta_n \leq 1. \tag{7.18}$$

PROOF. We consider the sequence $\{x_n\}$ given by

$$x_n = \frac{n!}{n^n e^{-n} \sqrt{n}} \quad \text{for } n = 1, 2, 3, \ldots. \tag{7.19}$$

We observe that the sequence $\{x_n\}$ is decreasing because

$$\frac{x_n}{x_{n+1}} = \frac{1}{e}\left(1 + \frac{1}{n}\right)^{n+1/2} > 1 \tag{7.20}$$

or

$$\left(n + \frac{1}{2}\right)\ln\left(1 + \frac{1}{n}\right) - 1 > 0$$

holds. Indeed, we shall at once show the double inequality

$$0 < \left(n + \frac{1}{2}\right)\ln\left(1 + \frac{1}{n}\right) - 1 < \frac{1}{4}\left(\frac{1}{n} - \frac{1}{n+1}\right). \tag{7.21}$$

Let A_1 denote the area of the region between the x-axis and the curve $y = 1/x$ in the interval $n \leq x \leq n + 1$; we know that

$$A_1 = \ln\frac{n+1}{n}.$$

Let A_2 denote the area of the region between the x-axis and the line segment connecting the points $(n, 1/n)$ and $(n + 1, 1/(n + 1))$ in the interval $n \leq x \leq n + 1$; clearly

$$A_2 = \frac{1}{2}\left(\frac{1}{n} - \frac{1}{n+1}\right)$$

and $A_1 < A_2$ because the curve $y = 1/x$ is concave up in the interval $n \leq x \leq n + 1$ and to get the region with area A_2 we replaced the curve $y = 1/x$ over the indicated interval by its chord. Finally, let A_3 denote the area of the region between the x-axis and the tangent line to the curve $y = 1/x$ at the point $x = n + \frac{1}{2}$ in the interval $n \leq x \leq n + 1$; we have

$$A_3 = \frac{1}{n + \frac{1}{2}}$$

and $A_3 < A_1$ in view of Proposition 4.16. Thus, $A_3 < A_1 < A_2$ and so

$$\frac{1}{n + \frac{1}{2}} < \ln\left(1 + \frac{1}{n}\right) < \frac{1}{2}\left(\frac{1}{n} + \frac{1}{n + 1}\right)$$

which is apparently equivalent to (7.21). But (7.21) implies (7.20).

Since $x_{n+1} < x_n$ and $x_n > 0$ for all n, we see that

$$\lim_{n \to \infty} x_n = \sigma \tag{7.22}$$

exists and σ is nonnegative. In fact, (7.21) shows that

$$1 < \frac{x_i}{x_{i+1}} < \exp\left(\frac{1}{4}\left(\frac{1}{i} - \frac{1}{i + 1}\right)\right), \tag{7.23}$$

where we use the notation $\exp(t)$ to signify e^t. For $i = n, n + 1, \ldots, n + k - 1$ the inequality (7.23) generates a system of inequalities which, upon multiplication, produces the inequality

$$1 < \frac{x_n}{x_{n+k}} < \exp\left(\frac{1}{4}\left(\frac{1}{n} - \frac{1}{n + k}\right)\right) < \exp\left(\frac{1}{4n}\right)$$

for any two positive integers n and k. Keeping n fixed and letting k become arbitrarily large, we see that $\sigma \neq 0$ and that

$$1 \leq \frac{x_n}{\sigma} \leq \exp\left(\frac{1}{4n}\right).$$

We may therefore put

$$x_n = \sigma \exp\left(\frac{\theta_n}{4n}\right) \quad \text{for } n = 1, 2, 3, \ldots, \tag{7.24}$$

where θ_n are suitable constants satisfying $0 \leq \theta_n \leq 1$. By (7.19) and (7.16)

$$\frac{x_{2n}}{x_n^2} = \frac{(2n)!}{(2n)^{2n}e^{-2n}\sqrt{2n}} \cdot \frac{n^{2n}e^{-2n}n}{(n!)^2}$$

$$= \frac{1 \cdot 3 \cdots (2n - 1)}{2 \cdot 4 \cdots (2n)} \cdot \frac{\sqrt{2n}}{2} \to \frac{1}{\sqrt{2\pi}} \quad \text{as } n \to \infty$$

and thus by (7.22) and because $\sigma > 0$ we have

$$\sigma = \lim_{n \to \infty} x_n = \sqrt{2\pi}. \tag{7.25}$$

From (7.25) and (7.24) we finally get

$$n! = n^n e^{-n}\sqrt{2\pi n}\,\exp\left(\frac{\theta_n}{4n}\right) \quad \text{for } n = 1, 2, 3, \ldots,$$

where θ_n denote constants satisfying $0 \leq \theta_n \leq 1$. $\qquad\square$

COMMENTS. From Proposition 1.5 in Chapter 1 we know that

$$\left(1 + \frac{1}{i}\right)^i < e < \left(1 + \frac{1}{i}\right)^{i+1}, \quad \text{where } i = 1, 2, 3, \ldots. \tag{7.26}$$

For $i = 1, 2, \ldots, n - 1$, (7.26) generates a system of inequalities which, upon multiplication, produce the inequality

$$e\left(\frac{n}{e}\right)^e \leq n! \leq en\left(\frac{n}{e}\right)^n. \tag{7.27}$$

While (7.27) represents only a coarse estimate of the order of magnitude of the number $n!$, it nevertheless suggests trying the number $(n/e)^n \cdot \sqrt{n}$ between $e(n/e)^n$ and $en(n/e)^n$; this is the clue for trying out the sequence (7.19).

We use Stirling's formula (7.18) to get some idea about the number 1000!. Since

$$\log_{10}(1000^{1000} e^{-1000} \sqrt{2000\pi}) = 2567.6046\ldots$$

and

$$\log_{10}(e)^{\theta*/4000}$$

is a number between 0 and 0.0001085..., it follows that

$$\log_{10}(1000!) = 2567.604\ldots$$

with either 6 or 7 being the digit in the fourth decimal place. This shows that the number 1000! has 2568 digits and that the number 1000! begins with the digits 402....

Another interesting question is: How many zeros are at the end of the number 1000!? Evidently, the number of terminal zeros of a number depends on how often the factor $10 = 2 \cdot 5$ occurs in its factorization. We must therefore find the exponent of the factors 2 and 5 in the prime factorization of 1000!. A moment's reflection shows that the prime 2 occurs to a much higher power in the prime factorization of 1000! than the prime 5 (to begin with, there are 500 even numbers between 1 and 1001). We shall see at once that there are exactly 249 "fives" occurring in the prime factorization of 1000!. Indeed, the enumeration of multiples of 5 gives us 200 terms, namely,

$$5, \quad 10, \quad 15, \quad \ldots, \quad 1000.$$

The enumeration of multiples of 25 ($= 5^2$) gives us 40 terms, namely,

$$25, \quad 50, \quad 75, \quad \ldots, \quad 1000.$$

The enumeration of multiples of 125 ($= 5^3$) gives us 8 terms, namely,

$$125, \quad 250, \quad 375, \quad \ldots, \quad 1000.$$

Finally, the enumeration of multiples of 625 ($= 5^4$) gives us a single term, namely,

$$625.$$

But $200 + 40 + 8 + 1 = 249$. Along the same lines of reasoning we can see that there are precisely 994 "twos" in 1000!. Thus, 1000! ends in 249 zeros.

In finding the number of zeros at the end of 1000! we have touched on an interesting subject pertaining to the theory of numbers. If p is a positive prime number, let $E_p(m)$ denote the highest power of the prime p that is a divisor of a given positive integer m. It can be shown that, if both the integer n and the prime number p are positive, the exponent of the highest power of p that divides $n!$ is given by the formula

$$E_p(n!) = \left[\frac{n}{p}\right] + \left[\frac{n}{p^2}\right] + \cdots + \left[\frac{n}{p^s}\right] \quad \text{with} \quad \left[\frac{n}{p^{s+1}}\right] = 0,$$

where $[t]$ denotes (here) the greatest integer less than or equal to t. Thus,

$$E_5(10001) = \left[\frac{1000}{5}\right] + \left[\frac{1000}{5^2}\right] + \left[\frac{1000}{5^3}\right] + \left[\frac{1000}{5^4}\right]$$

$$= 200 + 40 + 8 + 1 = 249,$$

$$E_2(1000!) = \left[\frac{1000}{2}\right] + \left[\frac{1000}{2^2}\right] + \left[\frac{1000}{2^3}\right] + \left[\frac{1000}{2^4}\right] + \left[\frac{1000}{2^5}\right]$$

$$+ \left[\frac{1000}{2^6}\right] + \left[\frac{1000}{2^7}\right] + \left[\frac{1000}{2^8}\right] + \left[\frac{1000}{2^9}\right]$$

$$= 500 + 250 + 125 + 62 + 31 + 15 + 7 + 3 + 1 = 994.$$

To see that the binomial coefficient $\binom{1000}{500} = 1000!/(500!)^2$ is not divisible by 7 we only need to note that $E_7(1000!) = 164$ and $E_7(500!) = 82$. Therefore, when 7^{164} is canceled out of the numerator and denominator of $\binom{1000}{500}$, no multiple of 7 remains in the resulting number.

We record for the convenience of the reader the number

$$25! = 15{,}511{,}210{,}043{,}330{,}985{,}984{,}000{,}000.$$

Stirling's Formula gives for 25! the approximate value 1.54596×10^{25}; the true value and the approximate value therefore differ by only about 0.3%. As n increases, the error between the true value and the approximate value for $n!$ decreases; it can be shown that this error is never more than $10/n$ percent.

Definition. Let $(a_1, a_2, \ldots, a_n)$ be a permutation of the numbers $1, 2, \ldots, n$ such that no element is back in its original place, that is, $a_1 \neq 1$, $a_2 \neq 2, \ldots$, $a_n \neq n$. Such a permutation is called a *derangement*. Let D_n be the number of derangements of the set $\{1, 2, \ldots, n\}$.

REMARKS. If we start with the $4! = 24$ permutations of the numbers 1, 2, 3, 4 and cross off all those with one in the first place, two in the second place, three in the third place, or four in the fourth place we have left the following nine arrangements:

$$2\ 1\ 4\ 3 \quad 2\ 3\ 4\ 1 \quad 2\ 4\ 1\ 3$$
$$3\ 1\ 4\ 2 \quad 3\ 4\ 1\ 2 \quad 3\ 4\ 2\ 1$$
$$4\ 1\ 2\ 3 \quad 4\ 3\ 1\ 2 \quad 4\ 3\ 2\ 1.$$

Therefore, $D_4 = 9$.

Proposition 7.18. *Let D_n be the number of derangements of the set $\{1, 2, \ldots, n\}$. Then*

$$D_n = n!\left(1 - \frac{1}{1!} + \frac{1}{2!} - \frac{1}{3!} + \cdots + (-1)^n\frac{1}{n!}\right).$$

PROOF. Let $D_0 = 1$ and $D_1 = 0$. Let us distinguish two kinds of derangements. We know that a_1 sits in the first position; suppose that 1 sits in the a_1th position, that is, a_1 and 1 just changed places. The rest of the $(n - 2)$ numbers must form a smaller derangement with each element moved from its initial position. This can happen in D_{n-2} ways. Since a_1 itself can be chosen in $(n - 1)$ ways the number of derangements of this kind is $(n - 1)D_{n-2}$. We can now count the number of derangements in which 1 is not in the a_1th position. First we can choose a_1 in $(n - 1)$ ways. Now add it to the front of any derangement of $\{2, 3, \ldots, n\}$ in which we have replaced the a_1 by 1. Since a_1 was not in place a_1, 1 will not be in place a_1. This process will produce all the derangements of the second kind. Clearly, there are $(n - 1)D_{n-1}$ of these.

Adding both together, we find

$$D_n = (n - 1)D_{n-1} + (n - 1)D_{n-2}. \tag{7.28}$$

Let us write (7.28) as

$$\frac{D_n}{n!} = \frac{n-1}{n}\frac{D_{n-1}}{(n-1)!} + \frac{n-1}{n(n-1)}\frac{D_{n-2}}{(n-2)!}.$$

We now introduce the notation

$$E_n = \frac{D_n}{n!}, \qquad E_0 = 1, \qquad E_1 = 0.$$

The E's satisfy the recurrence relationship

$$E_n = \left(1 - \frac{1}{n}\right)E_{n-1} + \frac{1}{n}E_{n-2}$$

or

$$E_n - E_{n-1} = \left(\frac{-1}{n}\right)(E_{n-1} - E_{n-2}). \tag{7.29}$$

Reiterating (7.29) for $(n - 1)$ instead of (n), we obtain the descent

$$E_n - E_{n-1} = \left(\frac{-1}{n}\right)\left(\frac{-1}{n-1}\right)(E_{n-2} - E_{n-3})$$

$$= \left(\frac{-1}{n}\right)\left(\frac{-1}{n-1}\right)\left(\frac{-1}{n-2}\right)(E_{n-3} - E_{n-4})$$

$$\vdots$$

$$= \frac{(-1)^n}{n!}.$$

We write this now in the form

$$E_n = \frac{(-1)^n}{n!} + E_{n-1}. \tag{7.30}$$

Reiterating (7.30) for $(n-1)$ instead of (n), we obtain another descent

$$E_n = \frac{(-1)^n}{n!} + \frac{(-1)^{n-1}}{(n-1)!} + E_{n-2}$$

$$\vdots$$

$$= (-1)^n\left(\frac{1}{n!} - \frac{1}{(n-1)!} + \frac{1}{(n-2)!} - + \cdots\right)$$

and so

$$D_n = n!\left(1 - \frac{1}{1!} + \frac{1}{2!} - \frac{1}{3!} + \cdots + (-1)^n\frac{1}{n!}\right), \tag{7.31}$$

finishing the proof.   □

DISCUSSION. For any n, the probability that any given permutation is a derangement is given by

$$\frac{D_n}{n!} = 1 - \frac{1}{1!} + \frac{1}{2!} - \frac{1}{3!} + \cdots + (-1)^n\frac{1}{n!}.$$

By (4.13) we know that

$$1 - \frac{1}{1!} + \frac{1}{2!} - \frac{1}{3!} + \cdots + (-1)^n\frac{1}{n!}$$

and $e^{-1} = 1/e$ can differ from each other by at most $3/(n+1)!$.

If we compare two packs of cards (one of them having been well shuffled), card by card, what is the probability that we shall get right through the packs without finding a single coincidence? The answer is $1/e$ (with an error of less than 10^{-69}, for packs of 52 cards, because 53! is approximately $4.2681617 \cdot 10^{69}$ by Stirling's formula). Many people are prepared to bet that no coincidence will occur, so an unscrupulous gambler might profit nicely by knowing that $e > 2$.

If 10 men check their hats in a cloakroom and the attendant gives each man back a hat at random then the probability that no one gets back his correct hat is about $1/e$. Surprisingly, if 100 men instead of just 10 checked their hats the probability that a random reassignment is totally wrong is still about $1/e$.

It should be noted that the expression for D_n in (7.31) can be rewritten in the simpler form

$$D_n = nD_{n-1} + (-1)^n, \quad \text{where } D_1 = 0. \tag{7.32}$$

Proposition 7.19. *Let*

$$T_n = \sum_{k=1}^{n} \frac{1}{k^2}.$$

Then $T_n \to \pi^2/6$ as $n \to \infty$.

PROOF. For any nonnegative integer n, let

$$J_{2n} = \int_0^{\pi/2} (\cos^{2n} t)\, dt \quad \text{and} \quad I_{2n} = \int_0^{\pi/2} t^2 (\cos^{2n} t)\, dt.$$

Applying integration by parts twice, we get

$$\int_0^{\pi/2} (\cos^{2n} t)\, dt = t(\cos^{2n} t)\Big|_0^{\pi/2} + 2n \int_0^{\pi/2} t(\cos^{2n-1} t)(\sin t)\, dt$$

$$= n\big[t^2 (\cos^{2n-1} t)(\sin t) \big]\Big|_0^{\pi/2}$$

$$- n \int_0^{\pi/2} t^2 \big[-(2n-1)(\cos^{2n-2} t)(\sin^2 t) + (\cos^{2n} t) \big]\, dt$$

$$= -2n^2 I_{2n} + n(2n-1) I_{2n-2}$$

or

$$J_{2n} = -2n^2 I_{2n} + n(2n-1) I_{2n-2}.$$

On the other hand,

$$\int_0^{\pi/2} (\cos^{2n} t)\, dt = \int_0^{\pi/2} (\cos^{2n-1} t)\, d(\sin t)$$

$$= (\cos^{2n-1} t)(\sin t)\Big|_0^{\pi/2} + (2n-1) \int_0^{\pi/2} (\cos^{2n-2} t)(\sin^2 t)\, dt,$$

that is, $J_{2n} = (2n-1)J_{2n-2} - (2n-1)J_{2n}$ or

$$J_{2n} = \frac{2n-1}{2n} J_{2n-2}.$$

Noting that $J_0 = \pi/2$, we see that

$$J_{2n} = \frac{(2n-1)(2n-3)\cdots 3 \cdot 1}{2n(2n-2)\cdots 4 \cdot 2} \cdot \frac{\pi}{2}.$$

We may thus conclude that

$$-2n^2 I_{2n} + n(2n-1)I_{2n-2} = \frac{(2n-1)!!}{(2n)!!} \frac{\pi}{2},$$

where we use the notation

$$(2n)!! = 2 \cdot 4 \cdots (2n-2)(2n), \qquad 0!! = 1,$$

$$(2n+1)!! = 1 \cdot 3 \cdots (2n-1)(2n+1), \qquad (-1)!! = 1.$$

(If n is a positive integer, then $n!!$ is the product of all positive integers less than or equal to n and of the same parity as n.)
Therefore,

$$\frac{(2n)!!}{(2n-1)!!}I_{2n} - \frac{(2n-2)!!}{(2n-3)!!}I_{2n-2} = -\frac{\pi}{4} \cdot \frac{1}{n^2}$$

and so

$$\frac{(2n)!!}{(2n-1)!!}I_{2n} - \frac{0!!}{(-1)!!}I_0 = \sum_{k=1}^{n}\left(\frac{(2k)!!}{(2k-1)!!}I_{2k} - \frac{(2k-2)!!}{(2k-3)!!}I_{2k-2}\right)$$

$$= -\frac{\pi}{4}\sum_{k=1}^{n}\frac{1}{k^2}.$$

This shows that

$$\frac{(2n)!!}{(2n-1)!!}I_{2n} = \frac{\pi^3}{24} - \frac{\pi}{4}\sum_{k=1}^{n}\frac{1}{k^2} = \frac{\pi}{4}\left(\frac{\pi^2}{6} - \sum_{k=1}^{n}\frac{1}{k^2}\right).$$

But

$$\lim_{n\to\infty}\frac{(2n)!!}{(2n-1)!!}I_{2n} = 0.$$

Indeed, since $(2/\pi)t \le \sin t$ for $0 \le t \le \pi/2$, we have

$$I_{2n} = \int_0^{\pi/2} t^2(\cos^{2n}t)\,dt \le \left(\frac{\pi}{2}\right)^2 \int_0^{\pi/2}(\sin^2 t)(\cos^{2n}t)\,dt$$

$$= \frac{\pi^2}{4}\left(\int_0^{\pi/2}(\cos^{2n}t)\,dt - \int_0^{\pi/2}(\cos^{2n+2}t)\,dt\right)$$

$$= \frac{\pi^3}{8}\left(\frac{(2n-1)!!}{(2n)!!} - \frac{(2n+1)!!}{(2n+2)!!}\right) = \frac{\pi^3}{8}\frac{(2n-1)!!}{(2n+2)!!}$$

and so

$$0 < \frac{(2n)!!}{(2n-1)!!}I_{2n} \le \frac{\pi^3}{8} \cdot \frac{1}{2n+2}.$$

The proof is complete. $\square$

REMARK. In Example 3 of Section 4 in Chapter 1 we derived the content of Proposition 7.19 in a different way.

3. Numerical Series

Definition. Let $\{x_n\}_{n=1}^{\infty}$ be a sequence of real numbers. The *infinite series* (or simply the series) generated by $\{x_n\}_{n=1}^{\infty}$ and denoted by

$$x_1 + x_2 + x_3 + \cdots \quad \text{or} \quad \sum_{n=1}^{\infty} x_n$$

is the sequence $\{s_k\}_{k=1}^{\infty}$ defined by

$$
\begin{aligned}
s_1 &= x_1, \\
s_2 &= s_1 + x_2 \quad (= x_1 + x_2), \\
&\vdots \\
s_k &= s_{k-1} + x_k \quad (= x_1 + x_2 + \cdots + x_k) \\
&\vdots
\end{aligned}
$$

If the sequence $\{s_k\}_{k=1}^{\infty}$ converges, that is, if there exists a (finite) real number S such that $s_k \to S$ as $k \to \infty$, then the series $\sum_{n=1}^{\infty} x_n$ is said to be *convergent* and S is called the *sum* of the series and we write

$$\sum_{n=1}^{\infty} x_n = S.$$

The elements x_n are called the *terms* and the elements s_k are called the *partial sums* of the infinite series $\sum_{n=1}^{\infty} x_n$. An infinite series is said to be *divergent* if its sequence of partial sums fails to be convergent.

REMARKS. As we have seen in the foregoing definition, the series $\sum_{n=1}^{\infty} x_n$ converges if and only if the sequence of partial sums of the series converges to a (finite) real number. Conversely, we observe that an arbitrary sequence $\{x_n\}_{n=1}^{\infty}$ of real numbers converges if and only if the series

$$x_1 + (x_2 - x_1) + (x_3 - x_2) + \cdots + (x_n - x_{n-1}) + \cdots$$

converges because the sequence of partial sums of this series coincides with the sequence $x_1, x_2, x_3, \ldots$. This means that convergence criteria for infinite series can also be used in the investigation of convergence of sequences.

Proposition 7.20. *The series $\sum_{n=1}^{\infty} x_n$ is convergent if and only if the remainder after m terms, that is,*

$$R_m = a_{m+1} + a_{m+2} + \cdots = \sum_{n=m+1}^{\infty} x_n,$$

is convergent. Moreover, if $\sum_{n=1}^{\infty} x_n$ is convergent, then $\lim_{m \to \infty} R_m = 0$.

PROOF. Let $R_{m,k}$ denote the *partial remainder*

$$R_{m,k} = x_{m+1} + x_{m+2} + \cdots + x_{m+k}.$$

Then

$$R_{m,k} = s_{m+k} - s_m, \qquad (7.33)$$

where s_{m+k} and s_m denote the partial sums

$$s_{m+k} = x_1 + x_2 + \cdots + x_{m+k} \quad \text{and} \quad s_m = x_1 + x_2 + \cdots + s_m$$

of the series $\sum_{n=1}^{\infty} x_n$. Assume now that the series $\sum_{n=1}^{\infty} x_n$ converges and has the sum S. Then, for m fixed and letting $k \to \infty$, $s_{m+k} \to S$. By definition, for fixed m and letting $k \to \infty$, $R_{m,k} \to R_m$. Hence, for fixed m and letting $k \to \infty$,

$$R_m = S - s_m \qquad (7.34)$$

by (7.33). Therefore, the series $\sum_{n=m+1}^{\infty} x_n$ is seen to be convergent. Moreover, (7.34) shows that $R_m \to 0$ as $m \to \infty$.

Conversely, if $\sum_{n=m+1}^{\infty} x_n$ is assumed to be convergent and having sum T, then $\lim_{k\to\infty} R_{m,k} = T$. By (7.33), $s_{m+k} = R_{m,k} + s_m$; keeping m fixed and letting $k \to \infty$, we obtain

$$\sum_{n=1}^{\infty} x_n = T + s_m,$$

showing that the series $\sum_{n=1}^{\infty} x_n$ is convergent. $\qquad\square$

REMARK. Proposition 7.20 shows that the initial terms of an infinite series have no effect whatsoever on the convergence or divergence of a series; it is the "tail end" of the series that matters as far as convergence or divergence is concerned.

Proposition 7.21. *Let $\sum_{n=1}^{\infty} x_n$ be a convergent series. Then $x_n \to 0$ as $n \to \infty$.*

PROOF. Let $\{s_k\}_{k=1}^{\infty}$ be the sequence of partial sums of $\sum_{n=1}^{\infty} x_n$. Then

$$x_n = s_n - s_{n-1} \to 0 \quad \text{as } n \to \infty$$

because s_n (and with it s_{n-1}) tend to S as $n \to \infty$, where S is the sum of the convergent series $\sum_{n=1}^{\infty} x_n$. $\qquad\square$

REMARK. Proposition 7.21 provides us with a very handy *necessary* condition for the convergence of an infinite series: If $\{x_n\}_{n=1}^{\infty}$ is not a null sequence, then the series $\sum_{n=1}^{\infty} x_n$ diverges. It is important to note, however, that the condition is *not sufficient*: If $\{x_n\}_{n=1}^{\infty}$ is a null sequence, it can still happen that $\sum_{n=1}^{\infty} x_n$ diverges. For example, the series

$$\sum_{n=1}^{\infty} \ln\left(1 + \frac{1}{n}\right) = \sum_{n=1}^{\infty} [\ln(n+1) - \ln n] \qquad (7.35)$$

satisfies the property that $x_n = \ln(1 + 1/n) \to 0$ as $n \to \infty$, but the series in (7.35) is divergent because

$$s_k = \sum_{n=1}^{k} [\ln(n+1) - \ln n] = \ln(k+1) \to \infty \quad \text{as } k \to \infty.$$

The series $\sum_{n=1}^{\infty} (1/n)$ exhibits the same kind of behavior; while $1/n \to 0$ as $n \to \infty$, the *harmonic series* $\sum_{n=1}^{\infty} (1/n)$ is known to diverge by Proposition 1.7. Another example is the divergent series $\sum_{n=1}^{\infty} (1/\sqrt{n})$; indeed,

$$s_k = 1 + \frac{1}{\sqrt{2}} + \frac{1}{\sqrt{3}} + \cdots + \frac{1}{\sqrt{k}} > k\frac{1}{\sqrt{k}} = \sqrt{k} \to \infty \quad \text{as } k \to \infty$$

and $1/\sqrt{n} \to 0$ as $n \to \infty$.

Proposition 7.22. *Let $\sum_{n=1}^{\infty} a_n$ and $\sum_{n=1}^{\infty} b_n$ be two convergent series with sums A and B, respectively. If α and β are any (finite) real numbers, then the series*

$$\sum_{n=1}^{\infty} (\alpha a_n + \beta b_n)$$

converges and has sum $\alpha A + \beta B$.

PROOF. The claim follows from Proposition 7.5. □

Proposition 7.23. *The series $\sum_{n=1}^{\infty} x_n$ converges if and only if for any $\varepsilon > 0$ there exists an integer n_0 such that*

$$\left| \sum_{k=n}^{m} x_k \right| < \varepsilon \quad \text{if } m \geq n \geq n_0.$$

PROOF. The claim is a direct consequence of Proposition 7.12. □

REMARK. Taking $m = n$ in Proposition 7.23, we get $|x_n| < \varepsilon$ for $n \geq n_0$ and Proposition 7.21 follows.

Proposition 7.24. *If $\sum_{n=1}^{\infty} x_n$ is a series of nonnegative terms and*

$$s_k = x_1 + \cdots + x_k,$$

then $\sum_{n=1}^{\infty} x_n$ is convergent if and only if the sequence $\{s_k\}_{k=1}^{\infty}$ is bounded.

PROOF. The claim follows from Proposition 7.8. □

REMARK. The harmonic series $\sum_{n=1}^{\infty} (1/n)$ is a series of nonnegative terms. Since

$$\frac{1}{n+1} + \frac{1}{n+2} + \cdots + \frac{1}{2n} > n\frac{1}{2n} = \frac{1}{2},$$

we see that

$$\frac{1}{2^{k-1}+1} + \cdots + \frac{1}{2^k} > \frac{1}{2}$$

and so the sums

$$\frac{1}{3}+\frac{1}{4}, \quad \frac{1}{5}+\cdots+\frac{1}{8}, \quad \frac{1}{9}+\cdots+\frac{1}{16}, \cdots$$

are larger than $\frac{1}{2}$ and therefore, letting $s_k = \sum_{n=1}^{k} 1/n$, we see that

$$s_{2^k} > k \cdot \tfrac{1}{2}$$

and so the sequence of partial sums $\{s_k\}_{k=1}^{\infty}$ is not bounded and $\sum_{n=1}^{\infty}(1/n)$ must therefore be divergent by Proposition 7.24.

Proposition 7.25 (Cauchy's Condensation Theorem). *Let $x_1 \geq x_2 \geq \cdots \geq 0$. Then the series $\sum_{n=1}^{\infty} x_n$ converges if and only if the following series converges*

$$\sum_{k=0}^{\infty} 2^k \cdot x_{2^k} = a_1 + 2x_2 + 4x_4 + 8x_8 + \cdots.$$

PROOF. By Proposition 7.24 it will be sufficient to consider the boundedness of the sequence of partial sums. Let

$$s_n = x_1 + \cdots + x_n \quad \text{and} \quad t_k = x_1 + 2x_2 + \cdots + 2^k \cdot x_{2^k}.$$

For $n < 2^k$,

$$s_n \leq x_1 + (x_2 + x_3) + \cdots + (x_{2^k} + \cdots + x_{2^{k+1}-1})$$

$$\leq x_1 + 2x_2 + \cdots + 2^k \cdot x_{2^k}$$

$$= t_k;$$

for $n \geq 2^k$,

$$s_n \geq x_1 + x_2 + (x_3 + x_4) + \cdots + (x_{2^{k-1}+1} + \cdots + x_{2^k})$$

$$\geq \tfrac{1}{2}x_1 + x_2 + 2x_4 + \cdots + 2^{k-1} \cdot x_{2^k}$$

$$= \tfrac{1}{2}t_k.$$

Thus, the sequences $\{s_n\}_{n=1}^{\infty}$ and $\{t_k\}_{k=1}^{\infty}$ are either both bounded or both unbounded. $\qquad\square$

Proposition 7.26 (Comparison Theorem). *We have the following comparison tests:*

(i) *If $|a_n| \leq c_n$ for $n \geq n_0$, where n_0 is some fixed integer, and if $\sum_{n=1}^{\infty} c_n$ converges, then $\sum_{n=1}^{\infty} a_n$ converges.*
(ii) *If $a_n \geq d_n \geq 0$ for $n \geq n_0$, where n_0 is some fixed integer, and if $\sum_{n=1}^{\infty} d_n$ diverges, then $\sum_{n=1}^{\infty} a_n$ diverges.*

PROOF. By Proposition 7.23, given any $\varepsilon > 0$, there exists some $\tilde{n}_0 \geq n_0$ such that $m \geq n \geq \tilde{n}_0$ implies $\sum_{k=n}^{m} c_k < \varepsilon$. Hence,

$$\left| \sum_{k=n}^{m} a_k \right| \leq \sum_{k=n}^{m} |a_k| \leq \sum_{k=n}^{m} c_k < \varepsilon$$

and (i) follows. Part (ii) follows from Proposition 7.24. $\qquad\square$

Proposition 7.27 (Limit Comparison Test). *Let* $\{a_n\}_{n=1}^{\infty}$ *and* $\{b_n\}_{n=1}^{\infty}$ *be sequences of positive real numbers.*

(i) *If* $\overline{\lim}_{n\to\infty} a_n/b_n < \infty$ *and* $\sum_{n=1}^{\infty} b_n$ *converges, then* $\sum_{n=1}^{\infty} a_n$ *converges.*

(ii) *If* $\underline{\lim}_{n\to\infty} a_n/b_n > 0$ *and* $\sum_{n=1}^{\infty} b_n$ *diverges, then* $\sum_{n=1}^{\infty} b_n$ *diverges.*

PROOF. (i) Choose β such that

$$\overline{\lim}_{n\to\infty} \frac{a_n}{b_n} < \beta < \infty.$$

Then there exists n_0 such that $a_n/b_n < \beta$ for all $n \geq n_0$; hence,

$$\sum_{n=n_0}^{\infty} a_n < \beta \sum_{n=n_0}^{\infty} b_n < \infty.$$

(ii) Choose α satisfying

$$0 < \alpha < \underline{\lim}_{n\to\infty} \frac{a_n}{b_n}.$$

Then, for some n_0, we have $a_n/b_n > \alpha$ for all $n \geq n_0$; hence,

$$\sum_{n=n_0}^{\infty} a_n > \alpha \sum_{n=n_0}^{\infty} b_n = \infty$$

and the proof is complete. $\qquad\qquad\qquad\qquad\qquad\qquad\qquad\qquad\qquad$ $\square$

Examples

1. If $|x| < 1$, then the *geometric series*

$$\sum_{n=0}^{\infty} x^n = \frac{1}{1-x};$$

if $|x| \geq 1$, the series diverges. Indeed, if $x \neq 0$, then

$$s_k = 1 + x + x^2 + \cdots + x^k = \frac{1 - x^{k+1}}{1 - x}$$

and the claim follows if we let $k \to \infty$.

2. The series

$$\sum_{n=1}^{\infty} \frac{1}{n^p}$$

converges if $p > 1$ and diverges if $p \leq 1$. Indeed, if $p \leq 0$, divergence follows from Proposition 7.21. If $p > 0$, Proposition 7.25 is applicable, and we are led to the series

$$\sum_{k=0}^{\infty} 2^k \frac{1}{2^{kp}} = \sum_{k=0}^{\infty} 2^{(1-p)k}.$$

But $2^{1-p} < 1$ if and only if $1 - p < 0$, and the claim follows by comparison with the geometric series taking $x = 2^{1-p}$. The case when $p = 1$ we have already considered in the Remark following Proposition 7.24 as well as in Proposition 1.7.

3. The series

$$\sum_{n=2}^{\infty} \frac{1}{n(\ln n)}$$

is divergent by Proposition 7.25 because

$$\sum_{k=1}^{\infty} \frac{2^k}{2^k (\ln 2^k)} = \frac{1}{\ln 2} \sum_{k=1}^{\infty} \frac{1}{k}$$

is divergent.

4. The series

$$\sum_{n=3}^{\infty} \frac{1}{n(\ln n)(\ln\{\ln n\})}$$

is divergent by Proposition 7.25 and the foregoing Example 3:

$$\frac{2^k}{2^k(\ln 2^k)(\ln\{\ln 2^k\})} = \frac{1}{k(\ln 2)(\ln[k\{\ln 2\}])} > \frac{1}{k(\ln k)}.$$

5. The series

$$\sum_{n=2}^{\infty} \frac{1}{n(\ln n)^2}$$

is convergent by Proposition 7.25 and the foregoing Example 2:

$$\frac{2^k}{2^k(\ln 2^k)^2} = \frac{1}{k^2(\ln 2)^2}$$

and $\sum_{n=1}^{\infty} (1/n^2)$ converges.

6. The series

$$\sum_{n=3}^{\infty} \frac{1}{n(\ln n)(\ln\{\ln 2^k\})^2}$$

converges by Proposition 7.25 and the foregoing Example 5:

$$\frac{2^k}{2^k(\ln 2^k)(\ln\{\ln 2^k\})^2} < \frac{1}{\ln 2} \frac{1}{k(\ln k)^2}.$$

7. The series

$$\sum_{n=2}^{\infty} \frac{1}{(\ln n)^{\ln n}}$$

is convergent. Indeed,

$$\frac{1}{(\ln n)^{\ln n}} = \frac{1}{(n)^{\ln(\ln n)}}$$

and $\ln(\ln n) > 2$ for all sufficiently large n. Hence,

$$\frac{1}{(n)^{\ln(\ln n)}} < \frac{1}{n^2}$$

for all sufficiently large n. But $\sum_{n=1}^{\infty} (1/n^2)$ is convergent and the claim follows by part (i) of Proposition 7.26.

8. The series

$$\sum_{n=3}^{\infty} \frac{1}{(\ln n)^{\ln(\ln n)}}$$

is divergent. Indeed,

$$\frac{1}{(\ln n)^{\ln(\ln n)}} = \frac{1}{e^{[\ln(\ln n)]^2}}$$

and $[\ln(\ln n)]^2 < \ln n$ for all sufficiently large n [note that $(\ln T)^2 < T$ for sufficiently large T as $(\ln T)^2/T \to 0$ when $T \to \infty$]. Hence, for all sufficiently large n we have

$$\frac{1}{(\ln n)^{\ln(\ln n)}} = \frac{1}{e^{[\ln(\ln n)]^2}} > \frac{1}{e^{(\ln n)}} = \frac{1}{n}.$$

But $\sum_{n=1}^{\infty} (1/n)$ diverges and the claim follows by part (ii) of Proposition 7.26.

9. The series

$$\sum_{n=3}^{\infty} \frac{1}{(\ln\{\ln n\})^{\ln n}}$$

is convergent. Indeed,

$$\frac{1}{(\ln\{\ln n\})^{\ln n}} = \frac{1}{n^{\ln(\ln\{\ln n\})}}$$

and $\ln(\ln\{\ln n\}) > 2$ for all sufficiently large n. Hence, for all sufficiently large n

$$\frac{1}{(\ln\{\ln n\})^{\ln n}} = \frac{1}{n^{\ln(\ln\{\ln n\})}} < \frac{1}{n^2}.$$

But $\sum_{n=1}^{\infty} (1/n^2)$ is convergent and the claim follows from part (i) of Proposition 7.26.

10. The series

$$\sum_{n=1}^{\infty}\left(1 - \frac{\ln n}{n}\right)^n$$

is divergent. Indeed, Proposition 7.27 is applicable. Let $b_n = 1/n$ and $a_n = [1 - (1/n)(\ln n)]^n$ in part (ii) of Proposition 7.27. In Example 13 of the worked examples following Proposition 7.15 we showed that

$$x_n = \frac{a_n}{b_n} = n\left(1 - \frac{\ln n}{n}\right)^n \to 1 \quad \text{as } n \to \infty. \tag{7.36}$$

REMARK. Another method for showing (7.36) is as follows: We have

$$\ln x_n = \ln n + n\ln\left(1 - \frac{\ln n}{n}\right).$$

By Taylor's Theorem (see Section 2 of Chapter 4)

$$\ln\left(1 - \frac{\ln n}{n}\right) = -\frac{\ln n}{n} - \frac{1}{2}\left(\frac{\ln n}{n}\right)^2 + \alpha_n\left(\frac{\ln n}{n}\right)^2,$$

where $\alpha_n \to 0$ as $n \to \infty$. Hence,

$$\ln x_n = -\frac{1}{2}\frac{(\ln n)^2}{n} + \alpha_n\frac{(\ln n)^2}{n} \to 0 \quad \text{as } n \to \infty$$

and (7.36) follows by the continuity of the logarithmic function.

11. The series

$$\sum_{n=1}^{\infty}\left(\frac{1}{n} - \ln\frac{n+1}{n}\right) \tag{7.37}$$

is convergent. Indeed, since $1/(n+1) < \ln\{(n+1)/n\} < 1/n$ [see (1.27)], we get

$$0 < \frac{1}{n} - \ln\frac{n+1}{n} < \frac{1}{n} - \frac{1}{n+1} \quad \text{for } n = 1, 2, 3, \dots.$$

But

$$\sum_{n=1}^{\infty}\left(\frac{1}{n} - \frac{1}{n+1}\right) = 1$$

because

$$\left(1 - \frac{1}{2}\right) + \left(\frac{1}{2} - \frac{1}{3}\right) + \left(\frac{1}{3} - \frac{1}{4}\right) + \cdots + \left(\frac{1}{k} - \frac{1}{k+1}\right) = 1 - \frac{1}{k+1} \to 1$$

as $k \to \infty$. Therefore, by part (i) of Proposition 7.26, the series (7.37) converges; by Proposition 1.7 its sum is Euler's constant $C = 0.5772156649\dots$.

12. The series

$$\sum_{n=1}^{\infty} \sin \frac{x}{n}$$

diverges if $x \neq 0$. Indeed, by part (ii) of Proposition 7.27, letting $b_n = 1/n$ and $a_n = \sin(x/n)$, we see that $a_n/b_n \to x$ as $n \to \infty$. However, the series

$$\sum_{n=1}^{\infty} \frac{1}{n} \sin \frac{x}{n}$$

converges for any (finite) value of x. Indeed, part (i) of Proposition 7.26 is applicable. Since $|\sin t| \leq |t|$ for any real number t, we see that

$$\left| \frac{1}{n} \sin \frac{x}{n} \right| \leq \frac{|x|}{n^2} \quad \text{for } n = 1, 2, 3, \ldots$$

and any real number x. But $\sum_{n=1}^{\infty} (1/n^2)$ is convergent; actually, we know that its sum is $\pi^2/6$ (see Proposition 7.19).

Proposition 7.28. *Let $\sum_{n=1}^{\infty} c_n$ and $\sum_{n=1}^{\infty} d_n$ denote, respectively, a convergent and a divergent series of positive terms. Then*

(i) *if the sequence $\{\gamma_n\}_{n=1}^{\infty}$ is positive and bounded, $\sum_{n=1}^{\infty} \gamma_n c_n$ converges;*
(ii) *if the sequence $\{\delta_n\}_{n=1}^{\infty}$ is positive and bounded below by the positive number δ, $\sum_{n=1}^{\infty} \delta_n d_n$ is divergent.*

PROOF. (i) If the partial sums of $\sum_{n=1}^{\infty} c_n$ remain constantly smaller than K and if the factors satisfy $\gamma_n < \gamma$ for $n = 1, 2, 3, \ldots$, then the partial sums of $\sum_{n=1}^{\infty} \gamma_n c_n$ obviously remain always smaller than γK and the claim follows by Proposition 7.24.

(ii) If $G > 0$ is arbitrary, then by assumption the partial sums of $\sum_{n=1}^{\infty} d_n$ are larger than G/δ from a suitable index n_0 onward. From the same index n_0 onward, the partial sums of $\sum_{n=1}^{\infty} \delta_n d_n$ are then larger than G and so $\sum_{n=1}^{\infty} \delta_n d_n$ is seen to diverge. $\qquad\qquad\qquad\qquad\qquad\qquad\qquad\qquad\qquad\qquad\qquad\qquad\quad\square$

Proposition 7.29. *Let $\sum_{n=1}^{\infty} c_n$ and $\sum_{n=1}^{\infty} d_n$ denote, respectively, a convergent and a divergent series of positive terms. If the terms of a given series $\sum_{n=1}^{\infty} a_n$ of positive terms satisfy, for every $n \geq n_0$ with n_0 fixed,*

(i) *the condition*

$$\frac{a_{n+1}}{a_n} \leq \frac{c_{n+1}}{c_n},$$

then the series $\sum_{n=1}^{\infty} a_n$ is also convergent. If however, for every $n \geq n_0$ with n_0 fixed, we have

(ii) *constantly*

$$\frac{a_{n+1}}{a_n} \geq \frac{d_{n+1}}{d_n},$$

then $\sum_{n=1}^{\infty} a_n$ must also diverge.

PROOF. The sequence of ratios $\gamma_n = a_n/c_n$ is, from a certain value of the index onward, monotone decreasing, and consequently, since all its terms are positive, it is necessarily bounded. Part (i) of Proposition 7.28 now establishes the claim.

(ii) We have, similarly, $a_{n+1}/d_{n+1} \geq a_n/d_n$, so that the ratios $\delta_n = a_n/d_n$ increase monotonically from a certain value of the index onward. But as they are constantly positive, they then have a positive lower bound. Part (ii) of Proposition 7.28 now proves the divergence. □

Definition. $\sum_{n=1}^{\infty} c_n$ is called *absolutely convergent* if $\sum_{n=1}^{\infty} |c_n|$ converges.

REMARK. Absolute convergence implies convergence by Proposition 7.26 (i).

Proposition 7.30 (Root Test). *Let $\sum_{n=1}^{\infty} c_n$ be a series of real terms and*

$$r = \overline{\lim_{n \to \infty}} \sqrt[n]{|c_n|}.$$

(i) *If $r < 1$, the series converges absolutely.*
(ii) *If $r > 1$, the series diverges.*

PROOF. Suppose that $r < 1$. Take a fixed β such that $r < \beta < 1$. Then there exists an integer $\tilde{n}$ such that $\sqrt[n]{|c_n|} < \beta$ for all $n \geq \tilde{n}$. Thus,

$$\sum_{n=\tilde{n}}^{\infty} |c_n| < \sum_{n=\tilde{n}}^{\infty} \beta^n = \frac{\beta^{\tilde{n}}}{1 - \beta} < \infty$$

and (i) follows.

If $r > 1$, then $\sqrt[n]{|c_n|} > 1$, hence $|c_n| > 1$, for infinitely many n, and so it is false that $c_n \to 0$ as $n \to \infty$. Therefore, (ii) follows from Proposition 7.21 and the proof is finished. □

Proposition 7.31 (Cauchy–Hadamard Theorem). *Let $\sum_{n=0}^{\infty} a_n x^n$ be a series of real terms and let*

$$\gamma = \overline{\lim_{n \to \infty}} \sqrt[n]{|a_n|}, \qquad R = \frac{1}{\gamma}.$$

(If $\gamma = 0$, $R = +\infty$; if $\gamma = +\infty$, $R = 0$.) Then $\sum_{n=0}^{\infty} c_n x^n$ converges absolutely whenever $|x| < R$ and diverges whenever $|x| > R$. (Series of the form $\sum_{n=0}^{\infty} a_n x^n$ are referred to as power series *and will be studied separately later on in this chapter.)*

PROOF. We let $c_n = a_n x^n$ and apply Proposition 7.30; we get

$$\overline{\lim_{n\to\infty}} \sqrt[n]{|c_n|} = |x| \overline{\lim_{n\to\infty}} \sqrt[n]{|a_n|} = |x| \cdot \gamma = \frac{|x|}{R}. \qquad \square$$

Lemma 1. *Let b denote a positive real number. Then $\lim_{n\to\infty} \sqrt[n]{b} = 1$.*

PROOF. If $b > 1$, then the claim follows from the Lemma preceding Proposition 1.10; if $0 < b < 1$, $\lim_{n\to\infty} \sqrt[n]{b} = \lim_{n\to\infty} 1/\sqrt[n]{b} = 1/1 = 1$. $\qquad \square$

REMARK. A much simpler way of proving Lemma 1 is to take the logarithm and to consider the sequence $\{(1/n)(\ln b)\}$.

Lemma 2, *Let $\{a_n\}_{n=1}^{\infty}$ be a sequence of positive real numbers. Then*

$$\underline{\lim_{n\to\infty}} \frac{a_{n+1}}{a_n} \leq \underline{\lim_{n\to\infty}} \sqrt[n]{a_n} \leq \overline{\lim_{n\to\infty}} \sqrt[n]{a_n} \leq \overline{\lim_{n\to\infty}} \frac{a_{n+1}}{a_n}. \qquad (7.38)$$

In particular, if $\lim_{n\to\infty} (a_{n+1}/a_n) = L$, where $0 \leq L \leq \infty$, then $\lim_{n\to\infty} \sqrt[n]{a_n} = L$.

PROOF. It is apparent that the last assertion follows from the first. The second inequality in (7.38) is obvious, while the first and the third have similar proofs. We only verify the first inequality in (7.38). Let

$$\underline{\lim_{n\to\infty}} \frac{a_{n+1}}{a_n} = a.$$

It is clear that $a \geq 0$. If $a = 0$, there is nothing to prove, thus we suppose that $a > 0$. Let $0 < \alpha < a$. Hence, there is an integer N such that

$$\frac{a_{k+1}}{a_k} > \alpha \qquad (7.39)$$

for all $k \geq N$. For $n > N$, multiply the inequalities (7.39) for $k = N, N + 1, \ldots, n - 2, n - 1$ to obtain

$$\frac{a_n}{a_N} > \alpha^{n-N} \quad \text{and} \quad \sqrt[n]{a_n} > \alpha(a_N \alpha^{-N})^{1/n}. \qquad (7.40)$$

Since (7.40) holds for all $n > N$, Lemma 1 shows that

$$\underline{\lim_{n\to\infty}} \sqrt[n]{a_n} \geq \lim_{n\to\infty} \alpha(a_N \alpha^{-N})^{1/n} = \alpha.$$

But α was an arbitrary (positive) number smaller than a, and thus

$$\underline{\lim_{n\to\infty}} \sqrt[n]{a_n} \geq a = \underline{\lim_{n\to\infty}} \frac{a_{n+1}}{a_n}. \qquad \square$$

Proposition 7.32 (Ratio Test). *Let $\sum_{n=1}^{\infty} c_n$ be a series of real terms with $c_n \neq 0$ for all n.*

(i) *If* $\overline{\lim}_{n\to\infty} |c_{n+1}/c_n| < 1$, *then the series converges absolutely.*

(ii) *If* $\underline{\lim}_{n\to\infty} |c_{n+1}/c_n| > 1$, *then the series diverges.*

PROOF. Put $a_n = |c_n|$ and apply Lemma 2 and Proposition 7.30. $\qquad\square$

COMMENTS. Strict inequalities are required in Propositions 7.30 and 7.32. If $c_n = 1/n$, then $\sum_{n=1}^{\infty} c_n$ diverges while $\lim \sqrt[n]{c_n} = \lim(c_{n+1}/c_n) = 1$. On the other hand, if $c_n = 1/n^2$, then $\sum_{n=1}^{\infty} c_n$ converges while $\lim \sqrt[n]{c_n} = \lim(c_{n+1}/c_n) = 1$.

The Ratio Test (see Proposition 7.32) is often easier to apply than the Root Test (see Proposition 7.30), but the Root Test has wider scope; our proof of the Ratio Test shows that any series that can be tested successfully by the Ratio Test can also be tested successfully by the Root Test. The following two examples show that the Root Test is strictly stronger than the Ratio Test. The Root Test detects that the series

$$\frac{1}{3} + \frac{1}{2^2} + \frac{1}{3^3} + \frac{1}{2^4} + \frac{1}{3^5} + \frac{1}{2^6} + \frac{1}{3^7} + \frac{1}{2^8} + \cdots \tag{7.41}$$

is convergent while the Ratio Test does not; the Root Test detects that the series

$$\frac{1}{2} + 2^2 + \frac{1}{2^3} + 2^4 + \frac{1}{2^5} + 2^6 + \frac{1}{2^7} + 2^8 + \cdots \tag{7.42}$$

is divergent while the Ratio Test does not. Moreover, the series (7.41) shows that $\underline{\lim}$ can not be replaced by $\overline{\lim}$ in (ii) of Proposition 7.32 and the series (7.42) shows that $\overline{\lim}$ can not be replaced by $\underline{\lim}$ in either (i) of Proposition 7.30 or (i) in Proposition 7.32.

Proposition 7.33 (Alternating Series Test of Leibniz). *If* $a_1 \geq a_2 \geq a_3 \geq \cdots \geq 0$ *and* $a_n \to 0$ *as* $n \to \infty$, *then the* alternating series

$$\sum_{n=1}^{\infty} (-1)^{n+1} a_n$$

converges. Moreover, if

$$s_n = \sum_{k=1}^{n} (-1)^{k+1} a_k \quad and \quad S = \sum_{k=1}^{\infty} (-1)^{k+1} a_n,$$

then

$$|S - s_n| \leq a_{n+1} \quad for \; n = 1, 2, 3, \ldots.$$

PROOF. For $n = 1, 2, 3, \ldots$ we have

$$s_{2n+2} - s_{2n} = a_{2n+1} - a_{2n+2} \geq 0,$$

$$s_{2n+2} - s_{2n+1} = -a_{2n+2} \qquad \leq 0,$$

$$s_{2n+1} - s_{2n-1} = -a_{2n} + a_{2n+1} \leq 0,$$

and therefore

$$s_{2n} \leq s_{2n+2} \leq s_{2n+1} \leq s_{2n-1}.$$

In addition,

$$\lim_{n\to\infty} (s_{2n-1} - s_{2n}) = \lim_{n\to\infty} a_{2n} = 0.$$

The subsequences $\{s_{2n}\}_{n=1}^{\infty}$ and $\{s_{2n-1}\}_{n=1}^{\infty}$ are bounded by 0 and a_1 and generate the nested sequence of closed intervals

$$\{[s_{2n}, s_{2n-1}]\}_{n=1}^{\infty}$$

and the lengths of these intervals, that is, $s_{2n-1} - s_{2n}$, tends to zero as $n \to \infty$. Evidently, S is the unique point common to all these intervals and

$$s_{2n} < S < s_{2n-1} \quad \text{for } n = 1, 2, 3 \dots.$$

Therefore, for $n = 1, 2, 3, \dots$,

$$|S - s_{2n}| \leq |s_{2n+1} - s_{2n}| = a_{2n+1} \quad \text{and} \quad |S - s_{2n-1}| \leq |s_{2n} - s_{2n-1}| = a_{2n}.$$

This completes the proof.   □

Proposition 7.34 (Kummer's Test). *Let $\{a_n\}_{n=1}^{\infty}$ and $\{b_n\}_{n=1}^{\infty}$ be sequences of positive numbers and put*

$$K_n = b_n - b_{n+1} \frac{a_{n+1}}{a_n}.$$

Then

(i) $\underline{\lim}_{n\to\infty} K_n > 0$ *implies the convergence of* $\sum_{n=1}^{\infty} a_n$, *and*
(ii) $K_n \leq 0$ *for all $n \geq N$ and the divergence of $\sum_{n=1}^{\infty} (1/b_n)$ implies the divergence of* $\sum_{n=1}^{\infty} a_n$.

PROOF. (i) Choose β such that $0 < \beta < \underline{\lim} K_n$. Then there exists an integer n_0 such that $K_n > \beta$ for all $n \geq n_0$. This implies that

$$0 < a_n < \frac{1}{\beta}(a_n b_n - a_{n+1} b_{n+1}) \tag{7.43}$$

for all $n \geq n_0$. Hence, the sequence $\{a_n b_n\}_{n=n_0}^{\infty}$ is a strictly increasing sequence of positive numbers, and so it converges to a nonnegative limit γ. Thus,

$$\sum_{n=n_0}^{\infty} \frac{1}{\beta}(a_n b_n - a_{n+1} b_{n+1}) = \lim_{k\to\infty} \sum_{n=n_0}^{k} \frac{1}{\beta}(a_n b_n - a_{n+1} b_{n+1})$$

$$= \lim_{k\to\infty} \frac{1}{\beta}(a_{n_0} b_{n_0} - a_{k+1} b_{k+1})$$

$$= \frac{1}{\beta}(a_{n_0} b_{n_0} - \beta) < \infty.$$

By (7.43) and Proposition 7.26 (i) we see that $\sum_{n=1}^{\infty} a_n$ converges.

(ii) For $n \geq N$ we have $a_n b_n - a_{n+1} b_{n+1} \leq 0$, and thus $\{a_n b_n\}_{n=N}^{\infty}$ is a non-decreasing sequence. Therefore,

$$a_n \geq \frac{a_N b_N}{b_n} \quad \text{for } n \geq N$$

and so Proposition 7.26 (ii) implies that $\sum_{n=1}^{\infty} a_n$ diverges. $\quad\square$

REMARK. Letting $b_n = 1$ for $n = 1, 2, 3, \ldots$, we have $K_n = 1 - a_{n+1}/a_n$ and $\underline{\lim} K_n = 1 - \overline{\lim}(a_{n+1}/a_n)$. Thus, in this case, Proposition 7.34 reduces to Proposition 7.32.

Proposition 7.35 (Raabe's Test). *If $a_n > 0$ for $n = 1, 2, 3, \ldots$ and*

$$R_n = n\left(1 - \frac{a_{n+1}}{a_n}\right),$$

then

(i) $\underline{\lim}_{n \to \infty} R_n > 1$ *implies the convergence of $\sum_{n=1}^{\infty} a_n$, and*
(ii) $R_n \leq 1$ *for all $n \geq N$ implies the divergence of $\sum_{n=1}^{\infty} a_n$.*

PROOF. In Proposition 7.34, take $b_n = 1 - n$ for $n > 1$ and $b_1 = 1$. Then $\sum_{n=1}^{\infty} (1/b_n)$ diverges and

$$R_n - 1 = (n-1) - n\frac{a_{n+1}}{a_n} = K_n \quad \text{for } n > 1$$

and the proof is complete. $\quad\square$

REMARKS. In Proposition 7.35, the assumption $\underline{\lim} R_n > 1$ is equivalent to the existence of some $\beta > 1$ such that $R_n > \beta$ for all large n; that is, $a_{n+1}/a_n < 1 - \beta/n$. Similarly, $R_n \leq 1$ is equivalent to $a_{n+1}/a_n \geq 1 - 1/n$. This shows that $a_{n+1}/a_n < \alpha < 1$ for all large n (hence $\sum_{n=1}^{\infty} a_n$ converges by Proposition 7.32), then $\sum_{n=1}^{\infty} a_n$ converges by Proposition 7.35; and if $a_{n+1}/a_n \geq 1$ for all large n (hence $\sum_{n=1}^{\infty} a_n$ diverges by Proposition 7.32), then $\sum_{n=1}^{\infty} a_n$ diverges by Proposition 7.35. Therefore, Raabe's Test tests successfully any series that the Ratio Test does. But Raabe's Test is stronger than the Ratio Test. Indeed, consider the *binomial series* $\sum_{n=0}^{\infty} \binom{\alpha}{n} x^n$ at $x = -1$, where α is a fixed real number and

$$a_n = \binom{\alpha}{n} = \frac{\alpha(\alpha - 1) \cdots (\alpha - n + 1)}{n!}(-1)^n \quad \text{for } n > 0.$$

Then

$$\frac{a_{n+1}}{a_n} = \frac{n - \alpha}{n + 1} \to 1 \quad \text{as } n \to \infty \tag{7.44}$$

and so the Ratio Test fails to decide the convergence or divergence of $\sum_{n=1}^{\infty} a_n$. But it follows from (7.44) that all terms a_n have the same sign if $n > \max\{\alpha, 0\}$.

Therefore, multiplying every term by -1 if necessary and noting that this affects neither convergence nor the ratios, we may assume that $a_n > 0$ for all large n. We have

$$R_n = n\left(1 - \frac{a_{n+1}}{a_n}\right) = \frac{n(1 + \alpha)}{n + 1} \to 1 + \alpha \quad \text{as } n \to \infty.$$

This shows that Raabe's Test gives the convergence of $\sum_{n=1}^{\infty} a_n$ for $\alpha > 0$ and the divergence of $\sum_{n=1}^{\infty} a_n$ if $\alpha < 0$ (the series $\sum_{n=1}^{\infty} a_n$ evidently converges if $\alpha = 0$).

Proposition 7.36 (Bertrand's Test). *If $a_n > 0$ for $n = 1, 2, 3, \dots$ and*

$$B_n = (R_n - 1)\ln(n) = (n - 1)\ln(n) - \left(\frac{a_{n+1}}{a_n}\right)n(\ln n),$$

then

(i) $\underline{\lim}_{n \to \infty} B_n > 1$ *implies the convergence of* $\sum_{n=1}^{\infty} a_n$, *and*
(ii) $\overline{\lim}_{n \to \infty} B_n > 1$ *implies the divergence of* $\sum_{n=1}^{\infty} a_n$.

PROOF. In Proposition 7.34 take $b_{n+1} = n(\ln n)$. Then $\sum_{n=1}^{\infty} (1/b_n)$ diverges and

$$K_n = (n - 1)\ln(n - 1) - n(\ln n)\frac{a_{n+1}}{a_n} = (n - 1)\ln\left(\frac{n - 1}{n}\right) + B_n.$$

Since

$$\lim_{n \to \infty} (n - 1)\ln\left(\frac{n - 1}{n}\right) = \lim_{n \to \infty} (n)\ln\left(\frac{n}{n + 1}\right) = \lim_{n \to \infty} \ln\left(1 + \frac{1}{n}\right)^{-n} = -1,$$

we have $\underline{\lim} B_n = 1 + \underline{\lim} D_n$ and the claim follows from Proposition 7.34. $\square$

Proposition 7.37 (Gauss' Test). *If $a_n > 0$ for $n = 1, 2, 3, \dots$ and if there exist a real number α, a positive number ε, and a bounded sequence $\{x_n\}_{n=1}^{\infty}$ of real numbers such that*

$$\frac{a_{n+1}}{a_n} = 1 - \frac{\alpha}{n} - \frac{x_n}{n^{1+\varepsilon}} \quad \text{for } n = 1, 2, 3, \dots,$$

then $\sum_{n=1}^{\infty} a_n$ converges if and only if $\alpha > 1$.

PROOF. We have

$$R_n = n\left(1 - \frac{a_{n+1}}{a_n}\right) = \alpha + \frac{x_n}{n^{\varepsilon}} \to \alpha \quad \text{as } n \to \infty$$

because $\{x_n\}_{n=1}^{\infty}$ is bounded. Thus, for $\alpha \neq 1$, the claim follows from Raabe's Test (see Proposition 7.35). For $\alpha = 1$ we use Bertrand's Test (see Proposition 7.36). We have

$$B_n = (R_n - 1)(\ln n) = \frac{x_n(\ln n)}{n^\varepsilon} \to 0 \quad \text{as } n \to \infty,$$

and so $\sum_{n=1}^{\infty} a_n$ diverges. $\qquad\qquad\qquad\qquad\qquad\qquad\qquad\qquad$ □

Worked Examples

1. Let α, β, and γ be real numbers, none of which is a negative integer or 0. Define $a_0 = 1$ and, for $n > 0$,

$$a_n = \frac{\alpha(\alpha + 1)(\alpha + 2)\cdots(\alpha + n - 1)}{n!}\frac{\beta(\beta + 1)(\beta + 2)\cdots(\beta + n - 1)}{\gamma(\gamma + 1)(\gamma + 2)\cdots(\gamma + n - 1)}.$$

The series $\sum_{n=0}^{\infty} a_n$ is called the *hypergeometric series*. Clearly,

$$\frac{a_{n+1}}{a_n} = \frac{(\alpha + n)(\beta + n)}{(1 + n)(\gamma + n)} = \frac{n^2 + (\alpha + \beta)n + \alpha\beta}{n^2 + (\gamma + 1)n + \gamma}.$$

Since $\lim_{n\to\infty}(a_{n+1}/a_n) = 1$, for all large n the terms a_n have the same sign and so Gauss' Test may be applied. Solving the equation

$$\frac{n^2 + (\alpha + \beta)n + \alpha\beta}{n^2 + (\gamma + 1)n + \gamma} = 1 - \frac{(\gamma + 1) - (\alpha + \beta)}{n} - \frac{x_n}{n^2}$$

for x_n, we find that $\{x_n\}_{n=1}^{\infty}$ converges and hence is bounded. Thus, Gauss' Test shows that $\sum_{n=1}^{\infty} a_n$ converges if and only if $\alpha + \beta < \gamma$.

2. Consider the series

$$\left(\frac{1}{2}\right)^p + \left(\frac{1\cdot 3}{2\cdot 4}\right)^p + \left(\frac{1\cdot 3\cdot 5}{2\cdot 4\cdot 5}\right)^p + \cdots.$$

Since

$$\frac{a_{n+1}}{a_n} = \left(\frac{2n + 1}{2n + 2}\right)^p \to 1 \quad \text{as } n \to \infty,$$

the Ratio Test (see Proposition 7.32) fails, and we turn to Raabe's Test (see Proposition 7.35). We find

$$n\left(1 - \frac{a_{n+1}}{a_n}\right) = n\left(1 - \left(\frac{2n + 1}{2n + 2}\right)^p\right) = \frac{2n}{2n + 2}\frac{1 - (1 - x)^p}{2x},$$

where $2n + 2 = 1/x$. But (by L'Hôpital's Rules)

$$\lim_{x\to\infty} \frac{1 - (1 - x)^p}{2x} = \lim_{x\to\infty} \frac{p(1 - x)^{p-1}}{2} = \frac{p}{2}.$$

Therefore, the given series converges for $p > 2$ and diverges for $p < 2$. It remains to consider the case $p = 2$. For $p = 2$ Raabe's Test fails but Gauss' Test (see Proposition 7.37) indicates divergence because

$$\left(\frac{2n+1}{2n+2}\right)^2 = 1 - \frac{1}{n} + \frac{5 - 4/n}{4n^2 + 8n + 4}$$

and so

$$\frac{a_{n+1}}{a_n} = \left(\frac{2n+1}{2n+2}\right)^2 = 1 - \frac{1}{n} + \frac{x_n}{n^2},$$

where $\{x_n\}_{n=1}^{\infty}$ is a bounded sequence.

3. The series

$$\sum_{n=1}^{\infty} \frac{1}{n!}\left(\frac{n}{e}\right)^n$$

diverges by Raabe's Test. Indeed,

$$n\left(1 - \frac{a_{n+1}}{a_n}\right) = n\left(1 - \frac{(1 + 1/n)^n}{e}\right)$$

and we only have to verify that

$$\lim_{x \to 0} \frac{1}{x}\left(1 - \frac{(1 + x)^{1/x}}{e}\right) = \frac{1}{2}.$$

4. Consider the series

$$\sum_{n=1}^{\infty} \tau(n)x^n,$$

where $x > 0$ and $\tau(n)$ denotes the number of divisors of the positive integer n. Since

$$x \leq \sqrt[n]{\tau(n)} \cdot x \leq \sqrt[n]{n} \cdot x$$

and (by the Lemma preceding Proposition 1.8)

$$\lim_{n \to \infty} \sqrt[n]{n} = 1,$$

we observe that the Root Test (see Proposition 7.30) is applicable. The series under consideration is convergent for $0 < x < 1$ and divergent for $x > 1$ (the series is apparently divergent for $x = 1$).

5. Let $0 < p \leq 1$. Then the series

$$\sum_{n=1}^{\infty} (-1)^{n+1} \frac{1}{n^p}$$

is convergent by Proposition 7.33; in particular, the *alternating harmonic series*

$$\sum_{n=1}^{\infty} (-1)^{n+1} \frac{1}{n}$$

is convergent and its sum is $\ln 2$ (see the Application following Proposition 1.6). The series

$$\sum_{n=1}^{\infty} (-1)^{n+1} \sin\frac{x}{n}$$

is convergent for arbitrary $x \neq 0$. While we can not apply Proposition 7.33 to the series directly, we can apply it to a certain remainder of the series. For sufficiently large n the signs of $\sin(x/n)$ and of x are the same and the absolute value of $\sin(x/n)$ decreases as n increases. From Example 12 following Proposition 7.27 we can see that the series $\sum_{n=1}^{\infty} (-1)^n \sin(x/n)$ is not absolutely convergent, however.

6. The series

$$1 - \frac{1}{3} + \frac{1}{5} - \frac{1}{7} + \frac{1}{9} - \frac{1}{11} + \cdots$$

is seen to be convergent by Proposition 7.33. Moreover, from (4.39) in Chapter 4 we can infer that

$$1 - \frac{1}{3} + \frac{1}{5} - \frac{1}{7} + \frac{1}{9} - \frac{1}{11} + \cdots = \frac{\pi}{4}. \tag{7.45}$$

The representation (7.45) is very interesting, but not suitable for the computation of the number π; to obtain an accuracy of six decimal places, for example, we would have to consider half a million terms of the series in (7.45), according to Proposition 7.33. Following equation (4.39) in Chapter 4 we considered Machin's method for computing the number π to a high degree of accuracy.

7. The series

$$\sum_{n=1}^{\infty} (-1)^{\sigma(n)} \left(\frac{n}{2n-1}\right)^n,$$

where $\sigma(n) = 1 + 2 + \cdots + n = n(n+1)/2$, is convergent. Indeed, observe that the Root Test (see Proposition 7.30) shows the absolute convergence of the series under consideration.

8. For $n = 1, 2, 3, \ldots$, let

$$a_n = \frac{2^n n!}{n^n}.$$

Then $a_n \to 0$ as $n \to \infty$. Indeed, since $a_{n+1}/a_n \to 2/e < 1$ as $n \to \infty$, the infinite series $\sum_{n=1}^{\infty} a_n$ converges by the Ratio Test (see Proposition 7.32) and therefore $a_n \to 0$ as $n \to \infty$ by Proposition 7.21.

Definition. The *Cauchy product* of two series $\sum_{n=0}^{\infty} a_n$ and $\sum_{n=0}^{\infty} b_n$ of real terms is the series $\sum_{n=0}^{\infty} c_n$, where

$$c_n = \sum_{k=0}^{n} a_k b_{n-k}.$$

Lemma. *Let $\{\alpha_n\}_{n=0}^{\infty}$ and $\{\beta_n\}_{n=0}^{\infty}$ be sequences of real numbers and α and β be (finite) real numbers.*

(i) *If $\sum_{n=0}^{\infty} |\beta_n|$ is convergent and $\{\alpha_n\}_{n=0}^{\infty}$ is a null sequence, then*

$$\gamma_n = \sum_{k=0}^{n} \alpha_k \beta_{n-k} \to 0 \quad as\ n \to \infty.$$

(ii) *If $\alpha_n \to \alpha$ and $\beta_n \to \beta$ as $n \to \infty$, then*

$$\frac{1}{n+1} \sum_{k=0}^{n} \alpha_k \beta_{n-k} \to \alpha\beta \quad as\ n \to \infty.$$

(iii) *If $\alpha_n \to \alpha$ as $n \to \infty$, then*

$$\frac{1}{n+1} \sum_{k=0}^{n} \alpha_k \to \alpha \quad as\ n \to \infty.$$

PROOF. (i) We put $\sum_{n=0}^{\infty} |\beta_n| = b$. Let $\varepsilon > 0$ and pick n_0 such that

$$|\alpha_n| < \frac{\varepsilon}{2b}$$

for all $n > n_0$. We set $\sum_{n=0}^{\infty} |\alpha_n| = a$. Since $\{\beta_n\}_{n=0}^{\infty}$ is a null sequence by Proposition 7.21, we can choose j_0 such that

$$|\beta_j| < \frac{\varepsilon}{2a}$$

for all $j > j_0$. Then, for $n > n_0 + j_0$,

$$|\gamma_n| \le \sum_{k=0}^{n_0} |\alpha_k||\beta_{n-k}| + \sum_{k=n_0+1}^{n} |\alpha_k||\beta_{n-k}| < \frac{\varepsilon}{2a} \cdot a + \frac{\varepsilon}{2b} \cdot b = \varepsilon$$

and so we have verified claim (i).

Claims (ii) and (iii) are mere reformulations of Propositions 7.15 and 7.13, respectively. ∎

Proposition 7.38 (Theorem of Mertens). *If at least one of the two convergent series*

$$\sum_{n=0}^{\infty} a_n = A \quad and \quad \sum_{n=0}^{\infty} b_n = B$$

*of real terms is absolutely convergent and if $\sum_{n=0}^{\infty} c_n$ is their Cauchy product,
then $\sum_{n=0}^{\infty} c_n$ converges and we have $\sum_{n=0}^{\infty} c_n = AB$.*

PROOF. Suppose that $\sum_{n=0}^{\infty} b_n$ is absolutely convergent. We put

$$A_n = \sum_{k=0}^{n} a_k, \quad B_n = \sum_{k=0}^{n} b_k, \quad C_n = \sum_{k=0}^{n} c_k, \quad \beta_n = b_n, \quad \alpha_n = A_n - A.$$

Then C_n is the sum of all the products $a_j b_k$ for which $0 \le j \le n$, $0 \le k \le n$,
and $0 \le j + k \le n$. Thus,

$$C_n = A_0 b_n + A_1 b_{n-1} + \cdots + A_n b_0$$

$$= (A + \alpha_0)\beta_n + (A + \alpha_1)\beta_{n-1} + \cdots + (A + \alpha_n)\beta_0$$

$$= AB_n + \sum_{k=0}^{n} \alpha_k \beta_{n-k}$$

$$\to AB + 0 = AB \quad \text{as } n \to \infty$$

by virtue of our assumptions and part (i) of the foregoing Lemma. $\square$

Proposition 7.39. *If $\sum_{n=0}^{\infty} a_n = A$ and $\sum_{n=0}^{\infty} b_n = B$ are convergent series and if
their Cauchy product $\sum_{n=0}^{\infty} c_n = C$ also converges, then $C = AB$.*

PROOF. Using the notation of the preceding proof, we have

$$C_k = A_0 b_k + A_1 b_{k-1} + \cdots + A_k b_0$$

and so

$$C_0 + C_1 + \cdots + C_n = A_0 B_n + A_1 B_{n-1} + \cdots + A_n B_0. \qquad (7.46)$$

Dividing both sides of the equation in (7.46) by $n + 1$ and then taking the
limit as $n \to \infty$, we obtain $C = AB$ in view of parts (ii) and (iii) of the Lemma
preceding Proposition 7.38. $\square$

REMARK. The Cauchy product of two convergent series may diverge. Take

$$a_0 = b_0 = 0, \quad a_n = b_n = \frac{(-1)^{n-1}}{\sqrt{n}} \quad \text{for } n = 1, 2, 3, \ldots .$$

Then $\sum_{n=0}^{\infty} a_n = \sum_{n=0}^{\infty} b_n$ converges by Example 5 of the worked examples
following Proposition 7.37, but $c_0 = c_1 = 0$ and, for $n = 1, 2, 3, \ldots$,

$$c_n = \sum_{k=0}^{n} a_k b_{n-k} = (-1)^n \sum_{k=1}^{n-1} \frac{1}{\sqrt{k}\sqrt{n-k}},$$

$$|c_n| = \sum_{k=1}^{n-1} \frac{1}{\sqrt{k(n-k)}} \ge \sum_{k=1}^{n-1} \frac{1}{\sqrt{(n-1)(n-1)}} = 1;$$

hence, $\{c_n\}_{n=0}^{\infty}$ is not a null sequence and so $\sum_{n=0}^{\infty} c_n$ diverges by Proposition
7.21.

Proposition 7.40 (Abel's Lemma). *If $\{a_n\}_{n=1}^{\infty}$ is a sequence of real numbers whose partial sums $s_n = a_1 + \cdots + a_n$ satisfy, for all $n = 1, 2, 3, \ldots, m \leq s_n \leq M$ for some real numbers m and M, and if $\{b_n\}_{n=1}^{\infty}$ is such that $b_1 \geq b_2 \geq b_3 \geq \cdots \geq 0$, then, for all $n = 1, 2, 3, \ldots$,*

$$mb_1 \leq \sum_{k=1}^{n} a_k b_k \leq Mb_1.$$

PROOF. An easy calculation shows that

$$\sum_{k=1}^{n} a_k b_k = \sum_{k=1}^{n} s_k(b_k - b_{k+1}) + s_n b_{n+1}.$$

Since $b_k - b_{k+1} \geq 0$ and $s_k \leq M$, we get

$$\sum_{k=1}^{n} a_k b_k \leq M \sum_{k=1}^{n} (b_k - b_{k+1}) + Mb_{n+1} = Mb_1.$$

The left-hand side inequality is verified in a similar fashion. □

Proposition 7.41 (Dirichlet's Test). *Let the sequences $\{a_n\}_{n=1}^{\infty}$ and $\{b_n\}_{n=1}^{\infty}$ be as in Proposition 7.40 and suppose, moreover, that $b_n \to 0$ as $n \to \infty$. Then the series $\sum_{n=1}^{\infty} a_k b_k$ converges.*

PROOF. By assumption, there is an $M > 0$ such that $|s_n| \leq M$ for all $n = 1, 2, 3, \ldots$. Hence, for any positive integers j and n with $j \leq n$,

$$\left| \sum_{k=j}^{n} a_k \right| = |s_n - s_{j-1}| \leq |s_n| + |s_{j-1}| \leq 2M.$$

By Proposition 7.40, applied to $\{a_k\}_{k=j}^{\infty}$ and $\{b_k\}_{k=j}^{\infty}$, this implies, for $n \geq j$,

$$\left| \sum_{k=j}^{n} a_k b_k \right| \leq 2Mb_j.$$

By assumption there is an integer n_0 such that $b_n < \varepsilon/2M$ for all $n \geq n_0$. Hence, $2Mb_j < \varepsilon$ for $j \geq n_0$; thus, for $n \geq j \geq n_0$,

$$\left| \sum_{k=j}^{n} a_k b_k \right| < \varepsilon.$$

The claim now follows from Proposition 7.23. □

Proposition 7.42 (Abel's Test). *If $\sum_{n=1}^{\infty} a_n$ is a convergent series and if the sequence $\{b_n\}_{n=1}^{\infty}$ is monotonic and bounded, then $\sum_{n=1}^{\infty} a_n b_n$ is also convergent.*

PROOF. Suppose $\{b_n\}_{n=1}^{\infty}$ is monotonically increasing and let $c_n = b - b_n$, where $b_n \to b$ as $n \to \infty$. By Proposition 7.41 the series $\sum_{n=1}^{\infty} a_n b_n$ converges. It is clear that the series $\sum_{n=1}^{\infty} ba_n$ also converges. But $a_n b_n = ba_n - a_n c_n$; thus,

$$\sum_{n=1}^{\infty} a_n b_n = \sum_{n=1}^{\infty} (ba_n - a_n c_n)$$

converges. □

4. Groupings and Rearrangements

Definition. A series $\sum A_k$ is said to arise from a given series $\sum a_n$ by *grouping of terms* (or by the *introduction of brackets*) if every A_k is the sum of a finite number of consecutive terms of $\sum a_n$, and every pair of terms a_m and a_n, where $m < n$, appear as terms in a unique pair of terms A_p and A_q, respectively, where $p \leq q$.

EXAMPLE. The grouping

$$(a_0 + a_1) + (a_2) + (a_3 + a_4 + a_5) + (a_6 + a_7) + \cdots$$

gives rise to the series $\sum A_n$, where $A_0 = a_0 + a_1$, $A_1 = a_2$, $A_2 = a_3 + a_4 + a_5$, $A_3 = a_6 + a_7, \ldots$.

Proposition 7.43. *The associative law holds for convergent infinite series without restriction in the following sense only:*

$$a_0 + a_1 + a_2 + \cdots = s$$

implies

$$(a_0 + a_1 + \cdots + a_{v_1}) + (a_{v_1+1} + a_{v_1+2} + \cdots + a_{v_2}) + \cdots = s,$$

if $v_1, v_2, \ldots$ denote any increasing sequence of distinct integers and the sum of the terms enclosed in each bracket is considered as one term of a new series

$$A_0 + A_1 + \cdots + A_k + \cdots$$

with

$$A_k = a_{v_k+1} + a_{v_k+2} + \cdots + a_{v_{k+1}}$$

for $k = 0, 1, 2, \ldots$ and $v_0 = -1$.

PROOF. The sequence of partial sums $\{S_k\}$ of $\sum A_k$ is evidently the subsequence

$$S_{v_1}, S_{v_2}, \ldots, S_{v_k}, \ldots$$

of the sequence of partial sums $\{s_n\}$ of $\sum a_n$. □

REMARK. The example

$$\left(2 - 1\frac{1}{2}\right) + \left(1\frac{1}{3} - 1\frac{1}{4}\right) + \left(1\frac{1}{5} - 1\frac{1}{6}\right) + \cdots = \frac{1}{1 \cdot 2} + \frac{1}{3 \cdot 4} + \frac{1}{5 \cdot 6} + \cdots$$

shows that grouping of terms may convert a divergent series into a convergent series. Equivalently, removal of brackets may destroy convergence.

Proposition 7.44. *If the terms of a convergent infinite series $\sum_{k=0}^{\infty} A_k$ are themselves finite sums (say, as in Proposition 7.43, $A_k = a_{v_k+1} + \cdots + a_{v_{k+1}}$; $k = 0$,*

$1, \ldots ; v_0 = -1$), *then we may omit the bracket enclosing these if and only if the new series* $\sum_{n=0}^{\infty} a_n$ *thus obtained also converges.*

PROOF. By Proposition 7.43, if $\sum a_n$ is convergent, then $\sum a_n = \sum A_k$. If $\sum a_n$ diverges, the equality between the series $\sum a_n$ and $\sum A_k$ becomes meaningless. $\square$

REMARK. The new series $\sum a_n$ obtained from $\sum A_k$ in Proposition 7.44 is certainly convergent if the numbers

$$A_k' = |a_{v_k+1}| + |a_{v_k+2}| + \cdots + |a_{v_{k+1}}|$$

form a null sequence. Indeed, if $\varepsilon > 0$ is given, pick m_1 so large that

$$|S_{k-1} - s| < \frac{\varepsilon}{2}$$

for every $k > m_1$ and pick m_2 so large that $A_k' < \varepsilon/2$ for every $k > m_2$. If m is larger than both these numbers m_1 and m_2, then we have

$$|s_n - s| < \varepsilon$$

for every $n > v_m$. For to each such n there corresponds a certain number k such that

$$v_k < n \leq v_{k+1}$$

and this number k must be larger than or equal to m. In that case, however,

$$s_n = S_{k-1} + a_{v_k+1} + \cdots + a_n, \qquad |s_n - S_{k-1}| \leq A_k' < \frac{\varepsilon}{2}.$$

And since

$$s_n - s = (s_n - S_{k-1}) + (S_{k-1} - s)$$

we then have in fact

$$|s_n - s| < \varepsilon \quad \text{or} \quad \sum_{n=0}^{\infty} a_n = s$$

which we wanted to show.

Examples

1. By Proposition 7.43 the following three series are convergent and have the same sum s:

$$1 - \frac{1}{2} + \frac{1}{3} - \frac{1}{4} + \frac{1}{5} - \frac{1}{6} + - \cdots, \tag{7.47}$$

$$\left(1 - \frac{1}{2}\right) + \left(\frac{1}{3} - \frac{1}{4}\right) + \left(\frac{1}{5} - \frac{1}{6}\right) + \cdots = \frac{1}{1 \cdot 2} + \frac{1}{3 \cdot 4} + \frac{1}{5 \cdot 6} + \cdots, \tag{7.48}$$

$$1 - \left(\frac{1}{2} - \frac{1}{3}\right) - \left(\frac{1}{4} - \frac{1}{5}\right) - \cdots = 1 - \frac{1}{2 \cdot 3} - \frac{1}{4 \cdot 5} - \cdots. \tag{7.49}$$

From (7.48) we see that

$$s > \frac{1}{1 \cdot 2} + \frac{1}{3 \cdot 4} = \frac{7}{12}$$

and from (7.49) we see that

$$s < 1 - \frac{1}{2 \cdot 3} = \frac{10}{12};$$

hence,

$$\frac{7}{12} < s < \frac{10}{12}.$$

Using more terms of the series (7.48) and (7.49), we can estimate s more closely; however, we know already from the Application following Proposition 1.6 that $s = \ln 2$. Again by Proposition 7.43, the series

$$\sum_{k=1}^{\infty} \left(\frac{1}{4k - 3} - \frac{1}{4k - 2} + \frac{1}{4k - 1} - \frac{1}{4k}\right) \tag{7.50}$$

has the same sum $s = \ln 2$ as the series

$$\sum_{k=1}^{\infty} \frac{(-1)^{k+1}}{k} \quad \text{and} \quad \sum_{k=1}^{\infty} \left(\frac{1}{2k - 1} - \frac{1}{2k}\right)$$

appearing in (7.47) and (7.48), respectively. By Proposition 7.22 we therefore have that

$$\sum_{k=1}^{\infty} \left(\frac{1}{4k - 2} - \frac{1}{4k}\right) = \frac{\ln 2}{2}. \tag{7.51}$$

Adding termwise series (7.50) and (7.51), Proposition 7.22 gives

$$\sum_{k=1}^{\infty} \left(\frac{1}{4k - 3} + \frac{1}{4k - 1} - \frac{1}{2k}\right) = \frac{3}{2}(\ln 2). \tag{7.52}$$

2. We once more consider the series (7.52), that is, we consider $\sum_{n=1}^{\infty} A_k$, where

$$A_k = \frac{1}{4k - 3} + \frac{1}{4k - 1} - \frac{1}{2k}.$$

Noting that A_k is positive and

$$A_k < \frac{2}{4k - 4} - \frac{1}{2k} = \frac{1}{2(k - 1)k} \leq \frac{1}{k^2} \quad \text{for } k > 1$$

the convergence of the series $\sum A_k$ follows by Proposition 7.26 from the convergence of the series $\sum (1/n^2)$. The Remark following Proposition 7.44 applies; indeed,

$$A'_k < \frac{2}{4k-4} + \frac{1}{2k} < \frac{1}{k-1}$$

and so $\{A'_k\}$ is a null sequence. Thus, the series

$$1 + \frac{1}{3} - \frac{1}{2} + \frac{1}{5} + \frac{1}{6} - \frac{1}{4} + + - \cdots \tag{7.53}$$

is seen to be convergent by Proposition 7.44. In view of (7.52) and Proposition 7.43 the sum of the series (7.53) is $\frac{3}{2}\ln 2$.

Definition. If $\sum_{n=1}^{\infty} a_n$ converges but $\sum_{n=1}^{\infty} |a_n|$ diverges, we say that $\sum_{n=1}^{\infty} a_n$ *converges conditionally.*

REMARKS. A series converges conditionally if it converges but does not converge absolutely. The alternating series

$$1 - \frac{1}{2^p} + \frac{1}{3^p} - \frac{1}{4^p} + - \cdots$$

converges absolutely if $p > 1$, converges conditionally if $0 < p \leq 1$, and diverges if $p \leq 0$. The alternating series

$$\frac{1}{2} - \frac{1 \cdot 3}{2 \cdot 4} + \frac{1 \cdot 3 \cdot 5}{2 \cdot 4 \cdot 6} - \frac{1 \cdot 3 \cdot 5 \cdot 7}{2 \cdot 4 \cdot 6 \cdot 8} + - \cdots$$

converges conditionally. Indeed, by Example 2 following Proposition 7.37 the series fails to converge absolutely. To verify the convergence of the series we can use Proposition 7.33. Since

$$1 \cdot 3 < 2^2, \quad 3 \cdot 5 < 4^2, \quad \ldots, \quad (2n-1)(2n+1) < (2n)^2,$$

it follows that

$$1^2 \cdot 3^2 \cdot 5^2 \cdots (2n-1)^2 (2n+1) < 2^2 \cdot 4^2 \cdot 6^2 \cdots (2n)^2$$

which is equivalent to the inequality

$$\frac{1}{2} \cdot \frac{3}{4} \cdot \frac{5}{6} \cdots \frac{2n-1}{2n} < \frac{1}{\sqrt{2n+1}}.$$

Thus,

$$\frac{1 \cdot 3 \cdot 5 \cdots (2n-1)}{2 \cdot 4 \cdot 6 \cdots (2n)} \to 0 \quad \text{as } n \to \infty.$$

The other assumptions in Proposition 7.33 are simple to check out.

Definition. Let $\{n_i\}_{i=1}^{\infty}$ be a sequence of positive integers such that each positive integer occurs exactly once among the n_i (i.e., if N denotes the set of positive integers, $\{n_i\}_{i=1}^{\infty}$ is a one-to-one function from N onto N). If $\sum_{n=1}^{\infty} a_n$ is a series of real numbers and if for $i = 1, 2, 3, \ldots$ we have the relation $b_i = a_{n_i}$, then $\sum_{i=1}^{\infty} b_i$ is called a *rearrangement of* $\sum_{n=1}^{\infty} a_n$.

DISCUSSION. Consider the alternating harmonic series

$$\sum_{n=1}^{\infty} \frac{(-1)^{n+1}}{n} = 1 - \frac{1}{2} + \frac{1}{3} - \frac{1}{4} + \frac{1}{5} - \frac{1}{6} + - \cdots;$$

it converges conditionally. We examine the following rearrangement of this series: Let the first p positive terms be followed by the first q negative terms, then the next p positive terms be followed by the next q negative terms, and so on. The resulting series converges to

$$\ln\left(2\sqrt{\frac{p}{q}}\right).$$

To see this, we let

$$H_n = 1 + \frac{1}{2} + \cdots + \frac{1}{n} = \ln n + C + \gamma_n,$$

where C is Euler's constant and $\gamma_n \to 0$ as $n \to \infty$ (see Proposition 1.7 in Chapter 1). Then

$$\frac{1}{2} + \frac{1}{4} + \cdots + \frac{1}{2m} = \frac{1}{2}H_m = \frac{1}{2}\ln m + \frac{1}{2}C + \frac{1}{2}\gamma_m$$

and

$$1 + \frac{1}{3} + \cdots + \frac{1}{2k-1} = H_{2k} - \frac{1}{2}H_k = \ln 2 + \frac{1}{2}\ln k + \frac{1}{2}C + \gamma_{2k} - \frac{1}{2}\gamma_k.$$

The foregoing shows that the partial sums of the series in question are of the form

$$\ln\left(2\sqrt{\frac{p}{q}}\right) + \delta_n,$$

where $\{\delta_n\}_{n=1}^{\infty}$ is a null sequence.

We note that, for $p = q = 1$, we get

$$1 - \frac{1}{2} + \frac{1}{3} - \frac{1}{4} + \frac{1}{5} - \frac{1}{6} + \cdots = \ln 2;$$

for $p = 2$, $q = 1$, we obtain [compare with (7.52)]

$$1 + \frac{1}{3} - \frac{1}{2} + \frac{1}{5} + \frac{1}{7} - \frac{1}{4} + \cdots = \frac{3}{2}\ln 2;$$

for $p = 1$, $q = 4$, we have

$$1 - \frac{1}{2} - \frac{1}{4} - \frac{1}{6} - \frac{1}{8} + \frac{1}{3} - \frac{1}{10} - \frac{1}{12} - \frac{1}{14} - \frac{1}{16} + \frac{1}{5} - \cdots = 0,$$

and so on.

Proposition 7.45. *Let* $\sum_{n=1}^{\infty} a_n$ *be a series of real terms and put*

$$p_n = \max\{a_n, 0\} = \frac{a_n + |a_n|}{2} \quad and \quad q_n = \min\{a_n, 0\} = \frac{a_n - |a_n|}{2}.$$

We have the following result:
(i) *If* $\sum_{n=1}^{\infty} a_n$ *converges absolutely, then both* $\sum_{n=1}^{\infty} p_n$ *and* $\sum_{n=1}^{\infty} q_n$ *converge.*
(ii) *If* $\sum_{n=1}^{\infty} a_n$ *converges conditionally, then both* $\sum_{n=1}^{\infty} p_n$ *and* $\sum_{n=1}^{\infty} q_n$ *diverge.*

PROOF. (i) If both $\sum_{n=1}^{\infty} a_n$ and $\sum_{n=1}^{\infty} |a_n|$ converge, then so does the series $\sum_{n=1}^{\infty}(a_n + |a_n|)$. Thus, $\sum_{n=1}^{\infty} 2p_n$ converges, implying that $\sum_{n=1}^{\infty} p_n$ converges. The convergence of $\sum_{n=1}^{\infty} q_n$ is established in a similar manner.

(ii) Suppose $\sum_{n=1}^{\infty} a_n$ converges but $\sum_{n=1}^{\infty} |a_n|$ diverges. Then the series $\sum_{n=1}^{\infty} p_n$ and $\sum_{n=1}^{\infty} q_n$ both diverge. For if both were convergent, then

$$\sum_{n=1}^{\infty} (p_n + q_n) = \sum_{n=1}^{\infty} a_n$$

would converge, contrary to assumption. Since

$$\sum_{n=1}^{k} |a_n| = \sum_{n=1}^{k} (p_n - q_n) = \sum_{n=1}^{k} p_n - \sum_{n=1}^{k} q_n,$$

divergence of $\sum_{n=1}^{\infty} p_n$ and convergence of $\sum_{n=1}^{\infty} q_n$ (or vice versa) implies divergence of $\sum_{n=1}^{\infty} a_n$, again contrary to assumption. □

REMARK. Note that the p_n are the positive and the q_n the negative terms of the sequence $\{a_n\}$.

Proposition 7.46 (Riemann's Rearrangement Theorem). *The terms of any conditionally convergent series can be rearranged to give either a divergent series or a conditionally convergent series whose sum is any preassigned number.*

PROOF. Let $\sum_{n=1}^{\infty} a_n$ be a conditionally convergent series with divergent nonnegative and nonpositive parts $\sum_{n=1}^{\infty} p_n$ and $\sum_{n=1}^{\infty} q_n$, respectively, and let c be an arbitrary (finite) real number. Let the rearrangement be determined as follows: First put down terms

$$p_1 + \cdots + p_{m_1}$$

until the total partial sum first exceeds c, then attach terms

$$q_1 + \cdots + q_{n_1}$$

until the total partial sum first falls short of c. Then attach terms

$$p_{m_1+1} + \cdots + p_{m_2}$$

until the total partial sum first exceeds c, then terms

$$q_{n_1+1} + \cdots + q_{n_2}$$

until the total partial sum first falls short of c, and so on. Each of these steps is possible because of the divergence of $\sum_{n=1}^{\infty} p_n$ and $\sum_{n=1}^{\infty} q_n$ [see Proposition 7.45 (ii)]. The resulting rearrangement of the given series $\sum_{n=1}^{\infty} a_n$ converges to c since $p_n \to 0$ and $q_n \to 0$ as $n \to \infty$ (because $a_n \to 0$ as $n \to \infty$).

To see that there are rearrangements of $\sum_{n=1}^{\infty} a_n$ the partial sums of which tend to $+\infty$, we consider a rearrangement of the series $\sum_{n=1}^{\infty} a_n$ in which we have alternately a group of positive terms followed by a single negative term. Since the series $\sum_{n=1}^{\infty} p_n$ diverges, its partial sums are unbounded and we can choose m_1 so large that

$$p_1 + \cdots + p_{m_1} > 1 - q_1,$$

then $m_2 > m_1$ so large that

$$p_1 + p_2 + \cdots + p_{m_1} + \cdots + p_{m_2} > 2 - q_1 - q_2$$

and, generally, $m_k > m_{k-1}$ so large that

$$p_1 + p_2 + \cdots + p_{m_k} > k - q_1 - q_2 - \cdots - q_k \qquad (k = 3, 4, \ldots).$$

Hence, the series

$$p_1 + \cdots + p_{m_1} + q_1 + p_{m_1+1} + \cdots + p_{m_2} + q_2 + p_{m_2+1} + \cdots$$

in which we have alternately a group of positive terms followed by a single negative term, is clearly divergent; its kth partial sum

$$p_1 + \cdots + p_{m_1} + q_1 + \cdots + p_{m_k} + q_k$$

exceeds k.

Analogously one may obtain rearrangements of $\sum_{n=1}^{\infty} a_n$ that tend to $-\infty$. $\square$

Proposition 7.47. *If $\sum_{n=1}^{\infty} a_n$ converges absolutely to A, then any rearrangement $\sum_{n=1}^{\infty} b_n$ of $\sum_{n=1}^{\infty} a_n$ also converges to A.*

PROOF. We first show that the proposition is true if $\sum_{n=1}^{\infty} a_n$ is a series of non-negative numbers.

For each positive integer k, let $s_k = b_1 + \cdots + b_k$. Since $b_i = a_{n_i}$ for some sequence $\{n_i\}_{i=1}^{\infty}$, we have

$$b_1 = a_{n_1}, \quad \ldots, \quad b_k = a_{n_k}.$$

Let $m = \max\{n_1, \ldots, n_k\}$. Then clearly

$$s_k \leq a_1 + \cdots + a_m \leq A.$$

Thus, $\sum_{n=1}^{\infty} b_n$ converges to some real number B (by Proposition 7.24). But

$$B = \lim_{n \to \infty} s_k$$

and so $B \leq A$ (i.e., $\sum_{n=1}^{\infty} b_n \leq \sum_{n=1}^{\infty} a_n$). However, since $\sum_{n=1}^{\infty} a_n$ is also a re-arrangement of $\sum_{n=1}^{\infty} b_n$, the same reasoning with the roles of the series $\sum_{n=1}^{\infty} a_n$ and $\sum_{n=1}^{\infty} b_n$ reversed shows that $A \leq B$. Hence, $A = B$.

We now consider the general case. If $\sum_{n=1}^{\infty} a_n$ converges absolutely, then both $\sum_{n=1}^{\infty} p_n$ and $\sum_{n=1}^{\infty} q_n$ converge by Proposition 7.45 (i); say $\sum_{n=1}^{\infty} p_n = P$ and $\sum_{n=1}^{\infty} q_n = Q$ (so that $Q \leq 0$). Then $A = P + Q$. For some $\{n_i\}_{i=1}^{\infty}$, we have

$$b_i = a_{n_i} = p_{n_i} + q_{n_i}. \tag{7.54}$$

Moreover, $\sum_{i=1}^{\infty} p_{n_i}$ is a rearrangement of the series $\sum_{n=1}^{\infty} p_n$ of nonnegative terms. Hence, by the first part of this proof, $\sum_{i=1}^{\infty} p_{n_i}$ converges and $\sum_{i=1}^{\infty} p_{n_i} = P$. Similarly, $\sum_{i=1}^{\infty} q_{n_i} = Q$. From (7.54) we see that $\sum_{i=1}^{\infty} b_i$ converges and

$$\sum_{i=1}^{\infty} b_i = \sum_{i=1}^{\infty} p_{n_i} + \sum_{i=1}^{\infty} q_{n_i} = P + Q = A.$$

It only remains to show that $\sum_{i=1}^{\infty} b_i$ converges absolutely. By (7.54), we have, however, that

$$|b_i| \leq |p_{n_i}| + |q_{n_i}| = p_{n_i} - q_{n_i}.$$

Thus, for any positive integer k,

$$|b_1| + \cdots + |b_k| \leq \sum_{i=1}^{k} p_{n_i} - \sum_{i=1}^{k} q_{n_i} \leq \sum_{i=1}^{\infty} p_{n_i} - \sum_{i=1}^{\infty} q_{n_i} = P - Q.$$

The partial sums of $\sum_{i=1}^{\infty} |b_i|$ are thus all bounded above by $P - Q$ and hence $\sum_{n=1}^{\infty} |b_i|$ is seen to be convergent (by Proposition 7.24). $\square$

Proposition 7.48. *If all rearrangements of $\sum_{n=1}^{\infty} a_n$ converge, then they all converge to the same sum.*

PROOF. Either $\sum_{n=1}^{\infty} a_n$ converges absolutely, in which case Proposition 7.47 applies, or the series converges conditionally, in which case there exists a divergent rearrangement by Proposition 7.46. $\square$

Definition. A series is said to *converge unconditionally* if every rearrangement of it converges.

REMARKS. To sum up, we have shown that a series converges unconditionally if and only if it converges absolutely. Also, if a series converges unconditionally, then every rearrangement of it converges to the same sum.

Definition. With an infinite matrix

$$
\begin{array}{cccc}
a_{11} & a_{12} & a_{13} & \cdots \\
a_{21} & a_{22} & a_{23} & \cdots \\
a_{31} & a_{32} & a_{33} & \cdots \\
\vdots & \vdots & \vdots &
\end{array}
$$

of real numbers we can associate a *double series*

$$\sum_{\mu, \nu=1}^{\infty} a_{\mu\nu}.$$

If $A\colon (\mu, \nu) \to \lambda$ is an enumeration of the pairs (μ, ν) of positive integers, we put $C_{\lambda}(A) = a_{\mu\nu}$ and call the series

$$\sum_{\lambda=1}^{\infty} C_{\lambda}(A)$$

the *ordering (relative to A) of* $\sum_{\mu, \nu=1}^{\infty} a_{\mu\nu}$ *into a single series.*

REMARK. Examples of ordering of double series into single series are *arrangement by diagonals*

$$a_{11} + (a_{12} + a_{21}) + (a_{13} + a_{22} + a_{31}) + \cdots$$

and *arrangement by squares*

$$a_{11} + (a_{12} + a_{22} + a_{21}) + (a_{13} + a_{23} + a_{33} + a_{32} + a_{31}) + \cdots;$$

in arrangement by diagonals we have grouped together the entries on successive diagonals and in arrangement by squares we have used rectangular grouping.

Proposition 7.49 (Main Rearrangement Theorem). *Let* $\sum_{\mu, \nu=1}^{\infty} a_{\mu\nu}$ *be a given double series. Suppose there exists a number M such that*

$$\sum_{\mu, \nu=1}^{m} |a_{\mu\nu}| \le M < \infty \quad \text{for all } m.$$

Then the following statements hold:
(a) *Any ordering of the double series into a single series is absolutely convergent and all single series have the same sum s.*
(b) *The series* $\sum_{\nu=1}^{\infty} a_{\mu\nu}$ *(column series) are absolutely convergent* $(\mu = 1, 2, 3, \ldots)$, *and so are the series* $\sum_{\mu=1}^{\infty} a_{\mu\nu}$ *(row series)* $(\nu = 1, 2, 3, \ldots)$.
(c) *The two series*

$$\sum_{\mu=1}^{\infty} \left(\sum_{\nu=1}^{\infty} a_{\mu\nu} \right) \quad \text{and} \quad \sum_{\nu=1}^{\infty} \left(\sum_{\mu=1}^{\infty} a_{\mu\nu} \right)$$

are convergent and their sums are equal (sum of the row series = sum of the column series), namely, equal to s.

PROOF. (a) Let n be a positive integer. We consider the pairs (μ, ν) which correspond to the numbers $\lambda = 1, \ldots, n$ under the enumeration A and let m be the largest of the μ and ν. Then

$$\sum_{\lambda=1}^{n} |C_{\lambda}(A)| \le \sum_{\mu, \nu=1}^{m} |a_{\mu\nu}| \le M.$$

By Proposition 7.24 we see that $\sum_{\lambda=1}^{\infty} C_{\lambda}(A)$ converges absolutely. If A^* is a second enumeration, then $\sum_{\lambda=1}^{\infty} C_{\lambda}(A^*)$ is a rearrangement of $\sum_{\lambda=1}^{\infty} C_{\lambda}(A)$; by the absolute convergence both series must have the same sum (see Proposition 7.47).

(b) If we put $m = \max\{\mu, n\}$, resp. $m = \max\{v, n\}$, then

$$\sum_{v=1}^{n} |a_{\mu v}| \le \sum_{\mu, v=1}^{m} |a_{\mu v}| \le M, \qquad \text{resp.} \ \sum_{\mu=1}^{n} |a_{\mu v}| \le \sum_{\mu, v=1}^{m} |a_{\mu v}| \le M;$$

thus, by Proposition 7.24, this point is also settled.

(c) For $m \le n$ we have

$$\sum_{\mu=1}^{m} \left| \sum_{v=1}^{n} a_{\mu v} \right| \le \sum_{\mu=1}^{m} \sum_{v=1}^{n} |a_{\mu v}| \le \sum_{\mu, v=1}^{n} |a_{\mu v}| \le M,$$

and passage to the limit as $n \to \infty$ gives

$$\sum_{\mu=1}^{m} \left| \sum_{v=1}^{\infty} a_{\mu v} \right| \le M,$$

showing the absolute convergence of the series formed by the column series. In the same way we get the absolute convergence of the series formed by the row series.

Finally, let $\sum_{\lambda=1}^{\infty} C_{\lambda}$ be some ordering of the double series into a single series. For any $\varepsilon > 0$ there exists a k_0 such that

$$\sum_{\lambda=k+1}^{\infty} |C_{\lambda}| < \varepsilon \quad \text{for all } k \ge k_0.$$

Corresponding to k we can find some n_0 such that $C_1, C_2, \ldots, C_k$ are among the $a_{\mu v}$ for $1 \le \mu \le n_0$, $1 \le v \le n_0$. For $m \ge n_0$, $n \ge n_0$ we then have

$$\left| \sum_{\mu=1}^{m} \sum_{v=1}^{n} a_{\mu v} - \sum_{\lambda=1}^{k} C_{\lambda} \right| \le \sum_{\lambda=k+1}^{\infty} |C_{\lambda}| < \varepsilon.$$

Upon passage to the limit as $n \to \infty$, $m \to \infty$ (resp. in the reverse order) and then followed by passage to the limit as $k \to \infty$, we get

$$\left| \sum_{\mu=1}^{\infty} \sum_{v=1}^{\infty} a_{\mu v} - \sum_{\lambda=1}^{\infty} C_{\lambda} \right| \le \varepsilon, \qquad \text{resp.} \ \left| \sum_{v=1}^{\infty} \sum_{\mu=1}^{\infty} a_{\mu v} - \sum_{\lambda=1}^{\infty} C_{\lambda} \right| \le \varepsilon.$$

But $\varepsilon > 0$ is arbitrarily small. $\qquad\qquad\qquad\qquad\qquad\qquad\qquad\qquad\qquad\Box$

REMARK. Double series satisfying the assumptions in Proposition 7.49 we shall call *absolutely convergent*; to such series we may assign the number s in part (a) of the proposition as the *sum*.

Proposition 7.50. *Let $\sum_{\mu=1}^{\infty} a_{\mu}$ and $\sum_{\lambda=1}^{\infty} b_v$ be absolutely convergent series. Then any of their products converges; moreover, all products converge to the same sum, namely,*

$$\sum_{\mu,\,\nu=1}^{\infty} a_\mu b_\nu = \left(\sum_{\mu=1}^{\infty} a_\mu\right)\left(\sum_{\nu=1}^{\infty} b_\nu\right). \tag{7.55}$$

In particular, the Cauchy product

$$\sum_{\omega=0}^{\infty}\left(\sum_{\mu=0}^{\omega} a_\mu b_{\omega-\mu}\right)$$

of the absolutely convergent series $\sum_{\mu=1}^{\infty} a_\mu$ and $\sum_{\nu=1}^{\infty} b_\nu$ is absolutely convergent and we have

$$\sum_{\omega=0}^{\infty}\left(\sum_{\mu=0}^{\omega} a_\mu b_{\omega-\mu}\right) = \left(\sum_{\mu=0}^{\infty} a_\mu\right)\left(\sum_{\nu=0}^{\infty} b_\nu\right). \tag{7.56}$$

PROOF. Since

$$\sum_{\mu,\,\nu=1}^{m} |a_\mu b_\nu| = \left(\sum_{\mu=1}^{m} |a_\mu|\right)\left(\sum_{\nu=1}^{m} |b_\nu|\right) \le \left(\sum_{\mu=1}^{\infty} |a_\mu|\right)\left(\sum_{\nu=1}^{\infty} |b_\nu|\right),$$

the assumptions of Proposition 7.49 are satisfied. Relation (7.55) is a consequence of part (c) of Proposition 7.49. Finally, the claim concerning the Cauchy product is an immediate consequence of the first part of the proposition. $\qquad\square$

EXAMPLE. Let $0! = 1$. By the Ratio Test (see Proposition 7.32) the series

$$\sum_{\mu=0}^{\infty} \frac{x^\mu}{\mu!} = E(x) \tag{7.57}$$

is absolutely convergent for any (finite) real number x. Hence, for any real numbers x and y the relation (7.56) gives

$$\left(\sum_{\mu=0}^{\infty} \frac{x^\mu}{\mu!}\right)\left(\sum_{\nu=0}^{\infty} \frac{y^\nu}{\nu!}\right) = \sum_{\omega=0}^{\infty}\left(\sum_{\mu=0}^{\omega} \frac{x^\mu}{\mu!}\frac{y^{\omega-\mu}}{(\omega-\mu)!}\right) = \sum_{\omega=0}^{\infty} \frac{1}{\omega!}\sum_{\mu=0}^{\omega}\binom{\omega}{\mu}x^\mu y^{\omega-\mu}$$

$$= \sum_{\omega=0}^{\infty} \frac{1}{\omega!}(x+y)^\omega,$$

that is, using the notation introduced in (7.57),

$$E(x+y) = E(x)\cdot E(y). \tag{7.58}$$

In particular, for $y = -x$ we get

$$E(x)\cdot E(-x) = E(0) = 1. \tag{7.59}$$

Taylor's Theorem (see Proposition 4.11 in Chapter 4) applied to the exponential function [note in particular (4.12) in Chapter 4] shows that $E(x)$ as defined in (7.57) is precisely the function e^x. Therefore, Proposition 7.50 provides us with an analytical method of verifying the familiar relations

$$e^{x+y} = e^x\cdot e^y \quad \text{and} \quad e^x\cdot e^{-x} = 1.$$

5. Uniform Convergence

Definition. Let $\{f_n\}_{n=1}^\infty$ be a sequence of real-valued functions defined on a set E of real numbers. We say that $\{f_n\}_{n=1}^\infty$ *converges pointwise on E* if, for every $x \in E$, the numerical sequence

$$\{f_n(x)\}_{n=1}^\infty$$

is convergent. In this case we define a function f on E by taking

$$f(x) = \lim_{n \to \infty} f_n(x)$$

for every $x \in E$. This function f is called the *pointwise limit of* the sequence $\{f_n\}_{n=1}^\infty$.

Definition. Let E be a set of real numbers and $\{f_n\}_{n=1}^\infty$ a sequence, each term of which is a real-valued function defined on E. We say that $\{f_n\}_{n=1}^\infty$ *converges uniformly on E to a function f* if, for every $\varepsilon > 0$, there exists an integer n_0 (depending only on ε) such that $n \geq n_0$ implies

$$|f_n(x) - f(x)| < \varepsilon \tag{7.60}$$

for every x in E. We denote this symbolically by writing

$$f_n \to f \quad \text{uniformly on } E.$$

REMARKS. Clearly, uniform convergence on E implies pointwise convergence on E; if the uniform limit exists, it is equal to the pointwise limit. The difference between pointwise convergence and uniform convergence is this: If $\{f_n\}_{n=1}^\infty$ converges pointwise on E, then there exists a function f such that, for every $\varepsilon > 0$, and for every x in E, there is an integer n_0, depending on ε *and* on x, such that (7.60) holds if $n \geq n_0$; if $\{f_n\}_{n=1}^\infty$ converges uniformly on E, we can find, for each $\varepsilon > 0$, *one* integer n_0 which will do for *all* x in E.

Since (7.60) is equivalent to

$$f(x) - \varepsilon < f_n(x) < f(x) + \varepsilon,$$

we see that if (7.60) is to hold for all $n \geq n_0$ and for all $x \in E$, then the graph of f_n (i.e., the set $\{(x, y): y = f_n(x), x \in E\}$) lies within a "band" of height 2ε situated symmetrically about the graph of f; see Figure 7.1.

Proposition 7.51. *Suppose, for every $x \in E$, $\lim_{n \to \infty} f_n(x) = f(x)$. Let*

$$M_n = \sup\{|f_n(x) - f(x)|: x \in E\}.$$

Then $f_n \to f$ uniformly on E if and only if $M_n \to 0$ as $n \to \infty$.

PROOF. The proposition is an immediate consequence of the definition of uniform convergence. $\square$

Proposition 7.52. *Let $\{f_n\}_{n=1}^\infty$ be a sequence of functions defined on a set E of real numbers. There exists a function f such that*

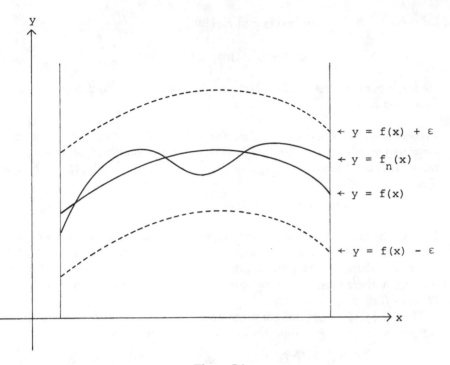

Figure 7.1

$$f_n \to f \quad uniformly \ on \ E$$

if and only if the following (called Cauchy's criterion for uniform conver-
gence) *is satisfied: For every* $\varepsilon > 0$ *there exists an integer* n_0 *such that* $m \geq n_0$,
$n \geq n_0$ *implies*

$$|f_m(x) - f_n(x)| < \varepsilon \qquad (7.61)$$

for every x *in* E.

PROOF. Suppose that $f_n \to f$ uniformly on E. Then, given $\varepsilon > 0$, we can find n_0
such that $n \geq n_0$ implies $|f_n(x) - f(x)| < \varepsilon/2$ for all x in E. Taking $m \geq n_0$, we
have $|f_m(x) - f(x)| < \varepsilon/2$, and hence we have

$$|f_m(x) - f_n(x)| \leq |f_m(x) - f(x)| + |f_n(x) - f(x)| < \varepsilon$$

for every x in E.

Conversely, suppose that condition (7.61) is satisfied. Then, for each x in E,
the sequence $\{f_n(x)\}_{n=1}^{\infty}$ converges by Proposition 7.12. Let

$$f(x) = \lim_{n \to \infty} f_n(x)$$

if $x \in E$. We must show that $f_n \to f$ uniformly on E. If $\varepsilon > 0$ is given, we choose
n_0 such that $n \geq n_0$ implies

$$|f_n(x) - f_{n+k}(x)| < \frac{\varepsilon}{2}$$

for every $k = 1, 2, \ldots$ and every x in E. Thus,

$$\lim_{k \to \infty} |f_n(x) - f_{n+k}(x)| = |f_n(x) - f(x)| \le \frac{\varepsilon}{2}.$$

Hence, $n \ge n_0$ implies $|f_n(x) - f(x)| < \varepsilon$ for every x in E and so $f_n \to f$ uniformly on E. $\qquad\square$

Proposition 7.53. *A sequence of functions* $\{f_n\}_{n=1}^{\infty}$ *defined on E does not converge uniformly on E to a function f defined on E if and only if for some $\varepsilon_0 > 0$ there exists a subsequence* $\{f_{n_k}\}_{k=1}^{\infty}$ *of* $\{f_n\}_{n=1}^{\infty}$ *and a sequence* $\{x_k\}_{k=1}^{\infty}$ *in E such that*

$$|f_{n_k}(x_k) - f(x_k)| \ge \varepsilon_0 \quad for\ k = 1, 2, 3, \ldots.$$

PROOF. The negation of uniform convergence: A sequence of functions $\{f_n\}_{n=1}^{\infty}$ defined on E fails to converge uniformly on E to a function f defined on E if and only if there exists a positive number ε_0 having the property that for any number N there exists a positive integer $n \ge N$ and a point x of E such that $|f_n(x) - f(x)| \ge \varepsilon_0$.

Thus, if $\{f_n\}_{n=1}^{\infty}$ does not converge uniformly on E to f, there are positive integers $n_1 < n_2 < \cdots$ and points $x_1, x_2, \ldots$ of E such that

$$|f_{n_1}(x_1) - f(x_1)| \ge \varepsilon_0, \qquad |f_{n_2}(x_2) - f(x_2)| \ge \varepsilon_0, \ldots$$

for some ε_0. $\qquad\square$

Examples

1. Let $f_n(x) = x^n$ for $|x| \le c$, where c is fixed and $0 < c < 1$. Then $\{f_n\}_{n=1}^{\infty}$ converges uniformly to $f(x) = 0$ (for $-c \le x \le c$) on the interval $|x| \le c$ because $|x^n| \le c^n$ and $c^n \to 0$ as $n \to \infty$. However, $\{f_n\}_{n=1}^{\infty}$ fails to be uniformly convergent to $f(x) = 0$ (for $-1 < x < 1$) on the open interval $|x| < 1$. In fact, for any positive integer n there is a point x in the open interval $(-1, 1)$ such that $|x^n|$ is not smaller than the positive number $\frac{1}{2}$; for example, let $x = 1 - 1/2n$ and observe that

$$\left(1 - \frac{1}{2n}\right)^n \ge 1 - \frac{1}{2n} \cdot n = 1 - \frac{1}{2} = \frac{1}{2}$$

by Bernoulli's Inequality [see (4.93) in Example 15 of Section 6 in Chapter 4]. By Proposition 7.53 $\{f_n\}_{n=1}^{\infty}$ thus fails to be uniformly convergent to 0 on $(-1, 1)$.

2. For $x \in [0, 2]$, let

$$f_n(x) = \frac{nx}{1 + n^2 x^2}.$$

It is clear that $f_n(0) = 0$ and $f_n(x) < 1/nx$ for $x > 0$; thus, the pointwise limit of $\{f_n\}_{n=1}^{\infty}$ is 0 on $[0, 2]$. Since

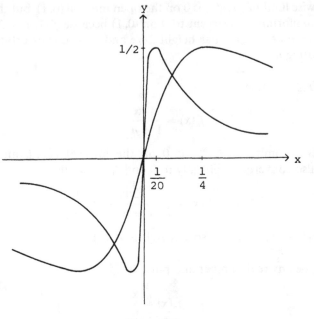

Figure 7.2

$$0 < f_n(x) = \frac{nx}{1 + n^2x^2} < \frac{nx}{n^2x^2} \leq \frac{1}{n}$$

for $1 \leq x \leq 2$, it is clear that $\{f_n\}_{n=1}^{\infty}$ converges uniformly to 0 on the interval $[1, 2]$. However, $\{f_n\}_{n=1}^{\infty}$ does not converge uniformly to 0 on the interval $[0, 1]$. Indeed, $f_n(1/n) = 1/2$ for $n = 1, 2, 3, \ldots$ and f_n assumes its largest value (namely, $1/2$) at the point $x = 1/n$; by Proposition 7.53 we therefore see that $\{f_n\}_{n=1}^{\infty}$ fails to converge uniformly to 0 on $[0, 1]$. The function f_n assumes not only its largest value at $x = 1/n$ but also has a local maximum there; we may thus speak of a *hump* at $x = 1/n$. Figure 7.2 illustrates the graph of

$$y = f_n(x) = \frac{nx}{1 + n^2x^2}$$

for the cases $n = 4, 20$. As we let n take the values $1, 2, 3, \ldots$, the hump glides toward the point $x = 0$.

3. For $0 < x < 1$, let

$$f_n(x) = \frac{1}{1 + nx}.$$

Evidently, the pointwise limit of $\{f_n\}_{n=1}^{\infty}$ is 0 on the open interval $(0, 1)$, but the sequence does not converge uniformly to 0 on $(0, 1)$ because, just as in Example 2, $f_n(1/n) = 1/2$.

4. For $0 < x < 1$, let

$$f_n(x) = 2n^2xe^{-n^2x^2}.$$

The pointwise limit of $\{f_n\}_{n=1}^{\infty}$ is 0 on the open interval $(0,1)$, but the sequence fails to be uniformly convergent to 0 on $(0,1)$ because $f_n(1/n) = 2n/e$. In this case the hump at $x = 1/n$ has height $2n/e$ and so becomes arbitrarily large with increasing n.

5. For $0 \leq x \leq 1$, let

$$f_n(x) = \frac{x}{1 + n^2 x^2}.$$

The pointwise limit of $\{f_n\}_{n=1}^{\infty}$ is 0 on the interval $[0,1]$; moreover, the sequence also converges uniformly to 0 on $[0,1]$ because

$$0 \leq f_n(x) = \frac{1}{2n} \frac{2nx}{1 + n^2 x^2} \leq \frac{1}{2n}$$

[note that $(1 - nx)^2 \geq 0$ and so $2nx \leq 1 + n^2 x^2$].

6. Let x be any real number and put

$$f_n(x) = \frac{x}{n}.$$

Then the pointwise limit of $\{f_n\}_{n=1}^{\infty}$ is 0 on the number line $(-\infty, \infty)$, but the sequence is not uniformly convergent to 0 on $(-\infty, \infty)$ because $f_n(n) = 1$. However, the sequence $\{f_n\}_{n=1}^{\infty}$ is seen to be uniformly convergent to 0 on any interval (a, b) of finite length.

The situation in the case of the sequence

$$h_n(x) = \frac{x^2 + nx}{n}$$

is entirely similar. The pointwise limit of $\{h_n\}_{n=1}^{\infty}$ is $h(x) = x$ on the number line $(-\infty, \infty)$, but the sequence is not uniformly convergent to $h(x)$ on $(-\infty, \infty)$ because $h_n(n) = n$; however, the sequence $\{h_n\}_{n=1}^{\infty}$ is uniformly convergent to $h(x) = x$ on any interval (a, b) of finite length.

7. Let x be any real number and put

$$f_n(x) = \frac{\sin(nx + n)}{n}.$$

Since $|\sin t| \leq 1$ for any real number t, we see that the pointwise limit of $\{f_n\}_{n=1}^{\infty}$ is 0 on $(-\infty, \infty)$. Since

$$|f_n(x) - 0| = \frac{1}{n}|\sin(nx + n)| \leq \frac{1}{n}$$

for any real number x, we see that $\{f_n\}_{n=1}^{\infty}$ converges uniformly to 0 on $(-\infty, \infty)$.

Proposition 7.54. *Suppose that $f_n \to f$ uniformly on an interval J. If each f_n is continuous at a point $\bar{x}$ of J, then the limit function f is also continuous at $\bar{x}$ and so*

$$\lim_{x \to \bar{x}} \lim_{n \to \infty} f_n(x) = \lim_{n \to \infty} \lim_{x \to \bar{x}} f_n(x).$$

PROOF. By assumption, for every $\varepsilon > 0$ there exists an integer N such that $n \geq N$ implies

$$|f_n(x) - f(x)| < \frac{\varepsilon}{3}$$

for every x in J. Since f_N is continuous at $\bar{x}$, there is a neighborhood $(\bar{x} - \delta, \bar{x} + \delta)$ with $\delta > 0$ such that $x \in (\bar{x} - \delta, \bar{x} + \delta) \cap J$ implies

$$|f_N(x) - f_N(\bar{x})| < \frac{\varepsilon}{3}.$$

But

$$|f(x) - f(\bar{x})| \leq |f(x) - f_N(x)| + |f_N(x) - f_N(\bar{x})| + |f_N(\bar{x}) - f(\bar{x})|.$$

If $x \in (\bar{x} - \delta, \bar{x} + \delta) \cap J$, each term on the right-hand side of the foregoing inequality is less than $\varepsilon/3$ and hence $|f(x) - f(\bar{x})| < \varepsilon$. ☐

REMARK. Uniform convergence is therefore sufficient, but not necessary, to transmit continuity from the individual terms of a sequence of functions to the limit function; for example, if $f_n(x) = x^n$, $f(x) = 0$, and J is the open interval $(-1, 1)$, the f_n and f are continuous on J but the convergence of f_n to f is only pointwise on J and not uniform on J (see Example 1 following Proposition 7.53). But there is a partial converse to Proposition 7.54 which we shall take up next.

Proposition 7.55 (Dini's Theorem). *Let $[a, b]$ be a closed interval of finite length and $\{f_n\}_{n=1}^{\infty}$ be a sequence of continuous functions on $[a, b]$ which converges pointwise to a continuous function f on $[a, b]$. If, moreover, the sequence is monotonic on $[a, b]$, then $f_n \to f$ uniformly on $[a, b]$.*

PROOF. We shall suppose that $f_n(x) \geq f_{n+1}(x)$ for $n = 1, 2, \ldots$, and for every x in $[a, b]$. We put

$$g_n(x) = f_n(x) - f(x).$$

Then $g_n \to 0$, and $g_n(x) \geq g_{n+1}(x)$ on $[a, b]$. We have to show that $g_n \to 0$ uniformly on $[a, b]$.

To show that $\{g_n\}_{n=1}^{\infty}$ is uniformly convergent to 0 on $[a, b]$, it is sufficient to establish that for any $\varepsilon > 0$ there is at least one index n such that the inequality

$$g_n(x) < \varepsilon \tag{7.62}$$

holds for all x in $[a, b]$ (since $\{g_n\}_{n=1}^{\infty}$ is a decreasing sequence of functions on $[a, b]$, the inequality (7.62) holds of course for all indices larger than n as well). Now, if the convergence were *not* uniform, by Proposition 7.53, there would exist a subsequence $\{g_{n_k}\}_{k=1}^{\infty}$ of $\{g_n\}_{n=1}^{\infty}$ and a sequence $\{x_k\}_{k=1}^{\infty}$ of points of $[a, b]$ such that

$$g_{n_k}(x_k) \geq \varepsilon_0 > 0$$

for some ε_0. By Proposition 7.10, the sequence $\{x_k\}_{k=1}^{\infty}$ contains a convergent subsequence which, for simplicity of notation, we shall again denote by $\{x_k\}_{k=1}^{\infty}$ and we assume that $x_k \to \bar{x}$ as $k \to \infty$. It is clear that $\bar{x}$ is in the interval $[a, b]$. By the continuity of each g_m with $m = 1, 2, 3, \ldots$, we have

$$\lim_{k \to \infty} g_m(x_k) = g_m(\bar{x}).$$

On the other hand, for each m and sufficiently large k with $n_k \geq m$ we have

$$g_m(x_k) \geq g_{n_k}(x_k) \geq \varepsilon_0.$$

Passage to the limit as $k \to \infty$ gives

$$\lim_{k \to \infty} g_m(_k) = g_m(\bar{x}) \geq \varepsilon_0.$$

But the inequality $g_m(\bar{x}) \geq \varepsilon_0$, which holds for every m with $m = 1, 2, 3, \ldots$, contradicts the assumption $g_m(\bar{x}) \to 0$ as $m \to \infty$ and the proof is finished. $\square$

REMARK. In Example 3 following Proposition 7.53 we have seen that

$$f_n(x) = \frac{1}{1 + nx}$$

fails to converge uniformly to 0 on the open interval $(0, 1)$. However,

$$f_n(x) \geq f_{n+1}(x)$$

for all x in $(0, 1)$, showing that closedness of the interval in Proposition 7.55 is essential. That the interval be of finite length in Proposition 7.55 is essential as well; the sequence of functions $\{h_n\}_{n=1}^{\infty}$, where

$$h_n(x) = -\frac{x^2}{n},$$

does not converge uniformly to 0 on $(-\infty, \infty)$, but $h_n(x) \leq h_{n+1}(x)$ for all x in $(-\infty, \infty)$.

Proposition 7.56. *Let $\{f_n\}_{n=1}^{\infty}$ be a sequence of functions, differentiable on a closed interval $[a, b]$ of finite length and such that $\{f_n(x_0)\}_{n=1}^{\infty}$ converges for some point x_0 of $[a, b]$. If $\{f_n'\}_{n=1}^{\infty}$ converges uniformly on $[a, b]$, then $\{f_n\}_{n=1}^{\infty}$ converges uniformly on $[a, b]$, to a function f, and, for $a \leq x \leq b$,*

$$f'(x) = \lim_{n \to \infty} f_n'(x). \tag{7.63}$$

PROOF. Let $\varepsilon > 0$ be given. Pick n_0 such that $n, m \geq n_0$ implies

$$|f_n(x_0) - f_m(x_0)| < \frac{\varepsilon}{2} \tag{7.64}$$

and, for $a \leq t \leq b$,

$$|f_n'(t) - f_m'(t)| < \frac{\varepsilon}{2(b-a)}. \tag{7.65}$$

Applying Proposition 4.5 of Chapter 4 to the function $f_n - f_m$, (7.65) shows that

$$|f_n(x) - f_m(x) - f_n(t) + f_m(t)| < \frac{|x-t|\varepsilon}{2(b-a)} < \frac{\varepsilon}{2} \tag{7.66}$$

for any x and t in $[a, b]$, if $n \geq n_0$, $m \geq n_0$. But

$$|f_n(x) - f_m(x)| \leq |f_n(x) - f_m(x) - f_n(x_0) + f_m(x_0)| + |f_n(x_0) - f_m(x_0)|$$

implies, by (7.64) and (7.66), that

$$|f_n(x) - f_m(x)| < \varepsilon$$

for $a \leq x \leq b$, $n \geq n_0$, $m \geq n_0$; thus, $\{f_n\}_{n=1}^{\infty}$ converges uniformly on $[a, b]$. Let, for $a \leq x \leq b$,

$$f(x) = \lim_{n \to \infty} f_n(x).$$

We now fix a point x on $[a, b]$ and define

$$h_n(t) = \frac{f_n(t) - f_n(x)}{t - x}, \qquad h(t) = \frac{f(t) - f(x)}{t - x} \tag{7.67}$$

for $a \leq t \leq b$, $t \neq x$. Then, for $n = 1, 2, 3, \ldots$,

$$\lim_{t \to x} h_n(t) = f_n'(x). \tag{7.68}$$

The first inequality in (7.66) shows that

$$|h_n(t) - h_m(t)| < \frac{\varepsilon}{2(b-a)}$$

for $n, m \geq n_0$; thus, $\{h_n\}_{n=1}^{\infty}$ converges uniformly, for $t \neq x$. Since $\{f_n\}_{n=1}^{\infty}$ converges to f, we get from (7.67) that

$$\lim_{n \to \infty} h_n(t) = h(t) \tag{7.69}$$

uniformly for $a \leq t \leq b$, $t \neq x$. Applying Proposition 7.54 to $\{h_n\}_{n=1}^{\infty}$, (7.68) and (7.69) give

$$\lim_{t \to x} h(t) = \lim_{n \to \infty} f_n'(x).$$

But this is (7.63), by the definition of $h(t)$. $\square$

REMARK. Note that the uniform convergence of $\{f_n\}_{n=1}^{\infty}$ implies nothing about the sequence $\{f_n'\}_{n=1}^{\infty}$; consider, for example, the sequence

$$f_n(x) = \frac{\sin nx}{\sqrt{n}}$$

for all real x. Then $\lim_{n\to\infty} f_n(x) = f(x) = 0$ for every real x and

$$f_n'(x) = \sqrt{n} \cos nx.$$

Hence, $\lim_{n\to\infty} f_n'(x)$ does not exist for any real x. The sequence $\{f_n\}_{n=1}^{\infty}$ converges uniformly on $(-\infty, \infty)$, but $\{f_n'\}_{n=1}^{\infty}$ does not even converge point-wise on $(-\infty, \infty)$. For example, $\{f_n'\}_{n=1}^{\infty}$ diverges since $f_n'(0) = \sqrt{n}$; but $f'(x) = 0$ for all real x.

Proposition 7.57. *Let* $\{f_n\}_{n=1}^{\infty}$ *be a sequence of continuous functions on a closed interval* $[a, b]$ *of finite length and suppose that* f *is a function on* $[a, b]$ *such that* $f_n \to f$ *uniformly on* $[a, b]$. *Then*

$$\lim_{n\to\infty} \int_a^b f_n(x)\, dx = \int_a^b f(x)\, dx. \tag{7.70}$$

PROOF. By Proposition 7.54 the function f is continuous and so the function $f_n - f$ are all Riemann integrable on $[a, b]$ (by Proposition 5.16 of Chapter 5). Let $\varepsilon > 0$. Since $f_n \to f$ uniformly on $[a, b]$, there exists a number n_0 such that

$$|f_n(x) - f(x)| < \frac{\varepsilon}{b - a}$$

for all x in $[a, b]$ and all $n \geq n_0$. Thus, $n \geq n_0$ implies

$$\left| \int_a^b f_n(x)\, dx - \int_a^b f(x)\, dx \right| = \left| \int_a^b [f_n(x) - f(x)]\, dx \right|$$

$$\leq \int_a^b |f_n(x) - f(x)|\, dx \leq \int_a^b \frac{\varepsilon}{b - a}\, dx = \varepsilon$$

(note that we have used Propositions 5.11 and 5.12 of Chapter 5). Thus, for any $\varepsilon > 0$, there exists a number n_0 such that $n \geq n_0$ implies

$$\left| \int_a^b f_n(x)\, dx - \int_a^b f(x)\, dx \right| \leq \varepsilon$$

and so we see that (7.70) holds. □

Examples

1. Let $f_n(x) = nx(1 - x)^n$. Then the pointwise limit of the sequence $\{f_n\}_{n=1}^{\infty}$ on $[0, 1]$ is 0, but the sequence does not converge uniformly on $[0, 1]$. Indeed, f_n has a hump at $x = 1/(n + 1)$ and

$$f_n\left(\frac{1}{n+1}\right) = \left(1 - \frac{1}{n+1}\right)^{n+1} \to e^{-1} \quad \text{as } n \to \infty$$

so that Proposition 7.53 applies. The sequence $\{f_n\}_{n=1}^{\infty}$ is not uniformly convergent on $[0, 1]$, yet term-by-term integration over $[0, 1]$ will give

$$\lim_{n \to \infty} \int_0^1 f_n(x)\, dx = \int_0^1 f(x)\, dx$$

in the case under consideration. Thus, uniform convergence is a sufficient, but not a necessary, condition for the validity of the interchange of limit and integration sign in relation (7.70).

2. Let $f_n(x) = nxe^{-nx^2}$, $n = 1, 2, \ldots, 0 \le x \le 1$. Then $f_n(x) \to 0$ as $n \to \infty$ for every x in $[0, 1]$. But

$$\int_0^1 nxe^{-nx^2}\, dx = \frac{1}{2}(1 - e^{-n})$$

and so

$$\lim_{n \to \infty} \int_0^1 f_n(x)\, dx = \lim_{n \to \infty} \frac{1}{2}(1 - e^{-n}) = \frac{1}{2}.$$

On the other hand, $\int_0^1 f(x)\, dx = 0$ and so

$$\lim_{n \to \infty} \int_0^1 f_n(x)\, dx \ne \int_0^1 f(x)\, dx. \tag{7.71}$$

The reason for the result in (7.71) is that the sequence $f_n(x) = nxe^{-nx^2}$ does not converge uniformly to 0 on $[0, 1]$. In fact, f_n has a hump at $x = 1/\sqrt{2n}$ and

$$f_n\left(\frac{1}{\sqrt{2n}}\right) = \frac{\sqrt{n/2}}{\sqrt{e}}$$

so that Proposition 7.53 is applicable.

3. Let $f_n(x) = (n + \sin x)/(3n + \cos^2 x)$. Then $\lim_{n \to \infty} \int_0^3 f_n(x)\, dx = 1$, by Proposition 7.57.

Indeed, the pointwise limit of $\{f_n\}_{n=1}^{\infty}$ is $\frac{1}{3}$ on $[0, 3]$. But $\frac{1}{3}$ is also the uniform limit of the sequence on $[0, 3]$ because

$$\left| \frac{n + \sin x}{3n + \cos^2 x} - \frac{1}{3} \right| = \left| \frac{3 \sin x - \cos^2 x}{3(3n + \cos^2 x)} \right| \le \frac{3|\sin x| + \cos^2 x}{3(3n + \cos^2 x)} \le \frac{3 + 1}{9n} = \frac{4}{9n}$$

for all x in $[0, 3]$.

Definition. Let E be a set of real numbers and $\{f_n\}_{n=1}^{\infty}$ a sequence, each term of which is a real-valued function defined on E. We say that the series $\sum_{n=1}^{\infty} f_n$ *converges uniformly on E* if the sequence of partial sums defined by

$$s_n(x) = \sum_{j=1}^{n} f_j(x)$$

converges uniformly on E.

DISCUSSION. Putting

$$\sum_{j=1}^{\infty} f_j(x) = s_n(x) + r_n(x),$$

where

$$s_n(x) = \sum_{j=1}^{n} f_j(x) \quad \text{and} \quad r_n(x) = \sum_{j=n+1}^{\infty} f_j(x),$$

we can reformulate in terms of series the condition (7.60) as follows: The series $\sum_{n=1}^{\infty} f_n$ is uniformly convergent on E if for every $\varepsilon > 0$ there exists an integer n_0 (depending only on ε) such that $n \geq n_0$ implies

$$|r_n(x)| < \varepsilon$$

for every x in E [we also say that the $r_n(x)$ tends uniformly to 0 on E].

Cauchy's criterion for uniform convergence considered in Proposition 7.52 reformulated for series reads: For every $\varepsilon > 0$ there exists an integer n_0 such that

$$|f_{n+1}(x) + f_{n+2}(x) + \cdots + f_{n+k}(x)| < \varepsilon$$

for all $n \geq n_0$ and every $k = 1, 2, 3, \ldots$ and every x in E.

Finally, we note that $\sum_{n=1}^{\infty} f_n$ converges uniformly on E if and only if for any sequence $\{x_n\}_{n=1}^{\infty}$ in E the corresponding remainders

$$r_n(x_n)$$

invariably form a null sequence. This is the formulation in terms of series of a statement equivalent to Proposition 7.53.

Proposition 7.58. *If a series $\sum_{n=1}^{\infty} f_n$ converges uniformly on a set E, then the general term f_n converges to 0 uniformly on E.*

PROOF. Let $s_n(x) = f_1(x) + \cdots + f_n(x)$ and $S(x) = \sum_{n=1}^{\infty} f_n(x)$. Then

$$|f_n(x)| = |s_n(x) - s_{n-1}(x)| = |[s_n(x) - S(x)] + [S(x) - s_{n-1}(x)]$$

$$\leq |s_n(x) - S(x)| + |s_{n-1}(x) - S(x)|.$$

Let $\varepsilon > 0$ be given. If n_0 is chosen such that $n \geq n_0 - 1$ implies

$$|s_n(x) - S(x)| < \frac{\varepsilon}{2}$$

for all x in E, then $n \geq n_0$ implies $|f_n(x)| < \varepsilon$ for all x in E. $\qquad\square$

Definition. A series of functions $\sum_{n=1}^{\infty} v_n$ is said to *dominate* a series of functions $\sum_{n=1}^{\infty} f_n$ on a set E if all terms are defined on E and that for any x in E we have $|f_n(x)| \geq v_n(x)$ for every positive integer n.

Proposition 7.59 (Comparison Test). *Any series of functions $\sum_{n=1}^{\infty} f_n$ dominated on a set E by a series of functions $\sum_{n=1}^{\infty} v_n$ that is uniformly convergent on E is uniformly convergent on E.*

PROOF. From Proposition 7.26 (i) we know that the series $\sum_{n=1}^{\infty} f_n(x)$ converges for every x in E. If $F(x) = \sum_{n=1}^{\infty} f_n(x)$ and $V(x) = \sum_{n=1}^{\infty} v_n(x)$, then

$$|[f_1(x) + f_2(x) + \cdots + f_n(x)] - F(x)| = |f_{n+1}(x) + f_{n+2}(x) + \cdots|$$

$$\leq v_{n+1}(x) + v_{n+2}(x) + \cdots = |[v_1(x) + v_2(x) + \cdots + v_n(x)] - V(x)|.$$

Let $\varepsilon > 0$ be given. If n_0 is such that $n \geq n_0$ implies

$$|[v_1(x) + \cdots + v_n(x)] - V(x)| < \varepsilon$$

for all x in E, then $|[f_1(x) + \cdots + f_n(x)] - F(x)| < \varepsilon$ for all x in E.   □

Proposition 7.60. *A series of functions converges uniformly on a set whenever its series of absolute values converges uniformly on that set.*

PROOF. The proposition is an immediate corollary to Proposition 7.59.   □

Proposition 7.61 (Weierstrass M-Test). *If $\sum_{n=1}^{\infty} f_n$ is a series of functions defined on a set E, if $\sum_{n=1}^{\infty} M_n$ is a convergent series of nonnegative constants, and if for every x in E we have*

$$|f_n(x)| \leq M_n \quad \text{for } n = 1, 2, 3, \ldots,$$

then $\sum_{n=1}^{\infty} f_n$ converges uniformly on E.

PROOF. Since any convergent series of constants (constant functions) converges uniformly on any set, we have that the proposition under consideration is merely a special case of Proposition 7.59.   □

COMMENTS. The Weierstrass M-Test (see Proposition 7.61) is an extremely useful test for uniform convergence. It should be noted carefully, however, that a series $\sum_{n=1}^{\infty} f_n$ of continuous functions on a closed interval $[a, b]$ of finite length can converge absolutely and uniformly on $[a, b]$, yet fail the Weierstrass M-Test. Here is an example: On the interval $[0, 1]$ let the function f_n be defined by

$$f_n(x) = 0 \quad \text{for } 0 \leq x \leq \frac{1}{2n + 1} \text{ and } \frac{1}{2n - 1} \leq x \leq 1;$$

$$f_n(x) = \frac{1}{n} \quad \text{for } x = \frac{1}{2n};$$

$\quad f_n(x)$ is defined linearly in the intervals $[1/(2n + 1), 1/2n]$
$\quad\quad$ and $[1/2n, 1/(2n - 1)]$.

It is easy to see that the series $\sum_{n=1}^{\infty} f_n$ satisfies the following conditions:

(a) The series is uniformly convergent on $[0, 1]$.
(b) Any rearrangement of the series is uniformly convergent (i.e., the series is uniformly and absolutely convergent).
(c) The series $\sum_{n=1}^{\infty} M_n$ diverges, where M_n is the upper bound of $|f_n(x)|$ as x ranges over the interval $[0, 1]$.

Definition. A sequence $\{f_n\}_{n=1}^{\infty}$ of functions is said to be *uniformly bounded on E* if there is a constant $M > 0$ such that $|f_n(x)| \leq M$ for all x in E and all $n = 1, 2, 3, \ldots$. The number M is called a *uniform bound for* $\{f_n\}_{n=1}^{\infty}$.

REMARK. If each individual function of a sequence $\{f_n\}_{n=1}^{\infty}$ is bounded on E, that is, if $\sup\{|f_n(x)|: x \in E\} \leq M_n$ for $n = 1, 2, 3, \ldots$, and $f_n \to f$ uniformly on E, then it is easily seen that $\{f_n\}_{n=1}^{\infty}$ is uniformly bounded on E.

Proposition 7.62 (Dirichlet's Test for Uniform Convergence). *Let $s_n(x)$ denote the nth partial sum of the series $\sum_{n=1}^{\infty} a_n(x)$, with a_n being a real-valued function defined on a set E of real numbers. Assume that $\{s_n\}_{n=1}^{\infty}$ is uniformly bounded on E by a constant M. Let $\{b_n\}_{n=1}^{\infty}$ be a sequence of real-valued functions which is monotonically decreasing in n for each fixed x in E [i.e., $b_{n+1}(x) \leq b_n(x)$ for each x in E and for every $n = 1, 2, 3, \ldots$], and suppose that $b_n \to 0$ uniformly on E. Then the series $\sum_{n=1}^{\infty} a_n(x)b_n(x)$ converges uniformly on E.*

PROOF. Since $b_n \to 0$ uniformly on E, for any $\varepsilon > 0$ there exists an integer j such that for all x in E,

$$|b_n(x)| < \frac{\varepsilon}{2M}$$

whenever $n \geq j$. Moreover, by the proof of Proposition 7.41,

$$|a_{j+1}(x)b_{j+1}(x) + \cdots + a_{j+p}(x)b_{j+p}(x)| < 2M|b_{j+1}(x)| < \varepsilon$$

provided that p is a positive integer and $x \in E$. The uniform convergence of the series $\sum_{n=1}^{\infty} a_n(x)b_n(x)$ on E now follows from Proposition 7.52. □

Proposition 7.63 (Abel's Test for Uniform Convergence). *Let $s_n(x)$ denote the nth partial sum of the series $\sum_{n=1}^{\infty} a_n(x)$, where each a_n is a real-valued function defined on a set E of real numbers. Assume that $\{s_n\}_{n=1}^{\infty}$ converges uniformly on E. Let $\{b_n\}_{n=1}^{\infty}$ be a sequence of functions which is uniformly bounded on E by a constant K. If $b_n(x)$ is either monotonically increasing in n for each fixed x in E, or monotonically decreasing in n for each fixed x in E, then the series $\sum_{n=1}^{\infty} a_n(x)b_n(x)$ converges uniformly on E.*

PROOF. Let $b_n(x)$ be monotonically increasing in n for each fixed x in E, that is, let $b_n(x) \leq b_{n+1}(x)$ for each x in E and for every $n = 1, 2, 3, \ldots$. Then the pointwise limit of b_n on E, say b, exists. Let

$$c_n(x) = b(x) - b_n(x).$$

Then $c_n(x)$ is positive (or zero) and monotonically decreasing in n for each fixed x in E. Also,

$$|b(x)| \leq K$$

for $x \in E$. Hence, if $n = 1, 2, 3, \ldots,$

$$c_n(x) \leq 2K$$

for $x \in E$. Since $\{s_n\}_{n=1}^{\infty}$ is uniformly convergent on E, there is a positive integer j such that

$$|a_{j+1}(x) + \cdots + a_{j+p}(x)| < \frac{\varepsilon}{2K}$$

whenever p is a positive integer and $x \in E$. Hence, by Proposition 7.40,

$$|a_{j+1}(x)c_{j+1}(x) + \cdots + a_{j+p}(x)c_{j+p}(x)| < \frac{\varepsilon}{2K}c_{j+1}(x) < \varepsilon.$$

By Proposition 7.52, $\sum_{n=1}^{\infty} a_n(x)b_n(x)$ is uniformly convergent on E.

Also, since $|b(x)| \leq K$ for all x in E, it follows from Proposition 7.52 that $\sum_{n=1}^{\infty} b(x)a_n(x)$ converges uniformly on E. The uniform convergence of $\sum_{n=1}^{\infty} a_n(x)b_n(x)$ on E is now a consequence of the relation

$$\sum_{n=1}^{\infty} a_n(x)b_n(x) = \sum_{n=1}^{\infty} [b(x)a_n(x) - a_n(x)c_n(x)]$$

and the proof is complete.   □

Proposition 7.64 (Tannery's Theorem). *Let $F(x) = \sum_{n=1}^{\infty} v_n(x)$, the series being uniformly convergent with regard to x for all positive x. Furthermore, for each fixed n, let*

$$v_n(x) \to w_n \quad \text{as } x \to \infty.$$

Then the series $\sum_{n=1}^{\infty} w_n$ is convergent and

$$F(x) \to \sum_{n=1}^{\infty} w_n \quad \text{as } x \to \infty,$$

that is, we have termwise passage to the limit

$$\lim_{x \to \infty} \left(\sum_{n=1}^{\infty} v_n(x) \right) = \sum_{n=1}^{\infty} \left(\lim_{x \to \infty} v_n(x) \right).$$

PROOF. The proof conveniently splits into two steps.

Step 1: We prove that the series $\sum_{n=1}^{\infty} w_n$ converges. Indeed, by the uniform convergence of $\sum_{n=1}^{\infty} v_n(x)$ for all positive x, for any $\varepsilon > 0$ there exists an integer N such that for all positive x we have

$$\left|\sum_{n=N+1}^{N+k} v_n(x)\right| < \varepsilon \quad \text{for } k = 1, 2, 3, \ldots.$$

As $x \to \infty$, $v_{N+1}(x) + \cdots + v_{N+k}(x) \to w_{N+1} + \cdots + w_{N+k}$. Hence,

$$\left|\sum_{n=N+1}^{N+k} w_n\right| \leq \varepsilon \quad \text{for } k = 1, 2, 3, \ldots$$

which (by Proposition 7.23) implies that $\sum_{n=1}^{\infty} w_n$ converges.

Step 2: The convergence of $\sum_{n=1}^{\infty} w_n$ having been established, we make a fresh start. Let $W = \sum_{n=1}^{\infty} w_n$. Then, since the series $\sum_{n=1}^{\infty} v_n(x)$ converges uniformly to its sum $F(x)$ for all positive x, and since $\sum_{n=1}^{\infty} w_n$ converges to the sum W, for any $\varepsilon > 0$ there is an integer N such that

$$\left|F(x) - \sum_{n=1}^{N} v_n(x)\right| < \varepsilon$$

for all positive x and

$$\left|W - \sum_{n=1}^{N} w_n\right| < \varepsilon.$$

Hence, for all positive x,

$$|F(x) - W| \leq \left|F(x) - \sum_{n=1}^{N} v_n(x)\right| + \left|\sum_{n=1}^{N} v_n(x) - \sum_{n=1}^{N} w_n\right| + \left|\sum_{n=1}^{N} w_n - W\right|$$

$$< 2\varepsilon + \left|\sum_{n=1}^{N} v_n(x) - \sum_{n=1}^{N} w_n\right|.$$

But as $x \to \infty$, each $v_n(x) \to w_n$ and so, since N is finite,

$$\sum_{n=1}^{N} v_n(x) \to \sum_{n=1}^{N} w_n.$$

Hence, there exists a number K such that

$$\left|\sum_{n=1}^{N} v_n(x) - \sum_{n=1}^{N} w_n\right| < \varepsilon \quad \text{when } x > K.$$

Finally, we have

$$|F(x) - W| < 3\varepsilon \quad \text{when } x > K.$$

But this means that $F(x) \to W$ as $x \to \infty$. $\qquad\qquad\square$

APPLICATION. We have

$$\lim_{n \to \infty} \left(\left(\frac{n}{n}\right)^n + \left(\frac{n-1}{n}\right)^n + \cdots + \left(\frac{1}{n}\right)^n\right) = \frac{e}{e-1}.$$

Indeed, let

$$A_n = \left(\frac{n}{n}\right)^n + \left(\frac{n-1}{n}\right)^n + \cdots + \left(\frac{1}{n}\right)^n.$$

Then

$$A_n = \sum_{k=0}^{n-1}\left(1 - \frac{k}{n}\right)^n = \sum_{k=0}^{n-1} b_k(n),$$

say. Now $b_k(n)$ is monotonically increasing in n and tends to e^{-k} as $n \to \infty$ so that

$$b_k(n) \le e^{-k} \quad \text{for all } n.$$

In addition, the series

$$\sum_{k=0}^{\infty} e^{-k}$$

converges, its sum being $1/(1 - e^{-1})$. But, when $k \to \infty$, we have also that $n \to \infty$ and so, by Proposition 7.64,

$$\lim_{n\to\infty} A_n = \sum_{k=0}^{\infty}\left(\lim_{n\to\infty} b_k(n)\right) = \sum_{k=0}^{\infty} e^{-k} = \frac{1}{1 - e^{-1}} = \frac{e}{e-1}.$$

Lemma. *Let f be a real-valued function defined on an interval $[a,b]$. If there are sequences $\{\alpha_n\}_{n=1}^{\infty}$ and $\{\beta_n\}_{n=1}^{\infty}$, where $a < \alpha_n < x < \beta_n < b$, and $\alpha_n \to x$, $\beta_n < x$ as $n \to \infty$, such that the limit*

$$\lim_{n\to\infty} \frac{f(\beta_n) - f(\alpha_n)}{\beta_n - \alpha_n}$$

does not exist, then f is not differentiable at x.

PROOF. We shall prove the contrapositive statement. Let f be a real-valued function on $[a,b]$ which is differentiable at a point x satisfying $a < x < b$. Let $\{\alpha_n\}_{n=1}^{\infty}$ and $\{\beta_n\}_{n=1}^{\infty}$ be any sequences such that $a < \alpha_n < x < \beta_n < b$, and $\alpha_n \to x$, $\beta_n \to x$ as $n \to \infty$. Then

$$\lim_{n\to\infty} \frac{f(\beta_n) - f(\alpha_n)}{\beta_n - \alpha_n} = f'(x). \tag{7.72}$$

Indeed, let $\lambda_n = (\beta_n - x)/(\beta_n - \alpha_n)$. Then $0 < \lambda_n < 1$ and

$$\frac{f(\beta_n) - f(\alpha_n)}{\beta_n - \alpha_n} - f'(x) = \lambda_n\left(\frac{f(\beta_n) - f(x)}{\beta_n - x} - f'(x)\right)$$

$$+ (1 - \lambda_n)\left(\frac{f(\alpha_n) - f(x)}{\alpha_n - x} - f'(x)\right).$$

The expressions within the braces both tend to zero as $n \to \infty$ and the sequences $\{\lambda_n\}_{n=1}^{\infty}$ and $\{1 - \lambda_n\}_{n=1}^{\infty}$ are both bounded. Hence, the limit in (7.72) exists and the equality holds.   □

Proposition 7.65. *There exists a real-valued continuous function on* $(-\infty, \infty)$ *that is nowhere differentiable.*

PROOF. Let

$$h(x) = x \qquad \text{for } 0 \leq x \leq 1$$
$$= 2 - x \qquad \text{for } 1 \leq x \leq 2,$$

and extend the domain of definition of h to all real x by requiring that

$$h(x + 2) = h(x);$$

in other words, h is *periodic* with period 2. It is clear that h is continuous on $(-\infty, \infty)$. Define

$$f(x) = \sum_{n=0}^{\infty} \left(\frac{3}{4}\right)^n h(4^n x). \qquad (7.73)$$

Since $0 \leq h(x) \leq 1$ for all real x, Proposition 7.61 shows that the series (7.73) converges uniformly on $(-\infty, \infty)$; hence, f is continuous on $(-\infty, \infty)$, by Proposition 7.54.

We fix a real number x and a positive integer m. There exists an integer k such that

$$k \leq 4^m x \leq k + 1.$$

We put

$$\alpha_m = 4^{-m}k, \qquad \beta_m = 4^{-m}(k + 1),$$

and consider the numbers $4^n \beta_m$ and $4^n \alpha_m$. If $n > m$, their difference is an even integer; if $n = m$, they are integers and their difference is 1; if $n < m$, no integer is situated between them. Hence,

$$|h(4^n \beta_m) - h(4^n \alpha_m)| = 0 \qquad \text{for } n > m, \qquad (7.74)$$
$$= 4^{n-m} \quad \text{for } n \leq m.$$

By (7.73) and (7.74),

$$f(\beta_m) - f(\alpha_m) = \sum_{n=0}^{m} \left(\frac{3}{4}\right)^n [h(4^n \beta_m) - h(4^n \alpha_m)].$$

Thus,

$$|f(\beta_m) - f(\alpha_m)| \geq \left(\frac{3}{4}\right)^m - \sum_{n=0}^{m-1} \left(\frac{3}{4}\right)^n 4^{n-m} > \left(\frac{1}{2}\right) \cdot \left(\frac{3}{4}\right)^m,$$

or

$$\left| \frac{f(\beta_m) - f(\alpha_m)}{\beta_m - \alpha_m} \right| > \frac{1}{2} \cdot 3^m. \qquad (7.75)$$

Since $\alpha_m \le x \le \beta_m$ and since $\beta_m - \alpha_m \to 0$ as $m \to \infty$, (7.75) shows that f is not differentiable at x, by the Lemma. $\square$

Worked Examples

1. The series

$$\sum_{n=1}^{\infty} \frac{x}{n^{\alpha}(1 + nx^2)} \quad \text{with } \alpha > 0$$

is uniformly convergent on $(-\infty, \infty)$. Indeed, the largest value of $x/(1 + nx^2)$ is attained for $x = 1/\sqrt{n}$ and so the given series is dominated by the convergent series

$$\sum_{n=1}^{\infty} \frac{1}{n^{\alpha+1}}$$

and the uniform convergence of the given series on $(-\infty, \infty)$ follows by Proposition 7.61.

2. The alternating series

$$\sum_{n=1}^{\infty} \frac{(-1)^{n+1}}{x^2 + n}$$

is uniformly convergent on $(-\infty, \infty)$ because, by Proposition 7.33,

$$\left| \sum_{n=m}^{\infty} \frac{(-1)^{n+1}}{x^2 + n} \right| \le \frac{1}{x^2 + m} \le \frac{1}{m}.$$

3. The series

$$\sum_{n=1}^{\infty} \frac{(-1)^{n+1}}{(1 + x^2)^n}$$

converges uniformly on $(-\infty, \infty)$ because, for $x \neq 0$,

$$\left| \sum_{n=m}^{\infty} \frac{(-1)^{n+1}}{(1 + x^2)^n} \right| \le \frac{x^2}{(1 + x^2)^n} = \frac{x^2}{1 + mx^2 + \cdots} < \frac{1}{m}.$$

However, the series

$$\sum_{n=1}^{\infty} \frac{x^2}{(1 + x^2)^n}$$

does not converge uniformly on $(-\infty, \infty)$ because

$$\sum_{n=1}^{\infty} \frac{x^2}{(1 + x^2)^n} = 1 \quad \text{for } x \neq 0$$

$$= 0 \quad \text{for } x = 0$$

and we know that the uniform limit of a sequence of continuous functions is a continuous function (see Proposition 7.54), but here the limit function has a point of discontinuity at $x = 0$.

4. The series

$$\frac{x}{1+x} + \frac{x}{(1+x)(1+2x)} + \frac{x}{(1+2x)(1+3x)} + \cdots$$

is not uniformly convergent on $(-\infty, \infty)$. Indeed, the nth partial sum

$$\frac{x}{1+x} + \frac{x}{(1+x)(1+2x)} + \cdots + \frac{x}{(1+\{n-1\}x)(1+nx)}$$

$$= \left(1 - \frac{1}{1+x}\right) + \left(\frac{1}{1+x} - \frac{1}{1+2x}\right) + \cdots + \left(\frac{1}{1+n-1x} - \frac{1}{1+nx}\right)$$

$$= 1 - \frac{1}{1+nx} = f_n(x)$$

and so $f_n(x) \to 1$ as $n \to \infty$, but $f_n(1/n) = \frac{1}{2}$. This shows that the convergence on $(-\infty, \infty)$ is not uniform.

5. Consider the series

$$\sum_{n=1}^{\infty} \frac{(-1)^n}{n} e^{-nx}$$

on $[0, \infty)$. Evidently, $|e^{-nx}| \leq 1$ for $x \geq 0$. Since the harmonic series $\sum 1/n$ is divergent, we can not apply Proposition 7.61. We can use Proposition 7.62 to show the uniform convergence of the given series on $[0, \infty)$ after checking that the partial sums of the series $\sum (-1)^n e^{-nx}$ are bounded. Alternately, Proposition 7.63 applies since the series $\sum (-1)^n/n$ is convergent and the bounded sequence $\{e^{-nx}\}$ is monotonically decreasing on $[0, \infty)$ (but not uniformly convergent to zero).

6. Power Series

Definition. Given a sequence $\{a_n\}_{n=0}^{\infty}$ of real numbers, the series

$$\sum_{n=0}^{\infty} a_n x^n \tag{7.76}$$

is called a *power series*; the numbers a_n are called the *coefficients* of the power series.

Proposition 7.66. *If $\sum_{n=0}^{\infty} a_n x^n$ is convergent for $x = x_0$, it is absolutely convergent for every value x such that $|x| < |x_0|$; if it diverges for $x = x_1$, it is divergent for any value of x such that $|x| > |x_1|$.*

PROOF. Since $\sum_{n=0}^{\infty} a_n x_0^n$ converges, the sequence $\{a_n x_0^n\}_{n=0}^{\infty}$ converges to zero by Proposition 7.21 and hence is bounded (by Proposition 7.1 or by Proposition 7.4). Thus, there is an M such that $|a_n x_0^n| < M$, for $n = 0, 1, 2, \ldots$. But

$$|a_n x^n| = |a_n x_0^n| \left|\frac{x}{x_0}\right|^n.$$

Therefore, if $|x/x_0| = c < 1$, the terms of the series $\sum_{n=0}^{\infty} a_n x^n$ are less than the corresponding terms of the series $M(\sum_{n=0}^{\infty} c^n)$.

From this the second part of the proposition follows easily. Suppose that $\sum_{n=0}^{\infty} a_n x_1^n$ diverges and that there were another value of x, say $x = x_0$, with $|x_0| > |x_1|$ for which $\sum_{n=0}^{\infty} a_n x_0^n$ converges. If this could happen, we would be in the following situation: There is an x_0 for which the series converges and an x_1 with $|x_1| < |x_0|$ for which it diverges. But this contradicts the first part of the proposition. $\qquad\square$

REMARK. It follows from Proposition 7.66 that only the following three cases can occur:

(i) The power series (7.76) converges for $x = 0$ and no other value of x; for example,

$$1 + 1!x + 2!x^2 + 3!x^3 + \cdots, \qquad 1 + x + 2^2 x^2 + 3^3 x^3 + \cdots.$$

(ii) The power series (7.76) converges for all values of x; for example,

$$1 + x + \frac{x^2}{2!} + \frac{x^3}{3!} + \cdots.$$

(iii) There is some positive number R such that, if $|x| < R$, the power series (7.76) converges and, if $|x| > R$, the power series (7.76) diverges; for example,

$$1 + x + x^2 + x^3 + \cdots, \qquad x - \tfrac{1}{2}x^2 + \tfrac{1}{3}x^3 - \cdots.$$

Definition. Given a power series $\sum_{n=0}^{\infty} a_n x^n$, let S be the set of x's for which this series converges. Then the number R, defined below, will be called the *radius of convergence of* $\sum_{n=0}^{\infty} a_n x^n$:

(i) $R = 0$ if $\sum_{n=0}^{\infty} a_n x^n$ converges only for $x = 0$.
(ii) $R = +\infty$ if $\sum_{n=0}^{\infty} a_n x^n$ converges for all x's.
(iii) $R = \sup\{|x|: x \in S\}$ if $\sum_{n=0}^{\infty} a_n x^n$ converges for some x's and diverges for others. The open interval $(-R, R)$ will be called the *interval of convergence*.

Proposition 7.67. *A power series* $\sum_{n=0}^{\infty} a_n x^n$ *converges absolutely for all* x's *inside the interval of convergence* $(-R, R)$ *and diverges for all* x's *satisfying* $|x| > R$.

PROOF. We restrict ourselves to case (iii), for clearly if $R = 0$ or $R = +\infty$ we have nothing to prove.

Let $x \in (-R, R)$. Then there is an x_0, $|x| < |x_0| < R$, for which the series converges (this is by the definition of supremum). By Proposition 7.66, the power series converges absolutely at x. Hence, it converges for all such x, that is, all such x for which $x \in (-R, R)$.

Suppose now that the power series does not diverge for some x_0, $|x_0| > R$. This means it converges for x_0. But then we have found a member of the set S larger than the supremum. This is a contradiction.   ☐

REMARK. Proposition 7.67 makes no statement about the endpoints $x = \pm R$ of the interval of convergence. It can happen that a power series converges at neither point, or at one point and not the other, or at both. Consider the power series

$$\text{(A) } \sum_{n=0}^{\infty} x^n, \quad \text{(B) } \sum_{n=1}^{\infty} \frac{x^n}{n}, \quad \text{(C) } \sum_{n=1}^{\infty} (-1)^{n+1} \frac{x^n}{n}, \quad \text{(D) } \sum_{n=1}^{\infty} \frac{x^n}{n^2},$$

$$\text{(E) } \sum_{n=0}^{\infty} (-1)^n \frac{x^{2n+1}}{2n+1}.$$

Using the Ratio Test (see Proposition 7.32) we see that each of these five power series is convergent for $|x| < 1$ and divergent for $|x| > 1$. Series (A) is divergent for $x = \pm 1$; series (D) is absolutely convergent for $x = \pm 1$; series (E) is conditionally convergent for $x = \pm 1$; series (B) is convergent for $x = -1$ and divergent for $x = 1$ while series (C) is convergent for $x = 1$ and divergent for $x = -1$.

DISCUSSION. If $\sum_{n=0}^{\infty} a_n x^n$ is any power series which does not merely converge everywhere or nowhere but at $x = 0$, then there exists a unique positive number R, namely, the radius of convergence of $\sum_{n=0}^{\infty} a_n x^n$, such that $\sum_{n=0}^{\infty} a_n x^n$ converges for every $|x| < R$ (in fact, absolutely), but diverges for every $|x| > R$.

Indeed, by assumption there exists at least one point of divergence, and one point of convergence $\neq 0$. We can therefore choose a positive number x_0 nearer 0 than the point of convergence and a positive number t_0 further from 0 than the point of divergence. By Proposition 7.66, the series $\sum_{n=0}^{\infty} a_n x^n$ is convergent for $x = x_0$, divergent for $x = t_0$, and therefore we clearly have $x_0 < t_0$. To the closed interval $J_0 = [x_0, t_0]$ we apply the method of successive bisection: We denote by J_1 the left or the right half of J_0 according to whether $\sum_{n=0}^{\infty} a_n x^n$ diverges or converges at the middle point of J_0. By the same rule we designate a particular half of J_1 by J_2, and so forth. The intervals of the nested sequence of closed intervals $\{J_n\}$ all have the property that $\sum_{n=0}^{\infty} a_n x^n$ converges at their left endpoint (call it x_n), but diverges at their right endpoint (call it t_n). For the sake of argument we assume that the sequence $\{J_n\}$ is infinite. The number R (necessarily positive because $R \geq x_0$), determined by the nested sequence of closed intervals $\{J_n\}_{n=0}^{\infty}$, is the radius of convergence. In fact, if $x = x'$ is any real number for which $|x'| < R$ (equality excluded), then we have $|x'| < x_k$, for a sufficiently large k, that is, such that the length of J_k is less than $R - |x'|$. By Proposition 7.66, x' is a point of convergence at the same time as x_k is; and indeed at x' we have absolute convergence. If, on the contrary, x'' is a number for which $|x''| > R$, then $|x''| > t_m$, provided m is large enough for the length of J_m to be less than $|x''| - R$. By Proposition 7.66, x'' is then a point of divergence at the same time as t_m is.

If beside R there were another number R' which had the properties claimed of R, then every point between R and R' would simultaneously have to be a point of convergence and a point of divergence for the power series $\sum_{n=0}^{\infty} a_n x^n$, but this is impossible in view of Proposition 7.66. Therefore R is unique.

This proves all that was desired and shows the *existence* of the radius of convergence of a given power series. The next proposition gives both existence and exact value of the radius of convergence of a given power series.

Proposition 7.68. *Let $\{a_n\}_{n=0}^{\infty}$ be a sequence of real numbers and let*

$$\gamma = \overline{\lim_{n \to \infty}} \sqrt[n]{|a_n|}. \tag{7.77}$$

(i) *If $\gamma = 0$, then $\sum_{n=0}^{\infty} a_n x^n$ converges absolutely for all real numbers x.*
(ii) *If $\gamma = L > 0$, then $\sum_{n=0}^{\infty} a_n x^n$ converges absolutely for $|x| < 1/L$ and diverges for $|x| > 1/L$.*
(iii) *If $\gamma = +\infty$, then $\sum_{n=0}^{\infty} a_n x^n$ converges only for $x = 0$ and diverges for all other real numbers x.*
(iv) *The number $1/\gamma$ is the radius of convergence of $\sum_{n=0}^{\infty} a_n x^n$, that is, if $R = 1/\gamma$, then $\sum_{n=0}^{\infty} a_n x^n$ converges for every x such that $-R < x < R$.*

PROOF. The proposition is a mere reformulation of Proposition 7.31. $\quad\square$

REMARKS. The radius of convergence of $\sum_{n=0}^{\infty} a_n x^n$ is also given by

$$\lim_{n \to \infty} \left| \frac{a_n}{a_{n+1}} \right| \tag{7.78}$$

provided that this limit exists; this fact is a simple consequence of the Ratio Test (see Proposition 7.32). Frequently, it is more convenient to use (7.78) than (7.77) to determine the radius of convergence of the power series $\sum_{n=0}^{\infty} a_n x^n$. Nevertheless, formula (7.77) is always applicable, while formula (7.78) is only applicable if the limit in (7.78) exists, be it finite or infinite. The power series

$$1 + \left(\frac{x}{2}\right) + \left(\frac{x}{4}\right)^2 + \left(\frac{x}{2}\right)^3 + \left(\frac{x}{4}\right)^4 + \cdots, \tag{7.79}$$

where

$$a_{2n-1} = \left(\frac{1}{2}\right)^{2n-1} \quad \text{and} \quad a_{2n} = \left(\frac{1}{4}\right)^{2n},$$

has radius of convergence $R = 2$ in terms of formula (7.77). Formula (7.78) is clearly not applicable to determine the radius of convergence of the power series (7.79) because $\lim_{n \to \infty} |a_n/a_{n+1}|$ does not exist in the case of series (7.79).

Proposition 7.69. *Suppose that the power series $\sum_{n=0}^{\infty} a_n x^n$ converges for $|x| < R$. Then $\sum_{n=0}^{\infty} a_n x^n$ converges uniformly on the closed interval $[-R + \delta, R - \delta]$, where δ is any assigned positive number less than R.*

PROOF. Let $\delta > 0$ be given. For $|x| \leq R - \delta$, we have

$$|a_n x^n| \leq |a_n (R - \delta)^n|;$$

and since

$$\sum_{n=0}^{\infty} a_n (R - \delta)^n$$

converges absolutely by Proposition 7.67, Proposition 7.61 shows the uniform convergence of $\sum_{n=0}^{\infty} a_n x^n$ on $[-R + \delta, R - \delta]$. $\square$

REMARK. In case the interval of convergence extends to infinity, the power series will be absolutely convergent for every value of x, but it need not be uniformly convergent on $(-\infty, \infty)$. However, it will be uniformly convergent on any interval $[-b, b]$, where b is a (finite) real number; a case in point is the power series

$$1 + x + \frac{x^2}{2!} + \frac{x^3}{3!} + \cdots.$$

Proposition 7.70 (Abel's Limit Theorem). *We have the following:*

(i) *If $\sum_{k=0}^{\infty} a_k$ converges, then $\sum_{k=0}^{\infty} a_k x^k$ converges uniformly on $[0, 1]$.*
(ii) *If $\sum_{k=0}^{\infty} a_k$ converges to L and $f(x) = \sum_{k=0}^{\infty} a_k x^k$, then $\lim_{x \uparrow 1} f(x) = L$.*

PROOF. Given $\varepsilon > 0$ we may choose n_0 such that

$$\left| \sum_{k=m+1}^{n} a_k \right| < \varepsilon$$

provided that $m, n \geq n_0$. That is,

$$-\varepsilon < a_{m+1} + a_{m+2} + \cdots + a_n < \varepsilon$$

for $m, n \geq n_0$. If $0 \leq x \leq 1$, then Abel's Lemma (see Proposition 7.40), applied to $\{a_k\}_{k=m+1}^{\infty}$ and $\{x^k\}_{k=m+1}^{\infty}$, gives

$$-\varepsilon x^{m+1} < a_{m+1} x^{m+1} + a_{m+2} x^{m+2} + \cdots + a_n x^n < \varepsilon x^{m+1} \qquad (7.80)$$

for $m, n \geq n_0$ and $0 \leq x \leq 1$. If $f_n(x) = \sum_{k=0}^{\infty} a_k x^k$, for $0 \leq x \leq 1$, then (7.80) implies that $|f_n(x) - f_m(x)| < \varepsilon$ for $m, n \geq n_0$, and $0 \leq x \leq 1$. By Proposition 7.52, the claim under part (i) now follows.

To verify part (ii), we use the fact that $\sum_{k=0}^{\infty} a_k x^k$ converges uniformly on $[0, 1]$, established in part (i), and so f is continuous on $[0, 1]$ (by Proposition 7.54). Thus, the one-sided limit $\lim_{x \uparrow 1} f(x) = f(1) = L$. $\square$

Lemma. *Let $\{u_n\}_{n=1}^{\infty}$ and $\{v_n\}_{n=1}^{\infty}$ be sequences of positive real numbers, the first sequence being bounded above, and in the second $v_n \to 1$ as $n \to \infty$. Then*

$$\overline{\lim_{n \to \infty}} \, u_n v_n = \overline{\lim_{n \to \infty}} \, v_n. \tag{7.81}$$

PROOF. It is clear that the sequence $\{u_n v_n\}_{n=1}^{\infty}$ is a positive sequence bounded above. If possible, let

$$\overline{\lim_{n \to \infty}} \, u_n v_n = s' > s = \overline{\lim_{n \to \infty}} \, u_n.$$

Take $2\varepsilon = s' - s$, and let $K = \sup\{u_n : n = 1, 2, \ldots\}$. Since $v_n \to 1$, as $n \to \infty$, there is a positive integer k_1 such that

$$|v_n - 1| < \frac{\varepsilon}{2K} \quad \text{for } n \geq k_1.$$

Thus,

$$|u_n v_n - u_n| = |u_n||v_n - 1| < K \frac{\varepsilon}{2K} = \frac{\varepsilon}{2} \quad \text{for } n \geq k_1.$$

Therefore,

$$u_n v_n < u_n + \frac{\varepsilon}{2} \quad \text{for } n \geq k_1.$$

But, since $\overline{\lim}_{n \to \infty} u_n = s$, there is a positive integer k_2 such that

$$u_n < s + \frac{\varepsilon}{2} \quad \text{for } n \geq k_2.$$

Therefore, $u_n v_n < s + \varepsilon$ for $n \geq k$, where $k = \max\{k_1, k_2\}$. But, since

$$\overline{\lim_{n \to \infty}} \, u_n v_n = s',$$

we know that $u_n v_n > s' - \varepsilon$, for an infinite number of values of n. Therefore, s' can not be larger than s.

Similarly, it can be shown that s' is not less than s; hence, $s' = s$. $\square$

Proposition 7.71. *The intervals of convergence of the two power series*

$$a_0 + a_1 x + a_2 x^2 + a_3 x^3 + \cdots \quad \text{and} \quad a_1 + 2a_1 x + 3a_3 x^2 + \cdots$$

are the same.

PROOF. Since $\sqrt[n]{n} \to 1$ as $n \to \infty$ (by the Lemma preceding Proposition 1.8 in Chapter 1), we have that

$$\varlimsup_{n \to \infty} \sqrt[n]{|a_n|} = \varlimsup_{n \to \infty} \sqrt[n]{n|a_n|}$$

by (7.81). The claim now follows by Proposition 7.68.   □

Proposition 7.72. *Let the power series $\sum_{n=0}^{\infty} a_n x^n$ have $(-R, R)$ for its interval of convergence and define*

$$f(x) = \sum_{n=0}^{\infty} a_n x^n \quad \text{for } -R < x < R.$$

Then

(i) $$\int_{x_0}^{x} f(t)\, dt = a_0(x - x_0) + \sum_{n=1}^{\infty} \frac{a_n}{n+1}(x^{n+1} - x_0^{n+1}),$$

   for $-R < x_0 < x < R$;

(ii) $$f'(x) = a_1 + 2a_2 x + 3a_3 x^2 + \cdots,$$

   where x is any point in the open interval $(-R, R)$;

(iii) $$f^{(k)}(x) = \sum_{n=k}^{\infty} n(n-1)\cdots(n-k+1)a_n x^{n-k}$$

   for any $x \in (-R, R)$ and

$$f^{(k)}(0) = k!\, a_k, \tag{7.82}$$

   where $k = 0, 1, 2, \ldots$.

PROOF. Part (i) follows from Propositions 7.67 and 7.57. Part (ii) is a consequence of Propositions 7.71 and 7.56. Part (iii) is a corollary to part (ii). This completes the proof.   □

REMARKS. Formula (7.82) shows that the coefficients of the power series development of f are determined by the value of f and of its derivatives at a single point. Also, if the coefficients are given, the values of the derivatives of f at the center of the interval of convergence can be read off immediately from the power series. Note, however, that although a function f may have derivatives of all orders, the series $\sum_{n=0}^{\infty} a_n x^n$, where a_n is calculated by (7.82), need not converge to $f(x)$ for any $x \neq 0$. In this case, f can not be developed in a power series of the form $f(x) = \sum_{n=0}^{\infty} c_n x^n$. For if we had $f(x) = \sum_{n=0}^{\infty} c_n x^n$, then

$$n!\, c_n = f^{(n)}(0)$$

would follows, implying that $c_n = a_n$. The function

$$f(x) = e^{-1/x^2} \quad \text{for } x \neq 0,$$

$$= 0 \qquad \text{for } x = 0$$

is a case in point; it has derivatives of all orders at $x = 0$ and $f^{(n)}(0) = 0$ for $n = 1, 2, \ldots$ (see Example 8 following Proposition 4.10 in Chapter 4).

Proposition 7.73 (Uniqueness Theorem for Power Series). *If* $\sum_{n=0}^{\infty} a_n x^n$ *and* $\sum_{n=0}^{\infty} c_n x^n$ *converge on some interval* $(-r, r)$, $r > 0$, *to the same function* f, *then*

$$a_n = c_n \quad \text{for } n = 0, 1, 2, \ldots .$$

PROOF. By (7.82) we have $n! \, a_n = f^{(n)}(0) = n! \, c_n$ for $n = 0, 1, 2, \ldots$. □

Proposition 7.74 (Multiplication Theorem for Power Series). *If* f *and* g *are given on the open interval* $(-r, r)$ *by the power series*

$$f(x) = \sum_{n=0}^{\infty} a_n x^n \quad \text{and} \quad g(x) = \sum_{n=0}^{\infty} b_n x^n,$$

then their product fg *is given on this interval by the series* $\sum_{n=0}^{\infty} c_n x^n$, *where the coefficients* $\{c_n\}$ *are*

$$c_n = \sum_{k=0}^{n} a_k b_{n-k} \quad \text{for } n = 0, 1, 2, \ldots .$$

PROOF. We have seen in Proposition 7.67 that if $|x| < r$, then the series giving $f(x)$ and $g(x)$ are absolutely convergent. If we apply Proposition 7.50, we obtain the desired conclusion. □

DISCUSSION. Consider the power series

$$\sum_{n=0}^{\infty} a_n x^n \tag{7.83}$$

and

$$\sum_{n=0}^{\infty} b_n x^n \tag{7.84}$$

having radii of convergence different from zero, and suppose that the smaller of these is denoted by r. As we know already, these series can be added, subtracted, and multiplied; the resulting series are convergent for $|x| < r$ and can be written as power series:

$$\sum_{n=0}^{\infty} a_n x^n \pm \sum_{n=0}^{\infty} b_n x^n = \sum_{n=0}^{\infty} (a_n \pm b_n) x^n,$$

$$\sum_{n=0}^{\infty} a_n x^n \cdot \sum_{n=0}^{\infty} b_n x^n = \sum_{n=0}^{\infty} (a_0 b_n + a_1 b_{n-1} + a_2 b_{n-2} + \cdots + a_n b_0) x^n. \tag{7.85}$$

Assuming that series (7.84) is identical with series (7.83), we see that a power series may be multiplied by itself in the interior of its interval of convergence as often as we please; we get

$$\left(\sum_{n=0}^{\infty} a_n x^n \right)^2 = \sum_{n=0}^{\infty} (a_0 a_n + a_1 a_{n-1} + a_2 a_{n-2} + \cdots + a_n a_0) x^n,$$

and generally, for every positive integral exponent m,

$$\left(\sum_{n=0}^{\infty} a_n x^n \right)^m = \sum_{n=0}^{\infty} a_n^{(m)} x^n, \tag{7.86}$$

where the coefficient $a_n^{(m)}$ depends on the coefficients $a_0, a_1, \ldots, a_n$ of the initial series (7.83), and comes about, as (7.85) shows, by addition and multiplication. Moreover, the series

$$\sum_{n=0}^{\infty} a_n^{(m)} x^n \quad \text{for } m = 2, 3, \ldots$$

are absolutely convergent as long as $\sum_{n=0}^{\infty} a_n x^n$ itself is.

We now consider the substitution of one power series into another power series. Let $y = f(x)$ be a function which can be represented by the power series $\sum_{n=0}^{\infty} a_n x^n$ in the open interval $(-R, R)$. Moreover, let $z = g(y)$ be a function which can be represented by the power series $\sum_{m=0}^{\infty} b_m y^m$ in the open interval $(-r, r)$.

If $|a_0| = |f(0)| < r$, then $|f(x)|$ is smaller than r for sufficiently small x (because f is differentiable and hence continuous on its interval of convergence) and so the functional composition $z = g[f(x)]$ has meaning for at least all those x satisfying $|x| < R$ and $\sum_{n=0}^{\infty} |a_n| |x|^n < r$.

Under the single condition that $|a_0| < r$ we can write the function $z = g[f(x)]$ in a neighborhood of the point $x = 0$ as a power series in x, if in the series $\sum_{m=0}^{\infty} b_m y^m$ we substitute for y the power series $\sum_{n=0}^{\infty} a_n x^n$ and then rearrange the terms in increasing powers of x. The details of proof for this result will be taken up in the next proposition.

Proposition 7.75 (Substitution Theorem for Power Series). *Let*

$$f(x) = \sum_{n=0}^{\infty} a_n x^n \quad \text{for } |x| < R,$$

$$g(y) = \sum_{m=0}^{\infty} b_m y^m \quad \text{for } |y| < r,$$

and

$$|a_0| < r.$$

Then the function $F(x) = g[f(x)]$ is defined and representable as a power series $\sum_{n=0}^{\infty} A_n x^n$ for at least all those x for which

$$|x| < R \quad \text{and} \quad \sum_{n=0}^{\infty} |a_n| |x|^n < r. \tag{7.87}$$

This is certainly the case for some neighborhood of $x = 0$.

PROOF. The last claim, namely, that (7.87) certainly holds for all x in a sufficiently small neighborhood of $x = 0$, follows simply for reasons of continuity:

$$\sum_{n=0}^{\infty} |a_n||x|^n \to |a_0| < r \quad \text{as } x \to 0.$$

By repeated multiplication of the power series $\sum_{n=0}^{\infty} a_n x^n$ with itself we get

$$\left(\sum_{n=0}^{\infty} a_n x^n \right)^m = \sum_{n=0}^{\infty} a_n^{(m)} x^n \qquad (7.88)$$

with

$$a_n^{(m)} = \sum_{k_1+k_2+\cdots+k_m=n} a_{k_1} a_{k_2} \cdots a_{k_m}, \qquad (7.89)$$

where the symbol $\sum_{k_1+k_2+\cdots+k_m=n}$ indicates that the summation is to be extended over all nonnegative integers $k_1, k_2, \ldots, k_m$ whose sum is n. We also adopt the notation in (7.88) for $m = 0$ and $m = 1$. In the same way we have

$$\left(\sum_{n=0}^{\infty} |a_n||x|^n \right)^m = \sum_{n=0}^{\infty} \alpha_n^{(m)} |x|^n$$

with

$$\alpha_n^{(m)} = \sum_{k_1+k_2+\cdots+k_m=n} |a_{k_1}||a_{k_2}|\cdots|a_{k_m}|.$$

Since $\alpha_n^{(m)}$ comes about by addition and multiplication of $|a_0|, |a_1|, \ldots, |a_n|$ in the same way as $a_n^{(m)}$ comes about by addition and multiplication of $a_0, a_1, \ldots, a_n$, it is clear that $|a_n^{(m)}| \le \alpha_n^{(m)}$. For brevity we put

$$\sum_{n=0}^{\infty} |a_n||x|^n = \rho$$

and assume that $\rho < r$. We have for every integer $M \ge 0$:

$$\sum_{n=0}^{M} |a_n^{(m)}||x|^n \le \sum_{n=0}^{M} \alpha_n^{(m)} |x|^n \le \sum_{n=0}^{\infty} \alpha_n^{(m)} |x|^n = \rho^m,$$

$$\sum_{m,n=0}^{\infty} |b_m||a_n^{(m)}||x|^n \le \sum_{m=0}^{M} |b_m|\rho^m \le \sum_{m=0}^{\infty} |b_m|\rho^m.$$

By the Main Rearrangement Theorem (see Proposition 7.49) it therefore follows that

$$F(x) = \sum_{m=0}^{\infty} \sum_{n=0}^{\infty} b_m a_n^{(m)} x^n = \sum_{n=0}^{\infty} \sum_{m=0}^{\infty} b_m a_n^{(m)} x^n.$$

We therefore have to set

$$A_n = \sum_{m=0}^{\infty} b_m a_n^{(m)}.$$

This completes the proof. □

COMMENT. Particularly important is the case $a_0 = 0$. Then the coefficients A_n are computed from the coefficients a_n and b_m in terms of finite sums rather than infinite series. In fact, if $a_0 = 0$, then $a_n^{(m)} = 0$ for all $m > n$, as can be seen from the definition of $a_n^{(m)}$ in (7.89) (in any system $k_1, k_2, \ldots, k_m$ of non-negative integers whose sum equals n with $n < m$, there must be at least one 0). Thus,

$$A_n = \sum_{m=0}^{n} b_m a_n^{(m)} \quad \text{(in the case } a_0 = 0\text{)}.$$

Proposition 7.76 (Taylor's Expansion Theorem). *Suppose that*

$$f(x) = \sum_{n=0}^{\infty} c_n x^n$$

with the power series converging for $|x| < R$. If $-R < a < R$, then

$$f(x) = \sum_{n=0}^{\infty} \frac{f^{(n)}(a)}{n!} (x - a)^n$$

for all x such that $|x - a| < R - |a|$ and we say that f can be expanded in a power series about the point $x = a$ which converges in $|x - a| < R - |a|$.

PROOF. We observe that

$$f(x) = \sum_{n=0}^{\infty} c_n [a + (x - a)]^n = \sum_{n=0}^{\infty} c_n \sum_{m=0}^{n} \binom{n}{m} a^{n-m} (x - a)^m$$

$$= \sum_{m=0}^{\infty} \left(\sum_{n=m}^{\infty} \binom{n}{m} a^{n-m} \right) (x - a)^m.$$

This is the desired expansion about the point $x = a$. To see its validity, we note that we have used Proposition 7.75 and merely substituted into the power series $\sum_{n=0}^{\infty} c_n x^n$ the simple power series $a + (x - a)$ with the variable of expansion $(x - a)$; the resulting expansion will be valid for at least those x which satisfy

$$|a| + |x - a| < R.$$

Finally, the form of the coefficients in (7.90) follows from part (iii) of Proposition 7.72. $\qquad\square$

DISCUSSION. An important example for the application of the Substitution Theorem for Power Series (see Proposition 7.75) is the *division of power series*.

Let the term a_0 of the power series $\sum_{n=0}^{\infty} a_n x^n$ be different from zero; we can then write $\sum_{n=0}^{\infty} a_n x^n$ in the form

$$a_0 \left(1 + \frac{a_1}{a_0} x + \frac{a_2}{a_0} x^2 + \cdots + \frac{a_n}{a_0} x^n + \cdots \right) = a_0 (1 + y)$$

provided we set

$$y = \frac{a_1}{a_0} x + \frac{a_2}{a_0} x^2 + \cdots + \frac{a_n}{a_0} x^n + \cdots.$$

Thus,

$$\frac{1}{a_0 + a_1 x + \cdots + a_n x^n + \cdots} = \frac{1}{a_0} \frac{1}{1 + y}$$

$$= \frac{1}{a_0} (1 - y + y^2 - \cdots + (-1)^m y^m + \cdots).$$

The last series plays the role of the series $\sum_{m=0}^{\infty} b_m y^m$ appearing in Proposition 7.75; here $r = 1$. By Proposition 7.75, the expression under consideration can be expanded in powers of x:

$$\frac{1}{a_0 + a_1 x + \cdots + a_n x^n + \cdots} = c_0 + c_1 x + \cdots + c_n x^n + \cdots$$

holds at least sufficiently small x, for example, for those which satisfy the inequality

$$\left| \frac{a_1}{a_0} \right| \cdot |x| + \left| \frac{a_2}{a_0} \right| \cdot |x|^2 + \cdots + \left| \frac{a_n}{a_0} \right| \cdot |x|^n + \cdots < 1.$$

We consider a second power series

$$h_0 + h_1 x + h_2 x^2 + \cdots + h_n x^n + \cdots$$

and suppose that it has a radius of convergence different from zero. Then the quotient

$$\frac{h_0 + h_1 x + \cdots + h_n x^n + \cdots}{a_0 + a_1 x + \cdots + a_n x^n + \cdots}$$

can be replaced by the product

$$(h_0 + h_1 x + \cdots + h_n x^n + \cdots)(c_0 + c_1 x + \cdots + c_n x^n + \cdots)$$

for sufficiently small x and can therefore be represented as a power series

$$d_0 + d_1 x + d_2 x^2 + \cdots + d_n x^n + \cdots.$$

The coefficients of this series can be determined most conveniently by commencing with the relation

$$(a_0 + a_1 x + \cdots + a_n x^n + \cdots)(d_0 + d_1 x + \cdots + d_n x^n + \cdots)$$

$$= h_0 + h_1 x + \cdots + h_n x^n + \cdots$$

in which the coefficients a_i and h_k are known. Multiplying the power series on the left-hand side of the foregoing equation by the general rule

$$\sum_{n=0}^{\infty} a_n x^n \cdot \sum_{n=0}^{\infty} d_n x^n = \sum_{n=0}^{\infty} (a_0 d_n + a_1 d_{n-1} + a_2 d_{n-2} + \cdots + a_n d_0) x^n$$

and comparing coefficients of like powers of x, we get the infinite system of equations

$$a_0 d_0 = h_0,$$

$$a_0 d_1 + a_1 d_0 = h_1,$$

$$a_0 d_2 + a_1 d_1 + a_2 d_0 = h_2,$$

$$\vdots$$

$$a_0 d_n + a_1 d_{n-1} + \cdots + a_{n-1} d_1 + a_n d_0 = h^n,$$

$$\vdots$$

Since $a_0 \neq 0$, the first equation yields $d_0 = h_0/a_0$, the second equation yields

$$d_1 = \frac{h_1 - a_1 d_0}{a_0} = \frac{a_0 h_1 - a_1 h_0}{a_0^2},$$

and so on. After having found the n coefficients $d_0, d_1, \ldots, d_{n-1}$, the $(n + 1)$th equation which contains the single unknown d_n puts us into a position to determine the value of d_n. In this manner we can determine (in fact, uniquely) the coefficients of the power series representing the quotient (7.91).

7. Some Important Power Series

The following series, convergent to the given function in the indicated intervals, are frequently employed in practice.

1. $(a + x)^m = a^m + m a^{m-1} x + \cdots + \dfrac{m(m - 1) \cdots (m - n + 1)}{n!} a^{m-n} x^n + \cdots$,

   where $-|a| < x < |a|$;

2. $a^n = 1 + \dfrac{x(\ln a)}{1!} + \dfrac{x^2 (\ln a)^2}{2!} + \cdots + \dfrac{x^n (\ln a)^n}{n!} + \cdots$,

   where $-\infty < x < +\infty$;

3. $e^x = 1 + x + \dfrac{x^2}{2!} + \cdots + \dfrac{x^n}{n!} + \cdots$,   where $-\infty < x < +\infty$;

4. $\sin x = x - \dfrac{x^3}{3!} + \cdots + (-1)^n \dfrac{x^{2n+1}}{(2n + 1)!} + \cdots$,   where $-\infty < x < +\infty$;

5. $\cos x = 1 - \dfrac{x^2}{2!} + \dfrac{x^4}{4!} - \cdots + (-1)^n \dfrac{x^{2n}}{(2n)!} + \cdots$,   where $-\infty < x < +\infty$;

6. $\tan x = x + \dfrac{x^3}{3} + \dfrac{2x^5}{15} + \dfrac{17x^7}{315} + \dfrac{62x^9}{2835} + \cdots$,   where $-\dfrac{\pi}{2} < x < \dfrac{\pi}{2}$;

7. $\cot x = \dfrac{1}{x} - \left(\dfrac{x}{3} + \dfrac{x^3}{45} + \dfrac{2x^5}{945} + \dfrac{x^7}{4725} + \cdots \right),$

where $-\pi < x < \pi$ except for $x = 0$;

8. $\sin^{-1} x = x + \dfrac{1 \cdot x^3}{2 \cdot 3} + \dfrac{1 \cdot 3 \cdot x^5}{2 \cdot 4 \cdot 5} + \dfrac{1 \cdot 3 \cdot 5 \cdot x^7}{2 \cdot 4 \cdot 6 \cdot 7} + \cdots$

$\quad\quad + \dfrac{1 \cdot 3 \cdot 5 \cdots (2n - 1)x^{2n+1}}{2 \cdot 4 \cdot 6 \cdots (2n)(2n + 1)} + \cdots, \quad$ where $-1 < x < 1$;

9. $\tan^{-1} x = x - \dfrac{x^3}{3} + \dfrac{x^5}{5} - \dfrac{x^7}{7} + \cdots + (-1)^n \dfrac{x^{2n+1}}{2n + 1} + \cdots,$

where $-1 \le x \le 1$;

10. $\sinh x = x + \dfrac{x^3}{3!} + \cdots + \dfrac{x^{2n+1}}{(2n + 1)!} + \cdots, \quad$ where $-\infty < x < +\infty$;

11. $\cosh x = 1 + \dfrac{x^2}{2!} + \dfrac{x^4}{4!} + \cdots + \dfrac{x^{2n}}{(2n)!} + \cdots, \quad$ where $-\infty < x + \infty$;

12. $\tanh x = x - \dfrac{x^3}{3} + \dfrac{2x^5}{15} - \dfrac{17x^7}{315} + \cdots, \quad$ where $-\dfrac{\pi}{2} < x < \dfrac{\pi}{2}$;

13. $\coth x = \dfrac{1}{x} + \dfrac{x}{3} - \dfrac{x^3}{45} + \dfrac{2x^5}{945} - \dfrac{x^7}{4725} + \cdots,$

where $-\pi < x < \pi$ except for $x = 0$;

14. $\sinh^{-1} x = x - \dfrac{x^3}{2 \cdot 3} + \dfrac{1 \cdot 3x^5}{2 \cdot 4 \cdot 5} - \dfrac{1 \cdot 3 \cdot 5x^7}{2 \cdot 4 \cdot 6 \cdot 7} + \cdots$

$\quad\quad + (-1)^n \dfrac{1 \cdot 3 \cdot 5 \cdots (2n - 1)x^{2n+1}}{2 \cdot 4 \cdot 6 \cdots (2n)(2n + 1)} + \cdots, \quad$ where $-1 \le x \le 1$;

15. $\tanh^{-1} x = x + \dfrac{x^3}{3} + \dfrac{x^5}{5} + \cdots + \dfrac{x^{2n+1}}{2n + 1} + \cdots, \quad$ where $-1 < x < 1$;

16. $\ln(1 + x) = x - \dfrac{x^2}{2} + \dfrac{x^3}{3} - \cdots + (-1)^n \dfrac{x^n}{n} + \cdots, \quad$ where $-1 < x \le 1$;

17. $\ln x = 2\left(\dfrac{x + 1}{x + 1} + \dfrac{1}{3}\left(\dfrac{x - 1}{x + 1} \right)^3 + \dfrac{1}{5}\left(\dfrac{x - 1}{x + 1} \right)^5 + \cdots \right.$

$\quad\quad \left. + \dfrac{1}{2n + 1}\left(\dfrac{x - 1}{x + 1} \right)^{2n+1} + \cdots \right), \quad$ where $x > 0$;

18. $\ln \dfrac{1 + x}{1 - x} = 2\left(x + \dfrac{x^3}{3} + \dfrac{x^5}{5} + \cdots + \dfrac{x^{2n+1}}{2n + 1} + \cdots \right), \quad$ where $-1 < x < 1$;

19. $\ln\dfrac{x+1}{x-1} = 2\left(\dfrac{1}{x} + \dfrac{1}{3x^3} + \dfrac{1}{5x^5} + \cdots + \dfrac{1}{(2n+1)x^{2n+1}} + \cdots\right),$

where $x < -1$ or $x > 1$;

20. $\sinh x + \sin x = 2\left(\dfrac{x}{1} + \dfrac{x^5}{5!} + \dfrac{x^9}{9!} + \cdots\right),\quad$ where $-\infty < x < +\infty$;

21. $\sinh x - \sin x = 2\left(\dfrac{x^3}{3!} + \dfrac{x^7}{7!} + \dfrac{x^{11}}{11!} + \cdots\right),\quad$ where $-\infty < x < +\infty$;

22. $\cosh x + \cos x = 2\left(1 + \dfrac{x^4}{4!} + \dfrac{x^8}{8!} + \cdots\right),\quad$ where $-\infty < x < +\infty$;

23. $\cosh x - \cos x = 2\left(\dfrac{x^2}{2!} + \dfrac{x^6}{6!} + \dfrac{x^{10}}{10!} + \cdots\right),\quad$ where $-\infty < x < +\infty$;

23. $\tanh x + \tan x = 2\left(x + \dfrac{2x^5}{15} + \dfrac{62x^9}{2835} + \cdots\right),\quad$ where $-\infty < x < +\infty$;

25. $\tanh x - \tan x = -2\left(\dfrac{x^3}{3} + \dfrac{17x^7}{315} + \dfrac{1382x^{11}}{155925} + \cdots\right),$

where $-\infty < x < +\infty$.

DISCUSSION. We shall now indicate how some of the foregoing expansions come about. In this connection, the matters presented in Section 2 of Chapter 4 are of particular interest.

Let f be a function having derivatives of any order in a neighborhood of the point $x = 0$. We consider the numbers

$$a_k = \frac{1}{k!} f^{(k)}(0), \qquad k = 0, 1, 2, \ldots$$

and formally write

$$s(x) = \sum_{k=0}^{\infty} a_k x^k = \sum_{k=0}^{\infty} \frac{1}{k!} f^{(k)}(0)x^k, \tag{7.91}$$

ignoring for the moment the convergence of the series, that is, ignoring for the moment the existence of the sum function s. By Proposition 4.11 we have

$$f(x) = \sum_{k=0}^{n} a_k x^k + R_n, \tag{7.92}$$

where $\sum_{k=0}^{n} a_k x^k$ is the Taylor polynomial of order n for f at the point 0 and R_n denotes the remainder (which is to be formed in terms of the function f). Comparison of (7.91) and (7.92) gives the following:

If for $|x| < \rho$, $\rho > 0$, we have

$$\lim_{n\to\infty} R_n = 0, \tag{7.93}$$

then

$$s(x) \equiv f(x),$$

that is, the power series

$$\sum_{k=0}^{\infty} a_k x^k \quad \text{with} \quad a_k = \frac{1}{k!} f^{(k)}(0)$$

represents on its interval of convergence the function f.

Indeed, by Proposition 7.73, the power series representation is unique. To see that condition (7.93) is not just necessary, but also sufficient for the convergence of the series (7.91), we note that (7.93) means that for any $\varepsilon > 0$ there is some $N > 0$ such that $|R_n| < \varepsilon$ provided $n > N$. But then

$$|a_{n+1} x^{n+1} + \cdots + a_{n+p} x^{n+p}| = |R_n - R_{n+p}| \le |R_n| + |R_{n+p}| < 2\varepsilon$$

for all $n > N$ and for arbitrary $p > 0$.

The series

$$f(x) = \sum_{k=0}^{\infty} \frac{1}{k!} f^{(k)}(0) x^k \tag{7.94}$$

is called the *Maclaurin series of f.* More generally, the series

$$f(x) = \sum_{k=0}^{\infty} \frac{1}{k!} f^{(k)}(x_0)(x - x_0)^k \tag{7.95}$$

is called the *Taylor series of f at the point* x_0. In the latter case of course we are dealing with a function having derivatives of any order in a neighborhood of the point $x = x_0$.

We note that condition (7.93) is not only necessary and sufficient for the convergence of the series (7.94) and (7.95), but also that *the given function f is represented by the series.*

A quick check of Section 2 of Chapter 4 will show that we have already proved a number of power series representations of elementary functions listed at the beginning of this section by proving that the remainder R_n tends to 0 with $n \to \infty$.

An efficient way to get the Taylor and Maclaurin series for elementary functions is to obtain a set of basic expansions only and then use the manipulative techniques of the last section to derive other expansions from the basic ones. For example, the series

$$\frac{1}{1+t} = 1 - t + t^2 - t^3 + \cdots$$

has radius of convergence $R = 1$. Therefore, for $|x| < 1$, integration from 0 to x gives

$$\ln(1 + x) = x - \frac{x^2}{2} + \frac{x^3}{3} - \cdots \quad \text{for } -1 < x < 1$$

and Proposition 7.70 gives that

$$\ln(1 + x) = x - \frac{x^2}{2} + \frac{x^3}{3} - \cdots \quad \text{for } -1 < x \le 1.$$

As a second example, note that the series

$$\frac{1}{1 + t^2} = 1 - t^2 + t^4 - t^6 + \cdots$$

has radius of convergence $R = 1$ and so, for $|x| < 1$, integration from 0 to x gives

$$\tan^{-1} x = x - \frac{x^3}{3} + \frac{x^5}{5} - \frac{x^7}{7} + \cdots \quad \text{for } -1 < x < 1.$$

By Proposition 7.70 we therefore get

$$\tan^{-1} x = x - \frac{x^3}{3} + \frac{x^5}{5} - \frac{x^7}{7} + \cdots \quad \text{for } -1 \le x \le 1.$$

As a third example, note that the series

$$(1 + t)^{-1/2} = 1 - \frac{1}{2}t + \frac{1 \cdot 3}{2 \cdot 4}t^2 - \frac{1 \cdot 3 \cdot 5}{2 \cdot 4 \cdot 6}t^3 + \cdots$$

has radius of convergence $R = 1$. Therefore, the same is true for

$$(1 - t^2)^{-1/2} = 1 + \frac{1}{2}t^2 + \frac{1 \cdot 3}{2 \cdot 4}t^4 + \frac{1 \cdot 3 \cdot 5}{2 \cdot 4 \cdot 6}t^6 + \cdots.$$

From this, by integration from 0 to x, where $|x| < 1$, we find

$$\sin^{-1} x = x + \frac{1}{2}\frac{x^3}{3} + \frac{1 \cdot 3}{2 \cdot 4}\frac{x^5}{5} + \frac{1 \cdot 3 \cdot 5}{2 \cdot 4 \cdot 6}\frac{x^7}{7} + \cdots \quad \text{for } -1 < x < 1.$$

Now, applying Proposition 7.70, we get

$$\sin^{-1} x = x + \frac{1}{2}\frac{x^3}{3} + \frac{1 \cdot 3}{2 \cdot 4}\frac{x^5}{5} + \frac{1 \cdot 3 \cdot 5}{2 \cdot 4 \cdot 6}\frac{x^7}{7} + \cdots \quad \text{for } -1 \le x \le 1.$$

As a fourth example we show that

$$\tan x = x + \frac{x^3}{3} + \frac{2x^5}{15} + \frac{17x^7}{315} + \cdots.$$

Indeed, let

$$\tan x = a_1 x + a_3 x^3 + a_5 x^5 + a_7 x^7 + \cdots,$$

noting that $f(x) = \tan x$ is an odd function, that is, $f(-x) = -f(x)$. Now, differentiation and the use of the identity

$$\sec^2 x = 1 + \tan^2 x$$

give

$$a_1 + 3a_3 x^2 + 5a_5 x^4 + 7a_7 x^6 + 9a_9 x^8 + \cdots$$
$$= 1 + [a_1 x + a_3 x^3 + a_5 x^5 + a_7 x^7 + \cdots]^2$$
$$= 1 + a_1^2 x^2 + 2a_1 a_3 x^4 + (2a_1 a_5 + a_3^2)x^6 + \cdots.$$

Equating coefficients of like powers products the recursion formulas

$$a_1 = 1,$$
$$3a_3 = a_1^2,$$
$$5a_5 = 2a_1 a_3,$$
$$7a_7 = 2a_1 a_5 + a_3^2,$$

and so on. Hence, $a_1 = 1$, $a_3 = 1/3$, $a_5 = 2/15$, $a_7 = 17/315$, and so on.

8. Miscellaneous Examples

EXAMPLE 1. We have

$$I = \int_0^1 \frac{\ln(1 + x)}{x} dx = \frac{\pi^2}{12}.$$

Indeed,

$$\frac{\ln(1 + x)}{x} = 1 - \frac{1}{2}x + \frac{1}{3}x^2 - \cdots + (-1)^{n-1}\frac{1}{n}x^{n-1} + \cdots$$

on the entire interval $[0, 1]$. Termwise integration gives

$$I = 1 - \frac{1}{2^2} + \frac{1}{3^2} - \cdots + (-1)^{n-1}\frac{1}{n^2} + \cdots = \sum_{n=1}^{\infty} (-1)^{n-1}\frac{1}{n^2}.$$

But

$$\sum_{n=1}^{\infty} \frac{1}{n^2} = \frac{\pi^2}{6}$$

(see Example 3 of Section 4 in Chapter 1) and so

$$I = \sum_{n=1}^{\infty} \frac{(-1)^{n-1}}{n^2} = \sum_{n=1}^{\infty} \frac{1}{n^2} - 2\sum_{n=1}^{\infty} \frac{1}{(2n)^2} = \frac{\pi^2}{12}.$$

EXAMPLE 2. We have

$$\sum_{n=0}^{\infty} \frac{(-1)^n}{3n + 1} = \frac{1}{3}(\ln 2) + \frac{\pi}{3\sqrt{3}}.$$

Indeed,

$$\sum_{n=0}^{\infty} \frac{(-1)^n}{3n+1} = \lim_{x \to 1-0} \sum_{n=0}^{\pi} \frac{(-1)^n}{3n+1} x^{3n+1}$$

$$= \lim_{x \to 1-0} \int_0^x \sum_{n=0}^{\infty} (-1)^n x^{3n} \, dx = \lim_{x \to 1-0} \int_0^x \frac{1}{1+x^3} \, dx$$

$$= \lim_{x \to 1-0} \left(\frac{1}{6} \ln \frac{(x+1)^2}{x^2-x+1} + \frac{1}{\sqrt{3}} \tan^{-1} \frac{2x-1}{\sqrt{3}} + \frac{\pi}{6\sqrt{3}} \right)$$

$$= \frac{1}{3} (\ln 2) + \frac{\pi}{3\sqrt{3}}.$$

EXAMPLE 3. To sum the series

$$\sum_{n=0}^{\infty} \frac{(-1)^n x^{2n}}{(n+1)(n+3)}$$

we start with the known series

$$\ln(1+x) = \sum_{n=1}^{\infty} \frac{(-1)^{n+1} x^n}{n} = \sum_{n=0}^{\infty} \frac{(-1)^n x^{n+1}}{n+1}.$$

Multiplying through by x, we obtain

$$x \ln(1+x) = \sum_{n=0}^{\infty} \frac{(-1)^n x^{n+3}}{n+1}.$$

Integration yields

$$\int \ln[x(1+x)] \, dx = C + \sum_{n=0}^{\infty} \frac{(-1)^n x^{n+3}}{(n+1)(n+3)}$$

or

$$\frac{1}{2}(x^2-1)\ln(1+x) - \frac{1}{4}(x-1)^2 = C + \sum_{n=0}^{\infty} \frac{(-1)^n x^{n+3}}{(n+1)(n+3)}.$$

Putting $x = 0$, we find that $C = -\frac{1}{4}$ and so

$$\frac{1}{2}(x^2-1)\ln(1+x) - \frac{1}{4}x^2 + \frac{1}{2}x = \sum_{n=0}^{\infty} \frac{(-1)^n x^{n+3}}{(n+1)(n+3)}.$$

Dividing by x^3, excluding $x = 0$, gives

$$\frac{x^2-1}{2x^3} \ln(1+x) - \frac{1}{4x} + \frac{1}{2x^2} = \sum_{n=0}^{\infty} \frac{(-1)^n x^n}{(n+1)(n+3)};$$

and, finally, replacing x by x^2, we obtain

$$\frac{x^4-1}{2x^6} \ln(1+x^2) - \frac{1}{4x^2} + \frac{1}{2x^4} = \sum_{n=0}^{\infty} \frac{(-1)^n x^{2n}}{(n+1)(n+3)} \quad \text{for } 0 < |x| < 1,$$

and we have what we set out to do. Note also the result that the left member, near $x = 0$, is an indeterminate form, as $x \to 0$, with limit $\frac{1}{3}$ (the $n = 0$ term on the right).

EXAMPLE 4. If s_r denotes the sum of the squares of the first r positive integers, then

$$\frac{s_1}{1!} + \frac{s_2}{2!} + \cdots + \frac{s_n}{n!} + \cdots = \frac{17e}{6}.$$

Indeed, since (see Example 4 of Section 4 in Chapter 1)

$$1^2 + 2^2 + \cdots + n^2 = \tfrac{1}{6}n(n + 1)(2n + 1),$$

we see that

$$\sum_{n=1}^{\infty} \frac{s_n}{n!} = \frac{1}{6} \sum_{n=1}^{\infty} \frac{n(n + 1)(2n + 1)}{n!}$$

and therefore it will be sufficient to show that

$$\sum_{n=1}^{\infty} \frac{n(n + 1)(2n + 1)}{n!} = 17e.$$

We do this now by starting with

$$e^x = \sum_{n=0}^{\infty} \frac{x^n}{n!}.$$

Multiplying both sides by x we get

$$xe^x = \sum_{n=0}^{\infty} \frac{x^{n+1}}{n!}.$$

Next, we differentiate twice with respect to x and obtain

$$e^x(2 + x) = \sum_{n=0}^{\infty} \frac{n(n + 1)x^{n-1}}{n!}.$$

Replacing x by x^2, we obtain

$$e^{x^2}(2 + x^2) = \sum_{n=0}^{\infty} \frac{n(n + 1)x^{2n-2}}{n!}.$$

Multiplying both sides of the foregoing equation by x^3 gives

$$e^{x^2}(2x^3 + x^5) = \sum_{n=0}^{\infty} \frac{n(n + 1)x^{2n+1}}{n!}.$$

Differentiation with respect to x gives

$$e^{x^2}(6x^2 + 9x^4 + 2x^6) = \sum_{n=0}^{\infty} \frac{n(n + 1)(2n + 1)x^{2n}}{n!}.$$

Setting $x = 1$, we get

$$17e = \sum_{n=0}^{\infty} \frac{n(n+1)(2n+1)}{n!} = \sum_{n=1}^{\infty} \frac{n(n+1)(2n+1)}{n!}$$

and we have obtained what we had set out to do.

EXAMPLE 5. We have

$$S = 1 - \frac{1}{2} - \frac{1}{3} + \frac{1}{4} + \frac{1}{5} - \frac{1}{6} - \frac{1}{7} + + - - \cdots = \frac{\pi}{4} - \frac{1}{2}(\ln 2).$$

Indeed, using the relation

$$\int_0^1 t^{n-1} \, dt = \frac{1}{n} \quad \text{for } n = 1, 2, 3, \ldots,$$

we see that

$$S = \int_0^1 (1 - t - t^2 + t^3 + t^4 - t^5 - t^6 + + - - \cdots) \, dt$$

$$= \int_0^1 [(1 - t^2 + t^4 - t^6 + \cdots) - (t - t^3 + t^5 - \cdots)] \, dt$$

$$= \int_0^1 \left(\frac{1}{1+t^2} - \frac{t}{1+t^2} \right) dt = \left(\tan^{-1} x - \frac{1}{2}\ln(1 + t^2) \right) \Big|_0^1$$

$$= \frac{\pi}{4} - \frac{1}{2}(\ln 2).$$

EXAMPLE 6. We have

$$H = \frac{1}{1 \cdot 2 \cdot 3} + \frac{1}{3 \cdot 4 \cdot 5} + \frac{1}{5 \cdot 6 \cdot 7} + \cdots = \ln 2 - \frac{1}{2}.$$

Indeed, using the identity

$$\frac{1}{k!} \int_0^1 t^{n-1}(1 - t)^k \, dt = \frac{1}{n(n+1)\cdots(n+k)},$$

with n and k denoting positive integers, we see that

$$H = \frac{1}{2!} \int_0^1 (1 - t)^2(1 + t^2 + t^4 + \cdots) \, dt = \frac{1}{2!} \int_0^1 \frac{(1-t)^2}{1 - t^2} \, dt$$

$$= \frac{1}{2!} \int_0^1 \frac{1-t}{1+t} \, dt = \frac{1}{2!} \int_1^2 \frac{2-s}{s} \, ds = \ln 2 - \frac{1}{2}.$$

EXAMPLE 7. We have

$$\sum_{n=1}^{\infty} \frac{1}{n(n+m)} = \frac{1}{m}\left(1 + \frac{1}{2} + \frac{1}{3} + \cdots + \frac{1}{m} \right)$$

because

$$\sum_{n=1}^{\infty} \frac{1}{(n+a)(n+b)} = \frac{1}{b-a} \int_0^1 \frac{x^a - x^b}{1-x} dx$$

for $a > -1$, $b > -1$, and $a \neq b$; taking $a = 0$ and $b = m$, we get the desired result on account of the relation

$$\frac{1-x^m}{1-x} = 1 + x + x^2 + \cdots + x^{m-1}.$$

EXAMPLE 8. We have

$$\int_0^1 \frac{x^{p-1}}{1+x^q} dx = \frac{1}{p} - \frac{1}{p+q} + \frac{1}{p+2q} - \frac{1}{p+3q} + \cdots,$$

where p and q are positive integers, because

$$\frac{x^{p-1}}{1+x^q} = x^{p-1}(1 - x^q + x^{2q} - x^{3q} + \cdots).$$

EXAMPLE 9. Let $x > 0$ and $x \neq 1$. Then

$$\frac{\ln x}{x-1} \le \frac{1}{x^{1/2}} \tag{7.96}$$

and

$$\frac{\ln x}{x-1} \le \frac{1+x^{1/3}}{x+x^{1/3}}. \tag{7.97}$$

Indeed, we start with the expansion

$$\ln \frac{1+t}{1-t} = 2\left(t + \frac{1}{3}t^3 + \frac{1}{5}t^5 + \frac{1}{7}t^7 + \cdots\right),$$

which holds if $|t| < 1$. To verify (7.96), we put $x = (1+t)^2/(1-t)^2$ and note that (7.96) becomes

$$\frac{1}{2t} \ln \frac{1+t}{1-t} - \frac{1}{1-t^2} \le 0 \quad \text{for } 0 < |t| < 1.$$

Using the power series expansions, we get

$$\sum_{n=1}^{\infty} \left(1 - \frac{1}{2n+1}\right) t^{2n} \ge 0 \quad \text{for } 0 \le |t| < 1,$$

which is evidently true.

To verify (7.97), we set $x = (1+t)^3/(1-t)^3$ and observe that (7.97) becomes

$$\frac{3}{2t} \ln \frac{1+t}{1-t} - \frac{t^2+3}{1-t^4} \le 0 \quad \text{for } 0 < |t| < 1.$$

Using power series expansions, we get

$$\sum_{n=0}^{\infty} t^{4n+2} + 3 \sum_{n=0}^{\infty} t^{4n} - 3 \sum_{n=0}^{\infty} \frac{t^{2n}}{2n+1} \ge 0 \quad \text{for } |t| < 1,$$

that is,

$$3 \sum_{n=0}^{\infty} \left(1 - \frac{1}{4n+1}\right) t^{4n} + \sum_{n=0}^{\infty} \left(1 - \frac{3}{4n+3}\right) t^{4n+2} \geq 0,$$

which is obviously true.

EXAMPLE 10. Let p and q be real numbers, and n be a positive integer. Then

$$\sum_{k=0}^{n} \binom{p}{k}\binom{q}{n-k} = \binom{p+q}{n}. \tag{7.98}$$

Indeed, we have $(1 + x)^p (1 + x)^q = (1 + x)^{p+q}$. If $|x| < 1$, then

$$(1 + x)^{p+q} = \sum_{n=0}^{\infty} \binom{p+q}{n} x^n$$

and

$$(1 + x)^p (1 + x)^q = \left(\sum_{n=0}^{\infty} \binom{p}{n} x^n\right)\left(\sum_{n=0}^{\infty} \binom{q}{n} x^n\right)$$

$$= \sum_{n=0}^{\infty} \left(\sum_{k=0}^{n} \binom{p}{k}\binom{q}{n-k}\right) x^n.$$

Since the two power series must be equal for $|x| < 1$, it follows that

$$\sum_{k=0}^{n} \binom{p}{k}\binom{q}{n-k} = \binom{p+q}{n}.$$

REMARK. Since

$$\binom{n}{k} = \binom{n}{n-k},$$

we have, by setting $p = q = n$ in (7.98),

$$\sum_{k=0}^{n} \binom{n}{k}^2 = \binom{2n}{k}.$$

EXAMPLE 11. We have

$$\cos^3 x - \frac{1}{3}\cos^3 3x + \frac{1}{3^2}\cos^3 3^2 x - \frac{1}{3^3}\cos^3 3^3 x + \cdots = \frac{3}{4}\cos x.$$

Indeed, we have the following succession of identities

$$4\cos^3 x = \cos 3x + 3\cos x,$$
$$4\cos^3 3x = \cos 3^2 x + 3\cos 3x,$$
$$4\cos^3 3^2 x = \cos 3^3 x + 3\cos 3^2 x,$$
$$\vdots$$
$$4\cos^3 3^n x = \cos 3^{n+1} x + 3\cos 3^n x.$$

If we multiply the first of these identities by 1, the second by $(-3)^{-1}$, the third by $3^{-2}, \ldots$, the nth by $(-3)^{-n}$, and denote by S_n the sum of the first n terms of the proposed series, we see that

$$4S_n = 3\cos x + (-3)^n \cos 3^{n+1} x.$$

Letting $n \to \infty$, we get that $S_n \to \frac{3}{4}\cos x$.

EXAMPLE 12. We have, for $|x| < 1$,

$$\frac{1}{1+x} + \frac{2x}{1+x^2} + \frac{4x^3}{1+x^4} + \frac{8x^7}{1+x^8} + \cdots = \frac{1}{1-x}.$$

Indeed, the nth partial sum of the series

$$\ln(1-x) + \ln(1+x) + \ln(1+x^2) + \ln(1+x^4) + \cdots$$

is equal to

$$\ln\{(1-x)(1+x)(1+x^2)\cdots(1+x^{2^{n-2}})\}$$
$$= \ln\{(1-x^2)(1+x^2)\cdots(1+x^{2^{n-2}})\}$$

$$= \cdots$$

$$= \ln(1-x^{2^{n-1}})$$

$$\to 0$$

as $n \to \infty$ for $|x| < 1$. Moreover, for $|x| \le \rho < 1$,

$$\left|\frac{2^n x^{2^n-1}}{1+x^{2^n}}\right| \le 2^n \rho^{2^n-1},$$

and the series $\sum 2^n \rho^{2^n}$ is convergent. Hence, the series

$$\frac{2x}{1+x^2} + \frac{4x^3}{1+x^4} + \frac{8x^7}{1+x^8} + \cdots$$

is uniformly convergent for $|x| \le \rho < 1$ and, by Proposition 7.56 (reformulated for series), its sum is the derivative of $-\ln(1-x) - \ln(1+x)$, that is, $1/(1-x) - 1/(1+x)$. The result required follows immediately.

EXAMPLE 13. (Theorem of Stolz). Let $\{x_n\}_{n=1}^{\infty}$ and $\{y_n\}_{n=1}^{\infty}$ be two sequences of real numbers. Suppose that $\{y_n\}_{n=1}^{\infty}$ is strictly increasing for all sufficiently large n and that $y_n \to \infty$ as $n \to \infty$. Then

$$\lim_{n\to\infty} \frac{x_n}{y_n} = \lim_{n\to\infty} \frac{x_n - x_{n-1}}{y_n - y_{n-1}} \tag{7.99}$$

provided that the limit on the right-hand side exists (be it finite or infinite).
  Indeed, we assume first that the limit is finite, that is,

$$\lim_{n\to\infty} \frac{x_n - x_{n-1}}{y_n - y_{n-1}} = r,$$

where r is a finite real number. Then, for any $\varepsilon > 0$, there exists a positive integer k such that, for $n \geq k$, we have

$$\left| \frac{x_n - x_{n-1}}{y_n - y_{n-1}} - r \right| < \frac{\varepsilon}{2} \quad \text{and} \quad y_n - y_{n-1} > 0.$$

We therefore see that all fractions

$$\frac{x_{k+1} - x_k}{y_{k+1} - y_k}, \quad \frac{x_{k+2} - x_{k+1}}{y_{k+2} - y_{k+1}}, \quad \ldots, \quad \frac{x_{n-1} - x_{n-2}}{y_{n-1} - y_{n-2}}, \quad \frac{x_n - x_{n-1}}{y_n - y_{n-1}}$$

are situated between $r - \varepsilon/2$ and $r + \varepsilon/2$. By Example 1 of the Worked Examples following Proposition 7.15, the fraction

$$\frac{x_n - x_k}{y_n - y_k}$$

must also be between $r - \varepsilon/2$ and $r + \varepsilon/2$ and hence

$$\left| \frac{x_n - x_k}{y_n - y_k} - r \right| < \frac{\varepsilon}{2}.$$

We now use the identity

$$\frac{x_n}{y_n} - r = \frac{x_k - ry_k}{y_n} + \left(1 - \frac{y_k}{y_n} \right) \left(\frac{x_n - x_k}{y_n - y_k} - r \right)$$

and we get

$$\left| \frac{x_n}{y_n} - r \right| \leq \left| \frac{x_k - ry_k}{y_n} \right| + \left| \frac{x_n - x_k}{y_n - y_k} - r \right|$$

(noting that $y_k/y_n \to 0$ as $n \to \infty$). We know already that the second summand on the right-hand side of the foregoing inequality is smaller than $\varepsilon/2$ for $n \geq k$; the first summand (whose numerator is a fixed quantity) also becomes smaller than $\varepsilon/2$ for $n \geq n'$ because $y_n \to \infty$. Choosing $n' \geq k$, we see that, for $n \geq n'$,

$$\left| \frac{x_n}{y_n} - r \right| < \varepsilon.$$

This then proves the claim for the considered special case.

To finish the proof, we observe that the case of an infinite limit easily reduces to the case of a finite limit. Take, for example, the case where

$$\frac{x_n - x_{n-1}}{y_n - y_{n-1}} \to \infty \quad \text{as } n \to \infty.$$

Then for all sufficiently large n, $x_n - x_{n-1} > y_n - y_{n-1}$; hence, $x_n \to \infty$ with $y_n \to \infty$ and the sequence $\{x_n\}_{n=1}^{\infty}$ must be strictly increasing for all sufficiently large n. Therefore, we can use what we have proved already and apply the result to the reciprocal expression, namely, y_n/x_n:

$$\lim_{n\to\infty} \frac{y_n}{x_n} = \lim_{n\to\infty} \frac{y_n - y_{n-1}}{x_n - x_{n-1}} = 0$$

and we conclude that $\lim_{n\to\infty} x_n/y_n = \infty$, finishing the proof.

EXAMPLE 14. If $a_n > 0$, $b_n > 0$, the series $\sum_{n=1}^{\infty} b_n$ diverges, and

$$\lim_{n\to\infty} \frac{a_n}{b_n} = L,$$

where L is finite or infinite; then

$$\lim_{n\to\infty} \frac{a_1 + a_2 + \cdots + a_n}{b_1 + b_2 + \cdots + b_n} = L.$$

Indeed, we only have to set $a_1 + \cdots + a_n = x_n$ and $b_1 + \cdots + b_n = y_n$ in (7.99) and everything is clear.

EXAMPLE 15. Let $\{t_n\}_{n=1}^{\infty}$ be a sequence with (finite or infinite) limit T. If $a_1, a_2, \ldots, a_n$ are positive numbers such that

$$a_1 + a_2 + \cdots + a_n \to \infty \quad \text{as } n \to \infty,$$

then

$$\lim_{n\to\infty} \left(\frac{a_1 t_1 + a_2 t_2 + \cdots + a_n t_n}{a_1 + a_2 + \cdots + a_n} \right) = T. \tag{7.100}$$

Indeed, (7.100) follows from (7.99) by setting $a_1 t_1 + \cdots + a_n t_n = x_n$ and $a_1 + \cdots + a_n = y_n$.

REMARK. Letting $a_1 = a_2 = \cdots = a_n = 1$ in (7.100), we obtain

$$\lim_{n\to\infty} \frac{t_1 + t_2 + \cdots + t_n}{n} = T, \tag{7.101}$$

provided that $\lim_{n\to\infty} t_n = T$, where T is finite or infinite.

Note that (7.101) is known to us from Proposition 7.13 already in the case where T is finite.

EXAMPLE 16. Let $a_1, a_2, \ldots, a_n, \ldots$ all be positive and $\sum_{n=1}^{\infty} a_n$ be a convergent infinite series with sum S. Moreover, let

$$b_n = \frac{a_1 + 2a_2 + \cdots + na_n}{n(n+1)}.$$

Then

$$\sum_{n=1}^{\infty} b_n = S.$$

Indeed, we know from (7.101) that the sequence

$$\frac{s_1 + s_2 + \cdots + s_n}{n}$$

tends to the same limit as the sequence $\{s_n\}$. We put $s_n = \sum_{k=1}^{n} a_k$. Then

$$\sum_{k=1}^{n} b_k = \frac{a_1}{1 \cdot 2} + \frac{a_1 + 2a_2}{2 \cdot 3} + \frac{a_1 + 2a_2 + 3a_3}{3 \cdot 4} + \cdots + \frac{a_1 + 2a_2 + \cdots + na_n}{n(n+1)}$$

$$= a_1\left(1 - \frac{1}{n+1}\right) + 2a_2\left(\frac{1}{2} - \frac{1}{n+1}\right) + \cdots + na_n\left(\frac{1}{n} - \frac{1}{n+1}\right)$$

$$= \frac{na_1}{n+1} + \frac{(n-1)a_2}{n+1} + \cdots + \frac{a_n}{n+1}$$

$$= \frac{s_1 + s_2 + \cdots + s_n}{n+1} = \frac{s_1 + s_2 + \cdots + s_n}{n} \cdot \frac{1}{1 + 1/n}.$$

Thus,

$$\lim_{n \to \infty}\left(\sum_{k=1}^{n} b_k\right) = S.$$

EXAMPLE 17 (Carleman's Inequality). Let $a_1, a_2, \ldots, a_n, \ldots$ all be positive and $\sum_{n=1}^{\infty} a_n$ be a convergent infinite series with sum S. Moreover, let

$$g_n = \sqrt[n]{a_1 a_2 \cdots a_n}$$

and the infinite series $\sum_{n=1}^{\infty} g_n$ have sum U. Then

$$U < eS. \tag{7.102}$$

Moreover, e is the *best possible constant* in (7.102).

Indeed, let b_n be as defined in Example 16. We have, by Proposition 4.17,

$$g_n = \sqrt[n]{a_1 a_2 \cdots a_n} = \frac{1}{\sqrt[n]{n!}}\sqrt[n]{a_1 \cdot 2a_2 \cdots na_n} \le \frac{n+1}{\sqrt[n]{n!}} b_n.$$

But

$$\left(1 + \frac{1}{1}\right)^1\left(1 + \frac{1}{2}\right)^2 \cdots \left(1 + \frac{1}{n}\right)^n = \frac{(n+1)^n}{n!} < e^n$$

and so

$$\frac{n+1}{\sqrt[n]{n!}} < e.$$

Thus,

$$\sum_{k=1}^{n} g_k < e \sum_{k=1}^{n} b_k,$$

and by letting $n \to \infty$ we deduce that the infinite series $\sum_{n=1}^{\infty} g_n$ converges (because, by Example 16, $\sum_{n=1}^{\infty} b_n = S$), and that its sum U can not exceed eS.

To verify that e is the best possible constant in (7.102), we write

$$a_n = \frac{1}{n} \quad \text{for } n = 1, 2, \ldots, m,$$

$$= 0 \quad \text{for } n > m;$$

we then have to prove that

$$\lim_{n \to \infty} \left(\frac{\sum_{n=1}^{\infty} g_n}{\sum_{n=1}^{\infty} a_n} \right) = e.$$

But

$$\frac{\sum_{n=1}^{m} g_n}{\sum_{n=1}^{m} a_n} = \frac{\sum_{n=1}^{m} \frac{1}{\sqrt[n]{n!}}}{\sum_{n=1}^{m} \frac{1}{n}} \tag{7.103}$$

and since $\sum_{n=1}^{\infty} (1/n)$ is divergent, by Example 14, if $m/\sqrt[m]{m!}$ tends to a limit as $m \to \infty$, the right-hand side of (7.103) tends to the same limit. However, we know from Example 10 of the Worked Examples following Proposition 7.15 that

$$\lim_{n \to \infty} \frac{\sqrt[m]{m!}}{m} = \frac{1}{e}$$

and so $m/\sqrt[m]{m!} \to e$ as $m \to \infty$, completing the proof.

EXAMPLE 18 (Bendixson's Test for Uniform Convergence). If $|\sum_{n=0}^{m} f_n'(x)|$ is less than a fixed number G at all points of an interval (a, b) of finite length and for all values of m, then if $\sum_{n=0}^{\infty} f_n(x)$ converges at all points of (a, b), it converges uniformly.

Indeed, divide the interval into j subintervals each of length $L = \delta/G$, where $\delta < \frac{1}{2}\varepsilon$, ε being an arbitrarily small positive number. Next, find m so that at the ends of each subinterval

$$g(x_r) = \sum_{n=m+1}^{m+p} f_n(x_r), \quad p = 1, 2, 3, \ldots,$$

is in absolute value less than δ. This can be done because the series converges at each of these points, and *they are finite in number* $(j + 1)$. Now if x is any point of the interval the distance to the nearest end of a subinterval (say x_r) is not larger than $L/2$; hence,

$$|g(x) - g(x_r)| < \left(\frac{L}{2} \right)(2G) = \delta,$$

because $|g'(x)| < 2G$. Thus,

$$|g(x)| < |g(x_r)| + \delta < 2\delta < \varepsilon,$$

and so the test for uniform convergence is satisfied.

EXAMPLE 19. Let $u_n \downarrow 0$ and $u_1 + u_2 + \cdots + u_n \to \infty$ as $n \to \infty$. Then

$$\lim_{n \to \infty} \frac{a_1 u_1 + a_2 u_2 + \cdots + a_n u_n}{u_1 + u_2 + \cdots + u_n} = \lim_{n \to \infty} \frac{a_1 + a_2 + \cdots + a_n}{n}$$

provided that the limit on the right-hand side exists.

[Observe that we are not assuming that $\lim_{n \to \infty} a_n$ exists and hence (7.100) does not apply; indeed, the foregoing result is of special value because it does not assume the existence of the limit $\lim_{n \to \infty} a_n$ as we shall see in Example 20.]

Indeed, put $a_1 + a_2 + \cdots + a_n = n A_n$ and assume that $A_n \to A$ as $n \to \infty$. Then

$$a_1 u_1 + a_2 u_2 + \cdots + a_n u_n$$

$$= A_1 u_1 + (2A_2 - A_1) u_2 + \cdots + [n A_n - (n-1) A_{n-1}] u_n$$

$$= A_1 (u_1 - u_2) + 2A_2 (u_2 - u_3) + \cdots + n A_n (u_n - u_{n+1}) + n A_n u_{n+1}.$$

Letting $c_n = n(u_n - u_{n+1})$ and noting that

$$c_1 + c_2 + \cdots + c_n = u_1 + u_2 + \cdots + u_n - n u_{n+1},$$

we get

$$a_1 u_1 + a_2 u_2 + \cdots + a_n u_n = A_1 c_1 + A_2 c_2 + \cdots + A_n c_n$$

$$+ A_n \{ (u_1 + u_2 + \cdots + u_n) \qquad (7.104)$$

$$- (c_1 + c_2 + \cdots + c_n) \}.$$

Since $\{ u_n \}$ is decreasing, c_n is positive. Moreover, $c_1 + c_2 + \cdots + c_n$ becomes arbitrarily large. Indeed, if L is any large number, we can determine k so that $c_1 + c_2 + \cdots + c_k$ exceeds L. And then we can determine n so that

$$u_{n+1} < \frac{u_1 + u_2 + \cdots + u_k - L}{k}$$

holds. These two consecutive determinations are possible since

$$u_1 + u_2 + \cdots + u_n \to \infty \quad \text{and} \quad u_n \downarrow 0 \quad \text{as } n \to \infty.$$

Noting that $\{ u_n \}$ is decreasing, we have

$$c_1 + c_2 + \cdots + c_n > u_1 + u_2 + \cdots + u_k + (n-k) u_{n+1} - n u_{n+1},$$

that is,

$$c_1 + c_2 + \cdots + c_n > u_1 + u_2 + \cdots + u_k - k u_{n+1} > L.$$

Hence, by (7.99),

$$\lim_{n \to \infty} \frac{A_1 c_1 + A_2 c_2 + \cdots + A_n c_n}{c_1 + c_2 + \cdots + c_n} = \lim_{n \to \infty} \frac{A_n c_n}{c} = \lim_{n \to \infty} A_n.$$

The second part on the right-hand side of (7.104) therefore consists of a factor which converges to zero, multiplied by

$$\frac{c_1 + c_2 + \cdots + c_n}{u_1 + u_2 + \cdots + u_n} = 1 - \frac{nu_{n+1}}{u_1 + u_2 + \cdots + u_n},$$

a quantity which is between 0 and 1. Hence, the considered expression tends to zero while the first part on the right-hand side of (7.104) converges to A. Thus,

$$\lim_{n\to\infty} \frac{a_1 u_1 + a_2 u_2 + \cdots + a_n u_n}{u_1 + u_2 + \cdots + u_n} = A.$$

EXAMPLE 20 (Theorem of Cesàro). The relative frequency of positive and negative terms in a conditionally convergent series $\sum \alpha_n$ for which $|\alpha_n|$ diminishes monotonically is subject to the following: The limit, if it exists, of the ratio P_n/Q_n of P_n, the number of positive terms, to Q_n, the number of negative terms α_k, for $k \leq n$, is necessarily 1.

Indeed, let

$$a_n = 1 \quad \text{if } \alpha_n \text{ is positive,}$$
$$= -1 \quad \text{if } \alpha_n \text{ is negative,}$$

and denote by u_n the absolute value of α_n. Then

$$\alpha_1 + \alpha_2 + \alpha_3 + \cdots = a_1 u_1 + a_2 u_2 + a_3 u_3 + \cdots.$$

The series $u_1 + u_2 + u_3 + \cdots$ is divergent and $u_n \downarrow 0$ as $n \to \infty$. By Example 19,

$$\lim_{n\to\infty} \frac{a_1 + a_2 + \cdots + a_n}{n} = \lim_{n\to\infty} \frac{a_1 u_1 + a_2 u_2 + \cdots + a_n u_n}{u_1 + u_2 + \cdots + u_n} = 0,$$

provided the first limit exists. Noting that $a_1 + a_2 + \cdots + a_n$ equals the excess of the number of positive terms over the negative terms, we see that the limit of the ratio of the two numbers satisfies

$$\lim_{n\to\infty} \frac{n + (a_1 + a_2 + \cdots + a_n)}{n - (a_1 + a_2 + \cdots + a_n)} = \lim_{n\to\infty} \frac{1 + (1/n)(a_1 + a_2 + \cdots + a_n)}{1 - (1/n)(a_1 + a_2 + \cdots + a_n)} = 1,$$

that is, the frequency of positive terms and negative terms is asymptotically the same.

EXAMPLE 21 (Theorem of Tauber). Suppose that the power series $\sum_{n=0}^{\infty} a_n x^n$ converges to $f(x)$ for $|x| < 1$ and that $\lim_{n\to\infty} na_n = 0$. If $f(x) \to A$ as $x \to 1-$, then the series $\sum_{n=0}^{\infty} a_n$ converges to A.

Indeed, it is desired to estimate differences such as $\sum_{n=0}^{N} a_n - A$. To do this, we write

$$\sum_{n=0}^{N} a_n - A = \left(\sum_{n=0}^{N} a_n - f(x)\right) + \{f(x) - A\}$$

$$= \sum_{n=0}^{N} a_n(1 - x^n) - \sum_{n=N+1}^{\infty} a_n x^n + \{f(x) - A\}.$$

(7.105)

Since $0 \le x < 1$, we have $1 - x^n = (1 - x)(1 + x + \cdots + x^{n-1}) < n(1 - x)$, so we can dominate the first term on the right-hand side by the expression

$$(1 - x) \sum_{n=0}^{N} na_n.$$

By assumption, $\lim_{n \to \infty} na_n = 0$; hence, by Proposition 7.13

$$\lim_{m \to \infty} \left(\frac{1}{m + 1} \sum_{n=0}^{m} na_n \right) = 0.$$

In addition, we have the relation $f(x) \to A$.

Now let $\varepsilon > 0$ be given and choose a fixed positive integer N which is so large that

(i) $\left| \sum_{n=0}^{N} na_n \right| < (N + L)\varepsilon$;

(ii) $|a_n| < \dfrac{\varepsilon}{N + 1}$ for all $n \ge N$;

(iii) $|f(x_0) - A| < \varepsilon$ for $x_0 = 1 - \dfrac{1}{N + 1}$.

We shall compute the magnitude of (7.105) for this value of N and x_0. From (i), (ii), (iii) and the fact that $(1 - x_0)(N + 1) = 1$, we get

$$\left| \sum_{n=0}^{N} a_n \right| \le (1 - x_0)(N + 1)\varepsilon + \frac{\varepsilon}{N + 1} \frac{x_0^{N+1}}{1 - x_0} + \varepsilon < 3\varepsilon.$$

Hence, $\sum_{n=0}^{\infty} a_n$ converges to A.

EXAMPLE 22 (Theorem of Bernstein). Let f be defined and possess derivatives of all orders on an interval $[0, r]$ and suppose that f and all its derivatives are nonnegative on the interval $[0, r]$. If $0 \le x < r$, show that

$$f(x) = \sum_{n=0}^{\infty} \frac{f^{(n)}(0)}{n!} x^n.$$

To prove this, we shall make use of formula (5.45); if $0 \le x < r$, then

$$f(x) = \sum_{k=0}^{n} \frac{f^{(k)}(0)}{k!} + R_n, \tag{7.106}$$

where

$$R_n = \frac{1}{n!} \int_0^r f^{(n+1)}(t)(r - t)^n \, dt = \frac{r^n}{n!} \int_0^1 (1 - s)^n f^{(n+1)}(rs) \, ds.$$

Since all terms in the sum (7.106) are nonnegative,

$$f(r) \ge \frac{r^n}{n!} \int_0^1 (1 - s)^n f^{(n+1)}(rs) \, ds. \tag{7.107}$$

Since $f^{(n+2)}$ is nonnegative, $f^{(n+1)}$ is increasing on $[0, r]$; therefore, if x is in this interval, then

$$R_n \le \frac{x^n}{n!} \int_0^1 (1 - s)^n f^{(n+1)}(rs) \, ds. \tag{7.108}$$

By combining (7.107) and (7.108), we have

$$R_n \le \left(\frac{x}{r}\right)^n f(r).$$

Hence, if $0 \le x < r$, then $\lim_{n \to \infty} R_n = 0$. This completes the proof.

EXERCISES TO CHAPTER 7

7.1. Let $0 < a_1 < a_2$ and define $a_{n+2} = \sqrt{a_{n+1} a_n}$ for $n = 1, 2, 3, \ldots$. Show that the sequence $\{a_{2n-1}\}$ is increasing and bounded above by a_2 and $\{a_{2n}\}$ is decreasing and bounded below by a_1 with both sequences tending to the common limit $(a_1 a_2^2)^{1/3}$.

[Hint: Note that a_{n+2}, being the geometric mean between a_n and a_{n+1}, lies between a_n and a_{n+1}. We also have $a_{n+2}\sqrt{a_{n+1}} = a_{n+1}\sqrt{a_n} = \cdots = a_2\sqrt{a_1}$.]

7.2. Let $0 < A \le a_1 < a_2 \le B$ and define the sequence $\{a_n\}$ as in Exercise 7.1. Show that

$$|a_{n+1} - a_n| \le \frac{B}{A + B}|a_n - a_{n-1}| \le \left(\frac{B}{A + B}\right)^{n-1}|a_2 - a_1|$$

and deduce that $\{a_n\}$ is a Cauchy sequence.

[Hint: We have $A \le a_n \le B$ for $n = 1, 2, 3, \ldots$ and $a_{n+1}^2 - a_n^2 = a_n(a_{n-1} - a_n)$ so that

$$|a_{n+1} - a_n| = \frac{a_n}{a_{n+1} + a_n}|a_n - a_{n-1}| \le \frac{B}{A + B}|a_n - a_{n-1}|.$$

Moreover, if $n > m$,

$$|a_n - a_m| = |a_n - a_{n-1} + a_{n-1} - a_{n-2} + \cdots + a_{m+1} - a_m|$$

$$\le |a_n - a_{n-1}| + |a_{n-1} - a_{n-2}| + \cdots + |a_{m+1} - a_m|.]$$

7.3. The sets of numbers $a_1, a_2, a_3, \ldots$ and $b_1, b_2, b_3, \ldots$ are defined in succession by the relations

$$a_{n+1} = \frac{a_n + b_n}{2}, \qquad b_{n+1} = \frac{2a_n b_n}{a_n + b_n},$$

where $0 < b_1 < a_1$. Show that for $n = 1, 2, 3, \ldots$

$$a_n > a_{n+1} > b_{n+1} > b_n > 0$$

and that both sequences $\{a_n\}$ and $\{b_n\}$ tend to the common limit $\sqrt{a_1 b_1}$.

[Hint: Note that b_{n+1} is the harmonic mean between a_n and b_n. Also observe that $a_{n+1} b_{n+1} = a_n b_n = \cdots = a_1 b_1$.]

7.4. Let $a_1 = 1$ and $a_{n+1} = 2(2a_n + 1)/(a_n + 3)$ for $n = 1, 2, 3, \ldots$. Show that $\{a_n\}$ is an increasing sequence bounded below by 0 and above by 2 and tending to the limit 2.

[Hint: We have

$$a_{n+1} - a_n = \frac{(2 - a_n)(1 + a_n)}{a_n + 3} \quad \text{and} \quad 2 - a_{n+1} = \frac{2(2 - a_n)}{a_n + 3}.]$$

7.5. Let $a_1 = \sqrt{2}$ and $a_{n+1} = (\sqrt{2})^{a_n}$ for $n = 1, 2, 3, \ldots$. Show that $\{a_n\}$ is an increasing sequence bounded above by 2. Find $\lim_{n \to \infty} a_n$.

[Hint: Apparently, $a_2 > a_1$. Moreover, $a_n > a_{n-1}$ implies $a_{n+1} > a_n$ because

$$\frac{a_{n+1}}{a_n} = (\sqrt{2})^{a_n - a_{n-1}}$$

and $2 > a_n$ implies $2 > a_{n+1}$ because $2/a_{n+1} = (\sqrt{2})^{2 - a_n}.]$

7.6. Let $a_{n+1} = \frac{1}{2}(a_n + a_{n-1})$ for $n = 1, 2, 3, \ldots$. Show that $\{a_n\}$ is a Cauchy sequence and $a_n \to (a_0 + 2a_1)/3$ as $n \to \infty$.

[Hint: We have $a_{n+1} - a_n = \frac{1}{2}(a_{n-1} - a_n)$ implying that

$$|a_{n+1} - a_n| = \frac{1}{2}|a_{n-1} - a_n| = \frac{1}{2^2}|a_{n-1} - a_{n-2}| = \cdots = \frac{1}{2^n}|a_1 - a_0|.$$

Moreover, $x_{n+1} + \frac{1}{2}x_n = x_n + \frac{1}{2}x_{n-1} = \cdots = x_1 + \frac{1}{2}x_0.]$

7.7. Let the sequence $\{a_n\}$ be as defined in Exercise 7.6. Show that

$$a_n = c_1 + c_2(-\tfrac{1}{2})^n,$$

where $c_1 = (a_0 + 2a_1)/3$ and $c_2 = 2(a_0 - a_1)/3$.

[Hint: If $a_n = t_n$ and $a_n = s_n$ satisfy $a_{n+1} = \frac{1}{2}(a_n + a_{n-1})$, then so does $a_n = c_1 t_n + c_2 s_n$ for any constants c_1, c_2. For $n \geq 2$, the equation is satisfied by $a_n = v^n$ if $v^2 = \frac{1}{2}(v + 1)$, that is, if $v = 1$ or $v = -\frac{1}{2}$ and so it is satisfied by $a_n = c_1 + c_2(-\frac{1}{2})^n$. Choose c_1, c_2 so that $a_n = c_1 + c_2(-\frac{1}{2})^n$ also for $n = 0$ and $n = 1$; we find $c_1 = (a_0 + 2a_1)/3$, $c_2 = 2(a_0 - a_1)/3$. Note also that $a_n \to (a_0 + 2a_1)/3$ as $n \to \infty$.]

REMARKS. The method of solution outlined in the Hint to Exercise 7.7 belongs to the theory of finite differences and difference equations. The interested reader may wish to consult S. Goldberg, *Introduction to Difference Equations with Illustrative Examples from Economics, Psychology and Sociology*, New York: Wiley, 1958.

7.8. The sequence defined by $a_0 = 0$, $a_1 = 1$, and $a_{n+1} = \lambda(a_n + a_{n-1})$ for $n = 1, 2, 3, \ldots$ converges provided $-1 < \lambda \leq \frac{1}{2}$.

[Hint: By the method outlined in the Hint to Exercise 7.7 we see that in this case we have the representation $a_n = c_1 v_1^n + c_2 v_2^n$, where v_1 and v_2 are the roots of the equation $v^2 = \lambda(v + 1)$, that is,

$$v_1 = \frac{\lambda + (\lambda^2 + 4\lambda)^{1/2}}{2} \quad \text{and} \quad v_2 = \frac{\lambda - (\lambda^2 + 4\lambda)^{1/2}}{2},$$

and c_1 and c_2 satisfy $c_1 = -c_2$ with $c_1 = 1/(\lambda^2 + 4\lambda)^{1/2}$. For $\{a_n\}$ to converge it is necessary that $|v_1| \leq 1$ and $|v_2| \leq 1$. But this is the case provided $-1 < \lambda \leq \frac{1}{2}$.

For example, if $\lambda \geq 0$ and $\lambda + (\lambda^2 + 4\lambda)^{1/2} \leq 2$, then $0 \leq \lambda \leq \frac{1}{2}$ and if $-4 < \lambda < 0$ and $|\lambda \pm i(-4\lambda - \lambda^2)^{1/2}| \leq 2$, where $i = \sqrt{-1}$, then it follows that $0 > \lambda > -1$.]

7.9. Let $3a_{n+1} = 2a_n + a_{n-1}$ with $a_0 = 7$ and $a_1 = 3$. Find $\lim_{n\to\infty} a_n$.

[Hint: See the method outlined in the Hint to Exercise 7.7 and note that in this case we have the representation $a_n = 4 - 9(-\frac{1}{3})^n$.]

7.10. If $a_{n+1} = +\sqrt{a_n + k}$, $a_0 > \frac{1}{2}$, $k > 0$, show that the sequence $\{a_n\}$ is convergent and that its limit is the positive root of the equation

$$x^2 - x - k = 0.$$

[Hint: We have $a_n > 0$ for all n. If $a_0^2 - a_0 = k$, then $a_n = a_0$ for all n, and so a_n is equal to the positive root of

$$x^2 - x = k.$$

If $a_0^2 - a_0 < k$ (so that a_0 is less than the positive root of $x^2 - x = k$), choose b_0 greater than the positive root of $x^2 - x = k$ so that $b_0^2 - b_0 > k$, and define b_n by $b_{n+1} = +\sqrt{b_n + k}$. Since $a_{n+1}^2 = a_n + k$, we have

$$a_{n+2}^2 - a_{n+1}^2 = a_{n+1} - a_n.$$

But $a_1^2 - a_0^2 = -(a_0^2 - a_0 - k) > 0$ and so $a_{n+1} > a_n$ for all n. Similarly, $b_{n+1} < b_n$ for all n. But $b_0 > a_0$ and $b_{n+1}^2 - a_{n+1}^2 = b_n - a_n$ so that $b_n > a_n$ for all n, and

$$b_{n+1} - a_{n+1} = \frac{b_n - a_n}{b_{n+1} + a_{n+1}} < \frac{b_n - a_n}{2a_1}, \qquad 2a_1 > 2a_0 > 1,$$

so that a_n and b_n tend to the same limit L (say). Since $a_{n+1}^2 = a_n + k$, we get by passage to the limit as $n \to \infty$ that $L^2 - L - k = 0$ and $L > 0$. If $a_0^2 - a_0 > k$, then the roles of a_n and b_n are interchanged.]

7.11. Let $a_{n+1} = k/(1 + a_n)$, $k > 0$, $a_0 > 0$. Show that the sequence $\{a_n\}$ is convergent and that its limit is the positive root of the equation

$$x^2 + x - k = 0.$$

[Hint: We have $a_n > 0$ for all n. If $a_0^2 + a_0 = k$, then $a_n = a_0$ for all n. If $a_0^2 + a_0 < k$, then

$$a_1 - a_0 = \frac{k}{1 + a_0} - a_0 = -\frac{a_0^2 + a_0 - k}{1 + a_0} > 0$$

and so $a_1 > a_0$. Since

$$a_{n+2} = \frac{k}{1 + a_{n+1}} = \frac{k(1 + a_n)}{1 + k + a_n},$$

we have

$$a_{2n+1} - a_{2n} = \frac{k(1 + a_{2n-1})}{1 + k + a_{2n-1}} - \frac{k(1 + a_{2n-2})}{1 + k + a_{2n-2}}$$

$$= \frac{k^2(a_{2n-1} - a_{2n-2})}{(1 + k + a_{2n-2})(1 + k + a_{2n-1})}$$

and so we see that $a_{2n+1} - a_{2n}$ has the same sign as $a_{2n-1} - a_{2n-2}$ and is therefore positive because $a_1 - a_0 > 0$. Also

$$a_{2n+1} - a_{2n} < \left(\frac{k}{1+k}\right)^2 (a_{2n-1} - a_{2n-2}).$$

Furthermore,

$$a_2 - a_0 = -\frac{a_0^2 + a_0 - k}{1 + k + a_0} > 0,$$

$$a_3 - a_1 = \frac{k(a_0^2 + a_0 - k)}{(1 + a_0)^2(1 + k) + (1 + a_0)k} < 0,$$

and

$$a_{2n+2} - a_{2n} = \frac{k^2(a_{2n} - a_{2n-2})}{(1 + k + a_{2n})(1 + k + a_{2n-2})},$$

$$a_{2n+3} - a_{2n+1} = \frac{k^2(a_{2n+1} - a_{2n-1})}{(1 + k + a_{2n+1})(1 + k + a_{2n-1})}$$

so that for all n, $a_{2n+2} > a_{2n}$, $a_{2n+3} < a_{2n+1}$, and so

$$a_{2n} < a_{2n+2} < a_{2n+3} < a_{2n+1},$$

that is, the closed interval $[a_{2n+2}, a_{2n+3}]$ is contained in the closed interval $[a_{2n}, a_{2n+1}]$ for all n.]

7.12. Let the sequence $\{a_n\}$ be the same as in Exercise 7.11. Show that, for any n, a root of $x^2 + x = k$ lies between a_{2n} and a_{2n+1}, provided $a_0^2 + a_0 \neq k$.
    [Hint: We have $a_{n+1} - a_n = -(a_n^2 + a_n - k)/(1 + a_n)$; if $a_0^2 + a_0 - k < 0$, then (see the Hint to Exercise 7.11)

$$a_{2n+1} > a_{2n}, \qquad a_{2n+1} > a_{2n+2}$$

and so $a_{2n}^2 + a_{2n} - k$ is negative and $a_{2n+1}^2 + a_{2n+1} - k$ is positive; hence, it follows since $x^2 + x - k$ is continuous that a root of $x^2 + x - k$ lies between a_{2n} and a_{2n+1}.]

7.13. If $a_{n+1} = 2k^2 a_n/(a_n^2 + k^2)$ and $b_{n+1} = (b_n^2 + k^2)/2b_n$, $a_0 > 0$, $b_0 > 0$, $k > 0$, then both a_n and b_n tend to k.
    [Hint: We have

$$\frac{k - a_{n+1}}{k + a_{n+1}} = \left(\frac{k - a_n}{k + a_n}\right)^2$$

$$= \left(\frac{k - a_0}{k + a_0}\right)^{2^{n-1}} \to 0.$$

Similarly,

$$\frac{b_{n+1} - k}{b_{n+1} + k} = \left(\frac{b_0 - k}{b_0 + k}\right)^{2^{n+1}} \to 0.]$$

7.14. If $|b_{n+2} - b_{n+1}| \leq k|b_{n+1} - b_n|$ and $k < 1$, verify that the sequence $\{b_n\}$ converges. Hence, show that, if for all n,

$$pc_{n+2} - (p + q)c_{n+1} + qc_n = 0, \qquad p > q > 0,$$

then $pc_{n+2} - qc_{n+1} = pc_1 - qc_0$ and c_n converges to $(pc_1 - qc_0)/(p - q)$.

[Hint: Since $|b_{n+2} - b_{n+1}|/|b_{n+1} - b_n| \le k < 1$, the series $\sum(b_{n+1} - b_n)$ is absolutely convergent, and so

$$\sum_{r=1}^{n-1} (b_{r+1} - b_r) = b_n - b_0$$

is convergent.

Hence, since $(c_{n+2} - c_{n+1})/(c_{n+1} - c_n) = q/p > 0$, we have

$$\frac{|c_{n+2} - c_{n+1}|}{|c_{n+1} - c_n|} = \frac{q}{p} > 1$$

so that c_n converges; let γ be the limit of c_n.

From $pc_{n+2} - qc_{n+1} = pc_{n+1} - qc_n$ it follows that $pc_{n+2} - qc_{n+1} = pc_1 - qc_0$ and so

$$pc_1 - qc_0 = \lim_{n \to \infty} (pc_{n+2} - qc_{n+1}) = (p - q)\gamma.]$$

7.15. If $\{a_n\}$ is a bounded sequence of real numbers and $2a_n \le a_{n-1} + a_{n+1}$, show that $\lim_{n \to \infty} (a_{n+1} - a_n) = 0$.

[Hint: We have $a_n - a_{n-1} \le a_{n+1} - a_n$, therefore $\{a_n - a_{n-1}\}$ is increasing; but bounded, therefore $a_n - a_{n-1} \to k$ as $n \to \infty$. If $k \ne 0$ suppose $k > 0$ (other case similar), then $a_n - a_{n-1} > \frac{1}{2}k$ for $n \ge N$, where N is an appropriate large integer. Thus,

$$a_M - a_{N-1} = \sum_{n=N}^{M} (a_n - a_{n-1}) > \frac{1}{2}(M - N)k.$$

This contradicts the boundedness of $\{a_n\}$. Thus, $k \le 0$. Similarly, $k \ge 0$. Hence, $k = 0$.]

7.16. Find the sum to n terms of the series

$$\sum_{k=1}^{\infty} \frac{k!}{(x + 1)(x + 2)\cdots(x + k)}$$

and hence show that the series converges for $x > 1$ and diverges for $x \le 1$.

[Hint: Let $u_{k+1} = k!/(x + 1)(x + 2)\cdots(x + k)$, $u_1 = 1$, then

$$(x + k)u_{k+1} = ku_k$$

and so $(x - 1)u_{k+1} = ku_k - (k + 1)u_{k+1}$; hence,

$$(x - 1)\sum_{k=1}^{n} u_{k+1} = 1 - \frac{(n + 1)!}{(x + 1)(x + 2)\cdots(x + n)}.$$

Therefore, if $x \ne 1$, then

$$\sum_{k=1}^{n} u_{k+1} = \frac{1}{x - 1} - \frac{(n + 1)!}{(x - 1)(x + 1)(x + 2)(x + n)}.$$

If $y > 0$, then

$$\left(1 + \frac{y}{2}\right)\left(1 + \frac{y}{3}\right)\cdots\left(1 + \frac{y}{n}\right) > 1 + y\left(\frac{1}{2} + \frac{1}{3} + \cdots + \frac{1}{n}\right) > \frac{Ny}{2}$$

if $n > 2^N$, and so

$$\frac{n!}{(x + 1)(x + 2)\cdots(x + n - 1)} = \frac{1}{(1 + y/2)(1 + y/3)\cdots(1 + y/n)} \to 0,$$

$y = x - 1 > 0$. Thus, for $x > 1$ the series converges to $1/(x - 1)$. If $x < 1$, put $1 - x = z > 0$, then

$$\frac{n!}{(x + 1)(x + 2)\cdots(x + n - 1)} = \frac{n!}{(2 - z)(3 - z)\cdots(n - z)}$$

$$= \frac{1}{(1 - z/2)(1 - z/3)\cdots(1 - z/n)}$$

$$> \frac{(1 + z/p)\{1 + z/(p + 1)\}\cdots(1 + z/n)}{(1 - z/2)(1 - z/3)\cdots\{1 - z(p - 1)\}}, \quad p > z,$$

$$\text{for } 1/(1 - z/r) > 1 + z/r, \quad r > z,$$

$$> \frac{z\{1/p + 1/(p + 1) + \cdots + 1/n\}}{(1 - z/2)(1 - z/3)\cdots\{1 - z/(p - 1)\}}$$

and so the sequence $\{n!/(x + 1)(x + 2)\cdots(x + n)\}$ diverges, and therefore the series $\sum u_k$ diverges.]

7.17. Show that, if $x > 1$, the series

$$\frac{1}{x + 1} + \frac{2}{x^2 + 1} + \frac{4}{x^4 + 1} + \frac{8}{x^8 + 1} + \cdots$$

has sum $1/(x - 1)$.
   [Hint: Note that $1/(x - 1) - 1/(x + 1) = 2/(x^2 - 1)$.]

7.18. Show that $k + 4$ terms of the series

$$\sum_{r=0}^{\infty} \frac{1}{r!}$$

suffice to determine the limit of the series to k decimal places.
   [Hint: For $n \geq 10$, $\frac{1}{2}(n!) > 3\cdot4\cdot5\cdot6\cdot7\cdot8\cdot9\cdot10^{n-9} > 10^{n-4}$; but

$$\frac{1}{(n + 1)!} + \frac{1}{(n + 2)!} + \cdots < \frac{1 + 1/2 + 1/2^2 + \cdots}{(n + 1)!}$$

$$< \frac{2}{(n + 1)!} < \frac{1}{10^{n-3}} \leq \frac{1}{10^{k+1}} \quad \text{if } n \geq k + 4.]$$

7.19. For what values of x is the series

$$\sum_{n=1}^{\infty} \frac{2^n x^n}{\sqrt{n}}$$

(i) absolutely convergent, (ii) convergent (but not absolutely), and (iii) divergent?

[Hint: Let $a_n = 1/\sqrt{n}$, then $a_n/a_{n+1} = \sqrt{1 + .1/n} = p_n$, say. Hence, $p_n^2 = 1 + 1/n$, therefore $p_n - 1 \le 1/k(p_n + 1) < 1/2k$, $n \ge k$, so that $p_n \to 1$. Thus, $\sum a_n t^n$ is absolutely convergent for $|t| < 1$, and divergent for $|t| > 1$. When $t = 1$, $\sum a_n t^n = \sum 1/\sqrt{n}$, which diverges since $1/\sqrt{n} > 1/n$, and when $t = -1$, $\sum a_n t^n = \sum (-1)^n/\sqrt{n}$, which converges since $1/\sqrt{n}$ steadily decreases with limit zero. Taking $t = 2x$, we have that

$$\sum_{n=1}^{\infty} \frac{2^n x^n}{\sqrt{n}}$$

is absolutely convergent for $|x| < \frac{1}{2}$, convergent for $x = -\frac{1}{2}$, and divergent for $x \ge \frac{1}{2}$ or $x < -\frac{1}{2}$.]

7.20. If $x_{n+1} = +\sqrt{x_n}$, $x_0 > 0$, then the transformation $y = \sqrt{x}$ gives

$$\int_1^{x_{n+1}} \frac{1}{y}\,dy = \frac{1}{2}\int_1^{x_n} \frac{1}{x}\,dx.$$

Hence, show that

$$2^n(x_n - 1) \to \int_1^{x_0} \frac{1}{x}\,dx = \ln x_0.$$

[Hint: We have

$$2^n \int_1^{x_n} \frac{1}{y}\,dy = \int_1^{x_0} \frac{1}{x}\,dx.$$

Since $x_{n+1} = \sqrt{x_n}$, we have $(x_{n+1} - 1)/(x_n - 1) = 1/(x_{n+1} + 1)$; but if $x_0 > 1$, then $x_n > 1$ for all n and so $(x_{n+1} - 1)/(x_n - 1) < 1/2$ and therefore $(x_n - 1)/(x_0 - 1) < 1/2^n$ so that $x_n \to 1$. If $x_0 < 1$, then $1/x_n > 1$ so that, by the foregoing consideration, $1/x_n \to 1$, that is, $x_n \to 1$. Finally, if $x_0 = 1$, then $x_n = 1$ for all n.
Next we observe that

$$\lim_{t \to 1} \frac{\displaystyle\int_1^t (1/x)\,dx}{t - 1} = \lim_{t \to 1} \frac{1}{t} = 1$$

and so

$$\frac{2^n \displaystyle\int_1^{x_n} (1/x)\,dx}{2^n(x_n - 1)} = \frac{\displaystyle\int_1^{x_n} (1/x)\,dx}{x_n - 1} \to 1.]$$

7.21. If $x_{n+1} = x_n/[1 + (1 + x_n^2)^{1/2}]$, then the transformation

$$y = \frac{x}{1 + (1 + x^2)^{1/2}}$$

gives

$$\int_0^{x_{n+1}} \frac{1}{1 + y^2}\,dy = \frac{1}{2}\int_0^{x_n} \frac{1}{1 + x^2}\,dx.$$

Hence, show that

$$2^n x_n \to \int_0^{x_0} \frac{1}{1+x^2} dx = \tan^{-1} x_0.$$

[Hint: Since $x_{n+1} = x_n/[1 + (1 + x_n^2)^{1/2}]$, we have $|x_{n+1}| < \frac{1}{2}|x_n|$ and so

$$|x_n| < \frac{1}{2^n}|x_0| \to 0.$$

Under the transformation

$$y = \frac{x}{1 + (1 + x^2)^{1/2}} = \frac{(1 + x^2)^{1/2} - 1}{x}$$

we have

$$y + \frac{1}{y} = \frac{2(1 + x^2)^{1/2}}{x}, \qquad y - \frac{1}{y} = -\frac{2}{x},$$

and so

$$\left(y + \frac{1}{y}\right)\frac{1}{y}\frac{dy}{dx} = \left(1 + \frac{1}{y^2}\right)\frac{dy}{dx} = \frac{2}{x^2};$$

hence,

$$\frac{1}{1 + y^2}\frac{dy}{dx} = \frac{1}{(y + 1/y)^2}\left(y + \frac{1}{y}\right)\frac{1}{y}\frac{dy}{dx} = \frac{x^2}{4(1 + x^2)}\frac{2}{x^2} = \frac{1}{2}\frac{1}{1 + x^2}.$$

Thus,

$$\int_0^{x_{n+1}} \frac{1}{1 + y^2} dy = \frac{1}{2}\int_0^{x_n} \frac{1}{1 + x^2} dx$$

and therefore

$$2^n \int_0^{x_n} \frac{1}{1 + y^2} dy = \int_0^{x_0} \frac{1}{1 + x^2} dx.$$

Since

$$\lim_{t \to 0} \frac{\int_0^t \frac{1}{1 + x^2} dx}{t} = \lim_{t \to 0} \frac{1}{1 + t^2} = 1,$$

we have

$$\frac{2^n \int_0^{x_n} \frac{1}{1 + y^2} dy}{2^n x_n} = \frac{\int_0^{x_n} \frac{1}{1 + x^2} dx}{x_n} \to 1$$

and so

$$2^n x_n \to \int_0^{x_0} \frac{1}{1 + x^2} dx = \tan^{-1} x_0.]$$

REMARK. In connection with the foregoing Exercises 7.20 and 7.21 the interested reader may want to look up Problems 85 and 86 in Chapter 3 of G. Klambauer, *Problems and Propositions in Analysis*, New York: Dekker, 1979.

7.22. The sequence $\{u_n\}$ is defined by $u_{n+1} = (6u_n^2 + 6)/(u_n^2 + 11)$ and by the value of the initial term u_0. Verify that if $u_n \to L$ as $n \to \infty$, then L is one of 1, 2, 3.
Show that, if $u_n > 3$, then

$$\text{(i)} \quad 3 < u_{n+1} < u_n, \qquad \text{(ii)} \quad \frac{u_{n+1} - 3}{u_n - 3} < \frac{9}{10},$$

and prove that, if $u_0 > 3$, then $u_n \to 3$ (monotonically) as $n \to \infty$.

[Hint: If $u_n \to L$ as $n \to \infty$, then $L = (6L^2 + 6)/(L^2 + 11)$ which reduces to $(L - 1)(L - 2)(L - 3) = 0$.
(i) If $u_n > 3$, then

$$u_{n+1} - u_n = \frac{6u_n^2 + 6 - u_n^3 - 11u_n}{u_n^2 + 11} = -\frac{(u_n - 1)(u_n - 2)(u_n - 3)}{u_n^2 + 11} < 0$$

and

$$u_{n+1} - 3 = \frac{3u_n^2 - 27}{u_n^2 + 11} > 0;$$

these two inequalities give $3 < u_{n+1} < u_n$.
(ii) From the foregoing expression for $u_{n+1} - 3$,

$$\frac{u_{n+1} - 3}{u_n - 3} = \frac{3(u_n + 3)}{u_n^2 + 11} = \frac{3}{10}\left(3 - \frac{(3u_n - 1)(u_n - 3)}{u_n^2 + 11}\right) < \frac{9}{10}$$

as $u_n > 3$. Multiplying the inequalities for $n = 0, 1, 2, \ldots, n + 1$, we get

$$0 < \frac{u_{n+1} - 3}{u_0 - 3} < \left(\frac{9}{10}\right)^{n+1}$$

(the first inequality holding as $u_{n+1} > 3$ and $u_0 > 3$). Since $\frac{9}{10} < 1$, this shows that $u_n \to 3$ as $n \to \infty$.]

7.23. Show that if k is a positive integer, then

$$\sum_{n=0}^{\infty} \frac{n^k}{n!}$$

is an integral multiple of e.
[Hint: Note that

$$n^k = A_1 n + A_2 n(n - 1) + A_3 n(n - 1)(n - 2) + \cdots + A_k n(n - 1)\cdots(n - k + 1),$$

where $A_k = 1$ and, for any r, $1 < r \le k$, A_r is the remainder when

$$n^{k-1} - A_1 - A_2(n - 1) - \cdots - A_{r-1}(n - 1)(n - 2)\cdots(n - r + 2)$$

is divided by $(n - 1)(n - 2)\cdots(n - r)$, so that each A_r is an integer. We have

$$\sum_{n=0}^{\infty} \frac{n^k x^n}{n!} = A_1 x \sum_{n=1}^{\infty} \frac{x^{n-1}}{(n-1)!} + A_2 x^2 \sum_{n=2}^{\infty} \frac{x^{n-2}}{(n-2)!} + \cdots + A_k x^k \sum_{n=k}^{\infty} \frac{x^{n-k}}{(n-k)!}$$

$$= (A_1 x + A_2 x^2 + \cdots + A_k x^k) e^x$$

and so

$$\sum_{n=0}^{\infty} \frac{n^k}{n!} = (A_1 + A_2 + \cdots + A_k)e.$$

By way of illustration, observe that $n^3 = n + 3n(n-1) + n(n-1)(n-2)$ and so

$$\sum_{n=0}^{\infty} \frac{n^3}{n!} = 5e.]$$

7.24. Show that

$$\sum_{n=1}^{\infty} \frac{1^3 + 2^3 + \cdots + n^3}{n!} x^n = (x + \tfrac{7}{2}x^2 + 2x^3 + \tfrac{1}{4}x^4)e^x.$$

[Hint: By formula (1.39),

$$1^3 + 2^3 + \cdots + n^3 = \frac{n^4 + 2n^3 + n^2}{4}$$

and so

$$1^3 + 2^3 + \cdots + n^3$$
$$= \tfrac{1}{4}[A_1 n + A_2 n(n-1) + A_3 n(n-1)(n-2) + A_4 n(n-1)(n-2)(n-3)],$$

where $A_1 = 4$, $A_2 = 14$, $A_3 = 8$, and $A_4 = 1$.]

7.25. Show that

$$\sum_{n=1}^{\infty} \frac{(n-1)x^n}{(n+2)n!} = \frac{(x^2 - 3x + 3)e^x + \tfrac{1}{2}x^2 - 3}{x^2}.$$

[Hint: We have

$$\sum_{n=1}^{\infty} \frac{(n-1)x^n}{(n+2)n!} = \frac{1}{x^2} \sum_{n=1}^{\infty} \frac{(n^2 - 1)x^{n+2}}{(n+2)!}.$$

But $n^2 - 1 = 3 - 3(n+2) + (n+2)(n+1)$ and so

$$\sum_{n=1}^{\infty} \frac{(n-1)x^n}{(n+2)n!} = \frac{1}{x^2}\left(3 \sum_{n=1}^{\infty} \frac{x^{n+2}}{(n+2)!} - 3x \sum_{n=1}^{\infty} \frac{x^{n+1}}{(n+1)!} + x^2 \sum_{n=1}^{\infty} \frac{x^n}{n!}\right)$$

$$= \frac{3(e^x - 1 - x - \tfrac{1}{2}x^2 - 3x(e^x - 1 - x) + x^2(e^x - 1)}{x^2}$$

$$= \frac{(x^2 - 3x + 3)e^x + \tfrac{1}{2}x^2 - 3}{x^2}.]$$

7.26. Show that

$$\sum_{n=1}^{\infty} \frac{n^3 + 2n^2 + n - 1}{n!} = 9e + 1.$$

[Hint: Let

$$n^3 + 2n^2 + n - 1 = A_0 + A_1 n + A_2 n(n - 1) + A_3 n(n - 1)(n - 2).]$$

7.27. If in the series

$$1 + \frac{1}{2} + \frac{1}{3} + \frac{1}{4} + \cdots$$

the signs of the terms are changed in such a way that p positive terms and q negative terms occur alternately, show that the resulting series converges or diverges according to whether $p = q$ or $p \neq q$.

[Hint: See the Theorem of Cesàro (in Example 20 of Section 8 in this chapter).]

7.28. Let $q_1, q_2, q_3, \ldots$ be the sequence of all positive integers whose decimal representations contain no digit 9. Does the series

$$\sum_{n=1}^{\infty} \frac{1}{q_n}$$

converge?

[Hint: There are $9^m - 9^{m-1}$ integers between $10^{m-1} - 1$ and $10^m - 1$ containing no nines at all. Hence, the sum in question is less than

$$\frac{9 - 1}{1} + \frac{9^2 - 9}{10} + \frac{9^3 - 9^2}{100} + \cdots = 80.]$$

7.29. Consider the power series $\sum_{n=0}^{\infty} a_n x^n$, where the coefficients satisfy the relation $a_{r+2} + c_1 a_{r+1} + c_2 a_r = 0$, where c_1 and c_2 are constants. Find the sum of this power series if $a_0 = 1$, $a_1 = -7$, $a_2 = -1$, and $a_3 = -43$ with x being such that the series in question converges.

[Hint: From the recurrence relation on the coefficients we get

$$-43 - c_1 - 7c_2 = 0 \quad \text{and} \quad -1 - 7c_1 + c_2 = 0$$

and so $c_1 = -1, c_2 = -6$. Letting S denote the sum of the series for values of x for which the series converges, we see that

$$S = 1 - 7x - x^2 - 43x^3 - \cdots$$

$$-xS = -x + 7x^2 + x^3 + \cdots$$

$$-6x^2 S = -6x^2 + 42x^3 + \cdots$$

and, by addition,

$$(1 - x - 6x^2)S = 1 - 8x$$

or

$$S = \frac{1 - 8x}{1 - x - 6x^2} = \frac{1 - 8x}{(1 + 2x)(1 - 3x)}.$$

We can also see that the series converges when $|x| < \frac{1}{3}$.]

7.30. Given that

$$\lim_{x \to 0} \frac{a \cos x + b \sin x + ce^x + d}{x^3} = 1,$$

find the values of a, b, c, and d.

[Hint: Using the series expansions of $\cos x$, $\sin x$, and e^x, we see that

$$\frac{a \cos x + b \sin x + ce^x + d}{x^3}$$

$$= \frac{a + c + d}{x^3} + \frac{b + c}{x^2} + \frac{c - a}{2x} + \frac{c - b}{6} + \left(\frac{c + a}{24}\right) x + \text{higher powers of } x.$$

If this is to tend to unity as $x \to 0$, it follows that

$$a + c + d = 0, \qquad b + c = 0, \qquad c - a = 0, \qquad c - b = 6.$$

Solving this system of equations, we get $a = 3$, $b = -3$, $c = 3$, and $d = -6$.]

7.31. Obtain the binomial series

$$z = (1 + x)^m = 1 + mx + m(m - 1)\frac{x^2}{2!} + m(m - 1)(m - 2)\frac{x^3}{3!} + \cdots$$

by taking

$$z = e^y = 1 + y + \frac{y^2}{2!} + \frac{y^3}{3!} + \cdots$$

and

$$y = m \ln(1 + x) = m(x - \tfrac{1}{2}x^2 + \tfrac{1}{3}x^3 - \cdots).$$

7.32. Let $x_n = (1 + a)(1 + a^2) \cdots (1 + a^n)$. Show that $\{x_n\}$ converges if $0 \le a < 1$.
[Hint: Use $1 + x \le e^x$.]

7.33. Let m be a positive integer. Show that

$$\sum_{n=1}^{\infty} \frac{1}{n(m + n)} = \frac{1}{m}\left(1 + \frac{1}{2} + \frac{1}{3} + \cdots + \frac{1}{m}\right).$$

[Hint: Let

$$S_n = \frac{1}{1(m + 1)} + \frac{1}{2(m + 2)} + \cdots + \frac{1}{n(m + n)}.$$

Then

$$mS_n = 1 + \frac{1}{2} + \frac{1}{3} + \cdots + \frac{1}{m} - \left(\frac{1}{n + 1} + \frac{1}{n + 2} + \cdots + \frac{1}{n + m}\right)$$

because

$$\frac{1}{k(m+k)} = \frac{1}{m}\left(\frac{1}{k} - \frac{1}{k+m}\right).]$$

7.34. Show that

$$\sum_{n=0}^{\infty} \frac{1}{(x+n)(x+n+1)(x+n+2)(x+n+3)} = \frac{1}{3x(x+1)(x+2)}$$

[Hint: Let

$$f(x) = \frac{1}{3x(x+1)(x+2)}.$$

Then

$$f(x) - f(x+1) = \frac{1}{x(x+1)(x+2)(x+3)},$$

$$f(x+1) - f(x+2) = \frac{1}{(x+1)(x+2)(x+3)(x+4)},$$

$$\vdots$$

$$f(x+n) - f(x+n+1) = \frac{1}{(x+n)(x+n+1)(x+n+2)(x+n+3)}.$$

Adding these equations and passing to the limit as $n \to \infty$ we obtain the desired result.]

7.35. Show that if $\sum_{n=1}^{\infty} a_n$ is absolutely convergent, then so are

(i) $\displaystyle\sum_{n=1}^{\infty} \frac{1}{1+a_n}$, where no $a_n = -1$, (ii) $\displaystyle\sum_{n=1}^{\infty} a_n^2$, (iii) $\displaystyle\sum_{n=1}^{\infty} \frac{a_n^2}{1+a_n^2}$.

[Hints: (i) Since $\sum a_n$ converges, we must have $a_n \to 0$ as $n \to \infty$ and so $1 + a_n > \frac{1}{2}$ for all sufficiently large n. Thus, $|a_n/(1+a_n)| < 2|a_n|$ for such n and so $\sum a_n/(1+a_n)$ converges absolutely.
(ii) Since $a_n \to 0$ as $n \to \infty$, $\{|a_n|\}$ is bounded. Let $M > |a_n|$ for all n. Then $a_n^2 = |a_n^2| = |a_n|^2 < M|a_n|$, and so $\sum a_n^2$ is absolutely convergent.
(iii) This follows from part (ii) together with part (i).]

7.36. Show that each of the following series has the indicated sum

(a) $\displaystyle\sum_{n=1}^{\infty} (-1)^{n+1} \frac{1 \cdot 3 \cdot 5 \cdots (2n-1)}{2 \cdot 4 \cdot 6 \cdots (2n)} \frac{1}{2n} = \ln\left(\frac{1+\sqrt{2}}{2}\right),$

(b) $1 + \displaystyle\sum_{n=1}^{\infty} \frac{1 \cdot 3 \cdot 5 \cdots (2n-1)}{2 \cdot 4 \cdot 6 \cdots (2n)} \frac{1}{2n+1} = \frac{\pi}{2},$

(c) $\displaystyle\sum_{n=1}^{\infty} (-1)^{n+1} \left(1 + \frac{1}{2} + \cdots + \frac{1}{n}\right) \frac{1}{n+1} = \frac{1}{2}(\ln 2)^2,$

(d) $\displaystyle\sum_{n=1}^{\infty} (-1)^{n+1} \left(1 + \frac{1}{2} + \cdots + \frac{1}{2n}\right) \frac{1}{2n+1} = \frac{\pi}{8}\ln 2,$

(e) $\sum_{n=1}^{\infty} (-1)^{n+1} \left(1 + \frac{1}{3} + \cdots + \frac{1}{2n-1}\right) \frac{1}{2n} = \frac{\pi^2}{32}$.

[Hint: Use Proposition 7.70 together with the expansions

(a) $\ln\left(\frac{1 + \sqrt{1+x}}{2}\right) = \sum_{n=1}^{\infty} (-1)^{n+1} \frac{1 \cdot 3 \cdots (2n-1)}{2 \cdot 4 \cdots (2n)} \frac{x^n}{2n}$,

(b) $\sin^{-1} x = x + \sum_{n=1}^{\infty} \frac{1 \cdot 3 \cdots (2n-1)}{2 \cdot 4 \cdots (2n)} \frac{x^{2n+1}}{2n+1}$,

(c) $\frac{1}{2}[\ln(1+x)]^2 = \sum_{n=1}^{\infty} (-1)^{n+1} \left(1 + \frac{1}{2} + \cdots + \frac{1}{n}\right) \frac{x^{n+1}}{n+1}$,

(d) $\frac{1}{2}(\tan^{-1} x)\ln(1+x^2) = \sum_{n=1}^{\infty} (-1)^{n+1} \left(\sum_{k=1}^{n} \frac{1}{k}\right) \frac{x^{2n+1}}{2n+1}$,

(e) $\frac{1}{2}(\tan^{-1} x)^2 = \sum_{n=1}^{\infty} (-1)^{n+1} \left(\sum_{k=0}^{n-1} \frac{1}{2k+1}\right) \frac{x^{2n}}{2n}$.]

7.37. Find the interval of convergence of the following power series and test for convergence at the endpoints:

$$\sum_{n=1}^{\infty} \frac{n^n}{n!} x^n.$$

[Hint: It is easy to see that the radius of convergence is $1/e$. To test at the endpoints $x = \pm 1/e$, we put $b_n = n^n/e^n n!$. From Stirling's Formula (see Proposition 7.17), it follows that $b_n \to 0$ as $n \to \infty$. Now

$$\frac{b_{n+1}}{b_n} = \left(\frac{1}{e}\right)\left(1 + \frac{1}{n}\right)^n < 1.$$

Thus, $\{b_n\}$ decreases to zero, and so $\sum_{n=1}^{\infty} (-1)^n b_n$ converges. On the other hand, by Stirling's Formula,

$$b_n > \frac{1}{2(2\pi n)^{1/2}}$$

for all sufficiently large n, and so $\sum_{n=1}^{\infty} b_n$ diverges.]

7.38. Let $\{f_n\}$ be a sequence of Riemann integrable functions defined on $[a, b]$ and assume that f is Riemann integrable on $[a, b]$ and

$$\lim_{n \to \infty} \int_a^b |f_n(x) - f(x)|^2 \, dx = 0.$$

Let g be Riemann integrable on $[a, b]$ and define, for x in $[a, b]$,

$$h(x) = \int_a^x f(t)g(t) \, dt \quad \text{and} \quad h_n(x) = \int_a^x f_n(t)g(t) \, dt.$$

Show that $h_n \to h$ uniformly on $[a, b]$.

[Hint: We have, by (5.41),

$$0 \le \left(\int_a^x |f(t) - f_n(t)| |g(t)|\, dt \right)^2 \le \left(\int_a^x |f(t) - f_n(t)|^2\, dt \right) \left(\int_a^x |g(t)|^2\, dt \right).$$

Given $\varepsilon > 0$, we can choose N so that $n \ge N$ implies

$$\int_a^b |f(t) - f_n(t)|^2\, dt < \frac{\varepsilon^2}{A},$$

where

$$A = 1 + \int_a^b |g(t)|^2\, dt.$$

Thus, $n \ge N$ implies $0 \le |h(x) - h_n(x)| < \varepsilon$ for every x in $[a, b]$.]

7.39. Find

(i) $\displaystyle \lim_{x \to 0} \frac{(1 + x)^{1/x} - e}{x}$ and (ii) $\displaystyle \lim_{x \to 0} \frac{(1 + x)^{1/x} - e + \frac{1}{2}ex}{x^2}$.

[Hint: Let $y = (1 + x)^{1/x}$. Then

$$\ln y = \frac{1}{x} \ln(1 + x) = 1 - \frac{x}{2} + \frac{x^2}{3} - \cdots$$

and so

$$y = e^{1 - x/2 + x^2/3 - \cdots} = e \cdot e^{-x/2 + x^2/3 - \cdots}$$

$$= e \left(1 + \left(-\frac{x}{2} + \frac{x^2}{3} - \cdots \right) + \frac{1}{2!} \left(-\frac{x}{2} + \frac{x^2}{3} - \cdots \right)^2 + \cdots \right)$$

$$= e \left(1 - \frac{x}{2} + \frac{11x^2}{24} - \cdots \right).$$

It is now readily seen that the answer to part (i) is $-e/2$ and the answer to part (ii) is $11e/24$.]

7.40. Show that

$$\lim_{x \to 0} \frac{(1 + x)^{1/x} - e + \frac{1}{2}ex + \frac{11}{24}ex^2}{x^3} = -\frac{7e}{16}.$$

[Hint: See Exercise 7.39.]

7.41. Show that

$$e^{-n} \left(1 + \frac{n}{1!} + \frac{n^2}{2!} + \cdots + \frac{n^n}{n!} \right) \to \frac{1}{2} \qquad \text{as } n \to \infty.$$

[Hint: See Problem 53 in Chapter 4 of G. Klambauer, *Problems and Propositions in Analysis*, New York: Dekker, 1979.]

Bibliography

Boyer, C. B. *The Concepts of the Calculus*. New York: Dover, 1949.

Hardy, G. H. *A Course of Pure Mathematics*, 10th ed. New York: Cambridge University Press, 1952.

Hewitt, E. and Stromberg, K. *Real and Abstract Analysis*. New York: Springer-Verlag, 1975.

Hille, E. *Analysis I & II*. New York: Blaisdell, 1964 & 1966.

Klambauer, G. *Problems and Propositions in Analysis*. New York: Marcel Dekker, 1979.

Klambauer, G. *Mathematical Analysis*. New York: Marcel Dekker, 1975.

Klambauer, G. *Real Analysis*. New York: American Elsevier, 1973.

Knopp, K. *Theory and Application of Infinite Series*, 2nd ed. New York: Hafner, 1951.

Stromberg, K. R. *An Introduction to Classical Real Analysis*. Belmont, CA: Wadsworth, 1981.

Toeplitz, O. *The Calculus: A Genetic Approach*. Chicago: University of Chicago Press, 1963.

Index

Undergraduate Texts in Mathematics

continued from ii

Martin: The Foundations of Geometry and the Non-Euclidean Plane.

Martin: Transformation Geometry: An Introduction to Symmetry.

Millman/Parker: Geometry: A Metric Approach with Models.

Owen: A First Course in the Mathematical Foundations of Thermodynamics.

Prenowitz/Jantosciak: Join Geometrics.

Priestly: Calculus: An Historical Approach.

Protter/Morrey: A First Course in Real Analysis.

Protter/Morrey: Intermediate Calculus.

Ross: Elementary Analysis: The Theory of Calculus.

Scharlau/Opolka: From Fermat to Minkowski.

Sigler: Algebra.

Simmonds: A Brief on Tensor Analysis.

Singer/Thorpe: Lecture Notes on Elementary Topology and Geometry.

Smith: Linear Algebra.
Second edition.

Smith: Primer of Modern Analysis.

Thorpe: Elementary Topics in Differential Geometry.

Troutman: Variational Calculus with Elementary Convexity.

Wilson: Much Ado About Calculus.